ISBN 978-0-266-82458-9
PIBN 10894910

# United States
# Circuit Court of Appeals
## For the Ninth Circuit.

1551

# Transcript of Record.

**(IN TWO VOLUMES.)**

DAVID G. LORRAINE,

Appellant,

vs.

FRANCIS M. TOWNSEND, MILON J. TRUMBLE and ALFRED J. GUTZLER, Doing Business Under the Firm Name of TRUMBLE GAS TRAP COMPANY,

Appellees.

**VOLUME I.**
**(Pages 1 to 320, Inclusive.)**

Upon Appeal from the United States District Court for the Southern District of California, Southern Division.

FILED
DEC 28 1922
F. D. MONCKTON,
CLERK.

Filmer Bros. Co. Print, 330 Jackson St., S. F., Cal.

United States

# Circuit Court of Appeals

For the Ninth Circuit.

# Transcript of Record.

(IN TWO VOLUMES.)

DAVID G. LORRAINE,

Appellant,

vs.

FRANCIS M. TOWNSEND, MILON J. TRUMBLE and ALFRED J. GUTZLER, Doing Business Under the Firm Name of TRUMBLE GAS TRAP COMPANY,

Appellees.

**VOLUME I.**

**(Pages 1 to 320, Inclusive.)**

Upon Appeal from the United States District Court for the Southern District of California, Southern Division.

Filmer Bros. Co. Print, 330 Jackson St., S. F., Cal.

# INDEX TO THE PRINTED TRANSCRIPT OF RECORD.

[Clerk's Note: When deemed likely to be of an important nature, errors or doubtful matters appearing in the original certified record are printed literally in italic; and, likewise, cancelled matter appearing in the original certified record is printed and cancelled herein accordingly. When possible, an omission from the text is indicated by printing in italic the two words between which the omission seems to occur.]

Index. Page

Index. Page

TESTIMONY ON BEHALF OF PLAINTIFFS—Continued:



TESTIMONY ON BEHALF OF DEFENDANT:

Index. Page

**Names and Addresses of Attorneys of Record.**

For Appellant:

WESTALL and WALLACE, Esqs. (JOSEPH F. WESTALL and ERNEST L. WALLACE), 902 Trust and Savings Building, Los Angeles, California.

For Appellee:

FREDERICK S. LYON, Esq., and LEONARD S. LYON, Esq., 312 Stock Exchange Building, Los Angeles, California, and FRANK L. A. GRAHAM, Esq., Higgins Building, Los Angeles, California.

---

**Citation.**

UNITED STATES OF AMERICA,—ss.

To Francis M. Townsend, Milon J. Trumble and Alfred J. Gutzler, Doing Business Under the Firm Name of Trumble Gas Trap Company, GREETING:

You are hereby cited and admonished to be and appear at a United States Circuit Court of Appeals for the Ninth Circuit, to be held at the City of San Francisco, in the State of California, on the 15th day of November, A. D. 1922, pursuant to an order allowing appeal filed and entered in the Clerk's office of the District Court of the United States, in and for the Southern District of California, in that certain suit being numbered E–113—Equity, wherein David G. Lorraine is defendant and you are plaintiffs to show cause, if any there be, why the decree rendered against the said appellant, as in the said order allowing appeal mentioned, should not be corrected, and speedy justice should not be done to the parties in that behalf.

WITNESS, the Honorable BENJAMIN F. BLEDSOE, United States District Judge for the Southern District of California, this 18th day of October, A. D. 1922, and of the Independence of the United States, the one hundred and forty-seventh.

BLEDSOE,

U. S. District Judge for the Southern District of California.

Due service of the above citation and receipt of a copy thereof is hereby admitted this 19th day of October, 1922.

FREDERICK S. LYON,
LEONARD S. LYON,

Solicitors and of Counsel for the Above-named Plaintiffs-Appellees. [1*]

[Endorsed]: E-113—Equity. In the United States Circuit Court of Appeals for the Ninth Circuit. Francis M. Townsend et al. vs. David G. Lorraine. Citation. Filed Oct. 20, 1922. Chas. N. Williams, Clerk. By L. J. Cordes.

*Page-number appearing at foot of page of original Certified Transcript of Record.

In the District Court of the United States, Southern District of California, Southern Division.

IN EQUITY—No. ——.

FRANCIS M. TOWNSEND, MILON J. TRUMBLE and ALFRED J. GUTZLER, Doing Business Under the Firm Name of TRUMBLE GAS TRAP CO.,

Plaintiffs,

vs.

DAVID G. LORRAINE,

Defendant.

**Bill of Complaint for Infringement of Letters Patent No. 1,269,134.**

Now come the plaintiffs in the above-entitled suit and complaining of the defendant above named allege:

I.

That plaintiffs, Francis M. Townsend, Milon J. Trumble and Alfred J. Gutzler are residents of the County of Los Angeles, State of California and citizens of said state.

II.

That defendant, David G. Lorraine, is a resident of the City of Los Angeles, State of California, and a citizen of said State.

III.

That the ground upon which the Court's jurisdiction depends is that this is a suit in equity arising under the patent laws of the United States.

IV.

That heretofore, to wit, on and prior to November 14th, 1914, said Milon J. Trumble was the original and first inventor of a certain new and useful invention, to wit, a crude petroleum and natural gas separator which has not been known or used by others in this country before his invention thereof, nor patented nor described in any printed publication in this or any foreign country before his said invention thereof, or more than two years prior to his application for a patent, nor was the same in public use or on sale in this country for more than two years prior to his application for a [2] patent in this country and being such invention, heretofore, to wit, on November 14th, 1914, said Milon J. Trumble filed an application in the Patent Office of the United States praying for the issuance to him of letters patent for said new and useful invention.

V.

That prior to the issuance of any patent thereon, said Milon J. Trumble, for value received, by an instrument in writing sold and assigned to Francis M. Townsend and Alfred J. Gutzler an undivided interest in and to aforesaid new and useful invention and in and to any and all letters patent that might be issued therefor on said application and in and by said assignment requested the Commissioner of Patents to issue said patent to said Milon J. Trumble, Francis M. Townsend and Alfred J. Gutzler, their heirs, legal representatives and assigns, which said assignment in writing was filed in the

Patent Office of the United States prior to the issuance of any letters patent on said application.

VI.

That thereafter, to wit, on June 11, 1918, letters patent of the United States for the said invention dated on said last-named day and numbered 1,269,134, were issued and delivered by the Government of the United States to the said Milon J. Trumble, Francis M. Townsend and Alfred J. Gutzler, whereby there was granted to Milon J. Trumble, Francis M. Townsend and Alfred J. Gutzler, their heirs, legal representatives and assigns for the full term of seventeen years from June 11, 1918, the sole and exclusive right to make, use and vend the said invention throughout the United States of America and the territories thereof, and a more particular description of the invention patented in and by said letters patent will more fully appear from the letters patent ready in court to be produced by the plaintiffs.

VII.

That plaintiffs ever since the issuance of said letters patent have been and now are the sole holders and owners of said letters patent and all rights and privileges by them granted, and have under the firm name of Trumble Gas Trap Co. [3] constructed, made, used and sold apparatus containing and embracing and capable of carrying out the invention patented by the said letters patent and upon each of said apparatus have stamped and printed the day and date of and the number of said letters patent and the same have gone into general use.

VIII.

That at divers times within six years last past in the Southern District of California the defendant herein, David G. Lorraine, without the license or consent of the plaintiffs has used the apparatus described, claimed and patented, and has made and used the apparatus described, claimed and patented in and by the said letters patent No. 1,269,134, and has infringed upon said letters patent and each and all of the claims thereof, and intends and threatens to continue so to do.

IX.

That by reason of the infringement aforesaid plaintiffs have suffered damages and plaintiffs are informed and believe that the defendant has realized profits but the exact amount of such profits and damages is not known to plaintiffs.

X.

That plaintiffs have requested the defendant to desist and refrain from further infringement of said letters patent and to account to the plaintiffs for the aforesaid profits and damages but the said defendant has failed and refused to comply with such request or any part thereof, and is now continuing and carrying on the said infringement upon said letters patent daily and threatens to continue the same and unless restrained by this Court will continue the same, whereby plaintiffs will suffer great and irreparable injury and damage for which plaintiffs have no plain, speedy or adequate remedy at law.

WHEREBY plaintiffs pray as follows:

I.

That a final decree be entered in favor of the plaintiffs, Francis M. Townsend, Milon J. Trumble and Alfred J. Gutzler and against the defendant David G. Lorraine perpetually enjoining and restraining the said defendant, his agents, servants, attorneys, workmen and employees, and each of them, [4] from using the apparatus described, claimed and patented in and by said letters patent No. 1,269,134, and from making, using or selling the apparatus described, claimed or patented in and by said letters patent and from infringing upon said letters patent or any of the claims thereof, either directly or indirectly or from contributing to any such infringement.

II.

That upon the filing of this bill of complaint or later on motion, a preliminary injunction be granted to the plaintiffs enjoining and restraining the defendant, David G. Lorraine, his agents, servants, attorneys, workmen or employees and each of them, until the further order of this Court from using the apparatus described, claimed and patented in and by said letters patent No. 1,269,134, and from making, using or selling the apparatus described, claimed and patented by said letters patent and from infringing upon said letters patent or any of the claims thereof either directly or indirectly, or from contributing to any such infringement.

III.

That plaintiffs have and recover from the defendant the profits realized by the defendant herein and the damages suffered by the plaintiffs and by reason of the infringement aforesaid, together with the costs of suit and such other and further relief as to the Court may seem proper and in accordance with equity and good conscience.

FRANCIS M. TOWNSEND.
MILON J. TRUMBLE.
ALFRED J. GUTZLER.

FREDERICK S. LYON,
LEONARD S. LYON,
FRANK L. A. GRAHAM,
Attorneys and Counsel for Plaintiffs,
504 Merchants Trust Bldg.,
Los Angeles, California. [5]

State of California,
County of Los Angeles,—ss.

Francis M. Townsend, Milon J. Trumble and Alfred J. Gutzler, being duly sworn, each for himself, deposes and says that he has read the foregoing complaint and knows the contents thereof; that the same is true of his own knowledge except as to matters therein stated on information and belief and as to those matters, he believes it to be true.

FRANCIS M. TOWNSEND.
MILON J. TRUMBLE.
ALFRED J. GUTZLER.

Sworn to before me this 31st day of December, 1920.

[Seal] LOUIS W. GRATZ,

Notary Public in and for Los Angeles County, State of California.

My commission expires July 21, 1924. [6]

[Endorsed]: No. E-113—Eq. United States District Court, Southern District of California, Southern Division. Francis M. Townsend, Milon J. Trumble and Alfred J. Gutzler, Plaintiffs, vs. David G. Lorraine, Defendant. In Equity. Bill of Complaint. Filed Jan. 3, 1921. Chas. N. Williams, Clerk. By R. S. Zimmerman, Deputy Clerk. Frederick S. Lyon, Leonard Lyon, Frank L. A. Graham, 504 Merchants Trust Bldg., Los Angeles, California, Attorneys for Plaintiffs. [7]

---

In the District Court of the United States, Southern District of California, Southern Division.

IN EQUITY—No. E-113.

FRANCIS M. TOWNSEND, MILON J. TRUMBLE, and ALFRED J. GUTZLER, Doing Business Under the Firm Name of TRUMBLE GAS TRAP CO.,

Plaintiffs,

vs.

DAVID G. LORRAINE,

Defendant.

**Answer to Bill of Complaint.**

Comes now David G. Lorraine, the defendant in the above-entitled cause, and, for answer to the bill of complaint of plaintiffs heretofore filed in this cause and to the several allegations therein contained, states:

I.

That he admits that the plaintiffs are residents and citizens of the County of Los Angeles and State of California.

II.

That he admits that this defendant is a resident and citizen of the city of Los Angeles and State of California.

III.

That he admits that this is a suit in equity arising under the patent laws of the United States.

IV.

That he denies that, on or prior to the 14th day of November, A. D. 1914, or at any other time, the said Milon J. Trumble was or now is the first or original inventor of any new or useful invention known or described as a crude petroleum and natural gas separator or that the same had not been known or used in this country before said alleged invention thereof by the said Milon J. Trumble, or that the same had not been patented or described in any printed publication in this or any foreign country for more than two years before said alleged invention thereof by the said Milon J. Trumble; or

[8] that the same was not in public use or not on sale in this country for more than two years prior to the alleged application of the said Milon J. Trumble for a patent therefor; this defendant states that he had not sufficient information to form a belief as to whether or not the said Milon J. Trumble did, on the 14th day of November, A. D. 1914, or at any other time, file an application in the Patent Office of the United States praying for the issuance to him of letters patent to any new and useful invention and, therefore, denies the same.

V.

That he has not sufficient information to form a belief concerning the truth of the allegations set out and contained in "Paragraph V" of said bill of complaint and, therefore, denies the same.

VI.

That he denies that, on June 11, 1918, or at any other time, letters patent of the United States for any new or useful invention were issued and delivered to the persons named in "Paragraph VI" of said bill of complaint or that said persons, or either, or any of them, or their heirs, personal representatives or assigns were granted the sole and exclusive right to make, use or vend any such alleged invention throughout the United States of America or the territories thereof. But this defendant states that the device described in the alleged letters patent granted to the plaintiffs is not a new or useful invention for the reason that said device is inoperative for the purpose stated in said letters patent and for which purpose said device is used and that, therefore, said device lacks utility and, for that reason, said patent is invalid.

VII.

That he does not have sufficient knowledge or information to enable him to form a belief as to the truth or falsity of the allegations contained and set out in "Paragraph VII" of said bill of complaint and, therefore, denies the same. [9]

VIII.

That he denies that, at any time or times within six years last past, in the Southern District of California, or elsewhere, this defendant, without the license or consent of the plaintiffs or otherwise, has used, manufactured or sold any apparatus or device described, claimed or patented in or by any letters patent issued or delivered to the plaintiffs, or either or any of them, or to any person or persons through, by or under whom said plaintiffs or either or any of them claim; or that this defendant has, at any time, infringed upon and such pretended or alleged letters patent or any claim or claims thereof; or that he intends or threatens so to do.

IX.

That he denies that, by reason of any infringement of any letters patent or any other rights, by this defendant, or his agents, employees, licensees or assigns, the plaintiffs, or either, or any of them, have suffered any damages or that this defendant has realized any profits whatsoever thereby.

X.

That he denies that the plaintiffs, or either or

any of them, have requested this defendant to desist or refrain from any infringement of any letters patent or to account to plaintiffs, or either or any of them, for any profits or damages, or that this defendant has failed or refused to comply with any such request or any part thereof; or that this defendant is now or, at any time, has been committing or carrying on any infringement upon any letters patent or threatens to continue so to do; or that the plaintiffs, or either or any of them, will suffer any great or irreparable injury or damage thereby; or that the plaintiffs have no plain, speedy or adequate remedy at law.

XI.

And this defendant, for another and further answer and defense to said bill of complaint, states: That the alleged patent No. 1,269,134, issued June 11th, A. D. 1918, [10] to the plaintiffs, in its very nature and by reason of the many patents which were previously issued for similar devices, is a secondary patent and must, therefore, be strictly construed and confined to the specific form of invention therein described and claimed and that any construction of said patent broad enough to include the separator made by this defendant, would render such alleged patent to the said plaintiffs invalid, because of anticipatory devices patented and disclosed long prior to the filing of the alleged application on which the said alleged patent to the plaintiffs was finally isued; that some of these prior patents here referred to are as follows:

| | | |
|---|---|---|
| No. 428,399, | W. Moore, | May 20, 1890. |
| No. 578,708, | W. J. Baldwin, | March 16, 1897. |
| No. 681,170, | C. R. Hudson, | August 20, 1901. |
| No. 815,407 | A. S. Cooper, | March 20, 1906. |
| No. 856,088, | A. T. Newman, | June 4, 1907. |
| No. 1,255,018, | P. Jones, | January 29, 1918. |
| No. 1,272,625, | W. H. Cooper, | July 16, 1918. |

that there are also other patents and printed publications, the titles, dates and places of publication are, at this time, unknown to this defendant, but for all of which this defendant is causing diligent search to be made and which, when known, this defendant prays leave to insert and set forth in this answer.

XII.

And this defendant, further answering said bill of complaint, states that this defendant, heretofore, and on or before the 8th day of July, A. D. 1915, invented a certain new and useful invention for separating natural gas from crude petroleum for which he has filed an application for letters patent; that the claims in such application have been duly passed by the Examiner in charge of the application and allowed and a patent for such invention will issue in due course, upon the payment of the final Government fee.

And this defendant states that the device for separating natural gas from crude petroleum which he has made is made according to the specifications and claims made and [11] so allowed in his said application for letters patent, and for which letters patent will be issued in due time and which said device the defendant states does not, in any manner whatsoever, infringe the alleged patent of the plaintiffs.

WHEREFORE, this defendant prays that the bill of complaint of the plaintiffs be dismissed and that this defendant be permitted to go hence with his costs in this behalf expended.

DAVID *S.* LORRAINE,
By CHAS. BAGG,
Solicitor for Defendant. [12]

[Endorsed]: No. E-113—Equity. U. S. District Court, Southern District of California, Southern Division in Equity. Francis M. Townsend et al. vs. David G. Lorraine. Answer. Received a copy of the within answer this the 28th day of January, 1921. Frederick S. Lyon, Leonard S. Lyon, Frank L. A. Graham, Attorneys for Plaintiffs. Filed Jan. 28, 1921. Chas. N. Williams, Clerk. Chas. Bagg, 632 Laughlin Bldg., Broadway 2555, Los Angeles, Attorney for Defendant. [13]

In the District Court of the United States, in and for the Southern Division of the Southern District of the State of California.

IN EQUITY—No. E-113.

FRANCIS M. TOWNSEND, MILON J. TRUMBLE and ALFRED J. GUTZLER, Doing Business Under the Firm Name of TRUMBLE GAS TRAP CO.,

Plaintiffs,

vs.

DAVID G. LORRAINE,

Defendant.

**Notice of Motion.**

To Francis M. Townsend, Milon J. Trumble and Alfred J. Gutzler, Plaintiffs in the Above-entitled Cause, and to Messrs. Frederick S. Lyon and Leonard S. Lyon, Their Solicitors:

You and each of you will please take notice that, on Monday, March 13th, 1922, at the hour of ten o'clock A. M. of said day, the defendant, David G. Lorraine, will call up for hearing the annexed motion for leave to amend defendant's answer, heretofore filed in this cause, before the Honorable Benjamin F. Bledsoe, Judge of the above-entitled court, at his courtroom in the Post Office Building, in the city of Los Angeles, State of California.

Dated March 2d, 1922.

CHAS. BAGG,
Solicitor for Defendant.

Service of the above and foregoing notice, together with a copy of the motion therein referred to, by copy, this the 3d day of March, A. D. 1922.

LYON & LYON,
GRAHAM & HARRIS,
Solicitors for Plaintiffs. [14]

---

In the United States District Court, in and for the Southern Division of the Southern District of the State of California.

IN EQUITY—No. E-113.

FRANCIS M. TOWNSEND, MILON J. TRUMBLE and ALFRED J. GUTZLER, Doing Business Under the Firm Name of TRUMBLE GAS TRAP COMPANY,

Plaintiffs,

vs.

DAVID G. LORRAINE,

Defendant.

**Motion for Order Permitting Defendant to Amend Answer.**

Comes now David G. Lorraine, the defendant in the above-entitled case and respectfully moves this Honorable Court for an order in this cause permitting this defendant to amend his answer heretofore filed in this cause in the following particulars to wit: By inserting in Paragraph XI of said answer the following list of patents in addition to those inserted therein at the time of the filing of the same:

No. 426,880, Walter Anderson Taylor, April 29, 1890.

No. 1,014,943, Eustace Vivian Bray, January 16, 1912.

No. 535,611, James S. Bougher, March 12, 1895.

No. 611,314, Joseph S. Cullinan, September 27, 1898.

No. 399,427, William Moore, March 12, 1889.

No. 1,226,913, C. C. Scharpenberg, May 22, 1917.

No. 1,182,873, Charles E. Fisher, May 9, 1916.

No. 1,095,478, Fritz Strohbach, May 5, 1914.

CHAS. BAGG,
Solicitor for Defendant. [15]

[Endorsed]: No. E-113. United States District Court, Southern District of California, Southern Division. Francis M. Townsend et al., Plaintiffs, vs. David G. Lorraine, Defendant. Notice of Motion and Motion to Amend Answer. Filed Mar. 4, 1922. Chas. N. Williams, Clerk. By Edmund L. Smith, Deputy Clerk. Chas. Bagg, 632 Laughlin Bldg., Los Angeles, Cal., Solicitor for Defendant. [16]

At a stated term, to wit, the January Term, A. D. 1922, of the District Court of the United States of America, within and for the Southern Division of the Southern District of California, held at the courtroom thereof, in the City of Los Angeles, on Monday, the 13th day of March, in the year of our Lord one thousand nine hundred and twenty-two. Present: The Honorable BENJAMIN F. BLEDSOE, District Judge.

No. E-113—EQ.

FRANCIS M. TOWNSEND et al.,

Plaintiffs,

vs.

DAVID G. LORRAINE,

Defendant.

**Minutes of Court—March 13, 1922—Hearing.**

This cause coming on at this time for hearing on motion to amend defendant's answer; now, both parties consenting thereto, it is by the Court ordered that this cause for hearing on said motion be continued to March 20th, 1922. [17]

---

At a stated term, to wit, the January Term, A. D. 1922, of the District Court of the United States of America, within and for the Southern Division of the Southern District of California, held at the courtroom thereof, in the City of Los Angeles, on Monday, the 20th day of

March, in the year of our Lord one thousand nine hundred and twenty-two. Present: The Honorable BENJAMIN F. BLEDSOE, District Judge.

No. E-113—EQ. S. D.

FRANCIS M. TOWNSEND et al.,

Plaintiffs,

vs.

DAVID G. LORRAINE,

Defendant.

**Minutes of Court—March 20, 1922—Hearing (Continued).**

This cause coming on at this time for hearing on motion to amend defendant's answer, and it appearing that both parties have consented to the hearing on said motion being continued to March 27th, 1922, it is thereupon ordered by the Court that said motion be and the same is hereby continued to said date. [18]

---

At a stated term, to wit, the January Term, A. D. 1922, of the District Court of the United States of America, within and for the Southern Division of the Southern District of California, held at the courtroom thereof, in the City of Los Angeles, on Wednesday, the 22d day of March, in the year of our Lord one thousand nine hundred and twenty-two. Present: The Honorable CHARLES E. WOLVERTON, District Judge.

No. E-113—EQ.

FRANCIS M. TOWNSEND et al.,

Plaintiffs,

vs.

DAVID G. LORRAINE,

Defendant.

**Minutes of Court—March 22, 1922—Hearing (Continued).**

This cause coming on at this time for final hearing; F. S. Lyon, L. S. Lyon and Frank L. A. Graham, Esqs., appearing as counsel for plaintiffs, and Charles Bagg and L. L. Mack, Esqs., appearing as counsel for defendant, and John P. Doyle being also present as shorthand reporter of the testimony and proceedings; and all parties having announced themselves as ready to proceed with the hearing and a statement having been made on behalf of the plaintiffs by F. S. Lyon, Esq., attorney for the plaintiffs and he having announced that the specifications of the patent against which infringements are charged are Nos. 1, 2, 3, 4, and 13, and a statement having been made on behalf of the defendant by Charles Bagg, Esq., attorney for the defendant, and the plaintiffs in support of the issues on their side having offered in evidence the following exhibits, to wit:

Plaintiff's Ex. No. 1—Letters Patent No. 1,269,134 to Milon J. Trumble and dated June 11, 1918.

Plaintiff's Ex. No. 2—Certified copy of File-wrapper and contents of same patent.

Plaintiff's Ex. No. 3—Copy of patent No. 1,373,664.

Plaintiff's Ex. No. 4—Copy of Reissue patent No. 15,220, dated November 8, 1921.

Plaintiff's Ex. No. 5—Small metal plate—Trumble Gas Trap Co.

which exhibits are admitted in evidence and filed; and [19]

Paul Paine, a witness herein, having been called, sworn and having testified on behalf of the plaintiff, and, in connection with his testimony, the following exhibit having been offered and admitted in evidence, as follows, to wit:

Plaintiff's Ex. No. 6—A Blue-print table of pressures, etc., and

Now, at the hour of twelve o'clock noon the Court declares a recess to the hour of two o'clock P. M. and

Now, at the hour of two o'clock P. M. the Court having reconvened and all parties being present as before except E. L. Kincaid, who is present as stenographic reporter of the testimony and proceedings; and

Paul Paine, a witness for the plaintiff, having resumed the stand and having given his testimony, and in connection with his testimony there having been offered and admitted in evidence the following exhibit, to wit:

Plaintiff's Ex. No. 7—Paine's Sketch of Stark Trap; and

Alfred J. Gutzler having been called, sworn and having testified in behalf of the plaintiff; and

O. W. Harris having been called, sworn and having testified in behalf of the plaintiff; and

In connection with his testimony there having been offered and admitted in evidence the following exhibits, to wit:

Plaintiff's Ex. No. $8^1$—1 Photograph,
Plaintiff's Ex. No. $8^2$—1 Photograph,
Plaintiff's Ex. No. $8^3$—1 Photograph,
Plaintiff's Ex. No. $8^4$—1 Photograph,

Two sketches having been admitted as Plaintiff's Exhibits Nos. $9^1$ and $9^2$ and

The following having been admitted in evidence, to wit:

Plaintiff's Ex. No. 10—A Model of Lorraine Trap; and

Leave is now granted to the plaintiffs to file a supplemental bill of complaint; and, good cause appearing therefor, this cause is continued to March 23, 1922, at the hour of ten o'clock A. M. for further hearing. [20]

At a stated term, to wit, the January Term, A. D. 1922, of the District Court of the United States of America, within and for the Southern Division of the Southern District of California, held at the courtroom thereof, in the City of Los Angeles, on Thursday, the 23d day of March, in the year of our Lord one thousand nine hundred and twenty-two. Present: The Honorable CHARLES E. WOLVERTON, District Judge.

No. E–113—Eq.

FRANCIS M. TOWNSEND et al.,

Plaintiffs,

vs.

DAVID G. LORRAINE,

Defendant.

**Minutes of Court—March 23, 1922—Hearing (Continued).**

This cause coming on at this time for further hearing; F. S. Lyon and Frank L. A. Graham, Esqs., appearing as counsel for the plaintiff and Charles Bagg and L. L. Mack, Esqs., appearing as counsel for the defendants and John P. Doyle, being also present as shorthand reporter of the testimony and proceedings; and

A supplemental bill having been filed and defendant herein having been given until Monday, March 27, 1922, to file answer to supplemental bill; and

O. W. Harris, a witness herein on behalf of the plaintiff, having resumed the stand and having given his testimony; and

In connection with his testimony there having been offered and admitted for Identification the following exhibit, to wit:

Plaintiff's Ex. No. 11 for Identification—Drawing of Lorraine Trap; and said O. W. Harris having been temporarily withdrawn; and

Thomas T. Davies having been called, sworn and having testified in behalf of the plaintiff; and

O. W. Harris, witness temporarily withdrawn as aforesaid, having resumed the stand having testified further; and

David Lorraine having been called, sworn and having testified as a witness for the Plaintiff; and in connection with his testimony there having been offered and admitted on behalf of the [21] plaintiff the following exhibits, to wit:

Plaintiff's Ex. No. 12—Blue-print drawing produced by David G. Lorraine.

Plaintiff's Ex. No. 13—Account sales; and

Wm. G. Lacy having been called, sworn and having testified in behalf of the plaintiff; and in connection with his testimony there having been offered and admitted on behalf of the plaintiff the following exhibits, to wit:

Plaintiff's Ex. No. 14—Letter of December 10, 1920, Townsend to Lacy Manufacturing Company.

Plaintiff's Ex. No. 15—Letter of December 13, 1920, Lacy Manufacturing Company to Townsend; and

Hans K. Hyrup having been called, sworn and having testified in behalf of the plaintiff; and

It is now by the Court ordered that the drawing heretofore marked Ex. No. 11 for Identification be admitted in evidence as Plaintiff's Ex. No. 11; and

Paul Paine having been recalled and having taken the stand on behalf of the plaintiff; and

Wm. C. Rae having been called, sworn and having testified in behalf of the plaintiff; and

Now, at the hour of twelve o'clock noon it is by the Court ordered that a recess be taken to the hour of two o'clock P. M. and

Now, at the hour of two o'clock P. M. the Court having reconvened and all being present as before; and

Wm. C. Rae having resumed the stand on behalf of the plaintiff; and

David G. Lorraine having been recalled to the stand; and

It is by the Court ordered that the drawing produced by said witness be admitted and filed in evidence on behalf of the plaintiff, to wit: [22]

Plaintiff's Ex. No. 16—Blue-print drawing produced by witness as aforesaid.

Plaintiff's Ex. No. 17—Sketch produced by witness David Lorraine.

Plaintiff's Ex. No. $18^1$, $18^2$, $18^3$—Three prints from drawings of Lorraine Gas and Oil Separator; and

Milon J. Trumble having been called, sworn and having testified for the plaintiff; and in connection with his testimony there are offered and admitted in evidence on behalf of the plaintiff the following exhibits, to wit:

Plaintiff's Ex. No. 19—Model of Trap "Lorraine 1922,"

Plaintiff's Ex. No. 20—Model Trumble Trap; and

Thereupon the plaintiff rests; and

Now, at the hour of 2:31 o'clock P. M. the Court declares a recess for ten minutes; and now

At the hour of 2:41 o'clock P. M. the Court having reconvened and all being present as before and counsel having announced their readiness to proceed with the trial of this cause and counsel having ordered that this cause be proceeded with; and

Defendant in support of the issues on his side having called to the stand the defendant, David G. Lorraine, who has heretofore been sworn, and said defendant having given his testimony; and

Upon motion of counsel for the defendant it is by the court ordered that Defendant's Ex. "A" heretofore offered for Identification be admitted in evidence, as follows, to wit:

Defendant's Ex. "A"—Model;

—and the following exhibit having been offered in evidence on behalf of the defendant, to wit:

Defendant's Ex. "B"—Sketch made by witness;

—and pamphlet marked Defendant's Ex. "C" for Identification, then offered in evidence and objection having been interposed on behalf of the plaintiff, is now admitted in evidence for limited purposes; and

Now, at the hour of 4:02 o'clock P. M. it is by the Court ordered that this cause be continued to the hour of ten o'clock A. M. [23] March 24th, for further hearing. [24]

---

United States District Court, Southern District of California, Southern Division.

IN EQUITY—No. E-113.

FRANCIS M. TOWNSEND, MILON J. TRUMBLE and ALFRED J. GUTZLER, Doing Business Under the Firm Name of TRUMBLE GAS TRAP CO.,

Plaintiffs,

vs.

DAVID G. LORRAINE,

Defendant.

**Supplemental Bill of Complaint.**

Come now plaintiffs above named and for their supplemental bill of complaint in the above-entitled suit, filed by leave of Court, further complaining, allege:

That on or about December 31, 1920, plaintiffs filed in this court their original bill of complaint against defendant alleging the invention by Milon J. Trumble of a certain invention in crude petroleum and natural gas separators and the grant, issuance and delivery to plaintiffs of letters patent of the United States numbered 1,269,134 on June 11, 1918, as in and by plaintiffs' original bill of complaint set forth, ready in court to be produced will more fully and at large appear; that plaintiffs are now the owners of the exclusive right, title and interest in, to and under said letters patent and so have been ever since the filing in this court of plaintiffs' said original bill of complaint.

That since the grant, issuance and delivery of said letters patent, and since the filing of plaintiffs' said original bill of complaint, and within the Southern District of California, and elsewhere, defendant has manufactured, sold and caused to be used, and has offered for sale, crude petroleum and natural gas separators embodying and containing the invention [25] patented in and by said letters patent, particularly as pointed out and claimed by claims 1, 2, 3 and 4 thereof, and defendant threatens and intends to continue to manufacture, sell, offer for sale and sell the same, all without the license or consent of plaintiffs and in infringement of said letters patent; that said crude petroleum and natural gas separators so manufactured, sold, offered for sale and caused to be used by defendant since the filing of plaintiffs' original bill of complaint herein differs only colorably from the crude petroleum and natural gas separators manufactured, sold, offered for sale and caused to be used by defendant in infringement of plaintiffs' patent at the time of the filing of plaintiffs' original bill of complaint herein and each thereof, contains the combination of elements and invention pointed out, claimed and patented in and by each of said respective claims, 1, 2, 3 and 4 of said letters patent; that one of said crude petroleum and natural gas separators cause to be manufactured by defendant for defendant by Lacy Manufacturing Company, of Los Angeles, California, and by it sold on March 17, 1922, to General Petroleum Company, Los Angeles, California, and now in the possession of plaintiffs, ready in court to be produced as may be required, will illustrate and exemplify the crude petroleum and natural gas separators now being manufactured, sold, offered for sale and caused to be used by defendant as aforesaid.

That plaintiffs do not know, and pray discovery thereof, as to how many such separators defendant has made, caused to be made, or sold, or used, and prays discovery thereof, that defendant is realizing and has realized large profits from the manufacture and sale thereof and plaintiffs have suffered great loss, injury and damage thereby and are suffering great and irreparable damage and injury therefrom. [26]

WHEREFORE, plaintiffs pray that defendant be enjoined both provisionally during the pendency of this suit and perpetually, from manufacturing, causing to be manufactured, selling, or offering for sale, or disposing of in any manner, or using, any crude petroleum and natural gas separators embodying or containing the invention patented in and by said letters patent and particularly set forth, pointed out and claimed in and by claims 1, 2, 3 and 4 thereof, and particularly from manufacturing, causing to be manufactured, selling or offering for sale, using or causing to be used, any crude petroleum and natural gas separator like or similar to the separator manufactured for defendant by said Lacy

Manufacturing Company and sold on March 17, 1922, to the General Petroleum Company aforesaid; and that defendant be required to account to and pay over unto plaintiffs all profits, gains and advantages realized by him from the manufacture, sale or use or such separators and all damages suffered by plaintiffs by reason of such infringing acts and that plaintiffs have judgment against defendant therefor and for their costs herein and for such other, further or different relief as may be just or equitable.

FRANCIS M. TOWNSEND,
MILON J. TRUMBLE,
ALFRED J. GUTZLER,
Plaintiffs.

FREDERICK S. LYON,
LEONARD S. LYON,
FRANK L. A. GRAHAM,
Attorneys for Plaintiffs. [27]

State of California,
County of Los Angeles,—ss.

F. M. Townsend, being first duly sworn, on oath says: That he is one of the plaintiffs in the above-entitled suit, that he has read the foregoing supplemental bill of complaint; that the same is true of his own knowledge except as to such matters and things as are therein stated on information and belief, and as to such he believes the same to be true.

F. M. TOWNSEND.

Subscribed and sworn to before me this 23d day of March, 1922.

[Seal] L. BELLE WEAVER,
Notary Public in and for the County of Los Angeles, State of California.

[Endorsed]: No. E–113. United States District Court, Southern District of California, Southern Division. Francis M. Townsend et al., Plaintiffs, vs. David G. Lorraine, Defendant. In Equity. Supplemental Bill of Complaint. Service, by copy, this the 23d day of March, A. D. 1922, at 9:16 A. M. Chas. Bagg, Solicitor for Defendant. Filed March 23, 1922. Chas. N. Williams, Clerk. Lyon & Lyon. Frederick S. Lyon, Leonard S. Lyon, 312 Stock Exchange Building, Los Angeles, Cal., Solicitors for Plaintiff. [28]

---

At a stated term, to wit, the January term, A. D. 1922, of the District Court of the United States of America, within and for the Southern Division of the Southern District of California, held at the courtroom thereof, in the city of Los Angeles, on Friday, the 24th day of March, in the year of our Lord one thousand nine hundred and twenty-two. Present: The Honorable CHARLES E. WOLVERTON, District Judge.

No. E-113—EQ.

FRANCIS M. TOWNSEND et al.,
Plaintiffs,

vs.

DAVID G. LORRAINE,
Defendant.

**Minutes of Court—March 24, 1922—Hearing (Continued).**

This cause coming on at this time for further hearing; F. S. Lyon, L. S. Lyon and Frank L. A. Graham, Esqs., appearing as counsel for the plaintiff and Charles Bagg and L. L. Mack, Esq., appearing as counsel for the defendant and G. J. Kaarnelly being present as stenographic reporter of the testimony and proceedings; and counsel for the respective parties having announced their readiness to proceed with the trial of this cause and the Court having ordered that this cause be proceeded with; and

David G. Lorraine, a witness for the plaintiff, resumes the stand and testifies further; and

In connection with his testimony there is offered and admitted on behalf of the defendant the following exhibit, to wit:

Defendant's Ex. "D"—Photograph of Lorraine Trap; and

Drawings produced by said witness having been offered in evidence by the plaintiff as Plaintiff's Ex. Nos. $19^2$, and $19^3$, and counsel for the defendant having interposed his objection to the admission into evidence of said exhibits and the Court having sustained defendant's objections, and the offer having been withdrawn by the plaintiff; and

Now, at the hour of twelve o'clock noon it is by the Court [29] ordered that a recess be taken to the hour of two o'clock P. M. to take autos to go to Trumble Plant at Alhambra to view a Lorraine Oil and Gas Separator and all parties having met at the said plant; and

Now, at the hour of 2:45 o'clock P. M. all appearances as before except L. L. Mack, one of the attorneys for the defendant, who is absent; and

The testimony of O. W. Harris, a witness heretofore sworn for the defendant, and of David G. Lorraine, the defendant, heretofore sworn in his own behalf, having been given; now,

At the hour of 3:55 o'clock P. M. the said cause is continued to be resumed at the courtroom on Monday, March 27th, 1922, at ten o'clock A. M. [30]

---

At a stated term, to wit, the January Term, A. D. 1922, of the District Court of the United States of America, within and for the Southern Division of the Southern District of California, held at the courtroom thereof, in the city of Los Angeles, on Monday, the 27th day of March, in the year of our Lord one thousand nine hundred and twenty-two. Present: The Honorable CHARLES E. WOLVERTON, District Judge.

No. E-113—EQ.

FRANCIS M. TOWNSEND et al.,
Plaintiffs,

vs.

DAVID G. LORRAINE,
Defendant.

**Minutes of Court—March 27, 1922—Hearing (Continued).**

This cause coming on at this time for further hearing; F. S. Lyon, L. S. Lyon and Frank L. A. Graham, Esqs., appearing on behalf of the plaintiffs and Charles Bagg and L. L. Mack, Esqs., appearing on behalf of the defendant and John P. Doyle, being present as shorthand reporter of the testimony and proceedings, and counsel for the respective parties having announced their readiness to proceed with the trial of this cause and the Court having ordered that this cause be proceeded with and

David G. Lorraine thereupon resumes the stand for further examination; and

In connection with his testimony there is offered and admitted in evidence on behalf of the plaintiff the following exhibit, to wit:

Plaintiff's Ex. No. 20—Cut of Lorraine automatic Oil and Gas Separator, being page 46 of the Oil Weekly, of February 25, 1922.

and the following exhibits having been offered and admitted in evidence on behalf of the defendant, to wit:

Defendant's Ex. "E"—Certified copy of patent to George L. McIntosh;

Defendant's Ex. "F"—Certified copy of patent to Walter Anderson Taylor;

Defendant's Ex. "G"—Certified copy of patent to Arthur W. Barker;

Defendant's Ex. "H"—Certified copy of patent to Eustace Vivian Bray;

[31]

Defendant's Ex. "I"—Certified copy of patent to Augustus Steiger Cooper;

Defendant's Ex. "J"—Certified copy of patent to Albert T. Newman; and

Walter P. Johnson and Luther L. Mack having been respectively called, sworn and having testified; and

Now, at the hour of 11:51 o'clock A. M. the Court declares a recess to the hour of two o'clock P. M.; and

Now, at the hour of two o'clock P. M. the Court having reconvened and all being present as before; and

Wm. A. Trout having been called, sworn and having testified in behalf of the defendant; and

In connection with his testimony there having been offered and admitted in evidence on behalf of the defendant the following exhibit, to wit:

Defendant's Ex. "K"—Sketch made by witness; and

A. A. Wharff having been called, sworn and having testified in behalf of the defendant; and

In connection with his testimony there having been offered and admitted in evidence on behalf of the defendant the following exhibit, to wit:

Defendant's Ex. "L"—Pencil sketch made by witness; and

W. H. Swope having been called, sworn and having testified in behalf of the defendant; and

In connection with his testimony there having been offered and admitted in evidence on behalf of the defendant the following exhibit, to wit:

Defendant's Ex. "M"—Pencil Sketch; and

Wm. G. Lacy having been recalled to the stand and having given his testimony; and

In connection with his testimony there having been offered and admitted in evidence on behalf of the defendant the following exhibit, to wit:

Defendant's Ex. "N"—Letter dated June 14, 1921, F. M. Townsend to Lacy Mfg. Co. [32]

Defendant's Ex. "O"—Letter dated June 15, 1922, Lacy Mfg. Co. to Townsend.

Defendant's Ex. "P"—Letter dated December 29, 1920, Dr. W. P. Keene to Lacy Mfg. Co.; and

The following exhibit having been offered and admitted in evidence on behalf of the plaintiff, to wit:

Plaintiff's Ex. No. 21—Photo; and

George H. Prout having been called, sworn and

having testified in behalf of the defendant; and

Now, at the hour of 3:30 o'clock P. M. it is by the Court ordered that a recess be taken for ten minutes; and now

At the hour of 3:40 o'clock P. M. the Court having reconvened and all being present as before; and

Robert W. Smith having been called, sworn and having testified in behalf of the defendant,

It is now by the Court ordered, good cause appearing therefor, that this cause be continued to Tuesday, March 28th, 1922, at the hour of 10 o'clock A. M. [33]

---

At a stated term, to wit, the January Term, A. D. 1922, of the District Court of the United States of America, within and for the Southern Division of the Southern District of California, held at the courtroom thereof, in the city of Los Angeles, on Tuesday, the 28th day of March, in the year of our Lord one thousand nine hundred and twenty-two. Present: The Honorable CHARLES E. WOLVERTON, District Judge.

No. E-113—EQ.

FRANCIS M. TOWNSEND et al.,

Plaintiffs,

vs.

DAVID G. LORRAINE,

Defendant.

**Minutes of Court—March 28, 1922—Hearing (Continued).**

This cause coming on at this time for further hearing; F. S. Lyon, L. S. Lyon and Frank L. A. Graham, Esqs., appearing as counsel for the plaintiffs and Charles Bagg and L. L. Mack, Esqs., appearing as counsel for the defendant, and John P. Doyle being also present as shorthand reporter of the testimony and proceedings; and counsel for the respective parties having announced their readiness to proceed with the trial of this cause and the Court having ordered that this cause be proceeded with; and

An answer to the supplemental bill of complaint having been filed and the defendant having rested; and

W. C. Ray and Paul Paine having been respectively recalled by the plaintiff in rebuttal; and

A portion of Lorraine trap cut from Lorraine trap heretofore exhibited to the Court, having been marked Plaintiff's Ex. No. 22 for Identification; and

W. L. McLaine having been called, sworn and having testified in behalf of the plaintiff; and

M. J. Trumble having been recalled to the stand on behalf of the plaintiff; and

It having been ordered by the Court, on motion of counsel for the plaintiff, that the Plaintiff's Exhibit No. 22 heretofore offered for Identification be admitted in evidence, to wit: [34]

Plaintiff's Ex. No. 22—Portion of trap; and

There having been offered and admitted on behalf of the plaintiff the following exhibit, to wit: Plaintiff's Ex. No. 23—Pressure Gage; and

O. W. Harris, having been recalled to the stand and having testified; and

There having been offered and admitted on behalf of the plaintiffs the following exhibit, to wit: Plaintiff's Ex. No. 24—Harris Pencil sketch.

And now the plaintiffs rest; and

The defendants having rested; and

A stipulation having been entered into by respective counsel that Judge Wolverton may sign decree in this cause outside of District, said written stipulation to be filed, it is by the Court ordered, at the hour of 11:22 o'clock A. M. that a recess be taken to the hour of two o'clock P. M.; and

Now, at the hour of two o'clock P. M. the Court having reconvened and all parties being present as before and the plaintiff having moved the Court to strike from the record Defendant's Exhibit "C," being a pamphlet issued by the Department of the Interior, and the motion having been taken under advisement with the allowance of an exception on behalf of either party; and

F. S. Lyon, Esq., having argued to the Court on behalf of the plaintiff; and

Charles Bagg, Esq., having argued to the Court on behalf of the defendant; and

F. S. Lyon, Esq., having argued in reply,—

Thereupon said cause was submitted to the Court upon the oral arguments and upon briefs to be filed by the plaintiff on or before Monday, April third,

1922, and by the defendant within seven days thereafter, and by the plaintiff in reply within seven days thereafter; and

At the request of John P. Doyle, shorthand reporter as aforesaid, it is by the Court ordered that said John P. Doyle [35] be permitted to temporarily take from the files such of the paper exhibits as he may need in transcribing his notes; and

Thereupon the Court on his own motion having directed that two enlargements of the drawings of the Trumble patent which had been used to illustrate the testimony and arguments, be filed as Plaintiff's Exhibits Nos. 25 and 26; and

Thereupon at the hour of 4:00 o'clock P. M. the Court declared a recess to the hour of ten o'clock A. M. Wednesday, March 29th, 1922. [36]

---

United States District Court, Southern District of California, Southern Division.

IN EQUITY—No. E-113.

FRANCIS M. TOWNSEND, MILON J. TRUMBLE and ALFRED J. GUTZLER, Doing Business Under the Firm Name of TRUMBLE GAS TRAP CO.,

Plaintiffs,

vs.

DAVID G. LORRAINE,

Defendant.

## Answer to Supplemental Bill of Complaint.

I.

Comes now the defendant, David G. Lorraine, and for answer to the supplemental bill of complaint heretofore filed in this cause on the 23d day of March, A. D. 1922, states:

That he admits that on or about December 31st, 1920, the plaintiffs filed in this court their original bill of complaint against defendant alleging the invention by Milon J. Trumble, admits, for the purpose of this suit, that the plaintiffs are now the owners.

This defendant denies that, at any time, since the grant, issuance or delivery of said letters patent or since the filing of plaintiff's said original bill of complaint, or at all, or within the Southern District of California or any other place, he has manufactured, sold or caused to be used, or has offered for sale, any crude petroleum and natural gas separators or any other device, in any manner embodying or emulating any invention patented in or by said letters patent, or of claims 1, 2, 3, or 4 thereof. Defendant denies that he has threatened or is now threatening or intends to continue to manufacture, sell, offer for sale and sell, any device or apparatus in any manner infringing upon the device described in said letters patent, or any of [37] the claims thereof. Defendant denies that any crude petroleum and natural gas separator manufactured or caused to be manufactured by this defendant or for the defendant by the Lacy Manufacturing Company, of Los Angeles, or any other person, firm or corporation, sold to or purchased by the General Petroleum Company or now in the possession of plaintiffs will show any device that in any manner or way infringes upon any letters patent granted to the said Milon J. Trumble or to any other person, other than to this defendant, or to any claim or claims thereof.

This defendant denies that he is realizing or has realized large profits from the manufacture or sale of any device whatsoever which, in any manner infringes upon any letters patent granted to the said Milon J. Trumble or to any other person, or to any claim or claims thereof; and this defendant denies that, by reason of any acts or conduct on the part of this defendant, the plaintiffs or either or any of them have suffered or are suffering any great or irreparable damage or injury.

II.

And this defendant for another and further answer and defense to said supplemental bill of complaint hereby adopts, reiterates and reaffirms each and every all and singular, the deni*ves* and allegations set out in paragraphs IV, V, VI, VIII, IX, X, XI, and XII, of his answer to plaintiff's original bill of complaint, to the same extent as though actually set out herein.

WHEREFORE, this defendant, having fully answered said supplemental bill of complaint, prays that said supplemental bill of complaint be dismissed and that this defendant be permitted to go hence with his costs in this behalf expended.

DAVID G. LORRAINE,
By CHAS. BAGG,
Attorney for Defendant. [38]

[Endorsed]: Equity—No. E-113. In the United States District Court for the Southern District of California, Southern Division. Francis M. Townsend et al., Plaintiffs, vs. David G. Lorraine, Defendant. Answer to Supplemental Bill of Complaint. Filed March 28, 1922. Chas. N. Williams, Clerk. Chas. Bagg, Solicitor for Defendant. [39]

---

In the District Court of the United States for the Southern District of California, Southern Division.

(Before Hon. CHARLES E. WOLVERTON, Judge.)

No. E-113—EQ.

FRANCIS M. TOWNSEND et al.,

Plaintiffs,

vs.

DAVID S. LORRAINE,

Defendant.

**Reporter's Transcript of Testimony and Proceedings.**

APPEARANCES:

F. S. LYON and LEONARD S. LYON, Esqs., for Plaintiffs.

CHARLES BAGG, Esq., for Defendant.

Los Angeles, California, March 22, 1922.

INDEX.



[40]

It is stipulated by and between counsel for the respective parties that the reporter's notes may be immediately transcribed and the original copy filed, and that the reporter provide each side with a copy thereof and that originally each side shall pay one-half of the cost of the transcript and copies will be taxable as costs in accordance with the court order for costs.

Mr. BAGG.—So stipulated.

Los Angeles, California, Wednesday, March 22, 1922, 10 A. M.

The COURT.—Are the parties ready for trial in the case of Townsend vs. Lorraine?

(Both sides announced ready.)

Mr. F. S. LYON.—If your Honor please, this is

a suit for infringement of patent. It is the ordinary suit for preliminary equitable relief and also for an accounting of profits and damages. The plaintiffs are the grantees of the patent issued to Milon J. Trumble, and is for an invention embodied in a device for separating oil and natural gas as the combination issues from an oil well.

In the production of crude oil many of the wells contain large quantities of natural gas, and in the earlier times that natural gas was dissipated and wasted. The time came here in California when it was desirable also to conserve that natural gas and the problem that faced Mr. Trumble was to produce a device by which the oil and gas might be economically separated and the gas conducted off to the source of either use or storage and the oil conducted to such storage as desired or to the point of transportation.

The problems that presented themselves to Mr. Trumble he solved in a very simple manner, and I shall present to your Honor a copy of the patent in suit and also a reproduction of the main view of that patent on a large scale, so that it may be more readily understood. [41]

The device in which Mr. Trumble actually embodied his invention is illustrated in Fig. 2 of the patent, a reproduction of which I now produce, as a large drum or vessel (1) having a large chamber into which the oil and gas from the well came in—in the particular embodiment shown in the drawings of the patent—at the top, and the intermingled oil and gas was then delivered under the pressure of the oil, reduced somewhat as they always reduce it at the flowing nozzle of the well which would deliver on to a surface in the form of a cone in this embodiment where it was directed; also that it flowed down a shelf or wall of the container in a thin body, thereby, as the oil went down, giving all parts of the oil in a thin body a chance to relieve itself, as you may say, of the gas contained in the oil, and the gas then could flow up through the outlet without going through the body of incoming oil and gas again.

The one great feature of that invention was this: that, the thin body of oil being delivered in thin and comparatively quiet condition, the oil could be separated from the gas without the gas taking up from the oil as much of the light part or gasoline vapors as would be taken up and carried off by it if it was just blown in in a mist and in an entire whirl; and also it is found that if it were simply allowed without separation first to issue into a solid body with a large depth then the gas bubbles would find some difficulty more or less, depending upon conditions, in getting through the body of oil, and then also as, apparently, the gas bubbles expanded they would take up larger amounts of the gasoline or lighter contents of the oil.

The specific invention, then, embodied a chamber and a means whereby the incoming stream of oil was directed upon a wall where it flowed down, and the outlet for gas being free at one side or at the center, so that the relieved gas would immediately escape, collecting the oil in the bottom of the chamber, where it, by means of a suitable float-operated valve, could be discharged intermittently, the device being entirely automatic in that regard. [42]

As the evidence that I shall produce shall further explain the invention to your Honor in detail I shall content myself in the opening with simply stating that after the commercial installation of a large number of these Trumble traps in the oil fields of California, they having become the standard equipment for that purpose, the defendant, who had been previously in other business and who had become familiar with the Trumble gas patent, entered the field as a competitor. He at first manufactured a device which is covered by a patent which he took out, and subsequently he has changed the form of that device to a second form. The defenses, as I understand them, are solely a defense of noninfringement.

With that statement I will content myself for the present.

Mr. BAGG.—If your Honor please, I think counsel should specify at this time in what respect or in what claims his contention is that we are infringing that patent, and that the burden is upon him to show that we are infringing; and in his opening statement he should state in what respect, so that the Court and counsel may know, and in what claim, the alleged infringing device manufactured and sold by the defendant, infringes the patent, and the claims set out by counsel for plaintiffs.

I think we are entitled to know that, and I think the Court is entitled to know it.

Mr. LYON.—We charge that the devices of the defendant infringe Claims 1, 2, 3, 4 and 13 of the patent in suit. I thought that information had already been given.

Now, I might state, for your Honor's convenience, if you wish it, that Claim 1 calls for this general combination, and I will use the specific embodiment of this invention which is shown in Fig. 2 in illustrating that:

"The combination of an expansion chamber"—which is the chamber inside of the vessel (1)—"arranged to receive oil and gas in its upper portion"—the oil and gas coming in at the top—"means for spreading the oil over the walls of such chamber to flow downward thereover"—the means being a baffle or other directing [43] device which will direct it onto the wall so that it will slide down the wall in a thin body and, in the specific embodiment here, these cones or deflectors. "Gas take-off means arranged to take off the gas from within the flowing film of oil"—and that is the outlet here through this pipe, and, as you will notice, it is in between the body of oil in there, or if you only use one-half of it it doesn't mix with the stream that is on the shelf as I stated. "An oil collecting chamber below the expansion chamber." There is, down below, oil collected in here so as to be above this valve and to be discharged to the bottom. "An oil outlet"—at the bottom—"from said collecting chamber and valve-controlled means arranged to maintain a submergence of the oil outlet." Here is the float on the valve. and that is the valve-control means he is speaking of (indicating). So that that is just a general claim of a general combination.

Now Claim 2 differs from that mainly in that it calls for a means for maintaining gas pressure upon the oil in here, that is, means of suitable regulating valve, which will hold back the pressure of the gas to the degree or the pressure desired. And Claim 3 differs in this: "In an oil and gas separator, the combination of an expansion chamber having a surface adapted to sustain a flow of oil thereover in thin body"—that is the wall or any other surface over which the oil is directed so that it spreads out. "Means for distributing oil onto such surface." And the specific embodiment of the cone might be a nozzle so long as it is carried over and directed on to, as you see, means for distributing oil on to such surface. "Pressure-maintaining means adapted to maintain a pressure on one side of the flowing oil." Now, of course this valve alone wouldn't do it if the other end was open, so that when you speak of valve means there must be a pressure-regulating valve somewhere on the gas outlet and the oil outlet must also be closed or we would not maintain pressure. "Withdrawing means arranged to take gas from the chamber." That is the outlet pipe. "And means for withdrawing the oil from the chamber."

Now Claim 4 is substantially the same general claim: (Reading same.) And Claim 13 is the general combination of—(Reading [44] claim.)That claim is not specifically limited by inclusion, by positive word, of the outlet means—of the valve mechanism here (indicating).

Mr. BAGG.—If your Honor please, I still think it is incumbent upon counsel in making his opening statement to make some statement as to in what respect or what kind of a device we have built or put on the market which infringes his device. I think he ought to, for the benefit of the court as well as counsel for defendant, make some statement which would indicate to the Court what kind of a device we were manufacturing and in what respects it infringed the patent.

The COURT.—Is that device set out in the pleadings?

Mr. BAGG.—No; it simply sets out the fact that we are infringing; and we did not know until counsel vouchsafed the information this morning as to exactly what claims he is contending we infringe. Of course now he has outlined that, but it seems to me it is incumbent upon him to make a statement for the benefit of the Court and counsel and describe what kind of an apparatus we are putting out and how it infringes and in what respect.

Mr. LYON.—I have not the slightest objection to stating what our position is, your Honor, in regard to what the defendant's device is.

The COURT.—Very well; make a statement.

Mr. LYON.—In the first place, the answer alleges that the defendant is manufacturing its devices in accordance with a certain invention for which an application for patent is pending, and I refer in that regard to paragraph 12 in the answer; and I suppose counsel for defendant will admit that the application referred to in the paragraph was the application for the defendant's patent No. 1,373,664, which issued April 5. 1921, will you (handing patent to counsel)?

Mr. BAGG.—Yes.

Mr. LYON.—And that is the device you admit in the answer you are manufacturing and using?

Mr. BAGG.—Yes. [45]

Mr. LYON.—And manufacturing and using it as described and diclosed in that patent?

Mr. BAGG.—Well, not exactly; no. We will not admit that we are making a trap just exactly as those drawings indicate.

Mr. LYON.—Well, substantially, so far as the mode of operation—

Mr. BAGG.—Substantially in conformity with the patent, yes.

Mr. LYON.—In the device, the Loraine trap, we have an oil and gas inlet from the wall which is indicated as 14. In this enlargement that I now present before you, your Honor, the oil enters into the general expansion chamber 2 at the side. Mr. Loraine instead of using the whole of the round chamber, has used only half of the chamber in this construction, cutting down the capacity as you will see in a moment, keeping the entire mode of operation the same. The oil and gas enters through the pipe 14, are diverted into a separator or directing device 17, the oil and gas being directed in a thin body along the wall 2 until the oil, with such sand or water, or whatever it contains drops into the bottom of the chamber. The gas in such a device flows up under the separator, through a port and out through the gas outlet. Part of it, may, perhaps, although the drawing is not clear, go through a little port I now point to. The specification of the patent says nothing about whether it does or does not. In the actual device that we have examined, the escaping gas went up underneath the separator, and corresponding in that regard and even in that detail of construction to the plaintiff's operation, where the oil and gas comes down and run on the side of the wall in a thin body; the gas goes up through underneath and through this outlet, corresponding in detail in that regard. The device is also provided with a float valve mechanism by which the amount of oil is retained in the device until it gets to a certain level and then automatically discharges in the same manner. The gas pressure regulating-valve 28 as shown in the specifications is not on this enlargement, because this enlargement is without figure 2, but it [46] is shown and described in the specifications and it is referred to in a number of places in the body of the description.

Now we contend, then, that in this device patented to *Loraine,* he has, as called by claim 1 of the patent the combination of an expansion chamber and it is immaterial for that purpose whether we refer to the whole of his chamber here or just this portion that is on the side, because that is an expansion chamber and the gas comes out there, some gas may come out here, depending upon the level of the oil. (Indicating.)

The COURT.—Is this a complete chamber?

Mr. LYON.—Well, it is complete to this extent: This is a wall which is cut right across on that (indicating). Mr. Harris, have you got a model? We have a model.

The COURT.—It cuts off a segment?

Mr. LYON.—It cuts off a segment of the chamber. I have got a glass model that it can be seen on very much more readily. This glass model will illustrate all except the float. As your Honor understands it is made of glass for convenience. This wall here represents that wall (indicating), and here is 17; here is the inlet pipe; here is the gas escape there—or some of it might escape through in there, we don't know, the patent doesn't tell us that. Now, then, after that was manufactured we brought this suit, and I show your Honor that each one of these four claims apply to that device, because it has, as called for in the claims, the expansion chamber adapted to receive gas; in its upper portion it has means for spreading the oil; it has got practically the same means, except he only uses half of it, over the wall of such chamber, in a thin flow downwardly, the oil flowing downwardly here, and we will demonstrate that by simply putting water in there. If your Honor wants to, you will see that the water under a head pressure will flow down because of the take-off means to take the gas from the thin flowing film of oil, an oil collecting chamber below the expansion chamber; and then he has his oil outlet which we have not shown in the model. The oil outlet is here (indicating), that is for the oil, water and gas. [47]

The COURT.—For the oil, water and gas.

Mr. LYON.—Yes. And the oil outlet is from this pipe here,—in this particular view it doesn't show—it comes right out to this pipe 32 which is connected with this here, and he maintains at least that much of a body of oil anyway, so that when he maintains that height of oil, the oil being up here (indicating), you see this is his expansion chamber (indicating), and he uses only this portion of the device for the operation. Possibly a small amount of interned gas here might escape if there was any; if so there is an open communication at the top and it can go over here to where the pressure that is maintained on this whole system by the valve that I pointed out is on the whole anyway. And here is his float and valve for the valve control means.

Now, I pointed out that he has the valve maintaining pressure, if desires, and to regulate it as desired, and when we take the third claim, "in an oil and gas separator," the combination of the expansion chamber having a surface adapted to sustain a flow of oil thereover in a thin body" there is his surface and there is his chamber as he uses it (indicating). There is the surface of the thin body of oil flows down so as it rolls over the interned gas can come out readily, and the gas is separated without picking up a large quantity of gasoline vapor and that it may readily get out, and another object, where the gas pressure is high, the purpose of maintaining the pressure here is to prevent formation of a heavy emulsion which cannot break.

Take the next element, means for distributing oil on to the surface, he has practically identically our same means only not quite so full circumference of cylinder, of full chamber—he cut it in two. He has a pressure-maintaining means adapted to maintain a pressure, that is this valve (indicating), and he has a withdrawing means to take the gas off, and he has means for withdrawing the oil. Those are the claims of claim 3 and they may be pointed out in claim 4 in the same manner. [48]

Now we come down to the more limited claim, 13, which is the most limited claims of the ones we are charged to infringe. We find it is "in an oil and gas separator, an expansion chamber; inlet means for feeding a foam composed of oil and gas through the central part of the expansion chamber; and imperforate spreader cone,"—without perforations in it—"having its apex pointing upwardly," it is up at this end (indicating), contending that that is the absolute equivalent, cutting off half of it, making no changes in the mode of operation or the effect, only cutting off the capacity, and that will be the evidence, we believe. (Continuing reading.) "located inside said chamber in such a manner as to spread a thin film of oil over the inner wall of said chamber," and the wording "inner" there means as distinguished from the outside, because in our device also we don't have a wall like that so that it is unnecessary to this wall we are speaking of, but it is the inside of a shell and not the outside that is mentioned by the word "inner" there—(Continuing reading:) "and means for taking gas from the central portion of said chamber." Mr. Harris, will you let me have a model of our device and the model 1922? I made a glass model. There would be the type of ours with two cones (indicating), and you see the gas outlet under that cone, or under that one (indicating)? The oil and gas comes through the top. Of course, on this model we haven't put in the valve mechanism, the float mechanism—of course you readily understand the patent is not limited to that.

Now Mr. Lorraine, we have discovered since the suit was brought, also has changed this device of his and attempted to retain our invention and yet get away from some of the appearance of it; so that instead of using the particular deflector which is shown in the patent to Lorraine and the particular gas outlet there he has put into the shell of his device two partition members or plates, and he has brought his gas and oil in at the top. He has put in a nozzle which directs the oil the same as our spreading device of our patent. There is this plate which corresponds to [49] his plate or partition 19 and also on to the inner face of the shell of his chamber, so that it flows down in a thin film down both the inner face of the chamber and to some extent to the face of the plate on the other side, which corresponds as a matter of fact to the other side, the other near side of our chamber, in effect.

The COURT.—Well, how is that deflected on this side, anyway?

Mr. LYON.—Well, I am going to give you that in just a minute. Now he takes off his gas from the top here, through this cross pipe, and it crosses over and goes down this chamber that is formed by this one down under there and up out through. He gives his gas a narrow passage up at the top there in order to get out. You asked me how the gas got out, didn't you?

The COURT.—No, I asked you how the oil was deflected on this side of the chamber, on the side of the chamber (indicating)?

Mr. LYON.—By this pipe. The oil comes into the pipe at the side here, and then this nozzle device, because it is flared in the manner it is distributes the oil in a thin film on both of those plates on this side. I can illustrate that very readily if your Honor desires by simply getting a can of water and attaching it here and you can see it poured down there, then, in that device. We maintain that this device infringes all of these claims with the possible exception of claim 13. Claim 13, as I have pointed out, called for the specific "imperforate spreader cone, having its apex pointed upwardly" and that has this cone here with its apex pointed upwardly. Now, perhaps, unless under a direct doctrine of equivalents, that is not found in this device. This device with the turn and the flare does exactly the same thing, but it is not necessary in this case that we find that that claim; claim 13, is infringed in this device, because if we apply it to the other four claims we find that they are in no manner limited to the particular spreading device that we have. On the contrary, claim 1 shows a "means for spreading oil over the wall of such chamber to flow downwardly thereover," but it is in claim [50] 2 and claim 3 which says "the combination of an expansion chamber having a surface adapted to sustain the flow of the oil thereover in a thin body, means for distributing oil onto such surface." It is not the cone it is limited to, and claim 4 is the same; so that the claims of the patent as they are allowed and issued do not limit us to the specific details of construction but are broad enough to cover the really broad invention of this man Trumble did make. I think that complies even with the request of counsel and I will offer in evidence, first, the patent in suit, being the Trumble patent number—

The COURT.—Did you have any more remarks to make?

Mr. BAGG.—I would like to make a statement for the defense.

The COURT.—Very well.

Mr. BAGG.—As counsel on the part of the defendant, we desire to state in the first place, this patent issued to Mr. Trumble is a combination patent, every one of whose elements are old, and that there is nothing new except perhaps in the combination of the old elements. This evidence will disclose the fact that the principle of all oil and gas traps is exactly the same, that they work on the principle of, I presume the best way, I might not be correct in my geography, but we will take the Great Lakes; Lake Huron and Lake Erie are connected by two rivers. The water flows from Lake Huron into Lake Erie and spreads out through that great space of water; then at the outlet of Lake Erie, as I understand it—as I say I might not be correct in my geography—but as I understand the Niagara Falls is there, and the water then rushes out there and as the water comes in it comes in with great force and great rapidity, and as it goes out it goes out with great force and great rapidity in the same way, but in the center, because of that large expanse of the lake there is practically no movement. Now the principle of all oil and gas separators is based upon that one principle, that you put a large quantity of oil and gas as it comes from the well, and as counsel on the other side has described it, it comes there thoroughly impregnated in a great many instances with various degrees of oil and gas and water [51] and sand. Now, the purpose of these oil and gas separators as he states is to allow these various elements to separate with the degree of their specific gravity, of course the sand falling to the bottom, the water next, the oil next and the gas coming off on top. This accomplished by this flowing down process. As the oil comes rushing into these oil and gas separators the capacity of them being so much larger than the inlet, it naturally flows down to practically no movement of the oil in this separator. That gives then the same a chance to drop to the bottom, the water next and the oil next and the gas comes out on top. Then there is a means of outlet; instead of having one outlet they have one outlet for the gas which is up high on top where the gas floats to; the next is the oil outlet which allows the oil to pass on into its proper receptacle, and then at the bottom there is a vent, a place where they can take out the sand and water which is waste, of course, saving the two valuable elements, the gas and the oil, and carrying it respectively to their places of use. Our contention will be that this principle in these oil and gas separators are almost as old as the oil industry itself; that the first oil and gas separator, our evidence will show, was made in 1856, and that there have been since that time many, many kinds of oil and gas separators made, and we don't know exactly how many, but there have been hundreds of them, probably thousands of them made; that there is nothing new in any of them so far as the principle of the construction of these various elements which controls the petroleum as it comes from the oil well. There is nothing new in the principle. The only thing that is new is that it may have different combinations such as counsel on the part of the plaintiff has described for his particular instrument. We are not going to contend that the patent granted to Mr. Trumble is insufficient or invalid. Now then, that being the case, his patent being a combination all of whose elements are old, our contention will be that unless he shows that we, by our apparatus do exactly the same thing that he does and exactly in the same way, we do not infringe, and that unless [52] we have all of the elements in our combination, if ours is a combination, that he has in his combination or their equivalent we do not infringe. Our first contention then will be that in every element or in every claim that he has he has what is called an expansion chamber. Our contention will be that we have no expansion chamber in our trap; that because of the history of his invention and because of the developments that took place in the patent office at the time his patent was pending, as shown by the file-wrapper which we will introduce, he is precluded from claiming anything except exactly what he describes in his patent. In other words, that his will be limited to cones in each and every instance, and a cone is not a *bassle* plate, and of course I take it that your Honor understands that just as well as we do.

The next contention will be that we do not spread the oil in any of our apparatus or any of the traps that we have put out in any thin film. We will contend that he is bound to spread the oil over the surface of his retaining wall or his retaining sides of his trap in a thin film. Our contention will be that the word "thin" and the word "film" mean exactly what they say and that the oil must be spread in a very thin film. Now, our contention will be that, while perhaps the word "thin" is a comparative term, what might be thin in one instance would be as grossly thick in another, but when you come down to describing a film, our contention will be that the cause of the interpretation placed upon it by his representatives at the time the patent was being granted, or being—at the time the patent was being carried through the patent office, that they are limited to a film of oil and not to any thin body of oil.

Our contention will further be that we do not spread a thin film or even a thin body of oil upon the face of this chamber; that this deflector, as he has described it here, is so wide in our trap as we make it, that is the space between it and the wall is so wide, that it is impossible for any film to be placed upon the side walls of the retaining chamber. The record will probably show that this is two inches thick, that is this space [53] around here is two inches between the edge of this cone or the edge of this baffle-plate or deflector, and the wall receptable into which this oil is pitched. We will contend that it is not a thin film in any sense of the word, is not even a thin body of oil compared with the size of the trap itself, and certainly not a film of oil. Our contention will be as stated before, that he is confined entirely, by reason of the state of the arc to the thin film of oil which means practically no body to it at all, no thickness. Now then, our further contention will be that this baffle-plate is nothing more than—serves no other purpose than to divert the oil to the side wall in order to prevent the stirring up of the body of the oil as it settles in the deflecting chamber. For instance, our contention will be that if this oil coming in here with the force that it does from a great many oil wells, or even from a pump, coming in here and allowed to shoot right down into this body of oil that is down here (indicating), would keep it so thoroughly stirred up that it would get no chance to settle, would just keep a stirring up process all the time and would keep that in such a shape that it would be impossible for the gas to come off from it or for the oil and water and sand to separate. It would just keep it stirred up and it would accomplish no purpose whatever; that the only purpose of this is simply like we have learned in drawing out soda-pop out of a bottle—of course we couldn't make any other illustration in the days of Volstead, these days—but you will notice that we as children learned that when we pour out soda-pop out of a bottle, when it foamed up the proper way to do to keep too much foam from coming on it, was to let it come out of the bottle and strike against the wall of the glass and then keep it from being stirred up. Now, that is our contention that this foam and stuff coming out from the oil with great force, if allowed to shoot right down into this body would keep it so thoroughly stirred up that it would never separate. Now those are our contentions.

The COURT.—I understand you to say, then, that the deflection of the oil on to the wall does not operate as an element in separating [54] the gas from the oil?

Mr. BAGG.—It operates in this way, that it keeps —no, it doesn't have anything to do with the separation of the oil from the gas. It simply prevents this oil—if it would shoot right down from here (indicating), keeping this body stirred up so that it wouldn't give an opportunity to stir up and separate the sand and water and oil and gas; it would simply keep it stirred up.

The COURT.—I understand you claim that that element that causes the oil to run close to the side of the chamber is not useful as an element in separating the gas from the oil?

Mr. BAGG.—We don't know; we are not prepared to say what effect it has if put in a very thin film like he has described it. We say we don't put any thin film on the side walls respectively. We just simply have a diverting process, and as a matter of fact if your Honor please, you will notice here according to their own description this tube here is split (indicating); then the oil coming out here would shoot over to this side and would shoot over to this side and wouldn't spread over here at all (indicating); it would just simply shoot. The only purpose of this whole thing is to spread this out so that it will not plunge right down into that body of oil in our case and keep it stirred up—simply a deflector just simply to change the direction of the oil as it comes in.

The COURT.—Now, you may proceed with your testimony.

Mr. BAGG.—If your Honor please, just a moment. Oh, yes, and another thing: Our contention will be that our oil simply comes down there in a more of a floating form, while theirs comes down with force. Our contention will also further be that in none of these traps in which we have this deflecting means is there any means for obtaining a pressure. Our testimony will further show that we have no pressure in our trap at all unless perhaps it might be back pressure resulting from some stoppage or restriction of the passage at the point where it is being used. Possibly, for instance, we are using it in the boiler, under a boiler; when the gas is being thrown into the boiler, it is restricted; the passage of course; that would necessarily throw a little back [55] pressure on his gas line and might cause a little pressure in there, but it would be very little. The testimony will show that practically none of the gauges—we have a gauge on our gas traps—show practically no pressure in our trap at all, that is in the one on which this deflector is has no pressure at all on it and that the gauge so shows. Our contention will further be that on the Trumble trap, in order to accomplish the purpose that he sets out it is absolutely necessary for him to maintain a pressure in there because his whole theory of the trap is based upon the fact that the oil comes down here and spreads out over the walls of this expansion chamber as he calls it; that there is a pressure put upon the outside which has a tendency to squeeze that gas and press that gas and oil—or gasoline, lighter hydrocarbons into the oil so that the gas as it comes out is not impregnated with the gasoline and what we call dry gas. Now, it is absolutely essential, our contention will be, for his trap to operate at all to have this pressure in it, and that is why he insists in all of his claims in having a gas pressure maintaining means which as he shows here in his—well, he showed it in his illustration what would correspond to the valve of an engine, that is in the gas line, a means by which when the pressure gets too great it closes or opens as the case might be and allows the gas to go out unrestricted; but he must have a restriction in his gas outlet line in order to produce and maintain a pressure on the inside of his trap against this flowing thin film of oil. Now, of course, a large portion of this will be brought out when we introduce the file-wrapper showing the history of his patent as it came through the office.

The COURT.—You may proceed, Mr. Lyon.

Mr. LYON.—Plaintiff offers in evidence the patent in suit 1,269,134 and asks that the same be marked Plaintiff's Exhibit 1. I think we had better use a numeral. We offer in evidence as Plaintiff's Exhibit 2 the filing-wrapper and contents of the Trumble application upon which exhibit 1 was issued. We offer in evidence under the admission and stipulation of counsel for defendant [56] made in open court this morning, copies of the Lorraine patent, 1373664, dated April 5, 1921, as Plaintiff's Exhibit 3; and as Plaintiff's Exhibit 4, a copy of the reissued letters patent No. 15220, dated November 8, 1921, as issued to Mr. Lorraine.

Mr. BAGG.—I think, if your Honor please, that hasn't been certified, has it?

Mr. LYON.—It is a printed copy. Well then, suppose, we don't bother the Court by putting in certified copies?

Mr. BAGG.—No, that is not necessary. That is all right.

Mr. LYON.—Now, will counsel for the defendant admit notice and demand from the plaintiff that defendant was infringing prior to the commencement of suit?

Mr. BAGG.—I think we deny that in the answer.

Mr. LYON.—Or will they admit that each one of the Trumble traps that were manufactured and sold by the plaintiffs bore the name of the Trumble Company's trap and the words "Patented June 11, 1918"? I ask these questions in order to possibly save formal proof.

Mr. BAGG.—All right, we will admit that.

Mr. LYON.—You will admit the notice and the markings?

Mr. BAGG.—Yes.

The COURT.—That is prior to what date?

Mr. LYON.—That is prior to the commencement of the suit, and the marking has been with the plate. We will just offer this plate in evidence as Plaintiff's Exhibit 5, and it shows the date "Patented June 11, 1918," and all the devices manufactured by the plaintiffs under the patent in suit have been so marked since the issuance of the patent. That is the stipulation. Mr. Paine, will you take the stand, please. [57]

**Testimony of Paule Paine, for Plaintiffs.**

PAULE PAINE, a witness called on behalf of the plaintiffs, being first duly sworn, testified as follows:

Direct Examination.

(By Mr. LYON.)

Q. Please state your name, age, residence, occupation.

A. Paul Paine; aged 40; residence, 607 Parkview Avenue, Los Angeles; controlling engineer.

Q. Are you connected with any oil companies at the present time?

A. I am not in the executive organization of any oil companies. I am a director in the Union Oil Company of California.

Q. And of what school are you a graduate, Mr. Paine?

A. I graduated—I have my training in, and in 1905 graduated from the Massachusetts Institution of Technology, more commonly called the "Boston Tech."

Q. As an engineer? A. As an engineer.

Q. After leaving that school did you have any connection whatever with the oil business?

A. I came west immediately and was in mining work for several years, and in 1909 entered the oil fields in the Midway field of California near Taft. I was engineer of the Honolulu Consolidated Oil Company. In 1911 I became assistant superintendent and in 1914 was made superintendent in charge

(Testimony of Paule Paine.)

of all the field operations for that company. The company was in the business of producing—of drilling oil wells, producing oil, producing gas, the sale of gas and the erection and operation of gasoline plants for the recovery of gasoline from the natural gas. In 1917 I left there and took charge of the operating departments in Oklahoma and Kansas of the Gulf Oil Corporation. In 1920 I entered business for myself as an independent engineer and have been occupied in that direction since then except for a nominal period when I was in the active organization of the Union Oil Company of California.

Q. Did you ever having anything whatever to do with the Trumble gas trap manufactured by the plaintiffs? [58]

A. Yes, I installed those on the property of the Honolulu Consolidated Oil Company. We were in the gas business, and at that time were supplying the major portion of the natural gas supply which came to Los Angeles.

Q. Approximately when did you install the first one of those Trumble gas traps?

A. In the latter part of February, 1915.

Q. Well, will you briefly describe that trap as installed at that time?

A. I had noted the operation of this trap on an adjoining property, and we had a well to which I considered it could be applied advantageously. This well was Well No. 3. Do you wish that detail?

Q. Yes.

(Testimony of Paule Paine.)

A. Well, Well No. 3 of the Honolulu Company on Section 10, township 32–24. This well was completed as a flowing oil well in August, 1914, and up to that time had been flowing this oil and gas out into an ordinary receiving tank, the oil being collected and moved from there to a shipping tank and the gas escaped into the atmosphere, and that was *was* entirely lost.

It was highly desirable to us that the gas be saved because we had several different uses for it, and these used obtained at different pressures. So this trap was installed at that well which was at that time flowing through a flow plug. The pipe-line which comes from a well in many instances has inserted in it a restricted opening, which is a solid piece of pipe of length from 6 inches to 12 inches with a small opening drilled through it—the opening in this case was five-eighths of an inch. The purpose of that is to restrict the flow of the well and to prevent it from flowing actively as it might. The reason for it being that we found that when the wells were wide open and flowed so vigorously they tended to bring in sand at the bottom of the well and stop the well altogether through this accumulation of sand at the bottom of the well. It had therefore become quite common to insert these flow plugs into the line. The effect of this flow plug, in addition to [59] maintaining probably a smaller initial production but a longer period of flow, was also to increase the pressure on the casing in the well through the fact that it backed up the pressure as

(Testimony of Paule Paine.)

we call it. After passing through this flow plug the oil and gas were conducted into a Trumble trap. The gas outlet was brought out to a near-by point and the gas permitted to escape in the atmosphere in the beginning. The oil outlet was extended to a receiving tank. The receiving end comprised two tanks each with a capacity of 5,000 barrels.

Q. What was the object at first in allowing the gas to escape into the atmosphere with that trap?

A. We were undecided at that time as to what effect upon the flow of the well the maintaining of the pressure on the trap might have. Obviously if the gas and oil outlets from the trap were closed everything would remain quiet and the well would not produce any oil or gas. Now, if it is allowed to flow wide open the effect as far as pressure is concerned would be practically the same as if it were going to a tank, and if the pressure on the trap were increased we were not entirely sure as to what the effect would be, therefore, the trap was allowed to operate in that manner for a period of a number of days, I don't recall how many. It was then decided that we would endeavor to save that gas and apply it to some beneficial use. The operating conditions on the property at that time were as follows: The fuel system of gas for the property operation operates under a pressure of from 20 to 40 or 50 pounds. If that gas could be saved at that pressure it could then be utilized for fuel purposes. If gas could be saved at a pressure of 75 pounds the gas could be delivered into the mains of the Midway

(Testimony of Paule Paine.)

Gas Company which was the company that obtained gas at Taft and brought it to Los Angeles for distribution in Los Angeles, and this vicinity. If the gas could be saved at a pressure of 175 pounds that gas could then be sold to the Southern California Gas Company, which was a local gas company engaged in the collection and distribution of natural [60] gas in Kern County. The price obtained in either case would have been 5 cents per 1,000 cubic feet, but because of contractural conditions with those two concerns it was desirable on our part that we save this gas at the higher pressure, if possible. The procedure was then started with the trap of allowing it to operate at these different pressure conditions. It was first allowed to operate with no pressure at all against the gas outlet. The gas escaped into the atmosphere. This continued for a three-day period. The reason for three days was that the well was producing about 1,200 barrels per day, and that provided about 3,600 barrels to this 5,000 barrel tank. At that time the oil was switched over into the second tank and the first tank of oil delivered to the Standard Oil Company. A valve on the gas outlet of the tank was then closed down slightly until the pressure on the inside of the Trumble trap registered, by means of a gauge attached to it, a pressure of 34 pounds. That was accomplished by just pinching the gas valve until it restricted the flow of gas sufficient to allow the gas to back up in the trap. We operated in that manner for a three-day period. Then when the

(Testimony of Paule Paine.)

tanks were switched the valve was closed down still further until the pressure held was 75 pounds. This continued for three days and the valve was closed down still further and the pressure of 177 pounds held upon the trap. No higher pressure was placed on the trap because there was no advantage to be gained by endeavoring to save the gas at a still higher pressure. During each of these three-day periods a number of observations were made of the volume of gas escaping, the quality of the oil which was obtained from the trap, the character of the gas which was escaping and the net loss. Observations were made on each day and then at the end of each period the average of these observations was ascertained so that we had those four periods of observations with no pressure on the gas outlet, with 34 pounds pressure, with 74 pounds pressure and with 177 pounds pressure.

Q. What was the gas pressure on the well itself?

A. At the beginning of the test the pressure was 380 pounds to the square inch, that is the pressure at the top of the casing [61] before the fluid passed through the flow plug. The oil was sold from the property under a contract providing for a graded payment dependent upon the quality of the oil. The quality of oil is registered in oil fields practice *ay* its gravity. The method of recording that is not in terms of specific gravity, but it is expressed in gravity baume as it is called. The weight of a fluid of 10 degrees gravity oil is the weight of water. As the fluids become lighter

(Testimony of Paule Paine.)
the gravity is expressed in higher terms of gravity baume. Thus 20 gravity oil is lighter than 15 gravity oil; 30 gravity oil is still lighter, and while the gravity of the oil is only a general index of its quality it is the expression customarily used in the oil field to register its value because the lighter oils, that is the oils of higher grade, as expressed in the Baume scale, usually contain the larger proportions of gasoline, and the lighter factors which have the greater value in commercial markets. So the price paid for this oil varied with its gravity. The higher the gravity the more money obtained for it. Now the oil ran from the tank to the Standard Oil Company. When the trap was operating with the gas escaping into the atmosphere it returned as having a gravity of 29.3 degrees. These determinations of the gravity of oil in the tank were not made by us, but were determined by the gauger of the Standard Oil Company who sampled the tanks and test the oil to ascertain its quality. We found that as we raised the pressure on the trap there was an increase in the gravity of the oil and accordingly in its quality. At 35 pounds pressure the gravity of the oil as delivered to the Standard Oil Company had increased from 29.3 to 30.9; at 74 pounds pressure the gravity of the oil had increased to 31.2. It so happened that at that time one price was effective for all from 29 gravity up to 31; and an increased price of 5 cents per barrel was effective for 31 and better. The application of this pressure therefore

(Testimony of Paule Paine.)

resulted in an increased value to the oil of 5 cents per barrel or approximately $60 per day. Now when the pressure was increased to 177 pounds it was found that, while the oil apparently had a still higher gravity when it went into the [62] shipping tank, that it didn't retain that still higher gravity but was 31.1 gravity when its quality was ascertained by the Standard Oil gauger. So much for the quality by increasing the pressure maintained in the trap.

Q. Well, why was this increased to 177 pounds and the increased gravity not fixed? When that oil was sold why did it drop back by the Standard Oil gauge?

A. Well, that will of course call for a conclusion on my part. The reason I will give for that will be this: That at the higher pressure, still higher pressure maintained upon that well certain gases were retained in the oil which went into the solution in the oil. Now, these gasses were always gasses. They were not gasoline vapors that had been retained in the oil in virtue of maintaining this higher pressure on the trap, and those gasses which are fixed gasses and are incondensable under the ordinary agreement of pressure or temperature, escaped from the oil in the shipping tank, and it is altogether possible—I observed it on occasion—that the escaping of those gasses carried along with them small quantities of gasoline vapor, so that I have had instances, similar instances where, through the maintaining of an unduly high pres-

(Testimony of Paule Paine.)

sure we arrived at quality of oil which was lower than that which would be obtained in a pressure range of from 75 to 100 pounds held on the trap.

Q. Now, proceed, Mr. Paine.

A. A collateral line of observations was made upon the quality of the gas escaping and upon its quantity. The quantity registered just about a million cubic feet per day. The weight of the gas was observed. The weight of gas is registered in specific gravity with reference to air as a unit, with the weight of the air at 1. The weight of gas is usually lighter of course and has smaller percentages which represent its weight as compared with air. The specific gravity of this gas with no pressure held upon the trap was .83. The gas at that time was escaping in a dense white cloud caused by the mist resulting from the gasoline vapors. At 36 pounds held on the trap the specific gravity of the gas declined to .79. At 74 pounds held on the trap the specific [63] gravity of the gas was .75 and 177 pounds held upon the trap the gravity of the gas was .70 resulting in a total decrease in the weight of the gas from .83 to .70. A further result of the installation of the trap was found to be an actual increase in the production from the well which was saved and sold. That was due to the large production of the well.

The COURT.—That was the gas that you saved and sold? A. No, sir; the oil.

Q. The oil?

(Testimony of Paule Paine.)

A. Yes, sir; due to the large production of the well, around 1200 barrels per day. It was impossible to ascertain exactly how many barrels were produced from it in every 24 hours due to difficulties of gauging and the fact that it would never be gauged at exactly the same time, and the further fact that wells frequently make a greater rate of flow at one time that they do at another. But the daily average production of the well had been observed over a long period, that is from August up until March, at that time. The well was of course declining in its rate of flow. It had by that time reached a very settled rate of decline, which was shown diagramatically, so that it was expected that during the ensuing months a certain production would be obtained from it, and it was found that the actual production from the well during that ensuing month was about 40 barrels per day in excess of that which had been expected. Following that month the rate of decline continued along the same rate as had obtained prior to the installation of the trap.

Mr. LYON.—You mean with the 40 barrel increase?

A. With the 40 barrel increase, as noted.

Q. Well, this 40 barrel increase then was what kind of a saving: was it gasoline vapor or what?

A. It must have been a saving of the gasoline vapors, because the gasoline vapors under these higher pressures, the dense white cloud at 36 pounds; the dense white cloud was not nearly as

(Testimony of Paule Paine.)
cloudy but it was still cloudy; at 74 pounds it had practically all disappeared, and at 177 pounds the color was all gone from the [64] gas and it was entirely colorless. These gasoline vapors had been prevented from escaping from the trap then with the gas and had been recondensed with the oil. As far as practical a check was made of that through ascertaining the difference in the weight of the gas at a specific gravity of .83 as compared with the same volume of gas at a specific gravity of .70 and the difference in weight of that gas was just about the weight of 40 barrels of gasoline of a gravity Baume, ranging from 70 to 75 which is the character of gasoline one would expect to be recondensed, or something on that order of magnitude. A further check could be provided, although the practical operations were not carried along with great enough degree of certitude to permit it to determine the kind of gasoline which would be necessary to raise 1200 barrels of oil of 29.3 degrees Baume gravity to a gravity of 31.1 degree. And it was in the same range. The ultimate effect therefore of the trap was first to conserve and utilize about one million feet per day of gas worth $50 per day, an increase in the value of the oil of about $60 per day, and an increase in actual oil saved from the well of about 40 barrels per day having a value of, at that time, of about $15 per day.

Q. Before passing from this general subject of the oil pressure, now why is it, Mr. Paine, that

(Testimony of Paule Paine.)

the gas may be readily separated from the oil as it is discharged through the flow plug of a well and yet not as readily separated, and I mean the gasoline vapor, in particular, from the natural gas if the gas and gasoline vapor are allowed to mix and then afterwards attempt to separate it?

A. That is due to a rather deep question of physics that has come into importance in connection with the manufacturing or increasing of gasoline from natural gas, the principle of "partial pressures" as it is called, which is this: If there are some vapors of gasoline in a gaseous state in gas and these vapors are there in a comparatively small proportion, such vapors if alone and not mixed with other gasses may condense at comparatively low pressure. One can have a gas for instance which is a gas vapor which is condensable at 10 pounds pressure if applied to it. Now, [65] if that gas is mixed with other gasses which are practically noncondensable in proportion of 5 per cent, say a 10 pound or 20 pound pressure then applied to the gas will not condense those vapors, but the pressure must be according to increase in the ratio of this dilution. If it were present there at 5 per cent then the pressure would have to be increased twenty times that which was required to condense it when it is alone and not intermingled with these noncondensable gasses. Now, if these gas vapors escape from the oil along with the gas they can then be carried along to a plant and compressed to a high pressure and cooled sufficiently to re-

(Testimony of Paule Paine.)

condense them. Then after they are once recondensed and separated from the other gas they can be obtained in a liquid state by the application of a comparatively low pressure, and that is a principle which applies to the preventative side of preventing the escape of gasoline vapors along with the gas; after they once escape into the other gas which is not condensable the separation and the saving of them is much more difficult and *expense.*

Q. Now, you have referred to the use on this Honolulu well in 1915 of a Trumble gas trap. Please identify that by its construction and mode of operation, if you can.

A. It was a cylindrical body that was—that had a conical shape at the bottom. It had an opening in the top through which the gas and oil gained access to the trap. That opening was, I think 6 inches in diameter or 8 inches in diameter. In the center of that opening was a smaller size of pipe which was the gas outlet. The oil and the gas from the well passed down inside of the larger sized pipe between the gas outlet pipe and this larger pipe into the body of the trap. Inside of the trap were some cones, either one or more, I don't know, cone-shaped bodies of metal over which the oil and the gas spread out, behaved like an umbrella. The oil then passed to the bottom of the trap. The gas came around up inside of the umbrella. The umbrella was extended from the bottom of this small pipe that came in at the top

(Testimony of Paule Paine.)

of the trap and through which the gas passes. On the side of the trap was an oil outlet, together with a valve and a control device attached to a float [66] inside of the trap by means of which the fluid level was maintained at practically the same point, so that no matter whether the well made larger quantities of oil at one time than at another, the fluid remained at about the same level. At the bottom of the trap was another opening through which any sand or water accumulated could be drawn off. At another point on the side of the trap was a water gauge glass for showing the height of the fluid in the trap.

Q. Then if I understand you correctly, in general construction and mode of operation the device is like that of Plaintiff's Exhibit 1, the Trumbel patent in suit.

A. It didn't have this side opening which comes out to the side and goes up to the top (indicating).

The COURT.—Which opening is that?

A. This here (indicating). That was absolutely, entirely off. It was not on the trap.

Q. Well, what is the purpose of this side opening?

A. I have never used a trap with that on there.

Mr. LYON.—Otherwise was it practically the same as in this drawing?

A. It had only one side outlet. This drawing shows two. It had only one side outlet through which the oil escaped.

(Testimony of Paule Paine.)

Mr. BAGG.—You don't know whether that No. 12 is the same?

A. Well, I infer that it was No. 12 because No. 12 shows a control valve on the inside whereas on this trap the control valve was on the outside.

Q. On the trap that you had?

A. Yes, on the trap which I always used at that time.

Q. There was not any inside?

A. There was none inside, no.

Q. But so far as the inlet of the oil and gas from the well and the umbrella cone, and so forth—

A. And the oil outlet, the gas outlet passing up through the trap and the outlet at the bottom, it was there.

Q. And where was the float for the valve in the Trumble trap that you had on in 1915 for the Honolulu Oil Company? [67]

A. It floated on the surface of the fluid level.

Q. Inside of the trap?

A. Inside of the trap, and was connected through a stuffing box to a lever on the outside of the trap, which actuated on an oil discharge valve situated on the outside of the trap on the oil discharge line.

Q. Subsequently to this administration and use of this first Trumble trap on the Honolulu Oil well in 1915, I believe you said February or March—

A. Yes, the latter part of February, second half of February, 1915.

Q. Did you thereafter use any more Trumble gas traps or know of any other person using them?

(Testimony of Paule Paine.)

A. Yes, we used a good many of them. We were completing quite a few nice wells at that time and it became our standard practice to put a Trumble trap on the well and to carry on a similar line of investigation, because it was found that no two wells behaved exactly alike, with respect to the back pressure that might be backed up against them and their behavior. Of course the prime object was to save the oil and then save the gas if we could, and wherever possible to save the gas at as high a pressure as possible because that eliminated the expense of later on recompressing the gas so that it could be carried to a market, and of course I observed the operation of these traps on a number of other properties in the Midway field.

Q. I notice from your testimony that evidently you found the greater saving of gasoline vapors in this Honolulu well in the range of pressures below and up to 34 pounds, is that correct?

A. We found in that case 36 pounds happened to be the point where we stopped; we found in general that the greater, by far the large bulk of gasoline saving was effective in the range up to about 25 pounds and that the additional saving of gasoline that would obtain from the higher pressures was very much smaller.

Q. Did you ever have any experience in equipping any other [68] wells with these Trumble gas traps, or assist in the operation of them?

A. Oh, I would assist operators sometimes when they were putting them in. It was a new thing

(Testimony of Paule Paine.)

at that time and we happened to have had some experience. In the North Midway field they developed some trouble with emulsion due to water coming along with the oil. When water occurs with the oil it may, especially in the pressure of gas, churn up into an emulsion of oil and water which is very difficult to handle and ordinarily will not be accepted by the producing agents. One property of the Maize Oil Company on section 28, I think it was 28, was a well that was flowing about 18 or 20 per cent of emulsion. We had had 82 per cent of clean oil and the balance was this emulsion of water and oil flowing through a flow plug, and I suggested to them that they put a trap on there in order to save the gas and reduce the proportion of emulsion that they had had there by maintaining a pressure on the trap. The effect of it was to reduce that proportion of emulsion down to about 3 per cent.

Q. You say "trap," what kind of a trap?

A. Trumble trap.

Q. And how was it that the use of the Trumble trap on that reduced that emulsion?

A. Probably the effect of that is that through maintaining the pressure on the trap the rate of the flow of the oil and gas as they pass through the flow plug is reduced, the velocity is reduced and the churning effect of the flow plug is accordingly reduced.

Q. Now, you refer to the building up of pressure within the Trumble gas trap by putting the valve

(Testimony of Paule Paine.)

on the gas outlet of the trap. What is it that builds up the pressure?

A. Why, the closing of the valve increases the friction factor of the gas passing through that opening to such an extent that the gas will not pass through as readily, and that backs up the gas in the trap until a pressure in the trap is reached [69] sufficiently high to force the quantity of gas coming through the well through that restricted opening.

Q. Then it finally comes down to a question of what is the pressure of the gas in the well; in other words, you must have pressure of gas in a well before you can get a pressure built up in the trap, must you not?

A. Oh, yes, if there is no gas. If production is coming from the well there must be a pressure to push it in there before one can build up a pressure in the trap.

Q. To your knowledge to what extent have these Trumble gas traps come into use since 1915?

A. Well, I don't know exactly the extent. They were widely used in the Midway field at that time and have been since then. Last week I observed traps in operation up there that I installed in 1915, and of course I have seen them operating in the mid-continent country, Texas and Oklahoma.

Q. You have no connection with the Trumble Gas Trap Company. A. None whatever.

Q. You have repeated from memory a number of figures in regard to tests that were made in this

(Testimony of Paule Paine.)
Honolulu Consolidated Oil Company's well in 1915. Have you any memorandum from which you can refresh your recollection on that?

A. I have a memorandum that I made at that time on a piece of tracing paper in order that I could take off blue-prints and give them to different people that were interested in it.

Q. Have you that tracing? A. Yes.

Q. If you have a blue-print from it please produce the blue-print along with it.

A. Well, I think I have a blue-print. I have got about everything else. Yes, that is it (producing paper).

Q. And this tracing that you have produced was made by you at the time? A. At that time, yes.

Q. And this blue-print is a true blue-print from it? [70] A. Yes.

Mr. LYON.—We offer the blue-print in evidence as Plaintiff's Exhibit 6. I don't think counsel will object to the blue-print.

Mr. BAGG.—No. I would just like to see it.

Mr. LYON.—We will give you a chance, and if there are any errors they will be corrected to correspond. You may cross-examine.

The COURT.—In obtaining the pressure inside of the tank that pressure comes from the pressure of the gas from the well, does it? A. Yes, sir.

Q. You maintain that by simply shutting off the pressure?

A. Yes, sir; closing down the valve part way.

The COURT.—Very well, go ahead.

(Testimony of Paule Paine.)

Cross-examination.

(By Mr. BAGG.)

Q. How long did you say you had been in what we know as the "oil game"?

A. Since 1909.

Q. Since 1909? A. Yes.

Q. You had seen prior to 1915 a number of other oil and gas separators besides the Trumble trap, hadn't you? A. Yes.

Q. As a matter of fact there were a large number on the market at that time, weren't they?

A. I knew of only one being marketed at that time.

Q. What one was that?

A. That was the McLaughlin trap.

Q. The McLaughlin trap?

A. The Stark trap was being made and used at that time, but they consisted substantially of a group of pipe fittings and nothing more at that time and it was not being marketed.

Q. Where was it you saw this Stark trap? [71]

A. On Section 36.

Q. On some well up there?

A. Yes, in the Midway field.

Q. In the Midway field? A. Yes.

Q. And the McLaughlin?

A. In the same district.

Q. How many of those McLaughlin traps did you say you had seen?

A. Well, I don't know, but there were quite a few of them in operation. They were the standard

(Testimony of Paule Paine.)
trap used on the properties of the Southern Pacific Company and were in use on some other properties as well.

Q. Well, in your study of the oil and the properties of oil and gas and the operation of oil and gas wells, you discovered that there were quite a number of devices some of them abandoned and some of them which were used for the purpose of separating the oil and the gas as it came from the well, did you not?

A. Well now, the only work I had done at that time was in the Midway field in California, and my knowledge didn't extend beyond the conditions there. In 1914 I moved over to the headquarters of the Honolulu Company and my work before that had been at the outlying properties where we were not confronted with the problem of high pressure gas, and that was my introduction to that problem.

Q. Well, you never had then any experience in oil fields outside of California?

A. Not at that time.

Q. Subsequently, however, you had?

A. Yes.

Q. You have been interested, I believe, in some oil fields down in Oklahoma? A. Oh, yes.

Q. Now, can you explain to the Court what the philosophy of these oil and gas traps is, why they operate?

Mr. LYON.—We object to that question as to the

(Testimony of Paule Paine.)

form. He [72] says "these gas traps." You mean the Trumble gas trap?

Mr. BAGG.—I am asking about all the gas traps he is familiar with. He has testified with reference to two that he knows and has seen.

The COURT.—I would suggest that he explain the Trumble trap first and then go to the others, because I would like to get the actual point in view first.

Mr. BAGG.—Very well, your Honor.

Q. Will you state to the Court, or explain to the Court if you can the philosophy or the principle which underlies the operation of the Trumble trap, in *ord*er words what makes it separate oil and gas?

A. For a consideration of the fundamental principles I would suggest that we dismiss from our mind this collateral device here which is not essential (indicating).

Q. Yes.

A. This portion (indicating). Now, the oil and gas from a well are flowing with a pressure, in other words if the valve at the top of the well were to be closed it would show a pressure on the pressure gauge. Of course the pressure declines and is smaller at times than it is at others, but the well flows in virtue of a prèssure; it passes, in the Trumble trap through this pipe line down through these openings shown at the top of the trap between the large pipe and the small pipe.

The COURT.—There are two pipes there, this pipe here and this inside (indicating)?

(Testimony of Paule Paine.)

A. Yes, sir; there is a larger pipe which we may assume to be 6 inches in diameter, and then inside of that is a smaller pipe that fits at the top here at No. 26 made tight so that the gas and oil will just go up into this small area above where the gas and oil come in. The oil and has gain outlet at No. 7 but pass through this opening and down through the space between the 6 inch and 2-inch pipe. It spreads out in this space at the top of the Trumble trap over this umbrella or cone-shaped piece of metal. The oil and gas comes down in the body of the trap. The [73] gas being lighter is separated from the oil in the body as it flows down over the umbrella and probably also separates, in minor degree, from the body of the oil as it has accumulated, passes upwards and out through the two-inch pipe which is situated inside of the 6-inch pipe, through the gas outlet shown as No. 10 and on to wherever the gas is to be used or into the atmosphere if it is to be wasted.

Q. Well, do you store that gas?

A. To a very limited degree. Natural gas is used in such large quantities that any storage in holders would not be commercially feasible. We must use pretty quick, must use it to-day or it is gone forever. Now, the oil accumuulated in the bottom of the trap and is withdrawn through an outlet. As I say, this is not the type, this drawing does not describe exactly the type of trap which was in use at that time, but is essentially similar except that the controlling outlet valve here is situated on

(Testimony of Paule Paine.)

the inside of the trap instead of the outside of the trap. The oil passes out of this opening. Now, wells surge at times, at times they may cease producing altogether for a period of a few weeks or a few hours and then surge up and make a violent flow. The result of such violent flow would be to raise the level of the oil in the trap. That brings about the function of the controlling outlet valve which by means of a float is regulated in the size of aperture through which the oil passes. As the fluid level rises, the fluid passes up and the float effects an opening in that outlet valve which lets more oil go through. That is cut down and it lessens and lessens until it reaches a certain level where it doesn't flow. The sand and water by reason of their greater weight settles to the bottom of the trap and may be withdrawn through an outlet at that point.

Q. Does the oil go down on the side of the tank in the operation?

A. I think—this is entirely conjecture, I have never been inside of one when it was performing—I think the oil flows down both on the inside of the shell of the tank and if this piece is sufficiently far back from the well (indicating), it may stop [74] and churn up the oil, but if this corrugation is directly up to the side on the inside of the shell of the trap my conclusion is that the oil in a large measure passes down as a film or small body of oil on the inside of the trap—that is what would be desirable to my mind rather than to drop the oil here into the body of the tank.

(Testimony of Paule Paine.)

Q. The object of the large tank is to give space so that the gas will be released from the oil as though you put it down in the open atmosphere?

A. To provide an opportunity for the gas to separate from the oil. Now the gas which is in oil, that is in large bulks, scattered throughout a large bulk of oil, very similar to these very large storage tanks, the gas takes quite a long time to escape from that oil altogether, because some of it is in the bottom of the body of the oil and it must gradually raise and it raises sometimes very slowly.

Mr. LYON.—The thinner the body of oil the quicker the gas gets out?

A. The more rapid is going to be its passage to the surface of the oil from which point it disassociates itself from the oil.

Mr. BAGG.—Well then, as a matter of fact the operation of this Trumble trap is that it affords a large space in the flow of the oil and gas which enables, or produces a slowing down of the velocity of the oil and gas in their flow and enables the gas and oil to separate?

A. I will agree with you if you will say "area" instead of "space."

Q. Well, "area"? A. Yes.

The COURT.—Your idea is that the spreading of the oil causes the condition under which the gas may escape from the oil?

A. It expedites the separation of the gas from the oil through the greater areas, not necessarily through the greater [75] space.

(Testimony of Paule Paine.)

Mr. BAGG.—But as a matter of fact the principle upon which all gas and oil separators are based is the fact that it affords, by reason of its enlarged space, and area, an opportunity for the oil and gas coming from the well to slow down or retard its velocity and give the oil and gas time and opportunity to separate.

Mr. LYON.—Now, wait a moment. I object to the form of the question, your Honor, although I would like to have it answered.

Mr. BAGG.—At this time I will restrict it to this one trap. That is the philosophy of it, isn't it?

The WITNESS.—Now, let's have the question.

(Last question is read by the reporter.)

A. Well, the function of any gas trap of course is to separate the gas from the oil and it is advantageous to have it larger than the size of pipe that is carrying the oil and gas to the trap.

Q. And they are all built upon that principle of velocity reducing means which gives the oil an opportunity to particularly become quiescent and then the oil and the gas separate, one going to the bottom and the other to the top.

A. No, no. Quiescence necessarily doesn't help it, Mr. Bagg, for this reason: If you have gas and oil mixed together in a body, quiescence, the gas will not separate as readily from the oil as it will if you churn it up, stir it up; you will then agitate it bringing particles of gas nearer to the surface, constantly accelerating the separation of the gas from the oil; so, while it is more or less speculative,

(Testimony of Paule Paine.)

I would not infer that its being quiescent would necessarily help the separation, but frequently a diminution in the velocity will help.

Q. I mean it slows it down. I presume no one would contend that the oil in any of these oil and gas separators was absolutely quiescent, but the velocity of it has been slowed down considerable as compared with its velocity as it comes from the well?

A. I would rather have the oil more active for an effective separation. [76]

Q. Well then, why wouldn't the oil and gas separate then in a pipe as it comes from the oil and gas well, why wouldn't it separate then?

A. Oh, because your oil is intimately mixed with the gas, and at a point where the gas, in virtue of its physical action of coming off would bring along with it particles of oil.

Q. And then when you slow it down then it gives it an opportunity to separate one from the other?

A. Of course, if a trap were high enough, no matter how much it were being agitated in the bottom, if the gas outlet were sufficiently far above where the particles of oil are being taken along in the stream of gas, why there would be an effective separation, even though the oil were active.

Q. Well, now, suppose just as an illustration that you had connected with that oil and gas well that was flowing at a considerable pressure such as you describe at the Honolulu well in 1915, and suppose that that was connected with a pipe-line that

(Testimony of Paule Paine.)

will cover miles long and the oil and the gas came through that pipe line at the same rate of speed that it came out of the oil and gas well, would the oil and gas be separated at the end of its journey, at the end of this pipe?

A. Oh no, without this pressure it wouldn't be?

Q. It wouldn't be separated to any appreciable extent whatsoever? A. No.

Q. That would be caused by the fact that the velocity was such going through that pipe that it didn't give the oil and the gas an opportunity to quiet down long enough to separate?

A. They would remain intermittently mixed throughout their passage there.

Q. The operation of the separation of oil and gas is just practically the same thing as though you were to take a bucket of water and mingle a lot of sawdust in it stir it up very thoroughly; then as long as you keep stirring it the sawdust and the water would not separate to any appreciable extent, would it? [77]

A. No.

Q. But then you allow that to settle or stop just a few minutes to take up something else and the first thing you know the sawdust will come to the top and the water drop to the bottom, wouldn't it?

A. Yes, sir.

Q. Now that is the same principle that obtains in all oil and gas separators?

A. Of course the fundamental principle of any

(Testimony of Paule Paine.)

gas trap is a separation of these fluids in virtue of their respective weights.

Q. And the principle upon which they operate is the slowing down process which gives them an opportunity to do that, is that correct?

A. The slowing down process to a minor degree. I don't believe that that is solely responsible, although of course a large settling area or space is desirable, but I cannot accept the theory that quiescence alone expedites the separation of the gas from the oil.

Q. Well, of course it is dependent upon the relative specific gravities of the oil and the gas; that is a fundamental principle which permits them to separate. If they were the same specific gravity they wouldn't separate under any circumstances, but the flowing down process which gives them an opportunity we will say to catch their breath and the gas to come out and the oil to settle, is practically the philosophy and basis upon which all oil and gas separators are based, isn't it?

A. To a certain extent, yes.

The COURT.—It is now the noon hour, and the Court will take a recess until 2 o'clock. [78]

AFTERNOON SESSION—March 22, 1922, 2 P. M.

PAUL PAINE resumed the stand.

The COURT.—You may proceed. Is there any further cross-examination?

Mr. BAGG.—Yes, if your Honor please.

Q. Mr. Paine, in that trap, that was installed back there in 1915 on the Honolulu well, do you

(Testimony of Paule Paine.)

know how many cones there were in that trap?

A. I do not. I did not open it up.

Q. You don't know whether there were any cones in there at all or not?

A. Well, I say I didn't open it up—we didn't take the whole works apart, but there was on that trap a manhole that could be taken off, through which one could gain entrance to the trap, and I took that off because I desired to replace the composition gasket in there with a lead gasket that would resist this light gravity oil, and I looked up there and I saw a conical shaped affair, but I could only see the one, and whether there was more than one or not I don't know.

Q. Did you observe how near the wall the edge of this cone came?

A. I didn't measure it or reach up to that point. It came very close to the wall.

Q. Very close to the wall? A. Yes.

Q. What was the estimated amount of space between the edge of the cone and the side wall?

A. Oh, I wouldn't like to estimate it. From that distance in looking, my recollection it might have been a quarter of an inch or it might have been three-quarters of an inch, but something of that degree.

Q. It was very close. Now, in the operation of that trap the pressure that you speak of came originally from the well, did [79] it not?

A. It is the pressure of the gas contained in the well that came along with the oil.

(Testimony of Paule Paine.)

Q. And that build up the pressure in the separator? A. Oh, yes.

The COURT.—Was there anything in the separator itself to measure this pressure?

A. No, sir, not inside of the trap.

Mr. BAGG.—Now, the pressure then in the trap came in at the upper portion with the oil and gas that came from the well?

A. No, pressure is not a definite thing that can come in; pressure is a condition, not a thing.

Q. Well, originally it started at the top; now for instance when the trap was first built up, when you first turned oil into it the pressure originally came in at the top, did it not, and then of course pressed down through the—

A. Possibly this will give you what you want: The fluid that was under pressure came in at the top of the trap.

Q. Well, that is about what we want. Now then the tendency of the pressure originally was downward, wouldn't it be on that flowing film of oil?

A. No, the tendency of the pressure in every case, Mr. Bagg, is in all directions.

Q. Yes, I understand that, but originally when you first started the pressure it would be on the top of the flowing film, or it would be above this film of oil as it flowed out of the top of this cone, wouldn't it?

A. No, the pressure as it exists in a container is equal at all points.

Q. Well, I understand, but then there was a

(Testimony of Paule Paine.)

pressure on the top of this film of oil anyway, wasn't there?

A. Yes, as there was all over the trap.

Q. Yes. And then it had a tendency to press that oil against the surface of that cone, didn't it?

A. Well, I wouldn't ascribe that to pressure, no. [80]

Q. Well, what would it be? A. Gravity.

Q. Simply the gravity. That was all there was on the oil surface of this cone, simply the gravity?

A. The oil fell on to the cone and moved down over it.

Q. Yes.

The COURT.—Well, did that release the pressure immediately when it entered the cone?

A. When it entered the trap?

Q. Yes. You would call this trap (indicating).

A. Yes, the whole trap.

Q. The whole thing. When it went in here did it release the pressure so that it wouldn't—

A. If the outlet were open, if the outlet to the trap were open it would reduce the pressure, yes, the gas outlet; it would vent it.

Mr. BAGG.—But there would be a pressure along here on top of this oil which would have a tendency to squeeze this oil in here, wouldn't it (indicating), squeeze it against the surface of this cone; it would have a squeezing effect, wouldn't it?

A. Well, the pressure would be exerted against every surface exposed there.

(Testimony of Paule Paine.)

Q. Yes, but it would be against the oil, having a tendency to press down toward this—

A. That tendency might be there.

Q. And it would be likewise the same, having a tendency to press this oil. Now, where would the pressure be on this side (indicating); where would the pressure be changed to?

A. Well, of course the pressure would be the same on the other side of this cone as it is against the oil.

Q. Yes, it would be underneath around here and up on the outside of this oil, wouldn't it?

A. The pressure, understand, must be the same at every point in this trap.

Q. So then when this oil came down here, flowed down this edge here, then the pressure would be behind it as well as on this [81] side of it, wouldn't it (indicating); the pressure would be along here (indicating); this is the film?

A. The pressure was there.

Q. The pressure was there. It would be on both sides of this film of oil, wouldn't it?

A. Now, what do you mean by "this film of oil"?

Q. Well, this film of oil, say, oil is coming in here now and flows down over this cone, strikes the edge of the cone drops down and starts to flow down the side wall of this receptacle. Now then, the pressure would be on the outside and on the inside, too, wouldn't it? A. Of what?

Q. Of this film of oil?

(Testimony of Paule Paine.)

A. Oh no, not if the film of oil is bearing tightly against the side of the iron.

Q. Well, wouldn't there be any pressure on the oil then at all?

A. It would be in the interior of the trap.

Q. It would be on the inside pressing toward this and having a tendency to push out then, wouldn't it?

A. Yes, the pressure is vertical in the trap.

Q. Well, it would have a tendency to push this oil this way, wouldn't it, out, in other words?

A. Yes, of course.

Q. If you were to take the side wall out the oil would shoot right out of the side, wouldn't it?

A. Out of the side of what?

Q. Just suppose you cut a hole right around the side wall of this receptacle and still have this force in there, then the oil would shoot right out of the side, wouldn't it?

A. Some of that oil that came up to that opening probably would pass out.

Q. Run all out and would go in the direction of the least resistance, wherever that least resistance might be?

A. Without the opening or with the opening?
[82]

Q. Yes, with the opening?

A. Why, a very small portion if the opening was small, a very small portion.

Q. Well, it would be just in proportion to the size of the opening, wouldn't it?

(Testimony of Paule Paine.)

A. Yes, probably.

Q. If the whole side were cut out all of it would jump out, wouldn't it? A. Yes, sir.

Q. It would be just in proportion to the size of that; in other words the pressure on this oil would have a tendency to carry the oil with it in the direction of the least resistance, wouldn't it?

The WITNESS.—What was that last question?

(The last question is read by the reporter.)

A. It would have that tendency, but the major influence on the oil of course would be gravity which is causing it to descend.

Q. Which is causing it to descend, but as it came down here, as it came past this hole, why it would shoot right out, wouldn't it?

A. A portion of it would.

Q. Practically all of it would flow over the hole?

A. Beg pardon?

Q. Practically all of it would be flowing down over that space where the hole is?

A. The oil which came opposite the opening and was not carried around the sides of the hole would obviously be—

Q. It is a general rule and a principle of pressure that it goes in the direction of the least resistance, that is ordinary pressure?

A. That is the principle of what moves flowing fluids, is the difference in pressure.

Q. And it has a tendency to go in whatever direction there is the least resistance. [83]

(Testimony of Paule Paine.)

A. Possibly. I don't want to lecture here, but I would like to differentiate on that.

Q. Go ahead.

A. Respecting those classes of fluids, the two classes of fluids, the liquids and the gasses.

Q. Yes.

A. Of course it is more or less simple, but in order that we may be clear on it, a liquid takes the shape of the vessel containing it, but maintains the same volume.

Q. Yes.

A. Gas takes the shape of the vessel containing it but occupies all the volume, all of the vessels containing it.

Q. Yes.

A. I speak of that here because we have two fluids before us.

Q. Yes.

A. One liquid, which is oil. That descends in virtue of gravity to the bottom of the trap. The gas responds to the law of gasses and fills all the remaining available space.

Q. Yes.

A. And applies the pressure equally in all directions.

Q. In all directions. Would there be any greater pressure above this top cone than there was below it?

A. There cannot be any greater pressure there unless the opening between the top cone and the bottom of the trap is so very small that it would

(Testimony of Paule Paine.)
introduce a slight friction factor with reference to the discharge of the gas and the oil through that opening.

Q. You mean the edge of the cone and the side of the wall?

A. The edge, yes. And I wouldn't judge that to be great enough to have any effect on that cone, no effect of consequence.

Q. That would depend altogether on, I guess you would call it viscosity of the oil, wouldn't it? If it were the heavier grade of oil the pressure would be greater and if not it would be [84] lighter?

A. Yes, sir; and the passage of the gas through a small opening.

Q. Otherwise it would be the same, wouldn't it?

A. The pressure would be the same. There is a friction factor there that would have to be overcome, but for all practical purposes that influence would be negligible.

Q. Now, when that oil comes in from the well and strikes the top of that cone, flows down the upper portion of the cone until it comes to the edge, now would the tendency of the oil be to remain against and flow down the side wall, or would it be the tendeney to kind of curve in or cup in?

A. I want to be sure that I understand your question.

Q. Well, I can illustrate it here. Now as this oil comes down and strikes this cone and comes down here to the edge, now would it have a tendency to continue right on down close up against this or

(Testimony of Paule Paine.)

would it have a tendency to kind of go in like that (illustrating).

A. My best judgment on that would be that if this shell were sufficiently close—if this cone were sufficiently close to the shell, the oil would continue from the edge of the cone over to the inside of the shell and then pass down.

The COURT.—Pass down along the—

A. The side of the shell.

Q. The side of the shell. But that wouldn't be an unbroken body of oil entirely around the cone because there might be passageways there through which the gas passes to the bottom of the trap, to the lower portion of the trap.

Mr. BAGG.—But it wouldn't cup in there or curve in there or whatever you would call it?

A. I don't think so.

Q. But if the oil was very heavy it might do that, mightn't it?

A. I doubt if weight would have as much effect on that as the viscosity of the oil to which you were referring a moment ago. [85]

Q. Well, the viscosity might have something to do with it, mightn't it?

A. The viscosity probably would.

Q. When I used the word "heavy" I meant heavy as distinguished from light oil?

A. Well, you see the viscosity of oil will vary greatly, even with oils of the same weight.

Q. Yes, but if this space in here was very nar-

(Testimony of Paule Paine.)
row do you think that the oil would continue down the side walls? A. I think so.

Mr. BAGG.—That is all.

Redirect Examination.

(By Mr. LYON.)

Q. Mr. Paine, in this Trumble trap to which you have been referring, what do you understand the function or object to be of conveying the oil down along the inner surface of the body of the chamber?

Mr. BAGG.—Now, if your Honor please, I think we will object to the form of that question because it is calling for a conclusion and it would probably better be answered by the inventor himself as to what—

The COURT.—This man is an expert.

Mr. BAGG.—Yes, sir; he might testify what the action would be, but I don't think he can tell what the object was. He can testify what the effect would be.

The COURT.—What we are getting at is the function.

Mr. BAGG.—Yes, he can testify what it did, what it actually did, but not what the object was.

The COURT.—There is no doubt about that.

The WITNESS.—The oil spreads out over this cone and as it passes down on the cone of course it becomes distributed over a wider and greater area. Then as it reaches the shell it passes down over the inside of this shell, at least probably the major [86] portion of it does, and in that manner exposes a large surface of fluid to this atmosphere of

(Testimony of Paule Paine.)

gas which is present in the trap. In that manner the gas more readily escapes from the oil than it would if contained in a solid body of great thickness. Just as I referred to before noon in the matter, gas contained in a large body of oil, in a very large tank, that does *does* not separate so readily as the gas which is in the bottom of that tank. It takes quite a while gradually to rise to the surface of the oil before it can escape from the oil, so that—

The COURT.—That is to say, if the oil was allowed to drop from there, straight down to the bottom without any outside hindrance the gas would not escape so readily as though distributed around there? A. No, sir; it wouldn't.

The COURT.—Very well. I understand it.

Mr. LYON.—Now, you have referred to viscosity. Just what do you mean by that as applied to an oil?

A. I don't know what the exact definition of viscosity is, but in simply language it is the readiness with which it flows.

Q. And when you say "of a higher viscosity" you mean the less readiness in flowing?

A. Good.

Q. And how does the viscosity of crude oil compare with the viscosity of water?

A. Well, it doesn't flow nearly as readily, as most crude oils. Occasionally a freak oil might have lower viscosity.

Q. Well then, for instance, if we were to take for example these glass models and flow water through them, the water would flow even more freely than

(Testimony of Paule Paine.)
ordinary crude oil mixed with gas? A. Yes.

Q. And would or would not such an illustration then where the water flowed down the inner surface of the glass be less of a flow down the surface than the oil?

A. Probably—I am not sure of this—probably the oil would tend to cling to the side of the vessel more readily than water [87] would.

Q. And that clinging would be due to friction?

A. Oh, that brings in the question of surface tension of fluid. There is another field.

Q. What I am after without leading you, the tendency of the oil then as it went down would be to roll over and over, wouldn't it, more than just simply to run down as water runs?

A. Oh, yes, the tendency particularly of California crude oils in their movements in all cases is to roll over and over in response to what we call turbulent flow. I speak of that because it is characteristic of Pennsylvania and Mid-continent oils to flow in what we call a straight line flow, each partiele going straight ahead.

Q. And then the oil flowing down over the inner surface of the chamber or tank, turning over and over, would expose more surface and body of the oil to the gas?

A. There would be that form of agitation in there which would bring the particles of gas more frequently to the surface of the fluid and allow them to escape into the body of the trap.

Q. Is that one of the reasons why in a trap of this

(Testimony of Paule Paine.)

Trumble type the separation is increased by the downward flow along the inner surface of the chamber, in your opinion? A. I think it is.

Q. What is it that makes the gas and oil flow from an oil well?

A. The pressure in different cases, different forms of pressure; sometimes it is the pressure due to the gas, the pressure under which the gas is contained in the well expanding; in other cases it is due to the pressure caused by water underlying the oil. I speak of that because in some field the wells produce large quantities of gas along with the oil. In other fields there are wells which produce, flow large quantities of oil and make almost no gas but which in their later lives produce large quantities of water; but it is a pressure of some form in either case. [88]

Q. Then the pressure that we have been referring to here in connection with this trap is all a natural pressure from the interior out?

A. That is the pressure I have been referring to, the gas pressure.

Q. You referred on cross-examination to a Stark trap. Briefly tell us what that was?

The COURT.—What is that, Mr. Lyon?

Mr. LYON.—(Spelling.) S-t-a-r-k.

A. The Stark trap as I knew it in the time under discussion comprised a long length of rather large sized pipe, inclined slightly. The oil was admitted to it at one end and flowed down in the pipe. Along the top of that pipe at a number of points

(Testimony of Paule Paine.)

were a number of small outlets situated above the level of oil which accumulated in this length of large sized pipe. The gas escaped through these many small openings. The oil was withdrawn from the end of the inclined pipe. It had no automatic feature. A man stood there, and by means of a gauge press on the side observed the level of fluid in the pipe and opened the oil valve on the oil outlet line faster at times when the oil level was rising, closed it slowly when the oil level was lowering and kept that level of oil at about a uniform point. The gas escaped from above the surface of the oil along that inclined length of large pipe.

Q. Well, then, you say you kept the oil in the Stark trap or pipe about a uniform level. How much of a body of oil did they usually keep in operating that trap?

A. The height of the oil? Oh, the pipe was inclined and so the height of the oil would vary at different points because it reached a level, but it would be, I would say it would occupy the entire space of the pipe at the lower end and gradually come down to nothing at the upper end. These traps would be a hundred or more feet in length in cases usually made out of five or six joints of large sized pipe and the pipe was usually about 20 feet in length, each joint was 20 feet in length.

Q. Will you make us a rough sketch, Mr. Paine, of that trap that you are referring to, the Stark trap? [89]

(The witness draws on paper with pencil.)

(Testimony of Paule Paine.)

Mr. LYON.—We offer the sketch made by the witness in evidence as Plaintiff's Exhibit No. 7.

The CLERK.—7 will be the next.

Mr. LYON.—Write the title "Paine sketch, Stark trap." (Addressing the clerk.)

Q. How as to the efficiency and operation did this Stark trap compare with the Trumble trap that you operated in 1915 on the Honolulu?

A. I never operated a Stark trap, but observed a number of them, studied them and considered the advisability of installing them. They appeared to be doing the work of separating the gas and the oil very efficiently. The disadvantageous features were there because of, first, their lack of flexibility, difficulty in moving them around quickly from place to place. What was desired by us was as cheap an article as we might attain that would do the work efficiently and one that could be put up very quickly so that there might not be unnecessary waste. The Stark trap required a great amount of pipe of both large pipe and small gas outlets, and a considerable number of valves and fittings of one kind and another and all of this equipment together with the considerable labor cost of installing them, making them prohibitive as compared with the Trumble trap.

Q. I believe you stated also that this Stark trap was nonautomatic; what did you mean by that?

A. Yes, I knew there was a third factor in there —I didn't recall it. The Trumble trap at that

(Testimony of Paule Paine.)

time required a man to be there to open and shut the valve on this oil outlet.

Q. You mean the Stark trap?

A. The Stark trap—to open and shut the valve on this oil outlet as the fluid level changed. If a well ceased flowing why he closed the valve altogether; if the well came in with a surge it was necessary to open it quickly and allow the oil to go out [90] and that was the means of operating the traps that I observed whereas the Trumble trap was absolutely as nearly completely automatic as one could expect to get.

Mr. LYON.—That is all.

Recross-examination.

(By Mr. BAGG.)

Q. Mr. Paine, I believe you have outlined here on this Plaintiff's Exhibit 7 the gas intake or the oil and gas intake and the oil outlet and the gas outlet. Now this Stark trap, the body, main body of the trap, is very much larger than both the outlet and the intake, isn't it? A. Yes.

Q. And the tendency, then, is to throw this oil in here and put it into this large barrel or whatever it is, has a tendency to slow down the speed of the oil or the velocity of the flowing of the oil and give it a chance to separate?

A. It has two tendencies; one to slow down that speed and the other to present a greater surface of fluid for the escape of gas.

Q. The oil comes in here (indicating); now you have indicated this oil level as being down here

(Testimony of Paule Paine.)
(indicating); this oil would be—there would be some oil up here as it came in here (indicating)?

A. There would be a stream of oil.

Q. There would be a stream of oil running down here. Would you mind illustrating that on your trap so it would show exactly how it would come in?

A. (Drawing on paper.) Now, this stream of oil coming in the form it would take and where the stream would enter the body of the fluid would depend entirely upon the velocity of the stream of fluid coming from the smaller pipe. It might come in there from the smaller pipe to the larger one at such a great velocity that it would in entirety be thrown out through the body of the [91] trap, or it would come down very slowly and trickle down and drop down. So I have indicated what might be called a composite picture occupying that whole body in there, but after it once gets down than it exposes the large surface for the escaping gas (indicating).

Mr. BAGG.—That is all.

Mr. LYON.—That is all.

## Testimony of A. J. Gutzler, for Plaintiffs.

A. J. GUTZLER, a witness called on behalf of the plaintiffs, being first duly sworn, testified as follows:

Direct Examination.

(By Mr. LYON.)

Q. Please state your name, age, residence and occupation.

(Testimony of A. J. Gutzler.)

A. A. J. Gutzler; residence, 1325 Fair Oaks Avenue; South Pasadena; age, 59 years.

Q. You are one of the plaintiffs in this action?

A. Yes, sir.

Q. How long have the plaintiffs been manufacturing and selling Trumble gas traps?

A. Since September.

Mr. BAGG.—Wait a moment. We object to that unless he testifies that he knows—no objection to that.

The COURT.—Well, of course if he doesn't know—

Mr. LYON.—I said he was one of the plaintiffs.

Mr. BAGG.—I beg your pardon. I thought you said "defendants."

Mr. LYON.—He is one of the plaintiffs.

The WITNESS.—Since September, 1914.

Mr. LYON.—You have heard Mr. Paine's testimony? A. Yes, sir.

Q. Where was the first Trumble gas trap installed, if you remember?

A. Out at Taft on the Northern Exploration Company, I believe.

The COURT.—Where is that?

A. In Taft, in this state. [92]

Mr. LYON.—On the Northern Exploration Company.

A. On the Northern Exploration Company.

Q. Can you state to what extent the plaintiffs have manufactured and sold Trumble gas traps embodying the invention here in issue?

(Testimony of A. J. Gutzler.)

A. I have got a memorandum of it. I will look it up. Do you want the total number?

Q. Yes, and the sale price.

A. 583; and the total sale price?

Q. Yes. A. $434.730.

Q. And can you give them also as to the states where they have been sold and installed, and by years?

A. I have got a copy of it. In California 281 traps were made and sold at a price of $170,005; in Texas, 161 at a price of $137,300; in Louisiana, 46, price $41,600; Arkansas 7 traps sold, total price, $7,650; Oklahoma, 41 traps sold at a price, $35,400; Wyoming, 4 traps sold at a price $1,350; exports 43 traps, $41,065.

Q. When you say "export," just what do you mean by that, Professor?

A. Well, in some cases we don't know where they go. We have export agents and in some cases we have shipped them to them and they build them up.

Q. In other words, they are exported from the United States to the old country?

A. To the old country. We know that some went to Mexico, and some to Bombay, India, but we are not sure of all of them.

Q. Now, can you give the different states by years for such traps? Have you that in the statement?

A. Yes. Do you want to follow down by years for all the states?

(Testimony of A. J. Gutzler.)

Q. Yes, by years in gross, if you have it that way.

A. I haven't got that. [93]

Q. Well, by years by states?

A. California in 1914, 3 traps. You just want the number of traps?

Q. That is all.

A. Not the amount of money?

Q. That is all that is necessary.

A. Three traps. California, 1915, 39; Wyoming, 1915, 2; California in 1916, 23; Oklahoma, 1, and Wyoming, 2. California, 1917, 30; Oklahoma, 2, in 1917; California, 1918, 23; export 1918, 1. California in 1919, 23; Texas, 1919, 53; export 4. California in 1920, 63. Texas, 1920, 94; Louisiana in 1920, 38; Arkansas in 120, 4. Oklahoma, 1920, 10, and export 35. 1921, California, 56—

Mr. BAGG.—Now, if your Honor please, I believe we will object to any testimony as to 1921 because that is after this suit had been filed. The suit was filed on the 3d day of January, 1921.

The COURT.—What is the object of this testimony?

Mr. LYON.—The only object of this testimany would be to show the general adoption and use of the device, unless it was also to show the effect of the infringement on the falling off of the sales, and we don't care whether we go into 1921 particularly or not, although it is competent to show the general adoption and use.

The COURT.—Very well, you may answer that question. I think probably it would be pertinent

(Testimony of A. J. Gutzler.)
to show the general use of it as compared with previously.

The WITNESS.—Shall I answer?

The COURT.—Yes.

The WITNESS.—1921, California, 56; Texas, 13; Oklahoma, 21; and export 3; Louisiana 8, and Arkansas 3. That is up to the 1st of January.

Mr. LYON.—You may take the witness.

Cross-examination.

(By Mr. BAGG.)

Q. That memorandum is just taken from your books, is it? [94]

A. That is taken from a register from our serial numbers and the journal.

Q. You didn't make the entries in those books yourself? A. I made a part of them.

Q. Well, you didn't make them all?

A. I looked after them all.

Q. That is not then taken from the books of original entry? A. Yes, sir.

Q. They are taken? That is merely a copy of that made by you.

A. This is a copy made by me of the original entries made under my direction.

Q. Did you make that yourself or did some one in your office make that?

A. I made this, but I didn't type it.

Q. Well, the memorandum you have there is not the copy that you made yourself?

A. Made from the books?

Q. Yes, sir. A. Yes, sir.

(Testimony of A. J. Gutzler.)

Q. Is it one you made from the books or is it one someone has typed? A. Oh—

Q. Under your instructions?

A. Typed under my instructions. I got it off the books and then had it typed. Of course, I didn't type it myself.

A. Oh no.

Mr. BAGG.—That is all.

Mr. LYON.—That is all. [95]

**Testimony of O. W. Harris, for Plaintiffs.**

O. W. HARRIS, a witness called on behalf of the plaintiffs, being first duly sworn, testified as follows:

Direct Examination.

(By Mr. LYON.)

Q. Please state your name, age, residence and occupation.

A. O. W. Harris, 45 years old; reside at 954 South Vermont, and I am attorney at law.

Q. Otherwise than your legal status of admission to the Bar what other education have you had, Mr. Harris?

A. Until I came to California in 1912 I was engaged entirely in engineering work, and all my educational experience up to that time had been wholly engineering work.

Q. With whom were you so employed in engineering work?

A. I was employed by the Westinghouse Electric

(Testimony of O. W. Harris.)

Manufacturing Company previous to coming to California.

Q. In doing what line of work?

A. In doing designing of electrical machinery of various kinds.

Q. How long were you with the Westinghouse?

A. I was with the Westinghouse Company for 8 years.

Q. Are you familiar with the Trumble gas traps?

A. I think I am, quite familiar with them.

Q. In what manner and under what circumstances did you acquire your knowledge of the Trumble gas traps?

A. Soon after coming to California I became associated with Mr. F. M. Townsend and Mr. Frank L. A. Graham, who at that time were in the business of soliciting patents and patent attorneys in general practice here. Mr. Townsend who is, as I understand it, one of the plaintiffs in this suit, was particularly interested with Mr. Trumble in other matters, and Mr. Trumble brought this invention to the office in which I was at that time employed, and an application for patent was made, a large portion or all of which I wrote and I had charge of the prosecution of the case before [96] the Patent Office. In other words, I wrote the most of the arguments previous to the allowance of the patent. I also saw the traps in operation and was familiar in general with the business in our office and our bookkeeper took care of a portion of the records, and so forth.

(Testimony of O. W. Harris.)

Q. Since the preparation and filing of the application for the Trumble patent in suit, have you seen and observed any of the Trumble traps in operation?

A. I have seen them from time to time when I would be in the oil fields. I have been in the oil fields a great deal of the time the last five or six years.

Q. To what extent, in accordance with your observation of such Trumble gas traps have they come into use in California?

A. In passing through the oil fields the Trumble traps are very readily identified through their peculiar shape and the way they are mounted, and very recently I had occasion to go through the oil fields and they are very noticeable—nearly every well around any field in this part of the state has a Trumble trap on it and they can be seen as you drive by on the road. They are in very common use on all of the wells.

Q. Are you acquainted with the defendant Lorraine? A. I have met Mr. Lorraine.

Q. Did you ever examine a trap on the Tonner lease? A. I did.

Q. And whose manufacture was that?

A. I believe it was of Mr. Lorraine's manufacture.

The COURT.—What trap was that? Excuse me.

Mr. LYON.—(Spelling.) T-o-n-n-e-r lease.

(Testimony of O. W. Harris.)

The WITNESS.—Of the General Petroleum Corporation at Brae, California.

Q. And did you identify that as of Mr. Lorraine's manufacture?

A. It had Mr. Lorraine's nameplate on it.

Q. Do you remember the number of the well?

A. The trap that I examined was on the Tonner No. 4 [97] well.

Mr. LYON.—All right. Will the defendant admit that was a trap of his manufacture and sold by the defendant?

Mr. BAGG.—No, we don't admit it.

Mr. LYON.—All right. When was it, Mr. Harris, that you examined, first examined that trap?

A. I don't remember the dates, but I have notes on which the dates appear, notes which I made at the time and also some photographs that I took.

Q. I show you four photographs and will ask you if you have ever seen them before (handing photographs to the witness)?

A. Yes, I made these photographs myself—that is, I made at least three of them; I think I made them all.

Q. And from them can you tell us approximately what the date was?

A. I was on the Tonner lease and made these photographs on the 15th day of December, 1920.

Q. I see "12-15-1920, O. W. H." appears at the bottom of two or more of these photographs. Did you scratch that on the negative?

A. I wrote that on the negative myself. It is

(Testimony of O. W. Harris.)

an autographic camera and I wrote the date on that at the time the photographs were taken.

Q. Do these four photographs correctly show the trap at the said Tonner lease at that time?

A. They do.

Q. And you said that on the trap it contained the name of the defendant "D. W. Lorraine," or "Lorraine trap" or what was it?

A. I have somewhere a sketch that I made on which that date appears if I can find it.

Q. Please produce such sketch?

A. I don't know that I have a memorandum with the exact reading of that nameplate, but I think it read "The Lorraine Gas Trap," and some other data below. The superintendent on the [98] lease—

Mr. BAGG.—Well, we object to what he said.

Mr. LYON.—All right. We will prove it before we get through. You are not mistaken that it was on Well No. 3 instead of 4 of the Tonner lease?

The COURT.—No. 3, you say?

Mr. LYON.—I think so.

A. Well, the two wells are side by side and it is possible I am mistaken. It was either Tonner No. said on No. 4.

Mr. BAGG.—We will admit it on No. 3, but you said on No. 4.

Mr. LYON.—Oh, you will admit it on No. 3?

Mr. BAGG.—Yes, we will admit it on No. 3.

Mr. LYON.—And that it was a trap of the defendant's manufacture and sold, is that correct?

(Testimony of O. W. Harris.)

Mr. BAGG.—Yes.

Mr. LYON.—We offer in evidence the four photographs as Plaintiff's Exhibit No. 8.

The CLERK.—Four photographs.

Mr. LYON.—Yes, $8^1$, $8^2$, $8^3$, $8^4$, I think that is the best way to mark them. Did you at the time of the examination of this trap have any opportunity to ascertain its interior construction?

A. Yes, I had an excellent opportunity.

Q. Please go ahead and detail what you did at that time find and what you ascertained it to be and what was the interior construction?

A. At the time that we visited the well there was also on the lease and at the well a Trumble gas trap which is shown I think in those photographs. It was possible to operate the well either on the Trumble or on the Lorraine, and the people in the oil company very kindly allowed us to see the Lorraine trap in operation, and they also shut it down and allowed us to get into the interior of it and take measurements, to investigate its interior construction. [99]

Q. Does this photograph $8^4$ show both the Trumble and Lorraine trap?

A. Yes, it shows it.

Q. Please take $8^4$ and mark one with the name "Lorraine" and the Trumble trap with the name "Trumble."

A. (Marking.) The Trumble trap is on the left and the Lorraine trap is on the right.

Q. Now, proceed.

(Testimony of O. W. Harris.)

A. In the top of the trap there was a manhole or cover, and we took this manhole off—after we shut the Lorraine trap down we took the manhole off, and were able to get inside and see how it was internally constructed. We couldn't see the entire internal construction of the trap and in some of it it was necessary to reach around down in and feel to see exactly how it was constructed. At that time I made sketches showing the interior construction as I found it, and made rough memorandums as to the internal construction, and I have here a sketch that was made on the lease at the time the dimensions were made, and also a second sketch that was made on the lease with the addition of some numerals that appear in ink which were made the next day.

Q. These correctly show, do they, the construction of that Lorraine trap, on the Tonner lease?

A. They correctly show the construction of that Lorraine trap as far as I was able to determine at the time.

Mr. LYON.—Well, we will ask that these two sketches be marked and offered in evidence as Plaintiff's Exhibits $9^1$ and $9^2$. Proceed.

The WITNESS.—The trap at the time it was operating—when we arrived on the lease the trap was operating under commercial conditions, and it had on the side of it the gauge, pressure gauge which I think is shown in the photographs—yes, it had a pressure gauge which is plainly shown in the photographs; and during its operation we ob-

(Testimony of O. W. Harris.)

served his gauge. The gauge appears on the side of the trap. During the entire operation of the trap [100] that gauge varied very little from 45 pounds on the trap, and during the time that I was on the lease 45 pounds was the pressure that was maintained on the interior of the trap as shown by the gauge which is a part of the trap.

Q. Well, please go ahead, by means of these sketches that you have produced and your knowledge gained from this inspection of this Lorraine trap on the Tonner lease, and tell us what its mode of operation was and its construction. And you may, at the same time, if you will, compare it with the device of the Trumbel patent, Plaintiff's Exhibit 1.

Mr. BAGG.—Now, if your Honor please, I think what he ought to testify to is what he actually saw there.

A. I will try to confine myself to that, Mr. Bagg. The gas and oil were let into the top of the trap through a pipe having the numeral 6 thereon, and down through the top of the trap into a spreader and was then spread on the baffle-plate, which extended across the trap with the exception of a space of two inches next to a partition which is marked 1; this partition marked 1 extended from the part near the center of the trap well down into the interior of the trap. The oil and the gas then came in, was from this pipe by this baffle-plate or spreader 4, passed down over the edge of the interior surface of the trap into this 15 inch

(Testimony of O. W. Harris.)

chamber on one side of the trap, and the gas which was released from the oil passed upwards and inside of the separator and through a two-inch throat device of the trap, being taken off by a gas pipe line marked 9.

Mr. BAGG.—Just a moment. I assume you didn't see that operation?

A. We didn't see the operation, couldn't get at it to see.

Q. All that is your theory?

A. Simply location.

Mr. BAGG.—We object to his theory, if the Court please, [101] because he hadn't shown himself to be an expert on the question of the action of oil and gas. He is qualified as an electrical expert but not as an oil and gas expert.

The COURT.—You think you have sufficient knowledge to testify as to the action of the oil in there?

The WITNESS.—I think I have, your Honor. I have had some additional experience which has not been detailed.

The COURT.—Very well. Do you want to examine him further about that?

Mr. BAGG.—Yes, I would like to test his—

The COURT.—Very well.

Mr. LYON.—Just generally, Mr. *Gutzler,* state what your experience has been in connection with the actual operation of the separation of oil and gas?

A. About 5 years ago—

(Testimony of O. W. Harris.)

Q. And oil separation, too?

A. About five years ago I became interested in the problem of taking water out of oil, not particularly taking gas out but taking water out of oil; and there was a company organized here in Los Angeles known as the National Dehydrating Company of which I was and am president, and during a period of three or four years I spent over half of my time in the oil fields, working with these dehydrating plants for taking water out of the oil and I lived on the leases and slept in bunk houses and I was with the oil men all the time and I think I have quite a fairly comprehensive knowledge of the oil business. At that time the problem we were working on was a somewhat similar problem to this. It was a problem of taking the gas out, very much as these take the water out. I saw a great deal of operators and talked with a great many men and I think I am competent to speak as to the method of those trap operations.

Mr. BAGG.—That is a matter for the Court to determine whether you are competent.

*The COURT.*—Well, the Court asked me about it. [102]

Mr. BAGG.—Whatever testimony you have given already with reference to the action of this oil and gas separator has been based purely upon your suspicion as to the action of it. You don't know positively how that did happen?

A. I have never seen the interior of that trap when it was in operation because it is a physical

(Testimony of O. W. Harris.)

impossibility to see it. You can't operate with the cover off very well.

Mr. BAGG.—Now, if your Honor please, we move to strike out the testimony of this witness as to the action of this oil and gas in this trap because of the fact that the witness has not qualified himself to testify as to the action of oil in this trap under these conditions; and from the further fact that the witness—that he testified he has never seen the gas trap in actual operation, that is so far as the internal action of the trap is concerned and is only speaking hypothetically.

The COURT.—Are you able to state what the operation is on your own machine?

Mr. BAGG.—Yes, sir; we will be able to furnish the Court with definite information as to how that operates.

The COURT.—How it operates?

Mr. BAGG.—Yes, and we will.

Mr. LYON.—From observation.

The COURT.—Well, this man is an expert and is more capable of testifying about those things than the general run of witnesses and knows more about it; he has been in this.

Mr. BAGG.—I don't think, your Honor, he testified he worked about oil and gas separators; he testified he slept out in the oil fields and has been out there a good deal. He didn't testify that he knew anything particularly about the action of oil and gas separators. He did testify he knew some-

(Testimony of O. W. Harris.)

thing about the dehydrating of oil which he says was a similar proposition.

The COURT.—Well, but his experience amounts to a good deal. I think I will allow the testimony.

Mr. BAGG.—Just give us an exception. [103]

The COURT.—You can take the exception.

A. In the center of the trap on the opposite side of the partition 1 from the chamber marked 2 in the sketch there were two floats.

Mr. LYON.—Point them out.

A. On the opposite side of the partition 1 (indicating), from this chamber 2 were two floats, long floats, and these floats were attached by a suitable lever system which can be distinguished in the photograph, particularly figure 1 in the photograph; a valve on the outside of the trap, one of which controlled the flow of oil and the other which controlled the flow of gas, and this linking system which extended outside of the trap terminated in a long lever on which there were weights, and on the operation of the trap as we saw it when we first went there these weights would move up and down very little evidently as the flow of oil in the trap varied—in other words there was a very slight movement, very little movement. However, by taking a hold of this weight it could be pulled down or could be pushed up, showing that it was balanced there, and had a capacity for moving in either direction, and by experimenting, pulling this up and down we determined that this lever and weight not only controlled the flow of gas but also

(Testimony of O. W. Harris.)

controlled the flow of oil, in other words that both the flow of oil and gas from this separator could be controlled by manipulation of the lever. I don't know what other information you might want. I looked it over carefully and Mr. Trumble was there and Mr. Townsend was there, Mr. Burrows of the General Petroleum was there, and we went into the interior of the trap. Mr. Trumble got up on top of the trap and looked in and we have his picture, figure 4 looking into the trap, shows the method of getting into it. We also have Mr. Trumble's picture moving the weight up and down showing how its action can be controlled.

The COURT.—What do you mean by the flow of oil, the gauge of the oil as it passed out?

A. No, I mean as the—evidently the well would flow more [104] or less as I would say by heads —in other words it was not an absolutely uniform flow, and if the flow of oil would increase—

Q. As it came into the tank?

A. As it came into the tank.

Q. That lever then regulated the flow of oil more or less as a gauge?

A. It seemed to regulate the pressure; it seemed to keep the pressure constant at all times.

Mr. LYON.—Mr. Harris, I show you a glass bottle with a top. What is it?

A. That is a model that shows the internal construction, that is, partially shows the internal construction of the Lorraine trap in regard to which I have been testifying.

(Testimony of O. W. Harris.)

Q. The one on the Tonner lease?

A. The one on the Tonner lease, No. 3 Well. It shows the oil in that pipe, the spreader, the throat inside of the spreader and the gas outlet pipe through which the gas was finally taken out. It doesn't show the oil outlet pipe nor does it show the two valves, one of which was on the oil outlet and the other of which was on the gas outlet. The gas outlet pipe comes from the top and contains the gas valve and the oil outlet comes down there (indicating) and contains the little handle for them to turn on this lever and both in turn operated to control the flow.

Q. I wish in giving your testimony you would speak just a little slower. I am afraid the reporter in this room where there is a good deal of noise cannot quite catch it all. Now this glass and brass model that you have last identified then, is a correct illustration of so much of the Tonner trap as it shows? A. Yes.

Mr. LYON.—We offer this model in evidence as exhibit—

The COURT.—Any objection to this model?

Mr. BAGG.—Yes, we have an objection.

Mr. LYON.—We offer it in evidence as Plaintiff's Exhibit 10. [105]

Mr. BAGG.—We object because it is only the testimony of a witness who went out there and examined the trap and he doesn't testify as to whether or not it is in exact proportion to the Trumble trap or is an exact duplicate, either drawn

(Testimony of O. W. Harris.)

to scale or drawn in any other way that makes it accurate.

The COURT.—Who made that model?

A. That model was made for me by a workman employed in Mr. Trumble's shop.

Q. Made according to size?

A. It was made from this sketch, plaintiff's Exhibit—it was made from a carbon copy of this sketch, Plaintiff's Exhibit No. 9, that is a sketch that I gave him.

Q. Have you got the dimensions there?

A. The dimensions are here, and that is proportional to the dimensions of this sketch, Plaintiff's Exhibit 9 as nearly as it could be duplicated.

Mr. BAGG.—Did you test that to see whether it was drawn to scale or not?

A. I checked up the main dimensions, that is all the dimensions that were at all essential in the thing. The importance of making a glass model, an exact duplicate of the model trap, is so you can see the internal processes of it.

Mr. BAGG.—Well, we renew our objection.

The COURT.—You say that that is a fair model in proportion to the size and dimensions and so forth of the original?

A. I think it is a fair representation.

The COURT.—The objection is overruled.

Mr. BAGG.—Note our exception.

Mr. LYON.—Will you take this and demonstrate by water and water can the action of the oil and gas in flowing, so as to show the distribution.

(Testimony of O. W. Harris.)

The COURT.—Are you testing that with water?

Mr. LYON.—With water. That was the object, if your Honor please, of my examination of Mr. Paine in which he said that [106] water had even less viscosity and the action of the oil in flowing down the sides of the chamber or trap would be even less than that of oil.

The COURT.—He said the oil would be—

Mr. LYON.—Roll over and over.

The COURT.—Would roll over and over, while the water would—

Mr. LYON.—In other words the water would go down with less action of rolling over and over, and just simply flow down while the oil would stop and roll over and over. Acts so as to throw all particles of its body outward so the gas could get out. But this demonstration just simply shows where the spreading of the oil is actually on what we term in our patent and claims the inner surface.

The COURT.—My impression was that the water followed the glass.

Mr. BAGG.—Yes.

The COURT.— —more readily, more evenly, than the oil would, and the water itself would not amount to very much.

Mr. LYON.—I would be very glad to use natural oil, but oil is black and the moment that we put oil in there, why we are through, we can't see through it. It is only an approximate test anyway and shows the action.

(Testimony of O. W. Harris.)

The COURT.—Have you any objections to that test?

Mr. BAGG.—Yes, sir, we do, because in the first place oil, as your Honor has suggested does not act the same as oil and gas, and in the next place that oil itself would not act the same, although he offers to produce oil here, it wouldn't produce the same effect because as testified by the witness and also specified in the specifications of the Trumble trap this material that comes into this gas trap is of a foaming substance, a mixture of oil gas, water and sand; until you get all that mixture bottled in foaming form as described this would not, even the oil itself give any idea of the action of the oil and gas when it comes in and the oil and gas separator. [107]

The COURT.—That doesn't appear to me to be a fair experiment.

The WITNESS.—It simply indicates, your Honor, how a fluid if it went through here this way the oil would go through; if you put water in it will work the same way.

Mr. LYON.—Now, you may answer the rest of the general question how this Lorraine trap on the Tonner No. 3 Well which has been stipulated to have been manufactured and sold by the defendant and which you say you examined and made these sketches of compares with the inventions of the patent in suit? I believe you are familiar you said with that patent? A. Yes.

Mr. BAGG.—Now, if your Honor please—

(Testimony of O. W. Harris.)

The COURT.—Just let me ask one question—what is the objection?

Mr. BAGG.—Now, if you Honor please,—I was going to object to that, being a matter for the Court to determine. The drawings are here, the description has been given and the Court will determine the action of those, the comparison of the action of the two traps. We have had expert testimony here to testify to the action of this oil as it comes into the Trumble trap. This present witness has testified as to the action of the Lorraine trap. Now it is for the Court to determine whether or not the action of the two are the same or similar or in any way one infringes the other.

The COURT.—That is the way we determine these things in patent suits is by expert testimony. The Court is not always an expert.

Mr. BAGG.—I know that, but in view of the fact that the action of both of these have been testified to by the experts, it is then for the Court to determine whether or not the action of the two is the same. The expert testified very carefully this morning as to the entire action of the Trumble trap. This witness testifies very carefully and very accurately as to his interpretation of the way the Lorraine trap acted. Now I take it that the Court having [108] seen and heard the explanation that these experts have as to the action of these two traps, it is for the Court to determine whether or not the action is similar.

The COURT.—The witness being an expert on

(Testimony of O. W. Harris.)

that subject I will take his testimony. There is one question I want to ask him further: Did you take note of the distance existing in this trap between the bottom of that incline and the wall itself (indicating)?

A. Well, your Honor, it was a very hard thing to determine. It was necessary—I might explain that our access, this being a commercial trap we couldn't discard in any way—we did cut into it the best we could. This hole which is marked 8 on the sketch, Plaintiff's Exhibit 9, normally has a cover on it and that cover was taken off. Now we could look in and we could see the top of the partition and the top of the baffle-plate and part of the baffle-floor as it went down into the trap. I was just about able by reaching in with my arm to reach to the bottom of the baffle-floor, and there was about room enough to get my hand down between the edge of the baffle and the wall of the trap—in other words, I would estimate the distance there might perhaps have been an inch.

The COURT.—What is it you call that piece?

A. The baffle or the spreader.

Q. Baffle?

A. Baffle. It might possibly have been an inch, possibly it might have been a little more, I wouldn't attempt to say accurately because we would only tell by feeling down in there.

The COURT.—Very well, that answers my question. Go ahead.

Mr. LYON.—And Mr. Harris in making this

(Testimony of O. W. Harris.)

comparison, if you are so inclined you may also have reference to the Lorraine trap, Plaintiff's Exhibit, Lorraine 3.

A. The patent, Lorraine patent, Plaintiff's Exhibit No. 3, is not exactly in accordance with the trap that I observed, in [109] that in the patent, particularly in figure 4, there is shown a sort of box marked with numeral 15, which has an opening in the front into which a pipe, 14, projects. On the trap which I observed on the Tonner lease there was no such box. There was a sort of spreader on the end of the pipe 14, but there was no box of this general type on there. The spreader came down and was supported by means of cleats or straps below here (indicating), and we are able to get the dimensions by taking the position of the rivets on the outside of the trap and determined that was 25½ inches down. Aside from that the patent—figure 4 of the patent is very similar to the trap that I saw; in other words, there is a partition in the patent marked 19, which extends from the top down to the bottom. There is a spreader marked 17 to the right of that partition as it is found in figure 4 of the patent and the oil comes in in the patent through a pipe 13, strikes upon a spreader and is, I assume, thrown outwardly by the spreader against the oil trap 2, moving down and there over.

Q. In the Lorraine specifications of the patent I call your attention to page 3, lines 5 to 12 in particular (handing document to the witness). Read that

(Testimony of O. W. Harris.)

to the Court in connection with this description of the spreader or duplicater you are referring to.

A. To make this portion of the patent intelligible we have to start on the line 129 of page 2. The Lorraine patent says: "The valve 28, having been set by its regulating means 29 to hold a given pressure in the receptacle 2 then as the emulsion is supplied to the receptacle by the oil supply pipe 12 the latter will be directed down the smaller compartment through the inlet or feed sleeve 15 by which the oil is showered onto the adjacent portion of the receptacle wall whence it flows downwardly between the wall and the partition 19, any gases being liberated rising to the top of this compartment and passing over the upper end of or through the partition 19 and accumulating in the upper end of the receptacle 2." [110]

Q. Will you please take this large reproduction and point out to the Court that portion of the description that you have been referring to as to showering action, and so forth?

A. The pressure, I might perhaps refer to lines 111 to 115 of the patent, in which it says: "as above mentioned, there is maintained in the receptacle a gas pressure as determined by the *by the* adjustment of the pressure regulating valve 28." In other words valve 28 regulates the pressure inside the receptacle, the pressure being maintained in the receptacle by that valve, the mixture of oil and gas which is under pressure; otherwise there would be no pressure in the receptacle and it is entirely dependent upon

(Testimony of O. W. Harris.)

the initial pressure to get any pressure at all. The oil and gas under pressure is delivered through the pipe 14 of the patent, passes downward through that pipe upon the baffle or spreader 17. It is thrown outwardly by that spreader on to the inner wall of the trap. It flows downwardly over that inner wall. The gas is released in its downward flow, passes upwardly through the throat between the vertical portion of the spreader 17 and the partition 19,—in other words, you see this throat up through here (indicating),—and it is finally taken off through a gas pipe which is not shown in this view because it is cut off. It is the gas pipe 22 of the patent.

Mr. LYON.—Now, what sentence or portions of specifications of this Lorraine patent, Plaintiff's Exhibit 3, do you rely upon for this statement that this incoming gas and oil is showered on to the inner surface of the chamber, 2.

A. Lines 5 and 6.

Q. Read it.

A. On page 3, which says: "by which the oil is showered on to the adjacent portion of the receptacle wall whence it flows downwardly between the wall and the partition 19."

Q. Now, then, proceed with your comparison with the invention of the patent in suit. [111]

A. Do you wish me to compare the claims of the Trumble patent?

Q. No, not the claims; the general mode of opera-

(Testimony of O. W. Harris.)

tion and the devices and the combination as it is expressed in the patent?

A. Well, so far as I can see the Trumble trap and the Lorraine trap operate exactly the same way. The Lorraine trap has certain structural differences—for instance, the Lorraine trap has very large, two very large floats which are placed on the other side of this partition. These floats operate two valves outside the trap one of which controls the oil and the other controls the gas. In the Trumble patent the float is placed in the bottom of the trap and operates a single valve; but so far as the method of getting the oil in the trap and getting the gas out of the oil, it seems to me that the Trumble and the Lorraine operate in exactly the same way, in other words, the oil and gas are brought in under pressure; they are allowed to flow downwardly through the trap; the gas is taken off through what might be called a separate conduit—in other words, it is taken away from the oil and it is taken out of it—I was somewhat led to that conclusion by the fact that if Lorraine didn't desire to get this operation it seems obvious to me that he would have placed this lower down in the chamber. He gains nothing by dropping that oil. He might better bring it down to the oil level and discharge it quietly, that is, and let it fall in there as it does. Then in the Lorraine trap you will find expansion chamber arranged to receive quietly rather than to let it fall in there as he does.

Q. Then in the Lorraine trap you will find an ex-

(Testimony of O. W. Harris.)

pansion chamber arranged to receive oil and gas, do you?

A. Yes. That may be considered as the entire interior of the chamber which is under uniform pressure or it may be considered as a partition between the partition 19 and the wall 2. In other words, you may consider this small portion in here as an expansion chamber (indicating), or may consider it all an expansion chamber, it is a matter of choice. [112]

Q. All above the liquid?

A. All above the liquid.

Q. Now, what means, if any, do you find in the Lorraine trap for spreading the oil over the wall of such chamber to flow downwardly thereover.

A. No. 17 of the patent.

Q. And where on this device are the gas take-off means?

A. The gas take-off means in the Lorraine is the pipe 22 in combination with this throat—this throat in here, (indicating)—in other words, that gas is taken off under this baffle, just as in the Trumble, is taken up in the Trumble and taken out, just as in the Trumble.

Q. In which figure is this? A. Figure 4.

Q. And you have valve control means arranged in the Tonner installation and in the Lorraine patent for controlling the submergence of oil in the chamber?

A. That is the oil flowing on the outside of the

(Testimony of O. W. Harris.)

trap which is worked from the floats instead of the top.

Q. And what means do you find for maintaining a pressure in the trap?

A. We find two means. In the patent he shows a valve 28 which he—may I see that trap—valve 28 which he describes in these words, page 2, line 32: "From the valve 26 is continued a delivery gas pipe 27 in which there is mounted a valve 26 controlling the flow from the receptacle 2." In other words that valve marked 28 in figure 3 of the patent is stated by Mr. Lorraine to be for the purpose of holding a pre-determined back pressure in the gas line and in the receptacle 2, in other words, in the interior of the trap.

Q. And do you find any other references to the maintenance of pressure in this patent, in this patent?

A. Yes, there are repeated references, line 111, page 2, it [113] is said: "There is maintained in the receptacle a gas pressure." On line 23, page 3, it is said: "The gas will be compressed to a pressure substantially equal to that determined by the valve 28." On line 46 it says: "By maintaining pre-determined pressure in the oil receptacle the latter is subjected to pressure." That is all that I have marked at the present time.

Q. Now, on the Tonner lease Lorraine trap, what means were there for maintaining that pressure?

A. There were the means in the patent corresponded to those you pointed out in the patent?

(Testimony of O. W. Harris.)

A. Corresponded to those in the patent?

Q. And what pressure was maintained at that?

A. There was a pressure of substantially 45 pounds. Unless we manipulate the trap the pressure remained very constant on that lease while I was there.

Q. Did you examine any other Lorraine trap within the last week? A. Yes, sir.

Q. Where?

Mr. BAGG.—We object to that, if your Honor please.

The COURT.—Isn't that matter in issue?

Mr. LYON.—This is the 1922 trap, another device they have been bringing out since the suit was filed.

The COURT.—Do you allege that?

Mr. LYON.—Yes, we allege in our bill of complaint that the defendant is—at the time it was filed, infringing, and that it intends to continue to infringe, and we are going to prove the intention, because it is about a year since the filing of the bill, by showing what they are now doing. They make a slight modification in the trap which is a device of this other model, and we will show we purchased one of their traps on last week in order to have the last evidence and not have to try two cases, and this is the device that we are now going to prove.

The COURT.—Did you file a supplemental complaint? [114]

Mr. LYON.—I don't know anything, your Honor, I could do, except to copy that bill, because a bill of

(Testimony of O. W. Harris.)

complaint in equity is in a form which brings that in—not like an action at law. We might be defeated in this case, having filed this bill of complaint in December, 1920, if we had only shown this installation in December, 1920, of this one trap, and they might say that they had ceased to manufacture and sell it and had no intention of infringing. Our allegation in that regard is this, "That at diverse times during the six years last past, in the Southern District of California, the defendant herein, David G. Lorraine, without the license or consent of plaintiffs, has used the apparatus described, claimed and patented and has made and used the apparatus described and infringed upon said letters patent and each and all of the claims thereof and intends and threatens to continue to do so." Now, all we could add to the general allegation if we filed a supplemental bill is that they are continuing to manufacture, and we would do that by producing the actual device that they are now making. They are continuing as is claimed to infringe and we say that is only a coverable change. We have never been in a patent case where the bill of complaint was in that general form that the court shut out any of their subsequent devices that the plaintiff elected to bring in as proof of intention and fact that they were continuing to infringe, because our whole injunction, which is the main relief, of course, that we are after, and the one that gives jurisdiction in equity, depends upon our showing that the infringing act was a continuing one.

(Testimony of O. W. Harris.)

The COURT.—Here is a device that was constructed from different principles involved in the other one?

Mr. LYON.—Not according to our theory, no different in principle.

The COURT.—I know. They might claim that on the other side. It might be well to file a supplemental complaint and get the [115] matter in issue.

Mr. LYON.—Well the whole matter is in issue because defendant denies an intent to infringe, and that is the only way we can prove it is to prove what they are actually doing, what they are doing now. Judge Trippet within the last few weeks defeated a litigant in equity because the testimony was two or three years old as to the actual infringement when there was nothing to show that during the intervening time they had used anything that did infringe, and defeated it on the ground of jurisdiction, so far as equity was concerned, and the defendant denied that he was longer infringing, just the same as they do here, And as to a supplemental bill our difficulty is this: Our supplemental bill would simply be what? A repetition of our allegation of the bill that they are now infringing and intend to infringe and we never would get through with the number of different devices that we could bring before the Court. And the situation is this, also: Supposing that the case is tried solely on the issue of this first device and an injunction is issued. Does it or does it not cover this device? If we had

(Testimony of O. W. Harris.)

gone to an injunction because we could try them in contempt proceedings; perhaps we would have to file a supplemental bill, but I know of no rule of equity which will shut us off from bringing in all the information up to the date of trial under the allegation of the intent to continue.

Mr. BAGG.—If your Honor please, the Court has held in Hammond Paving Company vs. Bryant, 113 Federal, 316, that a suit for infringement of patents cannot be sustained by proof of acts of infringement committed after the bill was filed. The court held on that very carefully. Now, then, of course with reference to what counsel on the other side said with reference to our intention to continue our infringement, that was merely based on an allegation in his petition for a bill of complaint for the purpose, if possible, securing a temporary restraining order. That is all that was put in there for. That is the customary allegation to put in a bill of complaint where one seeks to get a [116] temporary restraining order or temporary injunction as the case may be for the purpose of stopping further operations because of this intention to or expressed intention or alleged intention to continue, and ask the Court then by virtue of that allegation and that statement in the bill of complaint to grant a temporary injunction which would prohibit the party from continuing any act of infringement.

The COURT.—What have you to say about reaching this by supplemental complaint?

Mr. BAGG.—If they were to file a supplemental

(Testimony of O. W. Harris.)

complaint there is no doubt they could bring it in by supplemental complaint. I agree with the Court in that regard. If they would bring a supplemental bill here of course they could do that, but they can't do it in the absence of the supplemental bill because they stand upon the allegations in the bill of complaint at the time it was filed.

The COURT.—I think you had better file your supplemental bill

Mr. LYON.—We ask leave then to file a supplemental bill in the morning and bring this device in.

The COURT.—Have you one that has been operated since the commencement of this suit?

Mr. LYON.—I believe, your Honor, the practice in New York is when a question arises like this to dictate the amendment or supplemental bill right to the reporter. Now, maybe we might follow that practice here.

The COURT.—Well, you can file it in the morning.

Mr. LYON.—We can file it in the morning, that satisfies us. And shall we proceed with the evidence or hold that until morning?

The COURT.—You can hold that until morning. The Court is going to adjourn at 4 o'clock.

Mr. LYON.—Now, that is the next question I wanted to take up with Mr. Harris.

The COURT.—Very well. If you want to take this up you might proceed and if you don't get your bill in the evidence will go out.

Mr. BAGG.—Now, if your Honor please, I think

(Testimony of O. W. Harris.)

the supplemental bill—we ought to have some notice of what the allegations are and be prepared to meet it. [117]

Mr. LYON.—The allegation will be the same that you are continuing to manufacture and sell this trap like the one the Lacey Manufacturing Company is manufacturing. We can produce it for the purpose of evidence in this case and shall be pleased if the Court wishes and shall be willing to inspect the actual trap which we have opened up by cutting it with acetylene torch in order to show—we can show the entire interior of it.

Mr. BAGG.—If your Honor please, we came into this court upon the allegation of the bill of complaint originally filed, and we didn't come here to make a defense as to that particular trap and we certainly ought to have time to prepare for that; and I think that before the evidence should be taken in that case that we should have some notice of what the bill of complaint is going to be. We will probably have to prepare an answer to that supplemental bill and I think we ought to be allowed to see the supplemental complaint and will probably prepare a supplemental answer and then prepare for trial on that supplemental. Now, I think it is probably possible we could go right on with the trial, we could get ready and then take it up; but I wouldn't want to bind myself right now on that because of the fact I haven't consulted my client or my witnesses in regard to it.

(Testimony of O. W. Harris.)

The COURT.—You can serve that bill by nine o'clock in the morning?

Mr. LYON.—We could serve that bill a good deal earlier than that, your Honor. If we take an adjournment now and Mr. Bagg remains at his office we might serve it on him by 5 o'clock. We will file it in the morning. If you are not there we will serve it by nine in the morning.

The COURT.—Very well. The Court then will adjourn until to-morrow morning at 10 o'clock. [118]

[Endorsed]: Original. In the District Court of the United States for the Southern District of California, Southern Division. (Before Hon. Charles. E. Wolverton, Judge.) Francis M. Townsend et al., Plaintiff, vs. Davis *S*. Lorraine, Defendant. No. E–113—Eq. Reporter's Transcript of Testimony and Proceedings. Filed Apr. 7, 1922. Chas. N. Williams, Clerk. By R. S. Zimmerman, Deputy Clerk. Vol. I. Los Angeles, California, March 22, 1922. Reported by J. P. Doyle, E. L. Kincaid, Doyle & St. Maurice, Shorthand Reporters and Notaries, Suite 507, Bankitaly International Building, Los Angeles, California, Main 2896. [119]

## VOLUME II.

## INDEX.



## INDEX.

## VOLUME II.

113

Los Angeles, California, Thursday,
March 23, 1922, 10 A. M.

The COURT.—Are you ready to proceed, Gentlemen?

Mr. F. S. LYON.—Yes. If your Honor please, we were unable to serve the supplemental bill on Mr. Bagg until 9:15 to-day, but it is short and I might read it to your Honor.

The COURT.—Have you read it, Mr. Bagg?

Mr. BAGG.—It was only served on me at 9:15, your Honor, and I haven't had a chance to read it over thoroughly. I know what the contents of it are, however.

The COURT.—You will not want it read over then at this time?

Mr. BAGG.—No.

The COURT.—You will want time to answer?

Mr. BAGG.—Yes; and if it is necessary for us to make a discovery we will have to have time to go over our books.

The COURT.—And may the evidence go on in the meantime?

Mr. BAGG.—Oh, yes.

Mr. LYON.—We plead in this supplemental bill that they are infringing by this new type. We plead that that is illustrated by the machine or trap which the General Petroleum Company of California purchased from the Lacy Manufacturing Company, which is manufacturing the Lorraine traps for him, one of which we now have in our possession, and that that infringes Claims 1, 2, 3 and 4 of the patent in suit. The only discovery we ask is the usual discovery when we come to an accounting as to how many like that they have made. We bank our entire supplemental bill on this trap which we have in our possession, and we will prove that it was manufactured by the defendants. Unless, therefore, on the question of the validity of the patent, or on the question of infringement, the defendant has some other or different defense on the issues of validity or infringement that he wishes to put in, I would [122] suggest that we stipulate that the answers heretofore filed stand as the answer to the supplemental bill as well, unless Mr. Bagg has some other and additional defense, and if so he can state that and we can consider the answer so amended.

Mr. BAGG.—If your Honor please, we have no

objection to trying all the issues out in this one case, as to these various types of traps that we are alleged to infringe. We are perfectly willing to have them all thrashed out. But we would like to have a little time to study this trap. Now, he has described the trap as one that was sold by the Lacy Manufacturing Company. Of course that company is not a party to this suit, and we have to investigate their records to ascertain just what the type of trap was. Counsel are probably aware of the fact that we put out a number of different types of traps. They are adapted for different kinds of wells. Different wells have different characteristics, and you must have a trap of a certain model, adjusted in a certain way, to meet the requirements of a particular well. Now, in order to ascertain just exactly what the style of trap was that was sold to the General Petroleum Company we will have to go down there and investigate with the Lacy Manufacturing Company in order to ascertain what that is. It may be this type of trap,—we do not know. But we have never had any intimation that they were going to use that, until yesterday morning when they introduced that model, and consequently we don't know just exactly what the type of trap is. Now we are perfectly willing to thrash this out right now on any type of trap. All we want to know is just exactly what the type of trap is so that we can make the proper answer.

The COURT.—Well, can you accomplish that by Monday morning?

Mr. BAGG.—Oh, yes, we can answer easily by that time.

The COURT.—The usual practice is for an accounting to be taken up as a separate matter anyway. [123]

Mr. BAGG.—Yes, I understand that, but, as I understand, the rules require that we include this in our answer.

The COURT.—Yes.

Mr. LYON.—The only thing the answer requires now is a defense as to the validity of the patent and the denial of infringement. They can announce, under counsel's statement, that they do not contest the validity of the patent; that it is only a question of infringement. They can introduce the state of the art without pleading it. The only thing we have in that line, then, is that unless counsel has some other defense, they know what this trap is, and we are prepared to show that all we did in order to get this trap was to send an order down to the Lacy Manufacturing Company's plant where the Lorraine Gas & Oil Separator Company sold this trap for $1,300 to us, just the same as you would go out and buy a jack-knife or any other piece of mechanism, and as far as the statement that they have made a large number of different types of trap is concerned we have no knowledge of any such condition. We are, as I say, banking this allegation of infringement by our supplemental bill solely—

The COURT.—That includes this second trap?

(Testimony of Ford W. Harris.)

Mr. LYON.—Yes; that type.

The COURT.—Well, I don't think any injury can come from allowing the defendant to file its answer Monday, and in the meantime we can go on as though the answer were filed and the issues joined.

Mr. BAGG.—Very well.

Mr. LYON.—And if we get into the defendant's case we will reserve all questions of competency of the evidence and the admissibility of it under the pleadings until the answer is filed, without argument.

The COURT.—Yes. Call your next witness. [124]

**Testimony of Ford W. Harris, for Plaintiffs (Recalled).**

FORD W. HARRIS, recalled on behalf of plaintiffs, testified as follows:

Direct Examination (Resumed).

(By Mr. F. S. LYON.)

Q. Mr. Harris, you stated that you had observed this Lorraine trap that was on the Tonner lease, Well No. 3, and that it was being operated under a sustained pressure in the trap of 45 pounds. Is that correct? A. That is correct.

Q. Now, have you observed any other of the defendant's traps in operation? A. I have.

Q. Please state where they were and what the condition was with respect to such pressure on the traps.

(Testimony of Ford W. Harris.)

A. I have never seen many of the Lorraine traps. There is a trap across the road from the Tonner No. 3, on the Tonner No. 1 lease of the General Petroleum Company, and I observed that, and that was operating at a pressure of about 30 pounds. I also observed two traps on the lease of the Wonder Oil Company at Richfield; one of these traps apparently was shut down at the time and not operating and the other was operating under a pressure of your observation?

The COURT.—Those were Lorraine traps?

A. Lorraine traps. They had the Lorraine name on them. And they were constructed externally in accordance with these drawings here.

Q. (By Mr. F. S. LYON.) Is that the extent of your abservation?

A. That is the whole extent of my observation of the Lorraine traps.

Q. Can you give the date, approximately, when you made those observations?

A. I made my original observations before the suit was filed, but I made additional observations last Sunday. [125]

Q. And those traps were operating under pressure as you have indicated last Sunday?

A. Yes.

Q. And what means did you ascertain were on each of those respective traps for maintaining that pressure within the trap?

A. I didn't have an opportunity—or I didn't make any very careful investigation to see what

(Testimony of Ford W. Harris.)

the means were. I presume, however, there is—

Mr. BAGG.—Well, we object to any presumption, your Honor. He says he didn't observe.

The COURT.—I think you ought to state what you saw.

A. I simply observed that the gages on the traps showed this pressure. I don't know by what means the pressure was maintained.

Q. The pressure on the trap?

A. On the trap itself.

Q. What was the pressure?

A. Why, the pressure was in some cases 30 pounds and in some cases 25. It is very difficult to read these pressures on these particular gages, because they are high-pressure gages and they read down very low on the scale. In other words, they are not gages originally designed to read small pressures. They read up to several hundred pounds, and it is somewhat guesswork to read accurately on the gages, but they all show some pressure—at least 25 pounds.

Q. (By Mr. L. S. LYON.) Did you examine a Lorraine trap at the plant of the Trumble Company in Alhambra last Saturday?

A. Yes; I have examined that trap several times.

Q. Can you state the construction of that trap?

A. I think I can. [126]

Q. Please go ahead and state fully what was done while you were present with that trap and what you observed and how it is constructed and what is its mode of operation.

(Testimony of Ford W. Harris.)

A. I visited Mr. Trumble's shop or plant at Alhambra and this trap was in his yard there standing on the ground. We had a workman take—we first pried open the top of the trap, the manhole in the top of the trap, and we were able to see the entire internal construction of the trap, due to the peculiar construction of the trap inside. We could see a central chamber, but there were two side chambers which were blocked off by metal walls that we couldn't get into. We therefore had a workman take an acetylene torch and cut a hole inside of the trap adjacent to what we assumed to be the oil inlet. Upon cutting this hole inside of the trap we found a certain form of construction which is illustrated by a model which I have here (exhibiting). This model was made by a workman in that plant to illustrate the method of construction of the trap, and I won't vouch for the dimensions of it exactly, but I will vouch for it as correctly representing the general construction of the trap. On opening the manhole in the top of the trap we saw there was a central chamber between two partitions, and we could look down in this central chamber and see two floats in there similarly placed to the floats in Fig. 4 of the Lorraine patent. We couldn't see into the side chambers because there were walls across the top of it. In one side of the trap there was a pipe going in and we cut a hole to the right of that pipe—when you are looking at that side of the trap—and we saw that that pipe went in to the interior of the trap

(Testimony of Ford W. Harris.)

and it had firmly secured thereon an elbow, this elbow being inclined at an angle of approximately 45 degrees and being ground off so that it fitted tightly against the partition—which corresponds to the [127] partition 19 of the Trumble patent. We did not at that time—we later cut another hole on the other side so that we could get a much better view of the interior of the trap.

Q. Let me interrupt you just a moment, Mr. Harris, I present to you a drawing and ask you if you know what it is.

Mr. LYON.—I will state to the Court that I will prove the accuracy of this drawing by another witness a little later.

A. This is a drawing of the trap which is on the property of Mr. Trumble at Alhambra now and which I saw last Saturday and since.

Mr. LYON.—For identification we ask that this drawing be marked Plaintiff's Exhibit No. 11.

Q. Would it not be easier for you to use a blue-print of this? A. Yes; I think so.

(Blue-print presented to witness.)

Q. Now, as you proceed please take this drawing and point out on it the different parts you refer to and explain the construction to the Court.

A. In this drawing we have three figures. The figure on the left is a cross-section through the pipe on a central plane which passes through the oil inlet; Figure 2 is an elevation taken at right angles to Figure 1; and Figure 3 is a horizontal cross-section through what may be called the gas transfer

(Testimony of Ford W. Harris.)

pipe in the top of the trap. The oil inlet pipe which appears to the right of the first figure—that is, the figure to the left in the drawing—goes into the trap at right angles to the partition, which we may call a long partition, this partition being on the right-hand side of the trap, and there being a short partition on the left-hand side of the trap. Behind [128] these partitions there is an open space in which the floats are placed, and upon taking the manhole cover off the top of the trap we were able to see down into the trap, see the floats, see the top of the left-hand and right-hand chambers and see a pipe which connected these chambers. We were unable, looking into the top of the trap, to see what was inside of these chambers since the tops of the chambers were closed by a wall.

Referring to the right-hand figure, the oil inlet pipe is shown going into the center of the trap. We cut an irregular hole to the right of that little inlet pipe so that we could look in, and we found that the oil inlet pipe went into the trap a short distance and that it had secured thereon an elbow, this elbow being turned, as shown in the right-hand figure, at an angle of approximately 45 degrees, and being cut off so that it fitted very closely against the long partition—that is to say, the partition on the right-hand side of the trap. The floats were connected by levers inside.

Q. (By the COURT.) Do you call this a float (indicating)?

(Testimony of Ford W. Harris.)

A. Yes; these are floats, and the floats are shown here in this view. They were connected by a lever to two external valves—the two valves here, and shown on the end view in this figure here. Looking at the plan view which is on the upper right-hand corner of the sheet, it will be seen that the nipple is turned to this long and short chamber connected by a short pipe extending across the trap. The gas outlet pie is taken off the top of the trap, and is shown on the upper left-hand corner of the left-hand figure, being carried down back of the trap to a valve which is operated by the float. We later on cut another hole on the other side to give us better [129] access to the end of the trap to see how the nipple was placed and how it operated under operating conditions.

Q. Now, trace the action of the oil and the gas, and the separation of it, in this device.

Mr. BAGG.—To which questions we object for the reason that our contention is the witness is not qualified as an expert to describe the action of this oil and gas in the operation of the device.

Mr. F. S. LYON.—I just want to ask a preliminary question.

Q. You have observed a number of these similar devices in operation? A. Yes.

Q. And from your inspection of this particular trap last referred to, are you able to state how the oil and gas enter and are separated?

A. From my observation from certain experi-

(Testimony of Ford W. Harris.)
ments that I made upon the trap itself I think I can give a—

Q. (By the COURT.) How did you make those experiments?

A. By putting water through the trap to see how the water would flow through the interior of the trap.

Mr. BAGG.—We renew our objection to the testimony of this witness as to the operation of this trap for the reason that he has not qualified himself as an expert, and for the further reason that the test that he describes as having been made with water is not a competent test to show the action of oil and gas in an oil and gas trap.

Q. (By the COURT.) From your experience are you able to say what the action was?

A. I think I am; yes, sir.

The COURT.—The objection is overruled.

Mr. BAGG.—Exception.

A. Referring to the left-hand figure of this drawing, the oil and the gas enter the trap through a pipe in the top of the right-hand chamber, and being under some slight [130] pressure at least they naturally tend to follow their entering line; that is to say, they enter the trap horizontally on the central plane. If there were no elbow in the trap they would simply strike against the long partition, that is to say, the right-hand partition. The elbow being placed as it is tends to turn this oil and entrained gas, the oil and gas striking against the elbow and being turned around. Now upon putting

(Testimony of Ford W. Harris.)

the water through there we found that the water—

Mr. BAGG.—If your Honor please, we object to the water demonstration.

Q. (By the COURT.) Well, you may state your experiment there if you desire. I doubt whether the use of the water is a fair test.

A. We found that the water was strayed out over this partition and out through the elbow and being sprayed down across there; then with a very small stream of water—

Q. Well, what is your idea now as to the action of the oil?

A. I think the oil would have even a greater tendency to cling to the surface and cling to the partition because of its greater viscosity and its greater stickiness.

Q. Have you any idea why the pipe was put in there at an angle?

A. I think it was put in at an angle for the purpose of giving it a sweep and allowing it to spread around on the partition.

Q. (By Mr. LYON.) What mechanical feature in that particular elbow, if any, was there that gives you the opinion that it was intended to have the oil sprayed on to this partition and that does spray it on the partition?

A. The oil pipe entering the trap extends over toward the partition for a short distance. In other words, the elbow is not close up against the partition. It could [131] easily be. The end of the elbow which goes up against the partition is ground

(Testimony of Ford W. Harris.)

off to a knife edge so that it fits tightly against the partition and makes a close contact therewith. The lower end of the elbow forms a sort of sweeping curve that conducts the oil from over and tends to spurt it out in a very pretty form—a very efficient form—upon the partition.

Q. (By the COURT.) I wonder why that was not adjusted so that when the oil entered the whole of that . . . instead of in the pipe.

A. There is a reason for that, I think, your Honor. The gas is taken off through a connecting pipe as shown on the extreme right-hand corner of this figure and goes across into the other partition. The oil and gas being drawn away, it decreases the liability of gasoline vapors being carried away with the gas. In other words, to draw it away from the gas outlet, so that the gas can be taken out of that film and taken away on the other side.

Q. (By Mr. LYON.) Did you observe the mechanical construction of the faces of this elbow with relation to the direction of a fluid running on them?

A. The elbow is not an ordinary elbow. In other words, an ordinary elbow as you would purchase it would have the pipe threads in it because it is made for the purpose of making connection between the pipes. This elbow is very smoothly machined out to give it a flaring surface to assist in this direction of the oil over its surface and against the partition.

Q. And what was the object of running a small stream of water on that elbow?

(Testimony of Ford W. Harris.)

A. We wanted to see if the water under varying conditions and quantities would flow over that partition. We experimented by directing it in different directions in there to see that it would. It was possible by putting [132] the hose in far enough so that it wouldn't touch anything here (indicating) to have the water come through without touching the partition; in other words it didn't touch the elbow—it didn't touch the pipe. But if the water was directed against this pipe almost anywhere it invariably came out and fell down over the surface of this partition.

Q. In what form? A. In the form of a film.

Q. What was that due to?

A. That was due to the fact that the water naturally was directed toward the partition as it came in and had a certain velocity, and naturally was moving towards the partition, and this elbow didn't turn it exactly parallel to this partition but it turned it so that it struck the partition at an angle and, naturally, it adhered to the partition and flowed down over it.

Q. (By the COURT.) Wouldn't that elbow action go on—

A. No, it did not, in our experience, seem to go on. A little might flash off, but the great bulk of the oil flowed down over the partition. In our experiments we couldn't make it flow over the outer wall at all. It flowed over the partition. Now it might be, with a large supply, that some of it would go on the outer wall too. We had a garden hose.

(Testimony of Ford W. Harris.)
But being directed in this manner it not only tends to go on this partition but it is also directed towards the curved wall of the tank in case any of it gets off the partition. In other words, it is directed into that corner formed by the partition in the wall of the tank.

Q. (By Mr. LYON.) And how did the stream come out of the nipple—in the form of a column or how?

A. No; the stream came out of the nipple in the form of a sheet—not the nipple but the elbow—in the form of a [133] sheet—not the nipple but the elbow—in the form of a thin sheet. In other words, it went in in the form of a thin sheet and this thin sheet was continued over and went right on down over the partition.

Q. And then from your observation of this elbow what was it in the elbow that made it so come out in the form of a sheet rather than in a column?

A. It was the flaring of the elbow, the general shape of the elbow, the angle at which it was placed, and its relation with the entering oil. In other words, it is a very well worked out directing means for directing the oil over and spreading it out.

Q. Can you point that out on this drawing, what it is that does that?

A. Yes. The oil entering in this direction, that is to say, towards the partition, striking against that machined surface in the elbow, is simply turned and spread, and of course as it is turned it tends to spread out, and being spread it strikes on this parti-

(Testimony of Ford W. Harris.)

tion and flows down over in this thin body. I see no reason why oil would not act in exactly the same way.

Q. And you base that statement on what experience?

A. On a great deal of experience in handling oil at various times.

Q. Now in this device what means, if any, are provided for maintaining a pressure wthin the trap?

A. There are provided two valves, a gas valve and an oil valve, one of them being in the gas outlet and the other in the oil outlet, and by a suitable adjustment of these valve pressure could undoubtedly be maintained in the interior of the trap.

Q. Are those valves indicated on this drawing?

A. Yes; in all three figures of the drawing.

Q. Where? [134]

A. They are indicated in dotted lines on the left-hand figure, and near the bottom of the trap; they are indicated at the extreme right and bottom of the right-hand figure, and they are indicated in the plan view at the extreme right of the plan view which is at the upper right-hand corner of the sheet.

Q. And how, based upon your experience with such machines and with machinery in general, are those valves arranged mechanically to be operated upon this?

A. They are operated through the float and through a lever system which is not fully indicated on this drawing. So far as I can see, the arrange-

(Testimony of Ford W. Harris.)
ment for operating these valves is substantially like the other traps I have examined, with the possible exception of this independent adjustment on the gas and oil valve.

Q. And how does the operation of these valves, so far as maintaining pressure is concerned, and the discharge of oil, compare with the function and direction of the valves for these purposes in the Lorraine patent, Exhibit 3?

A. So far as I can see they are exactly similar. They are for the same purpose and operate in the same manner.

Q. And how do they compare with the valves for these purposes in the advice illustrated in the Trumble patent?

A. They are exactly similar to the two valves in the Trumble patent.

Mr. LYON.—You may inquire. Now, as I have stated to your Honor we will prove this drawing and then we will ask the court to inspect, before the close of the case, this particular device, and, if the court desires it, we will operate under actual oil.

Cross-examination.

(By Mr. BAGG.)

Q. Mr. Harris, you are a patent attorney, are you not? [135]

A. I am an attorney at law.

Q. Well, are you not a patent attorney also?

A. I am a patent attorney also.

Q. You are one of the attorneys regularly em-

(Testimony of Ford W. Harris.)

ployed by the Trumble Gas Trap Company or the plaintiffs in this case?

A. I don't know that the Trumble Gas Trap Company regularly employs an attorney. I am the attorney who got out their patent for them.

Q. Well, you are one of their attorneys in this case also, are you not?

A. I am not; no, sir.

Mr. L. S. LYON.—Will you stipulate that that trap we got last Saturday was purchased from and made by the defendant in this case? We have the driver here who delivered it, and as he is a busy man—

Mr. BAGG.—I probably will, but I don't want to until I have consulted my client.

Mr. L. S. LYON.—May we interrupt the examination of this witness, your Honor, so that we can put the driver on to testify that he took the trap out there?

The COURT.—If there is no objection to that procedure.

Mr. BAGG.—No objection, your Honor.

(Witness temporarily withdrawn.) [136]

**Testimony of Thomas T. Davis, for Plaintiffs.**

THOMAS T. DAVIS, called as a witness on behalf of plaintiffs, being first duly sworn, testified as follows:

Direct Examination.

(By Mr. F. S. LYON.)

Q. Your name is Thomas T. Davis? A. Yes.

(Testimony of Thomas T. Davis.)

Q. By whom are you employed?

A. By the General Petroleum Corporation.

Q. Los Angeles, California? A. Yes, sir.

Q. Last Friday did you take a gas trap to the yard of the Trumble Company in South Alhambra?

A. Yes, sir.

Q. Where did you get that trap?

A. At the Lacy Manufacturing Company.

Q. Under whose instructions?

A. Under Mr. Foster's.

Q. Who is Mr. Foster?

A. He is my transportation foreman.

Q. Transportation foreman of the General Petroleum Company? A. Yes, sir; at Vernon.

Q. And you delivered that trap at the yard of the Trumble Company in South Alhambra?

A. I did, sir.

Q. Did you make any changes of any kind in the trap from the time of its delivery to you until you delivered it to the Trumble Company? A. No, sir.

Mr. F. S. LYON.—That is all. Any questions, Mr. Bagg?

Mr. BAGG.—No. [137]

**Testimony of Ford W. Harris, for Plaintiffs (Recalled—Cross-examination).**

FORD W. HARRIS recalled.

Cross-examination (Resumed).

(By Mr. BAGG.)

Q. Now, you say that you are not one of the attorneys for the Trumble Gas Trap Company?

(Testimony of Ford W. Harris.)

A. I understand that I am not an attorney of record in this case.

Q. Well, you are, however, an attorney for them in another case that is similar to this, in which Mr. Moran is bringing a suit for infringement against the Trumble Gas Trap Company, are you not?

A. I wouldn't be sure of that. Mr. Graham, my partner, is attorney in both cases, and it really is—

Q. Well, your firm name is signed to the motion in that case, is it not?

A. It is possible that I am an attorney in the other case.

Q. And while your name does not appear, probably, as one of the attorneys in this case, your partner's name does appear on this supplemental bill of complaint?

A. I may not, technically, be an attorney in this case, but for all practical purposes I am an attorney in it.

Q. And as such attorney and representative of the Trumble Gas Trap Company you went out to see this trap on the lease that you described yesterday?

A. Yes, sir.

Q. And as their attorney you examined it with a view of testifying in this case? A. Yes, sir.

Q. Now you testified yesterday that you examined the pressure gauge on that trap and found that it showed a pressure of about 45 pounds. [138]

A. Yes, sir.

Q. That would indicate, as I understand it, the interior pressure on this trap?

(Testimony of Ford W. Harris.)

A. That is as I understand it; yes.

Q. Now, you testified yesterday that there were two valves operated by a float on the interior of this separator; is that correct?

A. There are two valves on the exterior which are operated by a float on the interior.

Q. Well, yes, that is what I intended to say.

A. Yes.

Mr. F. S. LYON.—Are you referring to the Tonner trap?

Mr. BAGG.—I am referring to the one he testified to yesterday.

Q. Now, those floats are operated by the height of the oil level in this larger chamber of the Moran trap, are they not? A. They are.

Q. Did you examine the mode of operation of these valves or these floats? A. I did.

Q. I wish you would explain to the Court if you can just how these valves worked as you observed them there that day.

A. My observations were confined to certain—what you might call rather obvious experiments. There is a long lever which projects out from the trap and carries weights on its outer end. This lever could be pushed up and pulled down, the pulling down of the lever corresponding to the condition which would occur when more oil came into the trap. In other words, as the float would rise the lever would be moved down, and *vice versa*. Upon pushing up the lever—which would correspond, of course, to the diminution of oil inside the trap—the

(Testimony of Ford W. Harris.)

flow of [139] oil into the tank into which the oil was delivered would be shut off. In other words, apparently, if the oil in the trap fell below a certain level the oil valve operated by these floats would close for the purpose of allowing more oil to collect, and as soon as the oil had collected and the flow had arisen a certain distance this oil would be released into the trap. So far as I could determine by an observation of the pressure gage and by opening a little auxiliary valve the gas valve operated in the opposite manner; that is to say, the gas valve apparently would close—or would open—you might see—the gas valve operated exactly opposite to the oil valve. In other words, when the oil valve was fully closed the gas valve was fully open, and *vice versa.*

Q. Well, then that is simply due to the opening in the oil exit, and the gas exit was governed altogether by the height of the oil in the trap?

A. That is so far as the two valves which were directly mounted upon the trap and operated by the float were concerned.

Q. So that it was the oil level that governed the operation of both of those valves?

A. So far as I could determine, yes.

Q. Now, you testified that the pressure on that trap was about 45 pounds to the square inch and remained more or less constant during that time. You don't know where that gas was going, do you, at that time?

A. Well, I know it was being delivered to a cer-

(Testimony of Ford W. Harris.)

tain pipe there. I don't know where the pipe went to. The pipe went down into the ground, and I don't know where it went from there.

Q. Did you say that observed a valve on that pipe line somewhere?

A. There was a valve on that pipe line just outside the trap, but this valve really has nothing to do with [140] this particular case because it was evidently an outlet valve to allow that gas to escape to the air and had nothing to do with controlling—

Q. It was what is known as a safety valve that is required by law in these cases, is it not?

A. Well, I don't know. It was simply a manually operated valve. I think it was there for some experimental purposes. I don't see that it had anything to do with this particular case.

Q. So that the only valve you observed, then, on this trap was this valve operated by this inferior float such as you have described?

A. These were the two valves that I observed, yes.

Q. Now, suppose this gas was being carried a long distance, there would be more or less pipe line friction there, would there not? A. Why, certainly.

Q. And that would have a tendency to cause a back-pressure in the gás trap which would show on the gas gage, would it not, or the pressure gage?

A. Yes.

Q. Now, if this gas was connected up and being delivered to some absorbing plant, and they had a reduction nozzle or some other means for reducing

(Testimony of Ford W. Harris.)

the flow, that would cause a back-pressure, would it not?

Mr. F. S. LYON.—We object to that as purely hypothetical and assuming a state of facts not shown to exist.

The COURT.—This is more or less theoretical, now, is it not?

Mr. BAGG.—I am just asking that to determine whether he is an expert—whether he is qualified.

A. So far as maintaining the pressure on this trap is concerned, it is quite immaterial; that is, a pressure could be maintained either by a valve put directly at the trap or by a valve on the end of the line or perhaps by [141] a contracted opening at the end of the line, which might be an absorption plant, or where it is delivered to the boilers, or anywhere. It is not necessary, to maintain the pressure on the trap, to put the pressure-maintaining valve at the trap; it can be placed at a distance.

Q. Now, do you know whether or not there was an absorption plant connected up with this—

A. I don't know that.

Q. As a patent attorney, the fact that this gas was being delivered to an absorption plant such as was described here a few minutes ago with a restricted passage for this oil, and this restriction was placed on this gas outlet line by the company purchasing the oil or gas or using it, that would not be attributed or charged up to the instruments of the oil and gas separator, would it?

(Testimony of Ford W. Harris.)

A. Well, as a patent attorney I think I ought to leave that to the Court.

Q. Well, you are a patent attorney.

A. Well, my opinion would be this, that if this trap as originally furnished had a pressure-reducing valve on it, and the purchaser was regularly using it and operating on a plant that also had a pressure-reducing valve I would be inclined to think he would take one of those valves off and not operate both of them. In other words, if he already had a means for maintaining the pressure he would not be apt to bother to put another in there which would certainly be without any function or purpose whatever.

Q. Then you observed no pressure-reducing means at all there at that trap?

A. Well, I observed a pressure, and I assumed that since there was a pressure there must be some means of producing it.

Q. But you don't know where that was?

A. I don't know where that was.

Q. You don't know whether it was down at the absorption plant or not? [142]

A. In the very nature of things I could hardly testify to where it was, because that pipe went down into the ground, and if some man had taken me down and said, This is the valve, I couldn't testify in regard to it.

Q. Now, you have examined the valve in the Trumble trap, have you not?

A. The construction of the valves.

(Testimony of Ford W. Harris.)

Q. Yes; and the valves generally?

A. No; not with any great particularity.

Q. Now, is the operation of those valves in the Trumble traps similar to those in the Lorraine trap?

A. The operation of the two valves shown in the Trumble patent is very similar.

Q. Now, the two valves in the Trumble trap are operated independently of each other, are they not?

A. You mean in the patent?

Q. Yes; in the patent.

A. Yes; one is operated by one float and another is operated by another float.

Q. And they bear no relationship one to the other?

A. Well, they are operated in a similar manner to the Lorraine valve.

(Last question read.)

A. I think they do.

Q. You say they do?

A. I think they do; yes. They are not mechanically connected, if that is what is you mean, Mr. Bagg.

Q. And the one that governs the opening of the oil outlet is in the bottom or in the lower portion of the trap and is operated altogether by the height of oil in that particular portion? A. Yes.

Q. Now, the valve which operates the gas outlet is considerably above the float that operates the oil outlet, [143] is it not? A. Yes, sir.

Q. And, as described in the specification for the patent, that valve is put in there for the sole pur-

(Testimony of Ford W. Harris.)

pose of preventing the mixture of the oil and gas as the gas passes out into the gas line?

A. I think the patent can perhaps speak for itself. I don't know as I would want to say just what that valve was put in there for. That is a matter I have not investigated.

Q. But as a matter of fact, before the gas valve would operate in the Trumble trap the valve in the oil line would have operated to its fullest extent, would it not?

A. Well, that would depend upon the dimensions of the trap and the way it is constructed.

Q. Well, the way it is described there—

A. I think it would as it is shown in the drawings.

Q. So they don't work, then, simultaneously at all?

A. Well, of course they are both operated by the oil level. First one opens and the other closes.

Q. That is in the Trumble trap? A. Yes.

Q. Now if the oil level were to get very high then it would close the upper valve, or the gas valve?

A. Yes.

Q. But it wouldn't unless it got to a dangerous height? A. No.

Q. But in the Lorraine trap they operate—one closes and the other opens. They are practically balanced, are they not?

A. Well, the difference is merely a matter of degree.

Q. Now, just answer my question, please. As you have described it, in the Lorraine trap, as one

(Testimony of Ford W. Harris.)

valve opens the other closes. They work in perfect harmony, do they not?

A. Well, I don't think I have described it that way. [144] I said that as the oil level rose in the Lorraine trap first the oil valve opened and then the gas valve closed.

Q. You mean, then, that there is a—

A. Not necessarily.

Q. —a period of time between those?

A. There may be. It depends upon the way in which it is constructed and the way in which it is adjusted.

Q. You didn't observe, then, how these valves are constructed?

A. It is my opinion—I couldn't check it up exactly—that on that trap that I examined at the Tonner lease the gas valve had very little function—had very little to do there. I don't think that gas valve would close until the oil level in the trap was very high, just as in the Trumble.

Q. Well, it probably wouldn't close, would it, but it might be reciprocal? A. Well, yes, it might.

Q. Now, is it not a fact, Mr. Harris, that those valves there work exactly in harmony and unison; as one closes the other opens just in exactly the same proportion, and as the opposite condition occurs, why, the other closes in the same way, so that they work practically in unison one with the other, operated by the same float?

A. I don't know that I am extremely competent

(Testimony of Ford W. Harris.)

to speak on the operation of those valves. This can be said, however: In this trap which is illustrated in this blue-print the two valves are each operated by a lever and these two levers may be adjusted so that the period between the complete closing of one valve and the complete opening of the other, for example, may be made anything you like. Now the two valves are, however, connected together and operated together.

Q. In the Trumble trap there is no outlet for the gas [145] above the upper cone, is there?

A. No.

Q. So that when the oil comes into the receiving chamber—as we will call this chamber above the cone—if there is any atmosphere in there that would have a cushioning effect, or if there was any gas in there it would have a cushioning effect, which would force the flow of the oil and gas as it passed from the oil well down past the edges and between the edges of the cone and the side wall of the expansion chamber, would it not? A. I don't think so.

Q. Well, where would it go?

A. I think it would flow over that surface naturally by gravity. I don't think the atmosphere has anything to do with it.

Q. In a small space, or proportionately small, such as is illustrated here by this model you have introduced, do you mean to say that that would flow without any pressure at all, and that no pressure would accumulate in there?

A. I don't see how it could accumulate. It is in

(Testimony of Ford W. Harris.)

open communication with the entire interior of the trap.

Q. Yes, but there is no escape for either the oil or gas from that receiving chamber except as it flows down the side wall of the trap?

A. That is true.

Q. Now, if there was a considerable flow or considerable pressure in that oil line as it comes from the well wouldn't there be a large amount of pressure inside of the trap? A. Yes, sir.

Q. Now, that would have a tendency to force that, because, there being no escape for the oil and gas in any other direction that would have a tendency to force that [146] oil and gas down past the edges of this cone and between the edges of the cone and the side wall of the expansion chamber, would it not?

A. No, sir.

Q. And that pressure in there wouldn't have any effect upon that at all?

A. No, sir; because you have the same pressure below that you have above. You have to have a difference in pressure to make anything flow. As long as the entire interior of the trap is under the same pressure it doesn't make any difference to the oil how it goes. It simply tends to go in the direction of gravity. If you have a difference in pressure then it will naturally move in the direction of the pressure.

Q. Suppose your well is flowing by heads, which is, as I understand it, what we call a gusher: there would be, at the times when the oil was flowing in

(Testimony of Ford W. Harris.)

heads, an enormous pressure coming with the flow of the oil into this receiving chamber, would there not? A. Yes.

Q. There would be an enormous pressure in there? A. Yes.

Q. Now, before that pressure could be communicated to the balance of the trap it would have to pass down through past the edges of these cones and the side walls of the trap, would it not?

A. Yes; the gas and oil would both pass.

Q. And that would have a tendency to force that oil down in there, would it not?

A. Momentarily, until the pressure equalized itself.

Q. Now, in Mr. Lorraine's trap such as you have described there is no such restricted chamber in his receiving chamber, is there?

A. You mean there is no dead space in the top? [147]

Q. Yes. A. No.

Q. There is a means by which they communicate over the top of this partition with the gas outlet and also with the oil outlet?

A. Yes, his pressures can immediately equalize themselves inside the trap.

Q. And therefore his oil never would, under any circumstances, be forced down over the side walls of this trap? A. No.

Q. It would always flow by gravity? A. Yes.

Q. In your opinion, is there ever any difference in

(Testimony of Ford W. Harris.)

pressure above the upper cone in the Trumble trap and below that cone?

A. In the event of a sudden increase in the pressure of the well or a sudden rise in the pressure of the well that rise in pressure would travel progressively through the pipe and through the trap. In other words, if your pressure was very suddenly increased in your well due to a difference in the rate of flow or a release of some formations below so that you had a sudden increase in the pressure, that increase in pressure would build up very rapidly in the pipe and in the top and bottom of the trap, but that would be an extremely momentary condition, only long enough for the gas to relieve itself.

Q. Now, is this space in the Trumble trap between the edges of the cone and the side walls of the expansion chamber, as you call it, sufficient to equalize that pressure almost immediately?

A. I think it is; yes. I don't know that I would want to testify as to that, because I really don't know what that distance is.

Q. You don't know what the distance is? [148]

A. No. Gas will go through a pretty small hole.

Q. Now, in the Lorraine trap if the well was flowing with a considerable amount of pressure and velocity coming into the receiving chamber, and if the oil and gas, as I understand,—or the oil particularly,—has to pass down the sides, as you have described it, has to pass down through this smaller chamber which is made by the partition and passes

(Testimony of Ford W. Harris.)
around under that partition and comes up on the other side, does it not?

A. The gas comes up on the other side.

Q. No; the oil. A. Yes.

Q. Some of the gas would be carried along too, would it not, and bubbled up?

A. Well, I wouldn't think very much would be carried along.

Q. Well, there would be some, though?

A. It is pretty hard to carry gas down through a body of liquid; I don't expect there would be any appreciable amount.

Q. But there might be some. A. Very little.

Q. Now, in the Lorraine trap as you observed it out there on that Tonner lease the oil level in the trap was considerably above the lower end of this partition, was it not?

A. We determined the oil levels and marked them on the sketches which are in evidence, I think.

Q. Well, you can answer that by saying that it was above it considerably, can't you?

A. Considerably above the—

Q. The lower end of this partition.

A. I was trying to see if I had some dimensions on that. I don't think we determined how far down that partition [149] extended, due to the difficulty in getting in there. We detemined where the oil level was and marked the dimension on it. But the oil level was about 18 or perhaps 20 inches below the bottom of the spreader.

Q. Well, you could tell by the rivets along the

(Testimony of Ford W. Harris.)

sides about how far this partition went down, could you not?

A. Yes, we could tell, but I don't think we put any dimensions on so that I could testify as to how far that went down.

Q. Well, you can testify as to whether or not, can you?

A. I think that partition extends down into the oil.

Q. Quite a little ways? A. Yes.

Q. Now, this partition runs clear across and is solid? A. Yes.

Q. So that there is no communication between the sides of the partition and either one of the two chambers?

A. Only a comunication over the top.

Q. Over the top and under the bottom?

Mr. LYON.—It is riveted to the sides, is it not?

Mr. BAGG.—Yes.

The WITNESS.—In the later traps it is not riveted,—it is welded. In this trap that is in Mr. Trumble's place now—

Q. (By Mr. BAGG.) Then the oil must go down underneath the bottom or lower portion of this partition and rise on the other side? A. Yes, sir.

Q. Now this, you say, would practically be pure oil, or oil with no gas in it?

A. Very little gas in it.

Q. And it would be considerably heavier and more compact than the oil and gas and foam as it came from the well? [150]

(Testimony of Ford W. Harris.)

A. Yes.

Q. Now, before this oil as it comes in from the oil and gas well can get down here into the other chamber it must pass around this partition?

A. Before any oil can get around?

Q. Yes.

A. Yes, sir.

Q. Now, if there was a considerable flow of oil into this trap with this space for communication between the trap—the two chambers above the partition—if there was a considerable flow of oil in there that would practically fill this chamber up, would it not, before it could get around?

A. Well, if you had enough to fill it up. I don't know how much that would be,—whether it would ever be practical in a trap.

Q. I know, but it could. There would be no pressure up here to force that down, would there, because of this communicating space up here?

A. Well, of course if your oil level got very high in that trap and you closed off your gas valve entirely you would have your full pressure of the well blowing out through the oil outlet.

Q. Yes, but suppose there was a great rush of oil or the oil was coming in in great quantities and at high velocity, with this communicating space above the partition which would enable the pressure to go on around and equalize on both sides, this could fill up in here, and probably would, away above this baffle-plate, would it not? A. No.

Q. You think not?

(Testimony of Ford W. Harris.)

A. I don't think it is probable at all; I think it is most unlikely. [151]

Q. Don't you think it is not only possible but probable that it would? A. No, I do not.

Q. That the oil and gas level in the smaller chamber would probably be away above the baffle-plate?

A. No; I don't think that condition ever occurs.

Q. Now if that would be true, then, the space between the edge of that baffle-plate and the wall of the chamber would have to be large enough to relieve that pressure?

A. It would have to be large enough to let the oil through.

Q. Now, I believe you stated that when you examined that trap there that day you reached in and felt the edge of that baffle-plate; and did you make any measurement as to how far that baffle-plate was away from the wall?

A. No. We couldn't make any measurement.

Q. But you did put your hand around there?

A. I reached my arm away down in there and felt of it.

Q. I believe you stated yesterday that you had had considerable experience in the oil and gas "game" as we call it.

A. I don't think I testified anything in regard to the gas game.

Q. Well, the oil game.

A. In the oil business I have had some experience.

Q. And that has extended over a period of how many years?

(Testimony of Ford W. Harris.)

A. Well, it has extended ever since I have been in California, about 10 years; but it was quite intensive over a period of about 3 years.

Q. The last 3 years?

A. No, the 3 years of 1918, 1920 and 1921.

Q. Now, you have seen a large number of other kinds of traps, have you not, oil and gas traps? [152]

Mr. F. S. LYON.—We object to that as not cross-examination.

Mr. BAGG.—Well, I will withdraw that. That is probably true.

Q. Do you know whether or not there are and have been for a number of years a large number of oil and gas separators on the market in different places?

Mr. F. S. LYON.—We object to that on the ground that it is not cross-examination.

Mr. BAGG.—I think probably that is correct, and I will withdraw it.

Q. You testified this morning with reference to that latest trap that you examined ,of the latest—we might possibly say the latest model trap that Mr. Lorraine is putting out. That doesn't any more than spread, as you call it, and—or spread out the oil over anything more than a small portion of this plate, does it (indicating)?

A. I would say that it spreads it over something less than half of it.

Q. All this elbow does is to deflect this flow of oil into this corner, is it not?

(Testimony of Ford W. Harris.)

A. No, that elbow not only deflects the oil but it spreads it and directs it against that partition.

Q. Well, now, the direction of that oil would be, as you stated this morning, coming almost in a straight line from the direction from which it started, would it not?

A. No, because gravity acts on it and as soon as it gets out it starts to flow.

Q. Well, if it were not for gravity it would shoot right out into this corner?

A. Yes, if you didn't have gravity.

Q. And if you had considerable pressure on there that would force it out, would it not? [153]

A. No, because the oil is partly spread before it leaves that elbow.

Q. Well, if there was a large amount of oil coming in there and coming out it would spread out, probably, but it would still shoot the greater portion of that down into that corner, would it not?

A. Well, it is not constructed so that it naturally will do that. The facts that it tends to fly out tends to relieve that pressure and at the same time tends to spread it in a film. If that were spread out as a nozzle it would tend to do that way, but being a direction elbow there I think there is very little tendency for it to do that.

Q. Well, it would work on the same principle as a shotgun, would it not?

A. I don't know, Mr. Bagg.

Q. Now, a shotgun, when it is fired, there is a large number of shot in it, and for a certain dis-

(Testimony of Ford W. Harris.)
tance as they go out of the muzzle of the gun they go practically together, but they begin to spread as soon as they leave the muzzle of the gun.

A. And fall.

Q. And begin to fall, yes, and also to go on; and after awhile they all fall to the ground in a large spray. A. Yes.

Q. Now, isn't that practically the same action as that of the oil in this case?

A. It is the action of anything that is momentarily acted upon by gravity. In other words, it tends to fall.

Q. And acted on by force?

A. Yes. The shape that that thing takes on that partition is somewhat similar to the shape of any falling body. In other words it is a sort of a trajectory curve.

Q. Well, the oil and gas coming into these various [154] traps come in either by virtue of natural pressure in the well or by pressure exerted upon them by the pumper. Is that so?

A. It is my understanding that these traps are not applied very often to pumping wells.

Q. Well, if they were applied to pumping wells they would come in in force? A. Yes.

Q. So that the oil always comes into these traps with more or less force? A. Yes, with some force.

Mr. BAGG.—That is all.

Mr. F. S. LYON.—That is all. [155]

**Testimony of David G. Lorraine, for Plaintiffs.**

DAVID G. LORRAINE, a witness called on behalf of the plaintiff, having been first duly sworn, testified as follows:

Direct Examination.

(By Mr. L. S. LYON.)

Q. You are the defendant in this case, are you not? A. Yes, sir.

Q. Were you served with a subpoena that you appear here with the drawings of the first trap furnished by you to the General Petroleum Company and installed on the Tonner lease at Brea?

A. Well, Mr. Lacy was.

Q. Weren't you served with a subpoena?

A. Not before I came into court; no.

Q. Well, have you been served with a subpoena at all? A. Yes, since that time.

Q. Have you that subpoena with you?

A. Yes, sir.

Q. When were you served?

A. Yesterday morning at ten o'clock (Producing subpoena).

Q. Will you produce the drawings of the first gas trap furnished by you to the General Petroleum Corporation and installed on the Tonner lease at Brea Canyon?

A. Well, after receiving the subpoena I didn't have the time to get that drawing out.

Q. When did you get the subpoena?

A. I got it yesterday after I came into court.

(Testimony of David G. Lorraine.)

Q. And you haven't had time since yesterday to get the drawing?

A. Not the correct drawing; no.

Q. Have you a drawing or a print of it?

A. Well, Mr. Lacy might have it there.

Q. Would you know if there was one here in court? [156] A. I would know it if I saw it, yes.

Q. Well, will you produce the drawing you are referring to?

A. (After examining papers with Mr. Lacy.) He doesn't happen to have it there. We can get it in about fifteen minutes. He has not the drawing of the trap that was installed on the Tonner lease, but he can get it in a short time. I didn't have time to get it from the time I received the subpoena. He made some changes in the tracing and he had to change it back again.

Q. (By the COURT.) How soon can you have it here? A. In about fifteen minutes, I believe.

Mr. BAGG.—We did have a model of that trap here.

Mr. L. S. LYON.—We would like a working drawing of it here.

Q. Have you also a working drawing of the trap that you sold to the General Petroleum Company on March 17, 1922, your order No. 23282?

A. Yes, we have that.

Q. Is that here in the courtroom?

A. Yes, that is here.

Q. Will you produce that drawing, please?

(Mr. Lacy hands drawing to witness.)

(Testimony of David G. Lorraine.)

Mr. L. S. LYON.—The drawing produced by the witness is offered in evidence as Plaintiff's Exhibit No. 12.

Q. Now, this Plaintiff's Exhibit No. 12 is not a complete working drawing, is it, Mr. Lorraine?

A. Well, it is an assembly, however.

Q. Well, you have complete working drawings of that trap, have you not?

A. We haven't any assembly, however, so that you can read it intelligently.

Q. But you have complete drawings of the trap?

A. Of different parts, yes. [157]

Q. Are those here in court?

A. Of the baffle features, that's all. You see it wouldn't have the floats in here, or the valves, or the pipe on it.

Q. What other drawings have you illustrating this trap here? Will you produce those?

A. Why, yes. Will you get that working drawing of this trap, Mr. Lacy, please? It is right in the roll there.

Mr. L. S. LYON.—Did you state, Mr. Lacy, that there were no working drawings here?

Mr. LACY.—There are no working drawings here, outside of the one drawing that you have produced there.

Q. (By Mr. L. S. LYON.) Well, you have working drawings of this trap, Exhibit 12, have you not?

A. Not anything more than that for the assembly view.

(Testimony of David G. Lorraine.)

Q. Have there never been any other drawings made? A. Only parts and pieces that is all.

Q. Well, that is all we want. Have you those drawings here? A. No, I have not.

Q. Can you produce those also?

A. That is the assembly view that we have used to assemble the trap with.

Q. Now, what drawings do you have to make the nozzle with—the elbow?

A. We have a separate drawing.

Q. You have a detail drawing, have you not?

A. Yes.

Q. We would like you to produce at two o'clock all of the drawings you have of the Tonner trap that we have been talking about, including your working drawings, and all of the drawings you have of this General Petroleum order trap, including the working drawings, and particularly the working drawing of the elbow. Can you do that? Will you agree to do that, or else we will—

A. I will try to do it; yes. That is all I can agree [158] to do.

Mr. L. S. LYON.—We would request a direction of the court that they be produced.

The COURT.—Very well; you will produce what you have, will you?

The WITNESS.—Yes.

Mr. BAGG.—Yes, your Honor, we will produce all that we have.

Mr. L. S. LYON.—Will you admit this sales invoice as being given by Mr. Lorraine in connection

(Testimony of David G. Lorraine.)
with the sale of the trap to the General Petroleum Corporation on March 17, 1922 (handing paper to counsel)?

Mr. BAGG.—Well, I don't know why we should not. Yes.

Mr. L. S. LYON.—That is offered in evidence as Plaintiff's Exhibit No. 13, meaning the sales invoice for the trap that the dray man, Mr. Davies, this morning testified to, connecting that particular trap with the defendant.

The COURT.—Very well.

Mr. L. S. LYON.—That is all, Mr. Lorraine; and I will ask Mr. Lacy to take the stand, please. [159]

**Testimony of William G. Lacy, for Plaintiffs.**

WILLIAM G. LACY, called as a witness on behalf of the plaintiff, having been first duly sworn, testified as follows:

Direct Examination.

(By Mr. L. S. LYON.)

Q. With whom are you associated in business, Mr. Lacy?

A. I am employed by the Lacy Manufacturing Company.

Q. That company manufactures traps for the defendant in this case, does it not? A. Yes, sir.

Q. You have been served with a subpoena to bring in certain records in this case, have you not?

A. Yes, sir.

Q. Have you with you the original letter or a copy thereof—I mean the original letter addressed

(Testimony of William G. Lacy.)
to William Lacy, President, under date of December 10, 1920, and signed by F. M. Townsend?

A. Yes, sir.

Q. Will you produce that letter, please?

A. Here is a copy of my reply attached to it (producing papers).

Q. And you have attached to it the reply letter under date of December 14, 1920, by Mr. Lacy to F. M. Townsend. A. Yes, sir.

Mr. L. S. LYON.—The letter and reply are offered in evidence as Plaintiff's Exhibits Nos. 14 and 15.

Mr. BAGG.—Now we would like to see the letter.

The WITNESS.—Now I am removing the Lacy Manufacturing Company replies from these letters (detaching copies and handing letters to Mr. Lyon).

Mr. L. S. LYON.—The letter of December 14 produced by the witness is offered in evidence as Plaintiff's Exhibit No. 14. [160]

Mr. BAGG.—We object to the introduction of that until we know what the contents of the letter is, your Honor. I haven't seen it.

The COURT.—What is the object of introducing that letter?

Mr. L. S. LYON.—The letter is a notice of infringement, may it please the Court, written prior to the filing of the suit.

The COURT.—A notice of—

Mr. L. S. LYON.—Of the claim of infringement, written by Mr. F. M. Townsend.

(Testimony of William G. Lacy.)

Mr. BAGG.—We admit that we received notice in a letter addressed to the Lacy Manufacturing Company, which is not a party to this suit. We admitted notice to ourselves. We stipulated that in the record, so I think there is no use in introducing this, and we object to it as incompetent, irrelevant and immaterial and not bearing upon any of the issues in this case.

Mr. L. S. LYON.—We want to show that the infringement in the case is deliberate, and we want to show the reply to this letter from Mr. Lorraine's manufacturing representative received by Mr. Townsend. The letter and its reply.

Q. Have you that reply, Mr. Lacy, the letter from Mr. William Lacy dated December 14, 1920, to F. M. Townsend? You took it off of this, you say.

A. I have the copy; you would have the original. Mr. Townsend would have the copy.

Q. You have the copy of it, have you?

A. I have the copy.

Q. Can you identify this letter written you as the original, from your copy (handing letter to witness)? A. Yes; that is the original.

Mr. L. S. LYON.—The letter is offered in evidence as Plaintiff's Exhibit No. 15.

Mr. BAGG.—To which we offer the same objection, your Honor. [161]

The COURT.—Very well; I will let it go in.

Mr. L. S. LYON.—That is all.

Mr. BAGG.—That is all. [162]

**Testimony of Hans K. Hyrup, for Plaintiffs.**

HANS K. HYRUP called as a witness on behalf of the plaintiff, having been first duly sworn, testified as follows:

Direct Examination.

(By Mr. F. S. LYON.)

Q. What is your name? A. Hans K. Hyrup.

Q. And what is your business, Mr. Hyrup?

A. Draftsman.

Q. How long have you been a draftsman?

A. About 10 or 12 years.

Q. Do you mean by that a draftsman who makes a business of making mechanical drawings?

A. Yes.

Q. And that has been your business for the last 10 or 12 years? A. Yes.

Q. By whom are you employed?

A. I am employed by Mr. Milon J. Trumble most of the time.

Q. At his shop in Alhambra?

A. At his shop in Alhambra; yes.

Q. Were you there last Friday? A. Yes.

Q. When Mr. Davies delivered the trap that he referred to? A. Yes, I was there.

Q. What was *down* with that trap?

A. It was taken off the truck and set up in place, and after that we opened up the trap.

Q. When was it opened up?

A. It was opened up Saturday morning some time. [163]

(Testimony of Hans K. Hyrup.)

Q. In the presence of whom?

A. In the presence of quite a few of the shop employees out there.

Q. Were there any changes at all made in that trap before it was opened up at the time that—

A. No; no changes.

Q. Now since the opening up of that trap did you make any drawings of it? A. Yes.

Q. I show you Plaintiff's Exhibit No. 12. Did you make that drawing?

A. Yes, that is the drawing I made.

Q. Does that correctly show this trap according to scale?

A. Yes, as nearly correct as I could scale it.

Q. You mean as near as you could scale it and reduce the dimensions? A. Yes.

Mr. F. S. LYON.—We offer the drawing marked Plaintiff's Exhibit 12 for identification in evidence.

That is all. [164]

**Testimony of Paul Paine, for Plaintiffs, (Recalled).**

PAUL PAINE, recalled as a witness on behalf of the plaintiff, testified as follows:

Direct Examination.

(By Mr. F. S. LYON.)

Q. Mr. Paine, will you examine the drawing of which the blue print, Exhibit No. 12, lying in front of you, is a copy and state whether you have ever examined the trap therein illustrated.

A. I saw a trap this morning which corresponds

(Testimony of Paul Paine.)

approximately to this. I haven't checked all the dimensions of the trap.

Q. Where did you see such trap?

A. At Alhambra in the yard of the experimental laboratory of Mr. Trumble.

Q. Did you make any particular observations of that trap?

A. I did of a portion of the top of it.

Q. What portion?

A. That portion which was visible from the outside and which was disclosed through two openings which have been cut in the side of it.

Q. Please explain to the Court where those openings were and what you observed.

A. Those openings were on that side of the trap which also contained a four inch inlet line. On the inside of the trap the oil inlet line had attached to it an elbow setting to which was attached a bell-shaped continuation of the elbow. I take it that this is the oil inlet portion of the trap (indicating). The oil inlet at the outside of the trap comprises a piece of four inch pipe welded to the trap and carried through. The extension [165] in the bell-shaped portion for the delivery of the oil has a diameter of eight inches at its terminus. This flare or bell-shaped extension bears on one side against the vertical baffle-plate; its other edge is at a distance of about two and a half inches from the inside of the trap, making the distance from the baffle to the edge of the trap a total of about ten and a half inches.

(Testimony of Paul Paine.)
The bell-shaped extension has an incline of about 32 degrees from the horizontal.

Q. Based upon your experience with these oil and gas separators and your knowledge of the flow of oil therein and of oil wells as heretofore stated by you, state what the progress of the intermingled oil and gas would be in this trap to which you have last referred?

A. The major tendency of the fluid, the mixture of oil and gas passing in there, would be to take the form of a sheet on the lower portion for the oil, and throughout the entire volume of the bell-shaped affair for the gas as it passes out of this bell, and to impinge against the baffle-plate and spread out over it as it descends in the trap.

Now, the degree to which that would extend would, in great measure, depend upon the proportions—the amounts—of oil and gas present, whether in small quantity or whether in large quantity.

Q. And as a mechanic what do you understand, then, is the purpose in the device of this nipple or elbow and its particular shape and the fact that it is brought into contact on the edge against the edge of the partition or wall?

A. Well, I would judge that to be for the purpose of causing oil to spread out over the baffle-plate in its descent.

Mr. F. S. LYON.—You may inquire. [166]

Cross-examination.

(By Mr. BAGG.)

Q. The character or proportion of that stream as

(Testimony of Paul Paine.)

it comes out of this elbow would be determined in a large measure by the force with which it was driven from the oil well or driven through this pipe?

A. The character of the stream would depend upon the capacity of the well to produce oil and gas; that is, its volume depends entirely upon what the well is producing.

Q. Well, suppose you had a gas well or an oil well that is flowing, we will say, with 200 pounds pressure, that would drive a stream of oil through this inlet pipe at a very high velocity, would it not?

A. That would depend entirely upon how fast it is moving; but as the film of oil, there are two factors involved there—the quantity of gas and the quantity of oil.

Q. Well, we will say just what we call the foam as it comes from the oil and gas well, and say with a pressure of 200 pounds, that would force this foam through this pipe at a very rapid rate, would it not?

A. I am unable to answer that until you tell me what you mean by foam because I have never thought of the production from a well in terms of foam.

Q. Well, that is the way it is described here in Mr. Trumble's patent. I don't know any more about that than you do, and probably not so much, but that is the way it is described in his patent, as a foam. Now if that was in the form of a foam, coming up from the oil well at a pressure of 200 pounds, that would shoot through here at a very rapid rate, would it not?

(Testimony of Paul Paine.)

A. I am unable to say how fast it would shoot.

Q. Well, pretty rapid? [167]

A. Because—it would be in motion. Now, if by foam is meant the spray of oil there, then there would be a possible division of the fluid coming from the well into the three classes of a solid body of liquid—which, of course, in response to the law of gravity would pass along the lower portion of the containing medium—and the spray, which would be in a condition of agitation above the surface of the liquid, and the gas, which is entirely about that. But through the fact that these are all moving with some degree of rapid*l*y they are to a great extent intermingled.

Q. And they would go through here at a rapid rate?

A. Well, rapid, of course, is a comparative term.

Q. I understand, but I couldn't tell you just exactly—

A. Well, they are in very active motion.

Q. Now, the way in which this oil and gas comes out of this elbow would be well determined by the rapidity or speed at which it travels as by the character of the fluid coming out, would it not?

A. Both of those factors would influence it.

Q. I see, one as well as the other.

A. Yes, sir.

Q. And if it were coming out at a very rapid rate and composed practically of oil and gas in a foamy form the chances are there wouldn't be much spreading over this baffle-plate, would there?

(Testimony of Paul Paine.)

A. It is possible in a discharge line to have two conditions there. By restricting the size of the opening one obtains the effect of a nozzle, of making it come out in a more direct line and with a greater piercing effect; on the other hand, if it is desired to slow down the motion and spread the fluid over a wider area the means to accomplish that would be to enlarge the opening at its discharge point. [168]

Q. But it would have a tendeney if it was coming in there at a very rapid rate and in the form of a foam to shoot that off into this corner here rather than to have it flow down the side of that wall, would it not?

A. Yes, but of course the tendency would be reduced through the agency of the bell-shape as compared with the effect it would have if the pipe had been continued in the original diameter that it had at the point where it entered the trap.

Q. Well, you have examined the gas trap that you have just described and the form of that elbow and the bell shape exit of it, and you would say that in that trap if that came as you saw it and examined it there—if that oil and gas in a foam-like form came through there at a very rapid rate it would have a tendency to shoot this oil and gas, the whole stream, into this corner over here instead of having it spread out over that plate to any considerable extent? A. No.

Q. You would not say that? A. No.

Q. You would say, then, it would spread over regardless of the amount of speed with which it

(Testimony of Paul Paine.)

came in there—you would say it would spread over there practically the same regardless of that speed?

A. No.

Q. Well, then what would you say?

A. I would say that the tendency for the fluid, of course, is to move in the direction of the aperture, in that direction; but for the large proportion of it to go over to the corner of the trap, I wouldn't say that that would be the result; that my judgment would be that it would spread over a considerable portion of the baffle-plate, [169] depending to no small degree upon the angle of inclination of that discharge opening as well as the extent to which that bell-shaped affair is flared out.

Q. Well, I am asking you to just describe it from the trap that you examined out there this morning, that bell shape and all, just under the same circumstances and the same trap that you examined this morning.

A. My judgment would be that the tendency in the trap I saw this morning would be for the fluid to spread over half of the baffle-plate.

Q. Regardless of the speed at which it was coming? A. No.

Q. And regardless of the character of the oil—whether it was oil or gas, and large proportions of gas, or a large proportion of oil?

A. It would make a difference.

Q. Then the speed would have a tendency to restrict that amount of surface on the baffle-plate than increase it, would it not?

(Testimony of Paul Paine.)

A. A greater speed would have a tendency to throw a greater volume of oil over towards the corner of the trap. Now there are two influences acting upon the body of fluid that comes into the trap—or on the body of liquid coming into the trap: one is the diversion over to the side of the trap caused by the angle of inclination of this bell-shaped affair; the second influence is the action of gravity which is pulling the oil down. Now, the extent to which those two will act upon each other will depend, in the major fact, on the speed with which that liquid enters the trap. If the liquid comes in at a high rate of speed a greater portion of that will be diverted to the side of the trap, and of course a high rate of speed assumes a large volume of liquid. Now if the liquid is coming in in smaller quantity [170] the major influence upon it will be that of gravity which is pulling the oil down, and in that case the oil will not go so far over to the side of the trap.

Q. Then the speed does have a tendency to restrict the amount of space that the oil flows over on this baffle-plate, does it not?

A. It has a tendency to determine the proportion of oil which goes over to one side of the bafflle-plate.

Mr. BAGG.—That is all. [171]

**Testimony of William C. Rae, for Plaintiffs.**

WILLIAM C. RAE, a witness called on behalf of the plaintiff, having been first duly sworn, testified as follows:

Direct Examination.

(By Mr. F. S. LYON.)

Q. Please state your name, age, residence and occupation.

A. William C. Rae; age, 50; Oak Knoll Circle, Pasadena; Sales Manager for the Trumble Gas Company.

Q. Have you ever observed in operation any of the defendant Lorraine's gas traps?

A. Yes, sir, I have.

Q. Please state where and approximately when, and what, if any, pressures you observed, and how you determined such pressures as were maintained within such traps.

A. The first trap I saw was on well No. 5 of the Tonner lease of the General Petroleum Company, the one you are discussing here.

Q. That is the one Mr. Harris spoke of?

A. Yes, that was running between 40 and 50 pounds pressure when it was first started and until it was connected up with the absorption plant when they had to maintain a pressure of about 27 pounds, I understand, and that registered on the gauge. That was the gasoline pressure to the absorption plant. But for many days before that they had run at 40 to 45 pounds.

(Testimony of William C. Rae.)

Q. (By Mr. BAGG.) You say you understood that. That is what someone told you?

A. That is so. The superintendent who operated the trap and who sends the gas to the compressor plant says that is the line pressure. [172]

Q. He told you that?

A. He told me that; and I saw with my own eyes that that trap had from 27 to 30 pounds. You couldn't tell definitely. They told me the line pressure was 27.

Mr. BAGG.—We object to that, your Honor.

The COURT.—Yes; that is hearsay.

Q. (By Mr. LYON.) How did you see with your own eyes that it had 27 pounds?

A. Well, the high pressure gauge he uses is about 300 pounds, and it is pretty hard to definitely tell within one or two pounds what he has on there. I can give you a reading for that later on if you want it.

Q. And according to your reading of that gauge what was the pressure?

A. Between 27 and 30 pounds.

Q. Now proceed with your other—

A. On Tonner No. 1 re-drilled well there was a test made of it. That trap had been working at different times at from 30 to 40 pounds. I never saw the gauge beyond 40 pounds. I haven't seen it now for about a month or a little more than a month.

Q. Well, you did observe that gauge?

A. I saw that gauge at the time it was installed.

(Testimony of William C. Rae.)

Q. And you did observe it at 40.

A. At 40 yes. At the start, yes.

Q. And at what other pressure did you observe it?

A. On a well at Hugo, No. 1 or 2, I don't just recall, but I think it was Hugo No. 1, the gauge showed no pressure. Said it was run without pressure. That was also a high pressure gauge. We knew that oil—or at least my service man would tell me—

Mr. BAGG.—Now we object to what the service man told him. [173]

A. All right, then, I know that that gas or oil, liquid, cannot be lifted without pressure. My company bought, and I had placed on the gas line, a pressure gauge, and that pressure gauge, an oil pressure gauge, showed 6 pounds, which lifted the oil 12 feet into the shipping tank. That also occurred on another Tonner well, but they say—the Lorraine people said there was no pressure on it. I went out to prove to my own satisfaction that I could see with that pressure gauge, and it did have a pressure to move it into the shipping tank.

Q. In other words, these high pressure gauges would not register at as low a pressure—

A. It didn't register anything; the needle was blank. The 30 pound gauge showed 6 pounds.

Q. Well, what other pressure?

A. On the Stearns, No. 4, I think, of the General Petroleum Company, that showed no pressure. We put the small gauge on that and it showed two and a half pounds.

Q. What other pressure?

(Testimony of William C. Rae.)

A. I have not observed the pressures down at Long Beach nor Huntington Beach. I have seeen a good many of these traps—

Q. Do you know what the gas pressures of those wells are, any of them? A. No, I don't know.

Q. Have you observed the pressure of any other wells that had gas traps put out by the Lorraine Manufacturing Company?

A. I have not observed the pressure on them.

Mr. LYON.—That is all.

(A recess was thereupon taken until 2:00 o'clock P. M.) [174]

AFTERNOON SESSION—2:00 o'clock.

The COURT.—You may proceed with the testimony.

**Testimony of William C. Rae, for Plaintiffs (Recalled).**

WILLIAM C. RAE, recalled.

Cross-examination.

(By Mr. BAGG.)

Q. Mr. Rae, you testified before lunch with reference to those various pressures on those traps.

A. Yes.

Q. I wish you would give me the pressure on each one of those traps that you have described. The first was on Tonner No. 3.

A. There were two pressures on that trap when I saw it at two different times—a pressure of between 40 and 45 in December, 1920; and later on, I should say perhaps three or four weeks, or around that, at

(Testimony of William C. Rae.)

any rate within the next month, it was reduced to between 27 and 30, the gas line pressure on the lease.

Q. That was indicated by the pressure gage on the Lorraine trap? A. As I saw it; yes, sir.

Q. Now what was the next one you saw?

A. Well, I can take— Q. Well, I don't care which one you take next.

A. Well, call it Tonner No. 1. That was last summer, in July or August. Around 40 pounds was the working pressure there for some time.

Q. And you never saw that again?

A. I have seen the trap quite often, but I have never looked at the gage. [175]

Q. That is the only time you looked at the gage?

A. Well, I saw it maybe a dozen times during three or four weeks test that was made there.

Q. And it averaged about 40 pounds?

A. 40 down to 27. When they got that down there they had to run it to the same pressure as they do all of the line pressures on that line.

Q. Now the next one was what?

A. Hugo No. 1 at Richfield.

Q. And what was the pressure in that?

A. It didn't show any pressure. I told you afterwards we had a 30-pound pressure gage on it afterwards, and it showed six pounds. That is the one we put on, on the gas line.

Q. Now, what is the next one?

A. On Stearns No. 4.

Q. And what was the pressure in that?

(Testimony of William C. Rae.)

A. No pressure on their gage. It registered two and a half pounds on the gage.

Q. Now what was the next one?

A. I have seen quite a number of them that I didn't—

Q. Now those are those four traps.

A. They are the specific traps that I paid a great deal of attention to. That was in the early part of the game when they came in.

Q. This was one of your gages that you put on this trap?

A. One that belonged to the Trumble Gas Trap Company, yes. I didn't put it on; it was our service manager that put it on.

Q. You couldn't vouch absolutely for the accuracy of that gage? A. I could not. [176]

Mr. BAGG.—That is all.

Q. (By Mr. LYON.) Do you know whether the pressure of gas from an oil well remains constant after the well has been in production for some time?

Mr. BAGG.—Now, I object to that because I don't think he has shown that he knows enough about oil wells to testify on that line.

Mr. LYON.—That is the question I asked him, if he knows.

A. I don't know.

Mr. LYON.—That is all.

Mr. BAGG.—That is all. [177]

**Testimony of David G. Lorraine, for Plaintiffs (Recalled).**

DAVID G. LORRAINE, recalled on behalf of plaintiffs.

Direct Examination (Resumed).

(By Mr. LYON.)

Q. Have you with you those drawings in regard to that Tonner trap?

A. Why, the drawings, Mr. Lacy tells me, the original drawings he is unable to find. They have been removed from the shop in some way. But we have a sketch there of the construction just as it was made, and we have another drawing here that hasn't the dimensions on it. It is an assembly view, but it was drawn true to scale.

Q. Now this assembly; is that a correct scale drawing?

A. Well, it is, with the exception of this right here (indicating). That should be two inches, and that measures up there one and a half inches.

Q. What is that?

A. That is that deflector?

Q. You say it should be two inches from the wall but in this drawing it is one and a half inches?

A. One and a half inches. Outside of that this is all true to scale.

Q. Now, who made this drawing?

A. Walter Lacy of the Lacy Manufacturing Company.

Q. Under your direction?

(Testimony of David G. Lorraine.)

A. Well, he copied it, but—

Q. When was it made?

A. I think the date is on here, isn't it? It was made after the trap went out.

Q. Was it made after the filing of this suit?

A. I believe it was, yes. But that sketch was made before the filing of the suit, that Mr. Graham has there. [178]

Mr. L. S. LYON.—We offer the print referred to by the witness as Plaintiff's Exhibit No. 16.

Q. Now, this sketch which you have produced, by whom was it made?

A. I sketched that myself.

Q. When was that made?

A. That was made prior to the filing of the suit.

Q. Well, was it made before the Tonner trap was put on or not? A. Yes.

Q. Before the Tonner trap was made?

A. Yes.

Q. How long before?

A. Oh, I should say a month.

Q. Was the Tonner trap made from this sketch? Were the drawings for the Tonner trap made from the dimensions on this sketch? A. Yes, sir.

Q. Now, I notice that the baffle-plate, there seems to have been some crosses on it as if you were going to take it out or remove it. The rest of the sketch is in ink, but those marks are in pencil. What was the purpose of those, and when were they put on?

(Testimony of David G. Lorraine.)

A. I don't know anything about these pencil marks. I haven't seen this for a year—over a year and a half.

Q. It would indicate as if *some* somebody was contemplating moving that over and taking it out, would it not?

A. Why, possibly. I couldn't say. I am no mind-reader as to that.

Q. Well, where has the sketch been since you made it?

A. Mr. Lacy brought it up from the factory just now.

Q. And the dimensions that are on the sketch are [179] correct, are they, and in accordance with the way the trap was actually made?

A. Well, I wouldn't say exactly. All I know about it is this here baffle-sheet.

Q. That is the only dimension that you know whether it is right or wrong?

A. Well, the size of the shell on that, and the position of the float-iron.

Q. When were those dimensions put on it?

A. Before the trap was constructed.

Mr. L. S. LYON.—This sketch is offered in evidence as Plaintiff's Exhibit No. 17.

Q. Was this sketch, Exhibit 17, made before or after you made the working drawings for the Tonner trap? A. Before.

Q. Was the Tonner trap the first trap you ever built? A. No, sir.

(Testimony of David G. Lorraine.)

Q. Had you ever built any traps before you made this sketch? A. Several.

Q. Several traps?

A. Yes; but not of that model.

Q. Is the Tonner trap the first one you ever made in the model that is drawn in that sketch?

A. Yes, sir.

Q. I show you three photostat or photographic prints and ask you to examine those and see if you ever have seen them before (handing same to witness). Not the prints, but the drawings of which these are photographs.

A. Yes, sir, I have seen these before.

Q. What are they?

A. Well, this shell here, and these dimensions on [180] this sketch here were never used to build a trap.

Q. Well, these are photographs of your drawings, are they not? A. Yes, sir.

Q. And what were those drawings made from?

A. Well, they were made to build a gas trap but it was changed to this sketch, this part of the shell and the baffle, before we started to construct the trap.

Q. Then these are photographs of the working drawings that you state you are unable to find, are they? A. No, they are not, exactly.

Q. But they are photographs of small working drawings? A. Yes, sir.

Q. And when were those working drawings made?

(Testimony of David G. Lorraine.)

A. Well, the date should be on those. I couldn't say exactly.

Q. Were they made before the Tonner trap was made?

A. Oh. yes. They were drawn the fourth month, third day, 1919.

Q. And when was the Tonner trap made?

A. Why, it was made some time in 1920, I believe.

Mr. L. S. LYON.—These three prints are offered in evidence as one exhibit—Exhibit No. 18.

Q. Now, were you able to produce the working drawing of this spreader or elbow that is used in the General Petroleum trap? A. The last one?

Q. Yes.

A. Mr. Lacy there says he couldn't get it. He went down after it.

Q. Well, you have that, haven't you?

A. No, he said he couldn't get it. He said he didn't have any down there. [181]

Q. Did you ever make any working drawings of that spreader, the elbow? A. I have not; no.

Q. Have any of them ever been made?

A. Well, we just simply bought that already made.

Q. Who did you buy it from?

A. Why, Mr. Lacy, I guess could say where it was bought.

Q. Well, after the purchase there was some machine work done on it, was there not?

(Testimony of David G. Lorraine.)

A. Just simply to cut the threads out of it, that is all.

Q. Whose idea was it as to the cutting out of those threads and how they were to be cut out?

A. That was mine.

Q. And who did you give the instructions to as to how the spreader was to be cut?

A. To the shop superintendent.

Q. And did you give him any sketches as to how it was to be done? A. No.

Q. What did you say to him?

A. Why, I just simply told him to take the threads out so that they wouldn't cut the oil flowing through there. To machine the threads out.

Q. And that is all you said? A. Yes.

Q. And you haven't any drawing as to how it was to be done at all?

A. No, not as yet. We are going to have a drawing made.

Q. What were you going to have a drawing made for?

A. Why, to show an assembly view of all the parts.

Q. Did you tell them to bell that separator out the [182] way it has been described by the witnesses?

A. I told them not to leave a sharp edge on it. It was immaterial how they would do it.

Q. Well, did you tell them to bell it out?

A. No, I didn't say bell it out.

Q. What did you say?

(Testimony of David G. Lorraine.)

A. I told them not to leave any sharp edge there so that it would obstruct the flow of the oil.

Q. Well, now, did you tell him anything esle? First you told him not to cut the threads out; now you say you told him not to leave any sharp edge. Did you say anything else at all?

A. Why, I never used any such word as belling it out.

Q. Well, then what did you say?

A. I told him to take the threads all out of there so that it wouldn't obstruct the flow of the oil—to machine it out.

Q. And that is all you said? A. That is all.

Q. You didn't say anything about a sharp edge?

A. Oh, no.

Q. What?

A. No. Never mentioned the words sharp edge.

Q. Now, how long ago was that?

A. Well, I couldn't say the exact day. Perhaps two months ago.

Q. How many of them have been made of that kind?

A. How many L's or how many traps?

Q. How many L's have been machined out for traps?

A. About 36. About that. I wouldn't say exactly, because we are making them every day.

Q. Are they all exactly the same?

A. Approximately, yes.

Q. And do you know what operations are gone

(Testimony of David G. Lorraine.)
through in [183] machining and shaping that spreader or L?

A. Why, the L is put into a lathe and they simply put a tool in the holder, and now I believe they are buying those L's without any threads in them.

Q. Now do they do anything to them?

A. Yes, there is a little groove left there to cut threads in and you cut that right out.

Q. And they bell them out, do they?

A. Yes, they bell them out.

Mr. L. S. LYON.—That is all.

Mr. BAGG.—That is all. [184]

**Testimony of Milon J. Trumble, for Plaintiffs.**

MILON J. TRUMBLE, called as a witness on behalf of the plaintiff, having been first duly sworn, testified as follows:

Direct Examination.

(By Mr. F. S. LYON.)

Q. Please state your name.

A. Milon J. Trumble.

Q. Where do you reside?

A. No. 100 North Stoneman Avenue, Alhambra, California.

Q. You are one of the plaintiffs in this case?

A. Yes, sir.

Q. And the Milon J. Trumble to whom, together with the other plaintiffs, the patent in suit was issued? A. Yes, sir.

Q. Now, have you ever had any experience in

(Testimony of Milon J. Trumble.)

the manufacture, installation or use of gas traps embodying your invention of the patent in suit?

A. I have.

Q. Did you ever observe the oil in such a trap?

A. I have.

Q. Please explain to us how you secured an opportunity to observe the flow of oil in such trap and what you observed.

A. I helped to install a trap on the McGinley lease at Montebello, and it was on what they call a head well; it flowed by heads; and after the main was installed we hadn't put the main head plate on and I had the oil turned into the trap several times to see what turn it would take in its course down the trap. This oil flowed in as a dead oil; there was no gas in it at that time. The trap was 200 feet away from the well, and was set up about 20 feet so that the oil would come out in the pipe and come up and go into the trap, so that I could see very plainly [185] the course of the oil, because the gas was not flowing through to interfere with it. The oil came into the trap and went down in the interior in a film. You could see it rolling in places, traveling down, just flowing down the interior of the trap. I could see perhaps three-quarters of the diameter of the trap.

Q. Have you examined the Lorraine gas trap which was delivered at your shop in South Alhambra last Friday? A. I have.

Q. Have you examined that carefully so as to

(Testimony of Milon J. Trumble.)

be able to state what would be the course of the oil in the use of such a trap? A. Yes, sir.

Q. Have you made any experiments to demonstrate such flow? A. I have.

Q. When did you make such experiments?

A. This morning, with oil.

Q. (By the COURT.) With dead oil?

A. With dead oil, that is, oil without gas mingled with it to spray it. It just simply flowed in slow and curved over and ran down; no gas with it.

Q. (By Mr. LYON.) Will you please now state what, in that experiment, you did, and what you observed of that Lorraine trap?

Mr. BAGG.—We object to this because of the fact that he has not testified that this experiment was conducted with oil and gas such as comes from the well, and I take it that your Honor will see that there is a considerable amount of difference between the foam composed of oil, gas, water and sand, a substantial difference, both in its action and in its general make-up, from the dead oil that he describes as having flowed through this trap. [186]

The COURT.—I heard you say that you were prepared to make an experiment in court.

Mr. F. S. LYON.—I don't know how much gas we can artificially put into the oil, but we can show your Honor the oil and the manner in which it operates. I think that so far as the evidence shows in this case to date it is entirely an assumption to say that the action would be any different

(Testimony of Milon J. Trumble.)
with the balanced pressures of gas in the trap and the intermingled gas.

Mr. LYON.—Mr. Trumble, from your experience with these gas traps are you able to state whether there would, in that regard, be any material difference between what you have referred to as dead oil and oil which contained a material amount of gas that comes from the well?

Mr. BAGG.—I think we will object to that, if the Court please. It is not shown that he is an expert in the matter or experienced in the matter of dealing with and handling oil as it comes—

The COURT.—Have you had any experience in that line?

The WITNESS.—I started in the oil fields in 1903 and I put in a good deal of my time in the developing of oil, drilling and with pick and shovel, too.

Q. Have you had occasion to observe the action of oil as you have denominated "dead oil" as it comes from the well in the foam?

A. I have, yes. The dead oil would flow along on the lower side of—

Q. I asked you if you had any experience in that line. Oh, you say you have? A. Yes.

Q. Then proceed?

A. The dead oil would flow along in a stream or body on the lower side of the pipe and would run down whichever way [187] it was deflected, and if it was mixed with oil or oil was mixed with gas, it would come in more of a spray and would whirl

(Testimony of Milon J. Trumble.)

the spray in the way it was deflected. Oil mixed with gas of course is better and travels very rapidly on account of being what we term a live oil.

Mr. LYON.—Then please explain to us this experiment that you made this morning with this Lorraine trap, what that showed with the dead oil, and how based upon your knowledge and observation of the action of the intermingling of oil and gas as it is delivered from a well, that would be in this Lorraine trap.

The COURT.—You insist upon the objection, do you?

Mr. BAGG.—Yes, sir.

The COURT.—I am inclined to hear this, because the Court would like to hear it.

Mr. BAGG.—Very well. We will withdraw our objection.

A. The dead oil as it comes into this trap flows in as I say in a quite slow stream in the lower side of the pipe. In this particular trap we have up there there is a four inch nipple that goes into the trap on this ell drawn out to a feather edge, cone-shaped, that sits up against a partition on which the ell has been ground off so it sits right against the partition pipe, and on this experiment out there the oil comes in slow and just takes the gravity and hits the baffle-plate and runs down a thin film.

The COURT.—That is the dead oil?

A. That is the dead oil. Now if it was live oil that comes in and was coming in it would go up

(Testimony of Milon J. Trumble.)
there and spread in a white fluid and pass down the plate only in a wider spray, that is the only difference.

Mr. LYON.—Mr. Paine was out there to that trap this morning, wasn't he?

A. Yes, sir. [188]

Q. Now, was it after Mr. Paine left that you made this experiment? A. It was.

Q. Mr. Harris wasn't there when you made that experiment? A. No, he was not.

Mr. LYON.—That is all. You may take the witness,—just a moment, I want to ask one question for the purpose of the record. You are prepared to reproduce that experiment to show the court, are you? A. Yes, sir.

Q. In the presence of counsel?

A. Yes, sir.

Q. And are all equipped for it, if desired?

A. Yes, sir.

Cross-examination.

(By Mr. BAGG.)

Q. Mr. Trumble, you testified that when you observed the action of your trap that you had the manhole off or the top off, or something of that kind? A. The manhole plate.

Q. There was not any gas in that hole then, was there?

A. No, sir; it was a dead oil because we opened it while it was dead in the pipe; that is it didn't contain gas and oil, it flowed in until the gas came in there and then we had to flow some of it back.

(Testimony of Milon J. Trumble.)

Q. Did you have your velocity reducing means of that well at that time? A. No, sir.

Q. Was that the oil as it came directly from the well? A. Yes, sir.

Q. It came in under pressure, did it?

A. Yes, sir.

Q. How much pressure?

A. The pressure on the well would go up to about 300 pounds I think before the well would head, start to flow.

Q. But there was no gas in that?

A. Not while we let it into the trap, not at this test, this particular test.

Q. So it just flowed in there quietly, then?

A. Yes, sir.

Q. Now, you didn't have an opportunity to observe how it acted on the side walls of the shell as you call it, after it passed off at the top of the cone or over the edge of the cone, did you?

A. I could see the oil coming down and around in the trap in a film, practically even—it seemed to be evenly divided over the surface.

Q. Over the surface of the trap? A. Yes.

Q. How could you see that?

A. I could look in through the manhole.

Q. Where was the manhole?

A. On the side of the trap nearer the bottom, nearer the cone-shaped bottom.

Q. Looking up that way? A. Yes.

Q. Now you never conducted any experiment with the Lorraine [189] with oil and gas and water and

(Testimony of Milon J. Trumble.)

sand in such a condition as it comes from the ordinary oil and gas well, did you?

A. I never carried on any experiments of those kind.

Q. The only experiments you made was simply putting oil into this trap or into this inlet and watching how it acted when it got out, as it came out of that elbow?

A. We pumped it up through an inch and a half pipe and it just filled this four inch elbow up and turned and ran over smooth and quiet. It was not a capacity of four inch or nowhere near it. It was a small pipe leading in there and it was sealed on the outside, and when we filled it to a level it ran over then and hit this curved cone, ran over the baffle-plate and ran down.

Q. Just a small stream of oil was all?

A. Yes, sir, it didn't matter. We had a valve there so it could be tested down to the size of your finger or a little larger stream.

Q. You didn't test it with anything like the capacity of the requirements that an oil and gas well would put on a trap?

A. I tested it first to the capacity of this pipe. The more oil we let in the more the spray was spread out in a film; the greater velocity the more the spray was spread out in the film and ran down the plate.

Q. Now there was no force upon this any more than just the natural gravity flow after it got up to this pipe?

(Testimony of Milon J. Trumble.)

A. No, we handled it with a centrifugal pipe.

Q. It then flowed up as it filled up this pipe and then flowed over? A. And flowed over.

Mr. BAGG.—That is all.

Mr. LYON.—The plaintiff rests. I want to offer in evidence before resting, however, this model which has been [190] used more visually than anything else.

Mr. BAGG.—If your Honor please, we object to the introduction of that because there has been no showing that that is constructed according to the dimensions of or drawn—constructed according to scale to show the dimensions of the original oil and gas trap.

The COURT.—Didn't somebody testify to that?

Mr. LYON.—Yes, he said it was substantially in accordance as nearly as he could get it. That is correct, isn't it, Mr. Harris? Mr. Harris is here.

The Witness HARRIS.—Yes.

The COURT.—Very well, it will be received.

The CLERK.—This will be 19.

Mr. LYON.—That is the Lorraine 1922 trap. Now I also have here an illustration that we have been using of the Trumble trap, simply for visualization and it might be made part of the record.

Mr. BAGG.—No objection.

Mr. LYON.—I offer that as exhibit 20.

The CLERK.—Exhibit 20.

Mr. LYON.—The plaintiff rests.

The COURT.—Has this been offered?

(Testimony of David G. Lorraine.)

Mr. LYON.—The first one was offered and accepted as Exhibit 10.

The COURT.—This one here (indicating).

Mr. LYON.—Yes. That is all.

Mr. BAGG.—Would your Honor give us about a 10 minute recess?

The COURT.—Yes, the Court will take a recess for 10 minutes.

(After recess.) [191]

DEFENSE.

**Testimony of David G. Lorraine, for Defendant.**

DAVID G. LORRAINE, called as a witness on behalf of the defendant, having been previously sworn, testified as follows:

The CLERK.—He has already been sworn.

Direct Examination.

(By Mr. BAGG.)

Q. You are the same David G. Lorraine who testified a few moments ago? A. Yes, sir.

Q. You are the defendant in this case?

A. Yes, sir.

Q. And are the designer and patentee of the Lorraine gas trap? A. Yes, sir.

Q. How long have you been in the oil and gas business?

A. Well, I haven't been steadily employed in the oil and gas industry, but I have been in and out of it since '96.

(Testimony of David G. Lorraine.)

Q. How much of that time have you been actively in the industry?

A. Well, I would say about 15 years.

Q. Do you understand the action of oil and gas separators and traps?

A. I do, a good many of them.

Q. I will ask you to state to the Court what is the principle upon which all of these oil and gas separators is based?

Mr. LYON.—One moment. We object to that as leading and suggesting and assuming a fact not appearing the evidence, that they all operate upon some one principle. [192]

The COURT.—Well, he has asked to explain the principle. I will hear the answer.

Mr. BAGG.—Read the question.

(The last question is read by the reporter.)

A. The principal operation, your Honor, of separating gas and oil is to reduce the velocity in a receptacle and to prevent the oil from going into the gas line, and to put all into a storage tank, a shipping tank so that there is no gas goes in with the oil into the tank and no oil goes with the gas into the gas line. And the principal way that is done in all types of gas separators is through a velocity-reducing means, in other words the oil is stopped for a certain period of time, and in a good many cases there is a certain amount of pressure put on to that oil for a certain period of time and in other cases there is no pressure upon that oil, but we must stop a certain volume of it for a cer-

(Testimony of David G. Lorraine.)

tain length of time in order to allow the oil to settle, to a settled state, and to allow the gas to come clean off of the oil, to rise to the top. That is the principle that they all work upon.

Q. Just be seated, Mr. Lorraine. Now, I wish you would state to the Court how in passing oil and gas as they come from the oil well, the gasoline or lighter hydro-carbons in the oil is made to remain in the oil after passing through the oil and gas separator, if you know?

Mr. LYON.—Now, we object to that upon the grounds that it is incompetent, no foundation laid, the witness not having qualified to answer the question.

The COURT.—I will hear the answer.

Mr. LYON.—That goes pretty heavily into chemistry.

The COURT.—Goes as I understand to the different elements of hydro-carbon in the oil.

Mr. BAGG.—Yes, sir, that is what is called the lighter contents or such as was testified to by Mr. Paine when he testified as to these, that the higher gravity oil first came off of the Trumble trap. [193]

The COURT.—Very well, I will hear the answer.

Mr. LYON.—Our objection is that the witness has not qualified, your Honor, on that phase of the case either by education or otherwise, in the chemistry of oils, in that regard.

The COURT.—Well, I will hear the witness until it appears that he is disqualified.

(Testimony of David G. Lorraine.)

A. I have made several tests in raising the gravity of oil in different types of gas traps, and I have found by carrying a certain volume of oil under a certain pressure for a certain period of time, regardless of how that oil flows into that trap or regardless of how it flows out, it will raise the same gravity, raise up the same gravity on the trap regardless of how it is made, on all devices or traps. That is a well-known fact and has been known for years. You don't have to spread it, don't have to baffle it in any way—just hold a certain pressure there and the oil absorbs the gasoline and the traps take off the dry gas, and if you take the pressure down on the trap, why, you get a wet gas. That takes place in all types of traps.

The COURT.—What do you mean by "wet gas"?

A. Well, gas that contains more gasoline in the gas.

Q. The result is that what we call hydro-carbons?

A. Yes, the lighter hydro-carbons.

The COURT.—Very well.

Mr. BAGG.—Then as I understand it from your testimony, if you want to retain the gasoline content or lighter hydro-carbons in the oil, you subject the oil while in the trap to a pressure?

Mr. LYON.—Now, wait a moment. We object to that as leading and suggestive.

The COURT.—How is that?

Mr. LYON.—We object to that as leading and suggestive.

Mr. BAGG.—Well, I am just getting at— [194]

(Testimony of David G. Lorraine.)

The COURT.—I think it is a little leading, but I will hear the answer.

A. Well, that is a system and it is well known and been known for a long time. If you put a pressure upon the traps while the oil and gas is flowing through the trap providing the gas contains a gasoline content. All gas that comes off from oil from all wells is not alike. For instance, it will run as high as three or four gallons to the thousand, while other gasses will only run as low as one per cent and five per cent. By putting this pressure on the oil as it goes through the trap it depends upon the volume of oil carried and the pressure carried whether you squeeze the gasoline content into the oil or whether you take the gass *of* at zero.

Mr. BAGG.—Now, then, if you desire to use the gas for the purpose of making what is called casing-head gas or extracting gasoline from it, *who* process do you use?

A. Why, they reduce the pressure down as low as possible. There is a rule.

The COURT.—That allows the gas to escape?

A. No, they get the gasoline then in the gas, and you take that in the extraction plant or the absorption plant, and you take the gasoline out of the gas there; and whether you use this for fuel or put it into a gas company's line like Mr. Paine spoke of here yesterday, it is desirable to get the dry gas as they only use it for fuel purposes, and then it raises the gravity of oil and raises the value

(Testimony of David G. Lorraine.)

of the oil, but in nine cases out of ten if you put a high pressure on to a gas trap, why you retard the production of the well. In a good many cases they think they are saving money, and they are losing money, in other words. The most desirable feature about a gas trap is to take the gas and oil and separate it and not put any back pressure on the well, that is to keep it down as low as possible. [195]

Mr. BAGG.—Now, in coming down to the trap that you installed at the Tonner lease, and which was described as the gas trap on Well No. 3, I will ask you to state to the Court or describe to the Court just exactly how that oil and gas separator was constructed?

A. Well, I can do it better from the model we had.

Q. Very well, if you have that model. I hand you herewith what purports to be a model of your oil and gas separator. I will ask the clerk to identify it as Defendant's Exhibit "A."

The COURT.—"A" for identification.

Mr. BAGG.—Now, will you explain to the Court, how that is constructed?

Mr. LYON.—Without an opportunity to examine this, without any foundation, of course, we will have to object to it. We will have to object to it on the ground there is no foundation laid.

Mr. BAGG.—Well, is that a model drawn practically to dimensions of the oil and gas separator installed on the Tonner lease at Well No. 3?

(Testimony of David G. Lorraine.)

Mr. LYON.—One moment. We object to that as leading and suggestive, your Honor. I don't like to be captious but the witness already testified that they have lost those drawings and they have nothing to make this from but memory. And I notice there is an exaggeration in the model already and I don't know how much more there is, and I would like to have it strictly proven, unless it is being used simply as a general illustration.

The COURT.—Who made that model?

The WITNESS.—Why, it was made according to my instructions.

Q. Is that made in accordance with the plans and specifications of the drawings that were originally adopted for [196] the purpose of constructing it?

A. Why yes, the whole principle is right here of the separating. As far as the baffle features are concerned they are identical. You can go up there and take it out and see the separator, the baffle feature. The float has fallen down in there, you will notice, but this baffle feature here, that is just exactly the same.

The COURT.—Very well. The Court will permit it.

The WITNESS.—The oil—

The COURT.—Set it up there on the table. I will see what I want.

A. The oil flows through here, through this pipe coming from the well (indicating), which as a rule consists of froth and foam, or foam and froth; it is divided here into two streams. The oil gravi-

(Testimony of David G. Lorraine.)

tates down through here. This is not a sealed top in any way; this don't come over the top here (indicating). It gravitates down here on each side. It is not forced down there in no thin film. The edge of the baffle to the wall is two inches.

Q. The baffle is two inches from the wall?

A. Yes, it is two inches and all the oil forced through. The oil comes up around this vertical partition and goes out here through this pipe here to the valve and the gas comes up through the top above this partition (indicating). That is all there is to the operation of that.

Q. How do you keep your pressure there is it *is* not sealed up at the top?

A. Well, what I mean it is not sealed over here, this separator chamber like the Trumble (indicating). The gas has a free exit at the top and the oil gravitates down. The difference between the Trumble trap and my separator is this: The oil and gas are forced down below this cone (indicating), from the pumping pressure or the flowing pressure [197] from the well, while my oil is allowed to gravitate down and the gas goes upwardly and up through this pipe (indicating).

Q. Well, the gas goes up in the other, doesn't it?

A. In the Trumble?

Q. Yes.

A. The gas and oil in the Trumble is forced down below this cone (indicating).

Q. I know. The gas goes up when it comes in the oil?

(Testimony of David G. Lorraine.)

A. Yes, you are correct there. It is all forced downwardly while my gas goes up and the oil goes down. You see the gas comes up from the top (indicating).

Q. I see.

A. And the oil gravitates down and deposits the sand and water at the bottom, and then the oil goes out of this outlet here (indicating).

Mr. BAGG.—Talk louder, Mr. Lorraine, so the stenographer can get you.

A. That there is practically the same only it hasn't got the large opening here (indicating). That would force some of the gas perhaps down below this here baffle-plate. If there was only that small—

The COURT.—Got a large space there?

A. Got a large space in the trap for the gas to go over the top. That is all the difference as far as that goes. That is constructed on the same lines. You see that the force forces the gas through the oil while ours goes up of its own volition. In this the gas would go up (indicating) and in that the oil would go down (indicating).

Mr. BAGG.—Now, Mr. Lorraine, just right there, when this gas and oil get in this trap and flow down through the pipe, before it strikes the baffle, describe to the Court what action or direction it takes before it strikes this baffle?

A. Here (indicating)? [198]

Q. Yes, sir.

(Testimony of David G. Lorraine.)

A. Why, it goes out in a divided stream, puts a stream out on each side.

Q. Go on and describe to the Court anything further.

A. This here in most cases on this side of this here partition (indicating), the oil is much higher than it is on the other side, because there is always a steady head still up here of oil and this froth or foam being lighter it is much higher here (indicating); it is just simply a settling process there. This here (indicating), if the well is flowing or pumping at all, that is filled with foam, this here chamber here, and gravitates down there and it holds the oil there and the gas goes upwards off of the top. The principal object in baffling the oil in my device is to prevent the oil from being stirred up with the water and to have it settle; it is not to run the oil down the side walls of the receptacle, to have any gasoline content squeezed into the oil. With the Trumble the oil and all the gas do come down below that cone there and that there action is positive, of course. I don't know whether that raises the gravity there or whether it raises it down there (indicating), but what I do know is that you can take any receptacle, the same size, retain the same pressure on it and the same amount of water for the same time, and your oil will be of this same gravity, because I have already tried it out here in the fields without the baffles.

The COURT.—What do you mean by "gravity"?

A. Why, as Mr. Paine was explaining yesterday,

(Testimony of David G. Lorraine.)
in order to raise the gravity of oil you put the pressure on the trap as the oil and gas passes through the trap, and of course it increases the value of the oil by raising the gravity.

Q. When that oil comes out it has a raised gravity? [199]

A. It makes a lighter oil and there is more gasoline squeezed into the oil. Of course, that is not desirable when you take the gasoline extraction plant; where they are using it for fuel purposes why, that is undesirable, because dry gas is just as good for fuel as the wet gas, and perhaps better.

Mr. BAGG.—On that trap that is installed on Well No. 3 on the Tonner Lease. I will ask you to state if there is any pressure-maintaining means used by you on that trap?

A. There is nothing—

Mr. LYON.—Wait a minute. We object to counsel leading witness. We don't think he ought to put the words in the mouth of the witness all the time. He hasn't even attempted to explain that device in full, your Honor.

The COURT.—Well, you can explain what if any pressure is maintained there and if so, how?

A. There is nothing connected with the trap that would maintain any pressure on a thin film of oil or on the receptacle. If we go to discharge the oil up in the tank, discharge it up, say higher than the tank, there is a pressure put on to the tank to discharge that oil up into that tank. There is

(Testimony of David G. Lorraine.)

a pressure put on the trap I should say, just enough pressure put there to discharge the oil. If the gas line is blocked in any way, for instance you have a small line, why of course it would build up a pressure, the friction on the line would do it; but as far as being a pressure means to sustain, to maintain a pressure upon the trap, there is nothing connected with it that would do it. This trap will handle a thousand barrels of oil and a million feet of gas with less than one pound pressure. I have done it out here with the Standard Oil Company and done it with the Shell Company. There is nothing there to maintain the pressure. The action of these valves that control this [200] separator is much different from anything that was ever constructed or invented before. They are both automatic; there is an automatic valve on the gas line and there is an automatic line on the oil discharge line. They leave the same sized opening at all times wherever they are set after they are once set; they don't hold any concrete pressure on the trap at all. We merely keep the oil level at one place in the trap. That is the purpose and that is the way they are, in every type of trap where the oil runs from the bottom to the top of the trap, that is most of them do. This acts different. The oil practically stays in one place in the trap, and that is what these valves are for to prevent the oil from going into the gas line or to prevent the gas from going out with the oil. It takes care of itself automatically. They can put a back pres-

(Testimony of David G. Lorraine.)

sure on the gas line and it doesn't make any difference; they can put a gas pressure on the oil line and it doesn't make any difference. The trap will take care of itself automatically, and this is the only trap that is constructed to-day in California I am sure that will do that.

Mr. BAGG.—I will ask you to state if there has ever been any pressure—well, that might be objected to as leading. What do you know about the pressure on this trap during its operation out there on this Tonner lease?

A. Well, when the absorption plant was taking the gas they had 55 pounds pressure on the whole line and also on the trap at one time that I know just as was stated here by one of the witnesses, but when they turned the gas loose why it just put enough pressure on the trap to put the oil into the storage tank, and they never touched the trap.

Q. And if anybody put that pressure on this trap—

A. We simply turn the gas into the gas line that went to the absorption plant and the absorption plant put the gas pressure on the line. [201]

Q. Was there any part or parcel of your trap that had anything to do with that pressure?

A. Absolutely not, because when we turned the pressure down—turned the gas out, the pressure went right down in the trap.

Q. Now, you say you have been in the oil industry more or less since '96?

A. More or less, yes.

(Testimony of David G. Lorraine.)

Q. And during that time you spent about 15 years in the oil industry?

A. I think so, just about.

Q. During that time have you had occasion to examine other oil and gas separators and traps?

A. Yes, I have.

Q. What, if any had you had occasion to examine?

Mr. LYON.—Wait a minute. May I ask what the purpose is of the examination?

Mr. BAGG.—The purpose, if your Honor please, is to show, ultimately to show that this baffle arrangement, the throwing of oil against the side walls of a trap or against a baffle-plate is old and it has been in use for years.

Mr. LYON.—Objected to as inadmissible under the pleadings. There is no defense of prior use and no defense of anything of the kind. The only defensive remark in the answer to the main suit is as follows, "And no defense of prior use or invalidity of the patent by anticipation is provable therefor."

Mr. BAGG.—If your Honor please, I will explain to the gentlemen, we are not attempting to show anticipation; we are simply showing, introducing this for the purpose of showing the state of the art, and we have a right to show that under the general showing. The gentleman conceded that in his argument yesterday about our filing an answer, by his supplemental plea here, conceded that we had a right under the [202] general is-

(Testimony of David G. Lorraine.)

sue to show the state of the art. We have plenty of authority to show that and we have a right to do that.

The COURT.—Well, I will overrule the objection.

Mr. LYON.—The testimony is received for the purpose of the state of the art?

Mr. BAGG.—That is all we expect.

Mr. LYON.—All right.

The WITNESS.—What was the question?

(Last question is read by the reporter.)

A. Well, I have examined numerous types of traps. My first experience with a trap was in Holgate, Ohio. It was nothing but a piece of pipe, and naturally I wanted to know what that trap was made like and one day I took it apart to see just what it was made like. It was nothing but a piece of pipe and there was an oil outlet, an oil inlet, an oil outlet and a gas outlet. That is what is called here in California a derrick. It is made out of casing, and you stand it up alongside of a derrick. They are built much larger here. And since that time I have built quite a few traps myself from different pipe, different models, and then the Tico, the Fisher trap, the Sharpenburg trap—

Mr. LYON.—Wait a moment. Your Honor, I think the witness ought to be required to give the actual dates of these things before he goes into them, for a large number of what he has already mentioned—at least one I noticed—are subsequent to our invention, and it is not the state of the art to show things which happened afterwards.

(Testimony of David G. Lorraine.)

Mr. BAGG.—We will admit that. Now, when was it that you examined this trap that you have just described?

A. In 1896.

Q. In 1896. Now then you mentioned the Tico trap. When did you have occasion to examine that?

A. Well, that was in Houston, Texas, or near Houston, Texas, Humble, Texas, near Houston, about 14 miles from [203] Houston. That was in 1905, in March, I believe.

Q. Now, state to the Court what the interior construction of that trap was?

A. Well, as near as I can remember it had an outlet at the bottom of the trap, and it had either three or four inlets at the top of the trap. They were connected to one pipe and they put the oil down the side walls of the trap.

Q. Now just describe to the Court—

A. I could make a sketch.

Q. Make a sketch upon that piece of paper now just showing the type of the trap and the inlet.

Mr. LYON.—I would like to ask the witness a question as to his competency in that regard, if I may?

The COURT.—Yes.

Mr. LYON.—In that Tico trap, Mr. Lorraine, that you say you saw near Humble.

A. It was in Humble; it was near Houston.

Q. Did you use that trap yourself?

A. No, I didn't.

Q. You saw it standing there under operation?

(Testimony of David G. Lorraine.)

A. No, I saw it when they put this trap together.

Q. Just explain fully what knowledge you have of it yourself?

A. Well, I was very much interested in the construction of it and I was not working there but I watched them put this trap up and I watched it operate.

Q. Did you assist in building it in any way?

A. No, sir, I did not.

Q. Have you any record of any kind that was made at the time that will show the construction of it?

A. No, I have not. I take it all from memory.

Q. What opportunity did you have for the measurements on the inside of the trap or its construction at that time? [204]

A. I had no opportunity only my own judgment.

Mr. LYON.—We object on the ground that the witness is not qualified.

The COURT.—I will overrule the objection.

Mr. BAGG.—Proceed, Mr. Lorraine, to make that drawing showing the interior of the trap.

A. (Drawing on paper.) This trap had an oil inlet in the top like this—either three or four of these, I wouldn't say which, around the top that spread the oil like that around the surface, and the oil discharged from the bottom and the gas out of the top in this manner (illustrating).

Q. Now, describe to the Court how the oil came in there and what the effect of it was.

A. Well, it splashed down the side wall here

(Testimony of David G. Lorraine.)

just the same as ours did, in order to prevent it from stirring at the bottom.

Q. Well, what would you say, from your experience as a builder of oil and gas traps and the workings thereof, was the effect of these several inlets constructed as you have described them? Describe them so that the stenographer can tell.

A. In order to reduce the velocity of the flow and at the same time spread the oil upon the side walls of the receptacle, to prevent it from foaming, to keep the foam from going into the gas line, that was the object of that baffle system.

Q. Now, how were those inlets constructed, so that the stenographer—the record will show a description, how they are constructed or were constructed.

A. Well, there was nothing but a pipe that was flanged in there and an elbow.

Q. Which way did the elbow turn? [205]

A. Turned toward the side wall.

Q. So the oil was discharged in which direction?

A. Toward the vertical wall. The oil would flow down and the gas would go up.

Q. Now, where was the gas taken off from?

A. Right in the top, as near as I can remember right at the center.

Q. That was the exit for the gas?

A. Yes, and the oil outlet was on the extreme bottom, and this set on a frame like this (illustrating).

Q. Now, what if any construction or structure

(Testimony of David G. Lorraine.)

was there in the interior of this trap to regulate the flow of oil and gas from the trap, if any?

A. Well, they used—at first on that trap they used a common ordinary gate valve to prevent that there trap from overflowing, and the float controlled the oil discharge valve.

The COURT.—What is that, the float?

A. The float, a vertical float controlled the oil discharge valve, and at first they used a common ordinary gate valve and the gas-discharge valve pinched it down a little and then afterwards they put in a glove valve, filed off the stem and put a weight across the stem and used it as a back pressure valve. That is how that trap—

Q. Now then, on the interior of this trap, describe to the Court now just the action of the oil and the gas after it got got into the trap?

A. Well—

Mr. LYON, Jr.—I don't believe the witness has shown he ever saw the inside of the trap work, if your Honor please. He hasn't testified to that. He said he stood around and watched them put it up.

The COURT.—I think he has made an explanation of it already that those flanges there was for the purpose of causing the oil to come down the side of the walls of the trap, and they made several of those intakes distributed around the traps so that the several intakes, as the oil came in it [206] would be distributed around the walls.

Mr. BAGG.—Yes, sir. Now, describe to the

(Testimony of David G. Lorraine.)

Court the action now after this oil got in here with reference to where the gas was taken off, and not from the trap but the interior, where did the gas separate from the oil, if you know?

A. Why, according to my judgment it separated from the oil as it come in here (indicating), and also some gas come off of the oil when it was in the bottom of the trap. This trap didn't seem to carry much oil. It had a gauge-glass here on one side and sometimes the oil would show in the glass and sometimes it wouldn't.

Mr. BAGG.—Now if your Honor please. I would like to have the clerk mark this Defendant's Exhibit "B."

The CLERK.—Defendant's Exhibit "B."

Mr. BAGG.—And we wish at this time to introduce that in evidence.

The COURT.—You say that was in 1905 when you saw this?

A. Yes, sir, it was in March, 1905.

Mr. LYON.—We object if the Court please to this exhibit on the same grounds. He has stated that it is only offered for the purpose of showing the state of the art, and that will be understood.

Mr. BAGG.—Yes, that is correct.

The COURT.—Yes.

Mr. LYON.—And we object generally to the exhibit on the ground that it is incompetent, not the best evidence, no foundation laid for the introduction of secondary evidence and incompetent, no foundation laid—not proven.

(Testimony of David G. Lorraine.)

The COURT.—The objection will be overruled.

Mr. LYON.—Note an exception.

Mr. BAGG.—What other traps, Mr. Lorraine, have you had occasion to examine other than the one, the Tico trap which you have just described?

A. Well, I examined the Brown trap. [207]

Q. You examined the Brown trap. When did you examine that?

A. The first Brown trap I seen was on what they called the K. T. & O. property near Taft, was in about 1915. It was an old trap.

Mr. LYON.—Wait a moment. We object on the ground that it is obvious that the witness' knowledge of that trap was subsequent to the date even of the filing of the application for the patent in suit.

The COURT.—When was that application?

Mr. LYON.—The application was filed in November, 1914.

The COURT.—And was issued in 1918.

Mr. LYON.—Yes, but the file-wrapper shows the date of the application. I think I have quoted it correctly, November 14th, 1914, so anything that he knows after at least November 14th is incompetent.

The COURT.—I think the objection will be well taken.

Mr. BAGG.—Yes, sir, we concede that. Prior to November 14, 1914, had you had occasion to examine any other traps than the ones you have described—the Tico trap?

(Testimony of David G. Lorraine.)

A. Well, I seen other traps. I seen the Cullum trap and the Washington trap, and I only observed them from their outside, not from their interior, but after that date of course I examined them.

Q. You examined them after that date?

A. Yes.

Q. Well, was the trap that you examined subsequent to November 14th, 1914, we will say—what did you call it, the Cullum trap?

A. The Cullum trap, yes.

Q. What can you say with reference to that being the same character of a trap as the trap you saw prior to that date?

Mr. LYON.—That is objected to as incompetent, no foundation [208] laid, calling for a mere guess and conclusion of the witness.

The COURT.—I think it is apparent from his own testimony that he couldn't testify to that because he didn't see the inside of that at all.

Mr. BAGG.—What other traps if any did you examine the workings of?

A. The McLaughlin traps.

Q. The McLaughlin traps. Was that prior to November 14th, 1914?

A. No, it was not. I seen them before that time but I never examined them.

Q. You never examined them until afterwards?

A. Until afterwards.

Q. You can't testify that the one you examined was like the one that you saw prior to that time?

Mr. LYON.—We object to that.

(Testimony of David G. Lorraine.)

A. No, I couldn't.

Mr. BAGG.—Mr. Lorraine, taking this model trap which has been introduced as model trap on the Tonner lease No. 3, I will ask you to state if you know where the level of the oil and gas would be in this trap during its operation in what was known as, what we might call the receiving chamber or the smaller chamber which contains a baffle-plate, as to whether or not it would be above or below the baffle-plate?

A. Well, that depends upon the flow and the volume of oil and gas that is coming into the receptacle, as the oil level varies on this side of this partition when it doesn't vary on the other side of the partition. If the oil should come in a little faster why it would build up here, sometimes build right up to the top and go over the top.

Q. I will ask you how you know that.

A. I drilled in here, not only in the Tonner trap but other traps and put cocks in here. [209]

Q. And what was the result when you opened those cocks?

A. Why, there was times there was a cock in there with a 5 inch nipple on it. That was all 5 inches deep.

Q. Where?

A. Right here, right by this baffle, right under the baffle here.

The COURT.—Under the baffle or over it?

A. In under it, right in this position, right here, just in under it (indicating).

(Testimony of David G. Lorraine.)

Q. I don't understand that.

A. Well, your Honor, this shell was tapped and drilled through here, and this here pipe was screwed in there (indicating).

Q. I understand that, but I don't understand why it should be piled up underneath but not above?

A. It was piled up above also. It shows that there was oil below there and also above.

Q. So that pile ran clear up? A. Yes, sir.

The COURT.—Oh, then, I understand you.

Mr. BAGG.—Now, Mr. Lorraine, I will ask you to state how many traps like the one you have just described as being located on the Tonner lease No. 3 you put out or built? A. Just one.

Q. After you had put that one out what did you do with reference to other models?

A. Why, we took this baffle out entirely, and I think Mr. Lacy there has the record. He can tell just how many we built without that baffle in there at all, but we still had this incoming baffle like that, but we didn't have this here deflector on the bottom (indicating).

Q. That is, you mean to say you didn't have this baffle at all or do you mean to say that it was not—

A. We used this vertical partition the same, just the same, but we abandoned the use of this baffle here and just [210] used this divider in the top.

Q. And there was no baffle there, they all struck nothing, then, after passing out of here (indicating)? A. No, sir.

(Testimony of David G. Lorraine.)

Q. After it passed out here it struck nothing?

A. Well, it may be some of it shot over this way with a big gas force (indicating).

Q. But it just dropped down? A. Yes.

Q. Now, why did you change from that form with the baffle-plate to the form without the baffle-plate?

A. Well, we found there were several reasons. We already decided not to use this baffle as we found that it was no good in there in that position as it held the oil up too high here (indicating). As soon as I installed the trap and put the trap into operation why I told Mr. Burrows and Mr. Swope that I wished that that baffle was out of there as it held the oil up too high on that side of the vertical partition.

Q. And you took it out?

A. Not that one. We left that one in.

The COURT.—That was held up because the space between the baffle-plate and the oil was not large enough to let the oil pass through, was that the reason for it?

A. Well, that might have been possible, but then it shot the froth and the foam, some of it over the top, and we had to carry out oil level too low on this side to make a complete separation of the oil when the oil reached the storage tank.

Q. Well the baffle-plate in there was the cause of the oil shooting over the top?

A. That was it.

(Testimony of David G. Lorraine.)

Q. Caused it to pile up there and it didn't have room to pass down by the wall, is that it? [211]

A. Well, yes, it held up the froth.

Q. I see.

A. I would say that is the resaon, yes, sir.

The COURT.—I see.

Mr. BAGG.—Now, what are the dimensions of that smaller chamber into which the oil and gas comes as it comes from the well?

A. The dimensions of this here is 11¾ inches I believe from this sheet to this wall (indicating).

Q. From the outside wall?

A. No, from the interior wall to the sheet would be 11¾ inches.

The COURT.—That is in the trap that you built?

A. Yes, sir. This baffle-plate here is 2 inches from the wall (indicating).

Mr. BAGG.—All the way down?

A. Yes, there is an area there of about 64 inches in the slot that comes down here (indicating).

Q. That well down there where you installed this trap upon the Tonner lease No. 3, what was the capacity of that well, if you know?

A. When we put the trap on it it was about 1200 barrels and about one million feet of gas.

The COURT.—A day? A. Yes, sir.

Mr. BAGG.—What was the size of that inlet pipe? A. Three inches.

Q. Now—

The COURT.—Did you pass that all through the tank, or, I mean, through the trap?

(Testimony of David G. Lorraine.)

A. Oh, yes, yes.

Q. Twelve— A. 1200 barrels.

Q. A day through that trap? [212]

A. Twenty-four hours, yes, sir. We have one trap that is handling 1800 barrels and about two million feet of gas, at Bolsa Chico.

Q. Same sized trap?

A. Same sized trap; yes, sir.

The COURT.—Very well.

Mr. BAGG.—Do you know what the pressure was on that well?

A. Why, the pressure for tubing I believe, was around about 200 pounds, that is on the inside of the flow nipple; the pressure at the top was about 55 pounds. That was regulated by the absorption plant.

The COURT.—You say at the trap or in the trap?

A. The pressure in the trap was about 55 pounds, but that was held back by the absorption plant. It was not taking the gas away fast enough so as to relieve the pressure.

Mr. BAGG.—Now, state whether or not, if you know, the flow from that well was constant formerly, or did the well flow in a head?

A. It flowed very steadily after we put this separator on it.

Q. I am speaking about formerly, before the separator was put on?

A. No, it flowed in heads.

Q. And after the separator was put on it?

(Testimony of David G. Lorraine.)

A. Flowed steadily.

Q. Flowed steadily. Now then, coming down to the last model which was described as having been sold to the General Petroleum Company, I will ask you to describe if you can how that trap is constructed (indicating)?

A. Why, I guess that there is about it, right there, from what I can see (indicating).

Mr. LYON, Jr.—Let the record show that the witness is [213] taking hold of and refers to Plaintiff's Exhibit—

Mr. LYON.—That is all right. Let us get the number, Mr. Lorraine. 19.

The WITNESS.—This here is setting a little closer to this wall (indicating), that is all the difference in this here. This elbow here, this is a common, ordinary reducing elbow that was bought I believe from the Pacific Supply Company by Mr. Lacy. It is from four inches to six inches. That is where the oil comes in at (indicating); that is the oil intake, and the oil goes down in this direction in a downwardly direction, and the gas passes through this channel over to this chamber (indicating); and this is a scrubbing chamber to clean the gas. The gas comes below that there partition and then out through this gas outlet here. As far as the valves are concerned they are just the same and this partition is just the same as the other trap. All the valves and all the mechanism hasn't been changed, but this baffling feature on the latest models has been changed and the height of the trap

(Testimony of David G. Lorraine.)

has been changed. The trap is a little bit longer, 18 inches.

Mr. BAGG.—Now, in the construction of an oil and gas separator what, if you know, is the most acceptable plan of construction with reference to the distance of the oil and gas exit from the oil and gas intake?

A. If I understand that question right, the question is where the oil comes in at and where the oil goes out?

Q. No, what is the most approved construction with reference to the distance between them?

Mr. LYON.—We object to that as a mere hypothetical question. There is no foundation laid that there is an approved or accepted distance.

The COURT.—You may state what you know about it.

A. Well, the way I built this separator was to make the gas travel as far as possible in the separator, to clean [214] itself, from the place where the oil and gas come into the trap, and also to make the oil travel a greater distance by allowing it to flow downwardly. This partition is not long enough here—and then deposit the sand and water in the bottom of the trap and then go for a distance upwardly to give the oil a longer distance to travel and give the gas a longer distance to travel.

The COURT.—How far would it travel there?

A. Well, if this partition comes down here a little lower it is something more like this one here

(Testimony of David G. Lorraine.)

and travels down this way (indicating), and then goes over to the outside.

Q. And goes out on the other side?

A. Yes, travels straight across; on our later models it travels straight across—it makes the oil travel a greater distance.

Q. That is on a principle that water will seek its level?

A. Exactly. The gas travels a greater distance to get to the outlet and that gives a chance to store the crude oil up.

Q. That is all to facilitate the separation of the gas from the oil?

A. That is it, in order to get the gas clean.

The COURT.—I understood that was the purpose of the instrument all the way through, from the beginning of this case.

Mr. BAGG.—You have heard the testimony of the witness Paine and the witness Harris in this case with reference to the character of the flow of this oil and gas in this trap after it leaves this T; I will ask you to state to the court what you know concerning the character of the flow as to whether or not it spreads out in a thin film on this baffle-plate, or whether it has a tendency to thicken and be more than a thin film? [215]

Mr. LYON.—Wait a moment. We object to the form of the question, your Honor, on the ground that the question involves mental reservation and is too indefinite and uncertain. The use of "thin" is a relative term. If the witness—

(Testimony of David G. Lorraine.)

Mr. BAGG.—Well, I will ask him to state how thick of any the oil will be on that baffle-plate, if he knows, under ordinary pressure?

A. In order to get oil—

Mr. LYON.—We object upon the ground that the witness is not qualified to answer the question.

The COURT.—I will hear the testimony. The witness appears to be an expert along that line.

A. In order to get oil or gas from an oil well into one of these gas traps it is either pumped in there or forced in there violently—it doesn't creep in there like a caterpillar—and then gravitate down—it must be pumped up there or it must flow in there with its own natural force; and there is only one way that oil and gas can come out of this unless you let it pour down there from a little bucket or pitcher, or something like that, or pump it up there real slow with not an oil well pump but a pump that you control by hand. There is only way that that can come out of there and you can take that out of there and put it on any well and that is in a round form. This oil and gas is composed of foam, just as Mr. Trumble says in his patent there and that reducing elbow will bring it out in a round form. As far as being a thin film on here it is impossible, or any kind of a film. You may put a stream down here, but the object of that is to keep stirring up the water and the sand from the oil so the oil will settle (indicating). The whole purpose of the construction of this trap (indicating), is to prevent any stirring action and to get the oil into a settled form

(Testimony of David G. Lorraine.)

so that the oil will stop and not lose any of [216] its gases. That is what makes this thing a good seller. It gets the gas clean. There is only one way that you can possibly bring that out of that round hole if there was any pressure there at all and that would be in that round form, the shape of that ell. If that was flat why it would bring it out into a thin film, or if it was conical and hit the side walls like that over there it would bring it down in a thin film, it would press right down there, but being that there is an elbow, why, this oil and gas can only fit that form, providing the oil has got gas in with it.

The COURT.—That is to say, after leaving it in the pipe it will spread out and expand in every direction until the elbow—

A. Why, it couldn't do anything else.

Q. I am asking for the facts and not arguing the question with you.

A. Yes. You see if you fill that, if there was two ounces of pressure on that there elbow would be filled all around with that two ounces of pressure. If there was one ounce of pressure on it it would be filled.

Q. What form does it take after it leaves the elbow?

A. It is directed to 45 degrees; that is to prevent it churning up the settled oil. Well, as I say, it might splash some on this wall—you couldn't prevent that; I couldn't build a gas trap without shooting some oil on the side walls of the receptacle, but

(Testimony of David G. Lorraine.)

it would be put on there in a film like form with this construction. In order to see how that would act you can take this out here, this same ell and we can put it on any oil well and put it right out in the open.

Q. Well, isn't it a fact that the more you spread your oil before it reaches the oil in the bottom of the tank won't it facilitate the separation of the gas from the oil?

A. I haven't found that to be so, although there are a good many experts that argue that. [217]

Q. I want to know as a fact. I am not telling you.

A. I understand that. There is a good many of them that believe that by stirring the oil or moving the oil they can sepaarte it that way, but I found out to get the oil in a settled condition just as quick as possible after it gets into the receiving chamber is the best way to separate oil and gas.

Mr. BAGG.—Now, Mr. Lorraine, I will ask you to state whether or not you have constructed an oil and gas separator exactly along the lines of the one that was sold recently to the General Pretroleum Company and delivered to the plaintiffs?

A. Yes, sir.

Q. Do you have such a trap, an exact duplicate of that? A. Yes, sir.

Q. And in a position where you can give to the court, if the Court so desires, an actual demonstration of its workings under normal conditions such as obtains in the actual separation of oil and gas?

(Testimony of David G. Lorraine.)

A. I believe we can with the Shell Oil Company's permission. They have one right alongside of a new Trumble.

Q. Where is that?

A. At Signal Hill, the Horsh Well.

Q. Will you make that demonstration to the Court at any time the Court may desire.

A. I would gladly do it with the Shell Oil Company's permission.

The COURT.—How far is that from here?

Mr. BAGG.—About 22 miles, I should say.

Mr. LYON.—While we are on that subject, you couldn't see anything inside of the trap with that demonstration, could you, Mr. Lorraine?

The COURT.—On the outside?

A. You could see how the traps work, and we could take [218] the bottom out of the trap.

Mr. LYON.—Just answer the question: If we went down to the Shell Oil Company we would have no opportunity to see how the oil or gas acted inside of the trap; we would have to content ourselves with just what we saw on the outside, wouldn't we?

A. I don't know, Mr. Lyon. I would like to answer your question. There is not any way that you could determine the action that took place in that separator unless you put it on to a well and made the receptacle out of glass, and then the oil would spread on the walls and you wouldn't be able to see the action at that.

Q. Well, is there one at the Shell Oil Company's, so that we could take it apart to any extent and see

(Testimony of David G. Lorraine.)

what is inside of it? Because I want to go down there and want to see it if we can.

The COURT.—The one of the Shell Oil Company I understand that is one you made and sold?

The WITNESS.—Not to the General Petroleum. They have another one. They bought another one.

Q. No, but that is a counterpart of the one that you sold?

A. Yes, sir, an exact duplicate.

Mr. LYON.—If there is one that we can open up I am perfectly willing to go down to the Shell Company, if we can see something there—in fact, I would be very glad to—but if all we can see is the outside of the trap it is a question in my mind whether we had better take the time.

The COURT.—That wouldn't disclose the action on the inside?

Mr. BAGG.—No, sir; I don't think it would. Mr. Lorraine, can you by opening a manhole in the trap show how the oil flows in the trap that you mentioned at the Shell Oil Company's? [219]

A. No, because there is no way that you can put a hole in that trap and have that trap function like it would function when it is sealed. That would change the course entirely. For instance, if this had no oil outlet here (indicating) and the gas outlet is over here (indicating), if the oil was not going out of that and gas at the same time it wouldn't function like a gas and oil separator. Neither would that model (indicating). As far as this baffling feature is concerned, why the best way

(Testimony of David G. Lorraine.)

to demonstrate that would be put it right on the end of a well, right in the open so you can see just how that acts. I will be perfectly willing to take this model right here just as it stands and attach that to a quarter inch line or whatever the line would be reduced to; and put this on the oil well line from any flowing well—that same well that we have mentioned—and watch the action right there. That is the way I test the baffling. I think that would be a fair demonstration, to take their own model.

The COURT.—That would be with dead oil?

A. No, take it right from the flowing well, tap the line and take it right from the flowing well.

The COURT.—Then if you admit that to the atmosphere then you have got the gas released at once, so I don't suppose that would be a good demonstration?

Mr. BAGG.—No, I think, if your Honor please, he means to tap on to the main line as it comes from the well and run a connection out so that the character of the oil as it comes into this model trap would be exactly the same as that which comes from the main line into the company's receptacle.

The COURT.—And you would have to have this trap full or you wouldn't get a demonstration at all.

Mr. BAGG.—Just simply to show the action of the oil as it came through. [220]

The COURT.—Yes, I see.

The WITNESS.—It would be as fair as making

(Testimony of David G. Lorraine.)

a hole in the trap to watch that flow down, because it would be open just the same.

Mr. BAGG.—Mr. Lorraine, I will ask you whether or not you have made any actual experiments with this method of construction to ascertain the effect of the oil as it comes into this trap through that inlet pipe and the elbow, as indicated?

A. Well, before I tested this baffle out—

Q. I am asking you if you made them; just answer the question yes or no. A. Yes, I have.

Q. Now, go on and state what those experiments consisted of when they were made, if you can remember.

A. Well, I have been testing baffles of different descriptions and different kinds for the last 12 or 13 years at least; and the way I test baffles is just the way I suggested that we test this. I test them out in the open and I watch their action right in the atmosphere. I watch the oil and see whether it is frothy or foamy when it comes from these different types of baffles. If it comes in a foam why the baffle is no good; if it discharges the oil without foam why it is considered a good baffle. That is the object, to do away with the foam. That is what the principle of all baffles have been constructed on.

Mr. BAGG.—Mr. Lorraine, I hand you herewith a pamphlet marked "Department of the Interior, Franklin K. Lane, Secretary. Bureau of Mines. Van H. Mankin, Director. Technical Paper No. 209," I think it is. "Technology 49."

Marked Defendant's Exhibit "C" for identification, and ask you to state what that is.

Mr. LYON.—Now, wait a moment. We object on the ground that the book must speak for itself. The witness is not [221] qualified to answer it, and I call the Court's further attention to the fact that the date of the publication on the inside is February, 1919, and it would have no pertinency or bearing; it is not a part of the prior art and in no manner can it have any effect as evidence in this case.

Mr. BAGG.—I am not offering it in evidence right now. I am simply asking him to identify it, and then I will.

Mr. LYON.—It is immaterial what it is as far as this case is concerned.

The COURT.—I think he has a right to identify it and then the court to determine whether it should be admitted as evidence.

Mr. BAGG.—Afterwards.

Mr. LYON.—Well, doesn't that book itself show what it is? That is the only question.

The COURT.—I presume the book would be its own proof.

Mr. BAGG.—I am not trying to prove its contents. I am simply asking him to identify it as to what it is and then after he has done it—

The COURT.—Doesn't it identify itself?

Mr. BAGG.—Beg pardon?

The COURT.—Doesn't the book identify itself?

Mr. BAGG.—I think not; it doesn't identify it

(Testimony of David G. Lorraine.)
to the extent of telling exactly what the contents of it are and I am just calling his attention to it.

The COURT.—He may identify it.

Mr. BAGG.—State what that is.

Mr. LYON, Jr.—We object upon the ground that the witness is not competent. He doesn't know anything about the book except what it shows.

The COURT.—What is the answer?

A. My answer is this pamphlet printed by W. R. Hamilton, Department of the Interior, describing different gas traps and their uses, the United States Government gives these away [222] free of charge. Technical paper 209.

Mr. LYON.—We move to strike out the answer from the record on the ground that it is incompetent.

The COURT.—I will let the answer stand.

Mr. LYON.—Note an exception.

The COURT.—You can have that identified.

Mr. BAGG.—If your Honor please, we desire to introduce this pamphlet in evidence for the purpose of showing the state of the art, the pamphlet giving a history of the oil and gas separators from the very beginning, gotten out and compiled at the instance of the Government and for free distribution and anyone can have it and it is sent out broadcast to all of the oil industry and so anybody should see it. These gentlemen themselves have one.

Mr. LYON.—We object on the ground it is incompetent, no foundation laid and not proper evidence; and on the further ground that if the date appearing upon the outside cover "1919" or in the statement on page 2 "First publication, 1919," is assumed to be the correct date of publication, it is not a proof of a prior art. It is not legal evidence of any kind and it is mere hearsay.

Mr. LYON, Jr.—We have had no opportunity to cross-examine whoever wrote that on any of the statements that are in there.

The COURT.—Does this show the date when these traps were used?

Mr. BAGG.—Some of them it does, if your Honor please. It doesn't tell them all.

The COURT.—Well, I think it ought to show the date when the traps were used or were invented.

Mr. BAGG.—It describes them as having been constructed, of course, prior to the time of the date of that pamphlet.

Mr. LYON.—Then under the law we would have a right to [223] cross-examine whoever wrote this article as to how much he knows about the original dates and so forth.

The COURT.—You haven't got the author here, have you?

Mr. LYON.—They haven't got the author here and know nothing about it.

The COURT.—Well, I will permit this to be admitted in evidence for the purpose of showing those traps where the date is apparent; and to all the others it will be no good at all.

Mr. BAGG.—Yes, sir, I understand.

Mr. LYON.—We note an exception to the ruling of the court.

The COURT.—Very well.

Mr. LYON.—Your Honor understands that there is no law that we are aware of that makes an instrument of this kind proof in court of the assertions that these things were known as of the date and that this instrument itself only speaks as to the date of its publication. That is our ground of objection.

The COURT.—I understand your ground of objection. The only thing in the mind of the court with reference to the introduction of that pamphlet is whether it is such a pamphlet, official pamphlet as a subject to be admitted in evidence.

Mr. BAGG.—I think it is, if your Honor please. That is the reason I am introducing it because it is an official document gotten out by the United States.

The COURT.—I won't do a *great of* harm anyhow.

Mr. BAGG.—It is now 4 o'clock, if your Honor please. I suppose it is adjournment time.

The COURT.—Yes. The court will adjourn until—

Mr. LYON.—We haven't got anything before us of course to show that even this Mr. Hamilton is a Government official or anything of that kind. [224]

The COURT.—It seems to have been put out by the Secretary of the Interior.

Mr. LYON.—As information, yes; but as information of 1919 rather than prior.

The COURT.—I will adjourn court until to-morrow morning at 10 o'clock. How long is it going to take, Mr. Bagg, to complete this case?

Mr. BAGG.—I don't think we will have very much more testimony. I think we are almost ready to close with Mr. Lorraine.

The COURT.—Very well.

Mr. LYON.—Is it your Honor's disposition to go out and see this Lorraine trap that we have at our shop?

The COURT.—Out 22 miles?

Mr. LYON.—No, out ten miles. It is right out here, the one at the plant where we have cut out a section and arranged it so we can make this demonstration with the oil and we can make it with oil and with gas in it, or without, as you desire, and with any pressure.

The COURT.—Any objection to that?

Mr. BAGG.—None whatever, if your Honor please.

The COURT.—Very well.

Mr. LYON.—I ask that because I want to take a nipple off of that after we have made the demonstration and offer it in evidence.

The COURT.—The Court will go out there. When can we go, to-morrow?

Mr. LYON.—Any time to-morrow.

Mr. BAGG.—Any time it suits your Honor.

Mr. LYON.—We will furnish machines for the Court, the reporter, and so forth.

The COURT.—I think the Court would rather go in the afternoon.

Mr. BAGG.—I think we have only one or two other witnesses, [225] whose testimony will be short.

The COURT.—We will adjourn, then, until to-morrow morning.

(Whereupon an adjournment was taken until 10 o'clock to-morrow, March 24, 1922.) [226]

[Endorsed]: Original. In the District Court of the United States for the Southern District of California, Southern Division. Before Hon. Charles E. Wolverton, Judge. Francis M. Townsend et al., Plaintiffs, vs. Davis *S.* Lorraine, Defendant. No. E-113—Equity. Reporter's Transcript of Testimony and Proceedings on Trial. Vol. II. Filed Apr. 7, 1922. Chas. N. Williams, Clerk. By R. S. Zimmerman, Deputy Clerk. Los Angeles, California, March 23, 1922. Reported by J. P. Doyle, E. L. Kincaid, Doyle & St. Maurice, Shorthand Reporters and Notaries, Suite 507, Bankitaly International Building, Los Angeles, California, Main 2896. [227]

INDEX.

VOLUME III.



[228]

Friday, March 24, 1922, 10 o'clock A. M.

The COURT.—You may proceed with the examination.

Mr. BAGG.—Mr. Lyon is not here yet.

The COURT.—Mr. Lyon is not here. You want to wait for him? We will wait a few minutes.

(Short recess.)

Mr. LYON.—It seems there was a mistake in time, your Honor, about ten minutes.

The COURT.—That is all right.

Mr. BAGG.—Are you ready to proceed?

Mr. LYON.—Yes, we are ready to proceed.

**Testimony of David G. Lorraine, for Defendant (Recalled).**

DAVID G. LORRAINE, recalled.

Direct Examination (Resumed).

(By Mr. BAGG.)

Q. Mr. Lorraine, I hand you Defendant's Exhibit "C" which was introduced in evidence yesterday, opened at page 12, and ask you to state whether or not the drawing contained on page 12 is a correct drawing of the Teco trap as you examined it in March—

A. 1905.

(Testimony of David G. Lorraine.)

Q. (Continuing.) —1905, and which you described yesterday upon the witness-stand?

Mr. FREDERICK S. LYON.—We object to that as leading and suggestive, and incompetent; no foundation laid, the witness not pretending to have made this drawing. He has already made a sketch of which he asserts he saw at that time, and we object further on the ground that this publication is not an official publication in any sense, and that is unproven here, a mere copy, not certified in any manner, and it is not pretended that this publication is, as a matter of act, any official investigation or [229] of any date. We do not think that a publication, and a mere essay written in 1919, simply upon the general statements collected by a man can be proof of a fact. It is a proof, of course, if otherwise competent, as far as a copy is concerned, of a publication in 1919, and as far as proving prior use, I call your Honor's attention to the high degree of proof that is required in that class of cases, and, as said by Justice Brown in the barbed-wire case, this oral testimony of this class is most satisfactory and least to be considered testimony of all kinds in law. The telephone cases, the barbed wire case, and a large number of cases have adverted to that fact, that every inventor is subjeet to attack by simply merely oral testimony that someone has seen a device at some previous time.

The COURT.—Let me see that pamphlet just a moment. (Receiving same.)

Mr. F. S. LYON.—The pamphlet itself does not pretend to give a date for this so-called Teco trap, and there is no pretense that it was made by general investigation. The author who writes the essay admits that he depends on the information of so and so, and so and so, naming different parties, even on the title page. Take for instance page 6, your Honor, under "Acknowledgments."

The COURT.—The Bureau of Mines—this seems to be printed in the Bureau of Mines, carrying out one of the provisions of the Act to disseminate information concerning investigations. The printing of each of its publications. "When this edition is exhausted, copies may be obtained at cost price only through the superintendent of documents, Government Printing Office, Washington, D. C." The superintendent of documents is not an official of the Bureau of Mines. He has an entirely separate office, and he should be addressed "Superintendent of Documents."

Mr. F. S. LYON.—Take, your Honor, page 6, under the heading "Acknowledgments" and you will see the author admits that it is under no knowledge of his. [230]

The COURT.—What was that?

Mr. F. S. LYON.—The bottom of page 6, under the heading "Acknowledgments."

The COURT.—Yes.

Mr. F. S. LYON.—Now, a patent issued by the Government itself can't be brought in unless there is a certain, a copy of another printed publication

(Testimony of David G. Lorraine.)

that is in the library of the Patent Office, the official library.

Q. (By the COURT.) I understand that you examined this device? A. Yes, sir.

Q. Personally? A. Yes, sir.

Q. And you know this to be a presentation of that device?

A. Well, it is not exactly the same, but the construction is the same. I should say that that trap is larger in diameter in proportion to its height of the trap that was constructed there, but outside of that the construction, as I know it to be, because I know we had an argument about the bottom being caved in in place of out.

The COURT.—I will admit this testimony.

Mr. F. S. LYON.—Note an exception.

The COURT.—You may have an exception.

Q. (By Mr. BAGG.) Now, you testify then, that this, so far as the internal construction is concerned, is a correct drawing of the trap as you examined it?

Mr. F. S. LYON.—Same objection and exception, your Honor.

The COURT.—Same ruling.

A. As far as this head here is concerned, and as far as this float is concerned. You can see the float through the gas outlet in the top, like that (indicating).

Q. (By Mr. BAGG.) Well, with reference to the intake and the means of bringing the oil, the gas into the chamber, is that—

(Testimony of David G. Lorraine.)

Mr. F. S. LYON.—The same objection and the same exception.

The COURT.—The same ruling all through.

Q. (By Mr. BAGG.) Is that the same as the way you examined it? [231]

A. One of the intakes, I took a piece of baling wire and looked through there, and also ran that baling wire through.

Q. (By the COURT.) Where is the oil brought in on that trap?

A. This shows four places. Now, I would not like to say that that particular trap had four places because I don't remember that part of it, but what I do remember is there was at least three inlets here in the top.

The COURT.—Yes, I understand.

Q. (By Mr. BAGG.) Now, did this—

The COURT.—Has that been patented?

Mr. BAGG.—Yes, sir.

The COURT.—Well, why not have the patent?

Mr. BAGG.—If your Honor please, I will explain to you why we do not have the patent: You will remember when we prepared for the trial of this case we only prepared for this particular patent, the trap that was installed down there on Well No. 3, and then they came in with the supplemental petition yesterday morning and we have not had time to get it.

The COURT.—Can you get that patent?

Mr. BAGG.—We will get it if your Honor will give us permission to do that.

Mr. F. S. LYON.—What patent do you refer to?

Mr. BAGG.—That is the Teco trap.

Mr. F. S. LYON.—That is the patent to Fisher that you referred to in your motion to amend your answer, isn't it?

Mr. BAGG.—No, no, it is not, not the Teco trap; it is a patent of a valve.

Mr. F. S. LYON.—Just a moment. Why, if your Honor please, I am not offering this in evidence or admitting its admissibility for the purpose of proving any of the allegations or recitals therein, but the defendant has in that motion to amend its original answer, set this up as a patent, and so far as we can see, it is identically the same as the Teco patent, and you will notice the [232] patent was issued in 1918, if I get it correct—

The COURT.—Patented May 9, 1916?

Mr. F. S. LYON.—1916.

Mr. BAGG.—If your Honor please, that, as it shows on its face, is a patent for that particular valve installed in there, and has no connection whatever with the trap. That patent is applied for by somebody else beside the Teco people. This is merely a patent for the—

Mr. F. S. LYON.—This very valve right here, that mechanism (indicating).

The COURT.—This one down here (indicating)?

Mr. F. S. LYON.—Yes. Take figure 1 there is the same view. Evidently the Bureau of Mines, so far as that is concerned, have taken the views of the drawing, practically.

Mr. BAGG.—If your Honor please, this is an application of Charles E. Fisher for a float operated drain valve and then the patent for which this is secured has nothing whatever to do with this trap. This is the Teco trap and we could not introduce this because it is a patent for a float operated drain valve. The Teco trap was patented before that.

Mr. F. S. LYON.—It seems to us, your Honor, that the parties have had a year and about three months in this case and this evidence is just as pertinent to the original complaint as the other, and they should have had any patents here that they wanted.

The COURT.—Well, I will hear this testimony with the understanding you get that patent.

Mr. BAGG.—Yes, sir. We will get the patent and submit it to your Honor.

Mr. L. S. LYON.—May we ask the defendant to prove the patent that you get, what it is?

Mr. BAGG.—We do not know at the present time.

Mr. L. S. LYON.—You do not ever know that there is— [233]

Mr. BAGG.—Yes, we do. It states on that pamphlet it is a patent.

Mr. F. S. LYON.—That is, all you know it states so here.

Mr. BAGG.—That is all we know, whether it is patented or not—

Mr. L. S. LYON.—We do not think the trial of the case, your Honor, should be—

(Testimony of David G. Lorraine.)

The COURT.—I am a little in doubt as to whether or not that patent is competent evidence in this case, but I am going to admit it and if I am wrong about it I will rule it out later.

Mr. F. S. LYON.—It don't seem to me that we ought to withhold the conclusion of the evidence in this case to now permit the defendant to make an examination and see if he can find another or different patent on this so-called trap.

The COURT.—Well, that is very good evidence that the patent has issued when it is referred to in that pamphlet.

Mr. F. S. LYON.—That is written in 1919, and here is this Fisher patent which shows identically and written in 1916.

Mr. L. S. LYON.—The patent the Government refers to is this Fisher patent, undoubtedly, because it is identically the same as they have in their patent.

The COURT.—That is a subsidiary patent.

Mr. BAGG.—Yes, a petition for an improvement.

Mr. F. S. LYON.—I do not know whether that is a fact or not in view of what counsel agreed to in his statement, your Honor.

The COURT.—I think I will adhere to what I said.

Q. (By Mr. BAGG.) Now, Mr. Lorraine, I will ask you to refer to page 15 of this same exhibit and ask you to state if the description of the Teco trap as outlined and set out beginning on page 15 is true and correct description of the Teco trap, as you examined back in 1905, and which you described yesterday?

Mr. F. S. LYON.—Well, we object to that, your Honor. [234] The pamphlet itself must stand as evidence so far as it is concerned, and certainly, where the high degree of proof in regard to oral testimony is required, that is required in the proof of prior use in patent cases, counsel ought not to be permitted to ask an omnibus question of that kind. We have had this witness' testimony as to his recollection and his knowledge. Now, if he simply be asked a leading question, "Is this an accurate description of what you saw" adds nothing to his evidence or anything else because it is not shown that he had anything to do with this publication at any time, and it would be simply, as a matter of fact, without any weight whatever.

The COURT.—As I understand, this is being offered as an exemplification of what the witness saw and examined himself?

Mr. BAGG.—Yes, sir. Not based upon the authenticity of that pamphlet at all but simply upon his recollection.

The COURT.—Very well. I will hear it.

Mr. F. S. LYON.—Our objection is that the witness should describe what he saw himself and not have an article allegedly describing the whole thing put before him and ask him if that is what he saw at that time. It is leading and suggestive.

The COURT.—Well, he could hardly ask that question in any other way. The witness intro-

(Testimony of David G. Lorraine.)

duced a drawing of his own here, and this now is to exemplify what he saw there, and he is now undertaking to describe that, undertaking to answer that question. I will hear the answer.

Mr. F. S. LYON.—Note an exception.

The COURT.—What is your answer?

A. Why, so far as I can remember, as I said in the beginning, I don't know whether there is three inlets at the top or four in this particular trap, but I am quite positive there are three, at least, and they were arranged so that they showered the oil upon the side walls of the trap as it entered. I saw the intake before they connected it up. I was very much interested in the construction of it and the base of this trap was turned in, [235] in place of out, and there was quite an argument there about them; they were afraid the trap would not stand the pressure without bulging. There was an argument about the action of the oil as it entered the trap, the stirring action, we were discussing it there at the time.

Q. (By the COURT.) You did not see the action of the oil as it entered?

A. I could not. After they turned the oil into it I could not see it.

Q. (By Mr. BAGG.) Mr. Lorraine, you heard the testimony of the witness Ray yesterday with reference to the various traps which he had examined and had been put out by you or your company. The first one was the Tonner lease No. 3. He testified that on two or three occasions he examined it

(Testimony of David G. Lorraine.)

along about December and there was between 40 and 45 pounds pressure. I will ask you to state what, if you know, was the disposition of the gas or what disposition was being made of the gas coming from that well at that time.

A. Why, the pumping of the gas into the absorption plant.

Q. How far away?

A. Well, the exact distance I could not say, but I would judge about 400 feet.

Q. Now, at that absorption plant what, if any, device do you know of that they had on the gas line for restricting the passage of the gas?

A. Well, just what they had to hold back the pressure, I don't know, but they evidently could not take care of the gas and was building up pressure on the line.

Q. Is that what caused this pressure of between 40 and 45 pounds at that time?

Mr. F. S. LYON.—Wait a moment. We object to that as leading and suggestive, and incompetent. The witness says he does not know. [236]

The COURT.—Well, do you know?

A. I know that the gas plant was holding the pressure back on the trap.

Q. (By Mr. BAGG.) I will ask you to state if there was any means whatever on that gas trap for the purpose of maintaining a gas pressure?

Mr. F. S. LYON.—We object on the ground the witness has already stated that he does not know. He says he knows the pressure was there. He has

(Testimony of David G. Lorraine.)

already stated he did not know what means it was.

The COURT.—What trap was that, now?

Mr. BAGG.—This trap was the trap down on No. 13 and No. 3.

The COURT.—Oh, yes.

Mr. BAGG.—The first trap they described.

The COURT.—I understand.

The WITNESS.—I know there was nothing connected with the trap or the valves that was maintaining a pressure on the trap that was installed.

The COURT.—They had the valves there?

A. It was in operation, yes.

Q. They had the meter there to determine the pressure? A. Oh, yes.

Q. All those appliances were there connected with the trap itself? A. Yes, sir.

Q. (By Mr. BAGG.) Was there any means on the trap,—I do not think I understood the witness, and I am asking that I may understand it myself, if your Honor please.

The COURT.—Yes.

Q. (By Mr. BAGG.) Was there any means upon the trap itself that you had installed or put on the trap or on the gas line for the purpose of maintaining any pressure within the trap?

Mr. F. S. LYON.—Well, we object to that, your Honor, as [237] leading and suggestive, and calling for a conclusion of the witness. He ought to be able to state what means, if any, there was there without simply giving a yes or no answer. He is building these traps and installing them, and he

(Testimony of David G. Lorraine.)
ought to know what he furnished, and so forth, in that line.

The COURT.—Well, the answer will probably call for an explanation. I will hear the answer.

A. Well, the valves are so constructed that go with the trap that they do not maintain any back pressure whatsoever on the trap. Only one valve moves toward the closing position, like the oil outlet valve, or gas valve, would move towards open position, and it leaves the same size of orifice or opening in the outlets at all times.

Q. (By the COURT.) Well, it is an automatic action, isn't it?

A. Automatic action, yes, sir, it is.

Q. Well, when the oil rises to a certain position, the oil valve acts and lets the oil out?

A. Yes, sir.

Q. And the same way with the other, with the gas?

A. Only they work in synchronism; they both work together.

The COURT.—I understand. What is the difference between that and the plaintiff's trap?

A. Well, the plaintiff's trap uses a back pressure valve that holds a certain pressure in the trap and maintains a pressure, not on all of his traps but, according to his patent, he has a back pressure valve. It does not show in this drawing (indicating); I do not notice that, and this gas outlet valve, shows in the patent.

Mr. BAGG.—Here is a copy of the Trumble patent.

(Testimony of David G. Lorraine.)

Q. I will ask you to state, Mr. Lorraine,—I am handing [238] you Plaintiff's Exhibit 1, and ask you to locate on the drawing in connection therewith the back pressure means or the gas pressure means described in that patent (handing document to the witness).

A. This is the back pressure valve right there (indicating).

Mr. F. S. LYON.—11 in the figure.

The WITNESS.—1, isn't it?

Mr. F. S. LYON.—1 of the Trumble patent.

The WITNESS.—The back pressure valve, the action is entirely different. This will hold the concrete pressure where my valves will hold no pressure at all, providing the gas can escape into the line.

Q. (By the COURT.) Is this where the oil enters?

A. No, sir, the oil enters through this line here, and the gas comes up through, below that (indicating).

Q. This is for the storage of the gas?

A. Yes, sir.

Q. This is the back pressure (indicating)?

A. That is the back pressure, the float control valve that prevents the trap from overflowing, but this is the back pressure valve here that maintains the pressure on the flowing volume of oil.

Q. (By Mr. BAGG.) What is the actual effect of this back pressure valve that you have described as—what was it?

(Testimony of David G. Lorraine.)

A. Figure 1.

Q. No. figure 11. What is the construction or the effect of that valve; what does it do?

A. It holds a constant pressure.

Q. Well, what does it do with reference to the amount of space in this passage line?

Mr. F. S. LYON.—Opens and closes the orifice.

Q. (By Mr. BAGG.) Does it have a tendency to restrict it?

A. Yes, it holds a steady pressure, where my valve is [239] an automatic valve, that does not hold a steady pressure and merely prevents the trap from overflowing.

Q. Then that has a tendency to restrict the passage of this gas out through the line; is that correct?

A. Yes, that is it.

Q. Now, then, I wish you would locate the other two valves in this patent, as shown on this drawing.

A. This lower valve here is the oil outlet, and this here upper valve here controls the gas outlet (indicating).

The COURT.—That is in there?

A. They do not work in synchronism as my valves work. My two valves work on the outside, the exterior of the trap, and one moves towards the closing position and the other moves towards opening position. These valves do not work together; the oil does up to this float before they close that valve and drop down again.

Mr. F. S. LYON.—To save time,—I do not want to interrupt you, your Honor, or Mr. Bagg, you do not contend that this valve that the witness has just last pointed out in the Trumble patent 40 in the center is the means for maintaining pressure?

Mr. BAGG.—No.

Mr. F. S. LYON.—Neither do we, your Honor, and that is a subsidiary feature which has only to do when there is an exceptional condition, and it is not the pressure means of the claims. I think you will agree with that.

Mr. BAGG.—Yes. What we are trying to—

Mr. F. S. LYON.—It has no comparison in any manner with the valves that the witness is referring to in his patented device.

The COURT.—But this upper valve is controlling?

Mr. F. S. LYON.—The valve 11 is part of the means for maintaining the pressure on the claim.

The COURT.—And that is in the claim? [240]

Mr. F. S. LYON.—Yes. And you have the Lorraine patent, just so that we may have the issues clear before your Honor, the Lorraine valves are valve 28, and so stated in these specifications and under the admission of the defendant that that is what they were used as.

Mr. BAGG.—If your Honor please, that is what the purpose of the testimony of this witness is, is to show that in the Trumble trap there are three valves, and that in the Lorraine trap, as described down there, and has been installed on that lease, only has the two valves which are worked internally. In other words, this valve here is not on that line

(Testimony of David G. Lorraine.)

and has never been used on that line, never was installed on it and never was used on it, and that the only valves that the plaintiff has in this particular trap are these two internal valves which regulate the work synchronously and opens the oil outlet and at the same time closes the gas outlet, and *vice versa.*

The COURT.—You mean the defendant?

Mr. BAGG.—Yes, sir, the defendant, and governed altogether by the height of the oil in the trap, and that we have not used that before. That is what we are trying to show.

Q. Now, Mr. Lorraine, I will ask you to state this: How many valves are used or required in the Trumble trap? A. Why, if the—

Q. Just answer that question.

Mr. L. S. LYON.—If the Court please, if that is the purpose of the testimony, we object to it as not raised by the pleadings. The pleadings state that he has made his traps in accordance with this particular patent. That is the issue raised by the answer. Now, he proposes to go back on his pleadings and states they are made some other way.

Mr. F. S. LYON.—Not only that, but that was the concession and stipulation on March 22d, the opening of this case.

Mr. BAGG.—If your Honor please, we stipulated—

Mr. F. S. LYON.—Just a minute, and I will read the record [241] on page 7 of the reporter's transcript: You will find this: "Mr. Lyon: In the first place the answer alleges that the defendant is manufacturing its devices in accordance with a certain . . . substantial conformity with that patent, yes."

Mr. L. S. LYON.—Paragraph 12 of the answer says "this defendant states the device for separating natural gas and crude petroleum, which he has made, is made according to the specifications and claims made and so allowed in his said application for letters patent." The issue is that they claim they have made it in accordance with the specifications. The only purpose of this testimony is to raise something that is not pleaded at all.

Mr. BAGG.—If your Honor please, in our answer we deny infringement of their patent, and we in our stipulation, we say that we are manufacturing this patent, or our machine, in conformity with the claim set out in our patent and specification. Now, then, we are not compelled—I do not think that learned counsel on the other side will say that we are compelled under any circumstances to make an exact duplication of that trap at any time. The only thing we are required to do under the patent would be to make one in substantial conformity with the drawings and the claims. Now that is all we are required to do. We can change those methods and means any way we want to.

The COURT.—Well, these valves are part of the claim, aren't they?

Mr. BAGG.—No, this pressure maintaining valve is not, no, sir.

Mr. F. S. LYON.—It is a part of your description, though, isn't it?

Mr. BAGG.—It is a part of our description, but it is no part of the claim. We are simply denying the infringement and we can show under that allegation,—we have plenty of authorities for that,—we can show that in this trap we have left out that particular valve, which is an actual fact, and we are still within the provisions of our claims and of our patents. [242]

The COURT.—Well, that patent is not here. The patents have not been issued.

Mr. BAGG.—The patent had not been issued at the time we filed our answer.

The COURT.—Your patent has been issued now?

Mr. BAGG.—Our patent has been issued since that time.

The COURT.—That contains that valve?

Mr. BAGG.—That provides for a valve of that kind, but does not necessarily provide that we shall use it, simply put in the drawing, and simply describes a part of the pressure regulating means, if we want to use it, but we do not have to use it if it is not necessary, and it would be a useless adjunct to our trap if not necessary, and we found it was not necessary here, and we never used it.

The COURT.—The only question is whether or not you are not bound by your answer and your stipulation.

Mr. BAGG.—I do not think so, if your Honor please, because of the fact that we have denied in-

(Testimony of David G. Lorraine.)

fringement. Now, then, on the rules of equity, as I understand it, they provide we can set up all the defenses that we want whether they are inconsistent or not. Now, then, we deny infringement. That means that we can say that our trap does not,—that we are putting out is anything other than what the patented device that the plaintiff is suing for. We can manufacture any kind of a trap we want to and prove any kind of a trap we want to, or any kind of a construction trap that we want to, providing it does not infringe the claim of the plaintiff. We are not bound by any patent, or anything else. Now, then, as a subsequent and additional defense we allege we are making these traps in accordance with the terms and provisions of our patent, but we can set up both of these defenses and prove either one of them under the rules of equity.

The COURT.—I will permit him to answer.

Mr. F. S. LYON.—Note an exception. [243]

The WITNESS.—Will you repeat that question he asked?

Mr. BAGG.—Well, to save time—

Mr. F. S. LYON.—Will you speak a little louder, please? I am afraid the reporter won't hear you and I know we can't at the table.

The COURT.—Are you speaking to me?

Mr. F. S. LYON.—No, I am speaking to Mr. Lorraine.

Q. (By Mr. BAGG.) In the Trumble trap, as disclosed in their exhibit—is this an exhibit in this case?

(Testimony of David G. Lorraine.)

Mr. F. S. LYON.—One.

Q. (By Mr. BAGG, Continuing.) Plaintiff's Exhibit 1 contains how many valves?

A. Well, I have noticed on these later traps—

Q. Well, I am talking about on this particular drawing, Mr Lorraine.

A. I never saw a Trumble trap with this floating valve in the interior.

Q. If you will just answer my question we will get along a lot faster.

According to the drawings in this trap, how many valves are there provided for?

A. Two or three here.

Q. All right. What are they?

A. Oil discharge valve, the valve to prevent oil from going into the gas line, a valve for maintaining a pressure means on the oil.

Q. Now, in the Lorraine trap as manufactured and installed and now being used on the Tonner lease No. 3, I will ask you to state whether or not there is any third valve or valves for maintaining that pressure on the gas line? A. No, sir.

Mr. F. S. LYON.—If your Honor please, it will be understood that our objection and exception goes to all the testimony with regard to an attempt to vary the admission and stipulation. [244]

The COURT.—Yes, so understood.

A. No, there is no valve on the gas line for maintaining any pressure upon the trap.

Q. (By Mr. BAGG.) Then I will ask you to state if there was any pressure at any time upon the

(Testimony of David G. Lorraine.)
Lorraine trap installed and used on Tonner lease No. 3 at the times described by the witness Ray—Will you read that question?

(Question read.) —was the same caused by any mechanism or device connected with the Lorraine trap.

A. No.

Mr. F. S. LYON.—Wait a moment. We object to that calling for a mere conclusion of the witness. It is perfectly obvious it must have some mechanism, your Honor, connected with it in some manner.

The COURT.—This was a question directed to the witness to say whether or not he has in that particular interfered with that patent. I think that is proper. You may answer the question—he has answered it, however.

Mr. BAGG.—Yes, sir.

Q. (By Mr. BAGG.) If there was such a pressure was the device causing such a pressure any part of the Lorraine trap?

Mr. F. S. LYON.—We object to that as calling for a conclusion of the witness, and not for a statement of fact; leading and suggestive. There was pressure there, and it was connected with a line and the pressure was maintained, the witness admits. Now, it is immaterial whether that valve was close or a long ways off, and it is simply an argument. The witness has already stated that they did not use the particular valve 28 close to the *tap*. Now, that is as far as he can go on that statement, and I think it is perfectly clear to the Court. He has also ad-

(Testimony of David G. Lorraine.)

mitted, by virtue of the connections there with the compressor plant that the pressure was maintained and that the necessary gaging, and so forth, were in the trap for reading it.

The COURT.—I will overrule that objection. You may answer. [245]

A. There was no valve or no means for maintaining any pressure on that trap for the simple reason when they turned the gas into the same line and did not put the gas into the absorption the pressure went right down on the trap.

Q. (By Mr. BAGG.) I will ask you to state if the Lorraine Gas Company, or you, had anything to do with the connections on the gas line to the absorption plant?

Mr. F. S. LYON.—We object to that, your Honor. It is immaterial whether they had anything to do with it or not. The question of infringement is what they sold these people and how it was intended to be used, and there is no denial that they sold them; the whole contention here, if the parties instead of using one valve used another pressure valve at another point, why, they have not changed it at all.

The COURT.—I will hear the answer. Of course, if two persons are violating,—are infringing, either one of them would be liable.

Mr. F. S. LYON.—Both of them would be liable.

The COURT.—Either one of them would be liable.

(Testimony of David G. Lorraine.)

The WITNESS.—What was the question, please?

(Question read.)

A. No, we had nothing to do with putting in the gas line. The gas line was already in when we put the trap out there, when we installed the trap.

Q. (By the COURT.) Did that gas line have a valve in it?

A. Not to my knowledge; no back pressure valve that I know of.

Q. (By Mr. BAGG.) Mr. Lorraine, I am handing you here a photograph which I will ask the clerk to kindly identify as Plaintiff's "D"—Defendant's Exhibit "D," and ask you to state what that photograph is (handing photograph to the witness).

Mr. F. S. LYON.—Wait a minute. We would like to see it.

Mr. BAGG.—Well, he can identify it. Just sit down and [246] identify it.

A. It is a photograph of the two separators at Tonner lease No. 3—this is the Lorraine and this is where the gas outlet was (indicating).

Q. Now, listen. Just identify that and tell us what that is.

A. This is the Lorraine separator installed on—

Q. What is that that I handed you there?

A. A photograph.

Q. When was that photograph taken?

A. It was taken in April, 1921; some time the latter part of April; I cannot say the exact date.

Q. 1921? A. Yes, sir.

(Testimony of David G. Lorraine.)

Q. Last year? A. Yes, sir.

Q. Who took that? A. The Putnam Studio.

Q. Under your direction? A. Yes, sir.

Q. And employed by you?

Mr. BAGG.—Now, if your Honor please, we offer this in evidence (handing photograph to Mr. Lyon).

Mr. LYON.—Do you offer this in evidence?

Mr. BAGG.—Yes.

Mr. LYON.—No objection. I want to ask one question:

Q. That photograph correctly shows the connections as made at that time, does it? A. Yes, sir

Q. (By Mr. BAGG.) Now, at the time this photograph was taken, was the Lorraine Oil and Gas separator shown therein connected up with the well and in operation?

A. Yes, sir. [247]

Q. I will ask you to examine the pressure gage on that and state to the Court what that reads.

A. Well, its reading here is about zero at the present time.

Q. (By the COURT.) It was in operation, you say, at the time.

A. Yes, sir.

The COURT.—Let me see that (receiving same).

Mr. F. S. LYON.—I suggest you take that photograph out.

Mr. BAGG.—Yes, we will take it out.

Mr. F. S. LYON.—I thought you could do it readily without loosening the whole thing.

Mr. BAGG.—That is all.

(Testimony of David G. Lorraine.)

Cross-examination.

(By Mr. L. S. LYON.)

Q. Now, Mr. Lorraine, as a matter of fact, in making your traps do you intend that it shall be used either with or without pressure, as the operator desires.

A. Well, if the pressure blows up in the gas line there is no way in the world that you could operate the trap without pressure. For instance, we had a trap on Wilbur well out here at the Shell Oil Company with a 4-inch line, and that line alone maintained a pressure of 80 pounds on that trap.

Q. You intend, then, that the operator shall, if the operator desires, use the trap with pressure?

A. All traps ever constructed had to have a pressure on them—

Q. That is not answering the question.

A. No, I do not intend to, no.

Q. Do you expect it will be used with pressure?

A. If the gas blows up the pressure it is unavoidable.

Q. Now, does it make any difference in your trap whether or not there is pressure in it?

A. No, it does not make any difference. [248]

Q. Is there any advantage in your trap at all to use pressure? A. No, it has no advantage.

Q. Well, now, I notice in Plaintiff's Exhibit 3,—which you state is your patent,—didn't you?

A. Yes, sir.

Q. And you are the David G. Lorraine to whom this patent was issued? A. Yes, sir.

(Testimony of David G. Lorraine.)

Q. And you signed the specifications, didn't you?

A. Yes, sir.

Q. And took the oath that it was correct?

A. Yes, sir.

Q. Well, now, how do you account for this statement: "From the above it will be seen that I have provided a method for separating or facilitating the separation of the gas and oil and separately discharging the same from emulsion; and furthermore, have provided a method in which, by maintaining a predetermined pressure in the oil receptacle, the latter is subjected to pressures having the effect of expressing the gaseous content from emulsions, the gaseous constituent in the emulsion being driven from the denser liquids by the increase in the pressure on the oil within the receptacle 2. This, therefore, prevents the loss of the valuable gaseous constituent such as occurs in apparatus in which the oil passes immediately from a well or other source to an apparatus in which it is subject only to atmospheric pressure." Did you write that? A. Yes, sir.

Q. Now, did you write this statement: "It has been found from practical experience in the operation of this apparatus that there is an increase in the production of oil from some wells because of the uniform flow from the separator which prevents the rapid increment in the quantity of sand ordinarily found in wells, [249] and which increment results in the clogging or jamming of the well and

(Testimony of David G. Lorraine.)

loss of production until the well is blown." Did you write that? A. Yes, sir.

Q. Now, then, you have there stated two advantages, haven't you, of keeping a pressure on the trap?

A. Well, that advantage is not for separating gas and oil.

Q. I thought you said there was no advantage in keeping a pressure in your trap?

A. I understood you to say it was in the separating, separating gas and oil.

Q. How about this first advantage where it says, "This, therefore, prevents the loss of the valuable gaseous constituent such as occurs in apparatus in which the oil passes immediately from a well or other source to an apparatus in which it is subject only to atmospheric pressure." There you have expressed an advantage of having pressure in your trap in regard to preserving these gaseous vapors, haven't you?

A. Well, do you understand the difference between emulsified oil and just separating gas and oil?

The COURT.—Do not argue the question.

Q. (By Mr. L. S. LYON.) What is that difference?

The COURT.—Just state the answer to these questions.

Mr. L. S. LYON.—Just read the question—Well, I will reframe the question to the witness.

The WITNESS.—What was the question?

Mr. L. S. LYON.—Read the question.

(Testimony of David G. Lorraine.)

(Question read.)

A. Well, that does not necessarily say that. Atmospheric pressure—you could take it off at zero in place of letting the vapor out— [250]

Q. Will you just answer the question?

A. Well, that is the way I would answer that.

Q. Well, you have expressed an advantage here of having pressure in your trap as an advantage in that it keeps your gaseous vapors from being lost? A. By atmospheric pressure, yes.

Q. They would be lost at atmospheric pressure, you said? A. Certainly.

Q. Now, then, by having a pressure in your tank you preserve some of them, don't you? Didn't you state that in your patent?

A. Well, you take them off the vacuum they would be preserved also.

Q. By having a pressure in your tank you do preserve them?

A. Not in this apparatus where we have no—

Q. In your patent you stated it was an advantage, didn't you?

A. Well, I do not think it is, though.

Q. Well, but you stated it was in your patent, didn't you? A. I guess I must have.

Q. In this photograph of the Tonner installation will you trace out the gas line of your trap? Just show so the Court can see it (handing photograph to the witness).

A. This is where the gas line—this is the gas

(Testimony of David G. Lorraine.)

outlet down to this valve and down over here, this is the outlet line.

Q. Now, where does that pipe go?

A. It goes in the ground, goes up to this absorption plant where you see those towers.

Mr. F. S. LYON.—Now, the witness has referred to a pipe which, for convenience, we will mark "X" in the photograph.

Q. (By the COURT.) Now this is the pipe here you refer to, running down here (indicating)?

A. Yes, sir.

Q. Comes down and joins to this (indicating)?

[251] A. Yes, sir.

Q. (By Mr. L. S. LYON.) Now, what is this right here next to this figure "X"?

A. That is a gate valve.

Q. Is that a valve? A. A gate valve; yes, sir.

Q. What is it, a gate valve?

A. It don't maintain any pressure; just simply closes and opens, a sliding valve, up and down.

Q. Well, if you would close it practically until it was entirely shut, what would happen? Would it change the pressure at all?

A. Yes, but it would not sustain a pressure; would not maintain a pressure.

Q. It could act as a back pressure valve, couldn't it? A. I would not say so; no.

Q. It would build up a pressure in the line, wouldn't it?

A. It could build it up providing you kept the gas liquid coming, yes, but this is not a back pres-

(Testimony of David G. Lorraine.)

sure valve and will not sustain or maintain a back pressure.

Q. Then, in this Tonner lease you did have, on the gas outlet pipe, a valve immediately adjacent the trap, did you not?

A. Not a back pressure valve.

Q. Well, you had a valve, didn't you?

A. Yes, sir.

Q. Why did you tell the Court there was no valve at all when you were describing that?

A. No back pressure valve.

Q. You did not say that. You said there was no valve. A. I said there was no valve?

Q. As I remember your testimony.

The COURT.—In its present construction?

Mr. F. S. LYON.—No, at the Tonner lease, I think, your Honor.

The WITNESS.—I said we sold them, I believe, —I did not [252] say those words. I said there was an oil discharge valve and a gas discharge valve that went with the trap and one moved toward the closing position and the other moved towards the open position. Those gate valves on the line have absolutely nothing to do with controlling the pressure or the functioning and the separating.

Q. What kind of a valve is this pictured as 28 in your patent, Plaintiff's Exhibit 3?

A. That is a back pressure valve, but that is not used at the Tonner lease.

(Testimony of David G. Lorraine.)

Q. Now, then, that would act as a means for maintaining pressure in the trap, would it not?

A. Yes, sir.

Q. Now, then, if you continue this pipe, pipe 27, gas outlet pipe to some other point where the plant itself maintains the pressure in the line either due to some restricting orifice or some valve at that plant, then the pressure would be maintained in the trap without this valve 28, wouldn't it?

A. It certainly would, if you keep the pressure on the line.

Q. Now then, doesn't this valve 28 open and close the gas outlet line 27?

A. We never had that type.

Q. I mean if it worked as in this patent, wouldn't it open and close it? A. Yes, sir.

Q. And would it maintain a back pressure then?

A. That weight on that bar there is connected with the stem, and their weight would hold just as much pressure, whatever it would weigh according to the distance from this fulcrum here to this fulcrum here (indicating).

Q. In other words, it would do it automatically, it would maintain a certain back pressure automatically.

A. Yes, sir; if that valve was used.

Q. Now then, how would it do that? [253]

A. Why, it would do that if the well would produce it.

Q. That weight would keep the valve closed a certain amount, wouldn't it? A. Yes, sir.

(Testimony of David G. Lorraine.)

Q. And by closing the orifice a certain amount it would build up the back pressure, would it not?

A. It simply would stay right on that much pressure and hold that much pressure back providing the well would produce that pressure.

Q. I mean in the line. A. Yes, sir.

Q. Now then, in a gate valve like in the photograph of this Tonner trap of yours, by turning the handle you would force down into the line a restriction, wouldn't you?

A. You would if that was done.

Q. And that restriction would build up a pressure, would it not, in the trap if there was gas coming through?

A. If you had such a well that would produce a certain amount of gas and a certain amount of oil, that restriction might maintain pressure, but there isn't any of those kind of wells.

Q. Now, did you know when you sold that trap for the Tonner No. 3 Well what line it was going to be put on? A. I did not.

Q. You did not know what was going to be done with it at all?

A. I knew that they were going to use it to separate gas and oil.

Q. And you did not know anything yourself about what they were going to do with it, where they were going to put it up?

A. Oh, I knew right where they were going to put it, yes, install it.

(Testimony of David G. Lorraine.)

Q. Did you know where they were going to make connections? A. No, sir; I did not.

Q. Didn't you tell them what they could connect it to? [254] A. No, we did not.

Q. Who did you talk to about it?

A. Mr. Burroughs, the superintendent.

Q. What is the conversation that you had with him? How did you get him to put the trap in?

A. Why, the trap that they were using at the time went out of operation two or three times in about five days; became totally inoperative.

Q. And what line was that trap on?

A. That was on the same line, I guess.

Q. The same line that is shown in this photograph? A. The same gas line; yes, sir.

Q. And that other trap that you referred to is shown in this photograph that you have produced, is it? A. Yes, sir.

Q. Then, all right. Go ahead and tell us what conversation you had concerning installing this trap.

A. Well, we installed the trap,—I told them I would ship the trap out there and let them put it up on trial; the trap was not sold. That was put in there on trial.

Q. Now, then, you understood all the conditions concerning this before you went out and sold the property, didn't you, or installed the trap, or to get them to take the trap?

A. To get them to try it?

Q. Yes. A. Yes, sir.

(Testimony of David G. Lorraine.)

Q. Now, did you know what pressure the Trumble trap was working under?

A. The Trumble trap was not working at that time.

Q. Well, when it was working.

A. No, I don't.

Q. You did not? A. No.

Q. Did you know what was being done with the gas? [255]

A. Yes, I knew it was going into the absorption plant.

Q. Now, do you understand anything about absorption plants at all?

A. Why, yes; I am very familiar with them.

Q. It requires the gas,—the gas has to come in under pressure to this absorption plant, doesn't it?

A. Yes, sir.

Q. And you knew that that trap of yours was going to be put up and the gas run.

A. Well, I wasn't positive what they was going to do with the trap. They was going to see whether the trap would work satisfactorily.

Q. You knew they were going to put it on that line, didn't you?

A. If it worked satisfactorily I thought they would, yes.

Q. That was the purpose you sent it there for, then. A. Yes, sir.

Q. Why do you put pressure gages on your traps?

(Testimony of David G. Lorraine.)

A. Why, to see whether there is any pressure built up on the trap.

Q. Why do you test them up to 250 pounds' pressure before you send them out?

A. Why, to be sure there are no leaks in them and you can't tell what any well is liable to do. Nobody knows how much of a volume of gas is liable to come out of one oil well.

Q. Now, haven't you represented that one of the advantages of your trap is that it can be run under a pressure up to as high as 250 pounds?

A. No, never have.

Q. You are willing to state that you never have represented that an advantage of your trap was that it could be run under pressure?

A. No, I would not say that, but I never set 250 pounds. [256]

Q. Well, what do you say?

A. I said it would operate at high or low pressure.

Q. (By the COURT.) Inside of the trap?

A. Yes, sir.

Q. (By Mr. L. S. LYON.) When did you first start in the business of making traps?

A. Here since this suit or before the suit?

Q. Well, before the suit, whenever you started.

A. Well, I have not had any on the market, that is, to do any gas trap business, but the first gas trap I made was sometime in about 1905, I should say, in about July, after I seen that there Teco trap that I just described, in Texas.

(Testimony of David G. Lorraine.)

Q. Now, when was it you started in here making traps?

A. About four years ago; three years ago.

Q. Just prior to that what business were you in?

A. Oil field work.

Q. What oil field work?

A. Why, I was at pump stations and engineering work and work like that.

Q. What did you do?

A. Run pump stations; worked in the oil fields; all kinds of work, general work.

Q. Were you employed by a company as a laborer of some kind?

A. Yes, sir—no, not a laborer; you could hardly call it a laborer. I worked at the pump stations and worked generally in oil fields.

Q. As a mechanic?

A. Well, as a mechanic, pumper and engineer.

Q. Then, about four years ago you became interested in starting to make a trap? A. Again, yes.

Q. Had you ever sold a trap before that?

A. Yes, sir. [257]

Q. How many?

A. Oh, I sold about four, I think; I guess I gave one away.

Q. How long before that?

A. Before I started out here?

Q. Yes.

A. 17 years now; about 17 years.

Q. Between that time you worked as a mechanic, did you? A. Yes, sir.

(Testimony of David G. Lorraine.)

Q. Now then, just before you started up in this locality about four years ago you had seen and observed the operation of the Trumble trap, hadn't you? A. I never saw the interior of it.

Q. And observed the operation of one, haven't you? A. Yes, sir.

Q. Now, the first traps that you made here, how were they constructed?

A. Well, one of the first traps that I made here was made something similar to this patent office drawing; that is, regarding this baffle feature.

Q. Did it have the baffle in it?

A. As is shown here; yes.

Q. As is shown on the patent?

A. Yes, had one float in it.

Q. Was that the Tonner? A. No, sir.

Q. What did you do with that trap?

A. Sold it to the K. T. & O.

Q. Did you make any traps with that baffle-plate in before you made the Tonner one?

A. No, sir; not that style.

Q. What other styles did you have?

The WITNESS.—Would you show me those drawings, please?

Mr. BAGG.—Are these the ones you want? [258]

The WITNESS.—Yes, sir. (Receiving same.) Do you want to look at these things?

The COURT.—Have those been offered?

Mr. BAGG.—If your Honor please, I do not see the purpose of this examination. In the first place it is not cross-examination of anything he said.

(Testimony of David G. Lorraine.)

Mr. L. S. LYON.—We offer the drawings produced by the witness in evidence.

The COURT.—What is the purpose?

Mr. L. S. LYON.—If we can just wait for a few minutes I will bring out the purpose, your Honor. I would rather not state it in front of the witness.

The COURT.—Very well.

Mr. BAGG.—Well, we object to the introduction of these until they have been—

Q. (By Mr. L. S. LYON.) What do those drawings show that you have produced, Mr. Lorraine?

A. Why they show gas traps I built before I started to build these traps, before I saw the Trumble trap.

Q. When were those made?

A. Why those were made while I was with the Canal Zone Oil Company.

Q. When? A. 1916.

Mr. L. S. LYON.—These are offered in evidence as Plaintiff's Exhibit—

The CLERK.—19.

Mr. BAGG.—We object to the introduction—

Q. (L. S. LYON.) Do any of these—

Mr. BAGG.—Wait just a moment. I would like to get in an objection.

Mr. L. S. LYON.—Pardon me.

Mr. BAGG.—I object to the offer on the ground they are incompetent, [259] irrelevant and immaterial; haven't any connection with this case, having been made,—testified by the witness hav-

(Testimony of David G. Lorraine.)

ing been made prior to his having seen the Trumble trap and should not be in evidence in this case as an infringing device.

The COURT.—Your purpose is to develop his trap from what he did.

Mr. L. S. LYON.—Yes, your Honor.

The COURT.—Is that prior to the Trumble trap?

Mr. L. S. LYON.—No, this is all admittedly after the Trumble patent came on, after the Trumble invention.

Mr. BAGG.—I think it was before he had ever seen the Trumble trap, wasn't it?

The WITNESS.—Yes, sir.

Q. (By Mr. L. S. LYON.) Where were you working when you made this? A. Made the—

Q. Yes, this drawing and this device that is in this last drawing.

A. Well, when I made the drawing first,—not that drawing, but the original drawing, was in San Francisco, in 1911.

Q. Well, but this trap here, this drawing here, where were you when you made this one?

A. I was in Taft.

Q. In Taft? A. Yes, sir.

Q. Who were you working for?

A. I wasn't working for anybody at that time.

Q. What were you doing?

The COURT.—This was in 1911, you said?

A. No, sir, when that drawing—the original drawing was made in 1911 and then I went to the Canal Zone Oil Company and I made a model of it

(Testimony of David G. Lorraine.)

and then later I had this Patent Office drawing made. I left the drawing there. This was just a photostatic [260] print of what was made in Washington.

Q. (By Mr. L. S. LYON.) When did you make the first trap like this last drawing?

A. About 1916, the model.

Q. Where were you

A. Canal Zone Oil Company near Santa Maria; Grazioza lease.

Q. At Santa Maria?

A. Yes, sir, Grazioza lease.

Q. What did you do at Taft? You started to say something about being at Taft when you did something. What was it you did there?

A. When I made this—fixed this sketch up and sent it to Washington?

Q. Yes.

A. I wasn't doing anything there at the time.

Q. You were at Taft, were you?

A. Yes, sir. I wasn't working for anyone.

Q. Now, then, how many traps like this did you make? A. I just made that one model.

Q. Then the next trap that you made was the Tonner trap? A. No, sir.

Q. What was the next trap you made?

A. The next trap I made was the K. T. & O. trap.

Q. And that K. T. & O. trap was just the same as your patent drawings, as far as you can tell, the patent drawing?

A. That is a good description, yes.

(Testimony of David G. Lorraine.)

Q. Now, there were Trumble traps at the Union property, right near that K. T. & O. installation, weren't there?

A. Yes, I believe there was.

Q. Now, didn't you testify yesterday that the only trap you had ever made that had a baffle in like the Tonner trap was the Tonner trap and you put it up and found it wasn't any good and took it out. [261]

A. Yes, sir, that is the only one I have made like that Tonner baffle.

Q. Well, Mr. Lorraine, but you made one before and put it up at the K. T. & O. just like the patent drawing there.

A. The one in the Tonner is not like the Patent Office drawing. That has a sheet down here below. This will force the oil down here where the Tonner would be compelled to gravitate.

Q. What would be the progress of the oil in this K. T. & O. trap?

A. Just as you see it here in this here drawing, to allow the gas to escape up here out of this opening and the oil to discharge here in the lower portion of that sleeve; that is a sleeve inside of another sleeve.

Mr. BAGG.—If your Honor please, it seems to me like *this all* incompetent, irrelevant and immaterial, and it seems to me like counsel on the other side should begin to develop their purpose by this time.

(Testimony of David G. Lorraine.)

Mr. F. S. LYON.—With the answer, Mr. Bagg, he is making traps in accordance with this patent and we are now asking him about a trap that he admits is made in accordance with the patent and how it works.

The COURT.—Well, that has no reference, then, to these other drawings.

Mr. F. S. LYON.—No, those early drawings, no. This is what he calls his K. T. & O. trap.

The COURT.—Well, I understand. We had better settle this matter about these other drawings.

Mr. F. S. LYON.—We will withdraw the earlier drawings, then, your Honor.

The COURT.—Very well.

Mr. L. S. LYON.—That is those that were marked Plaintiff's Exhibits.

The COURT.—19–1 and 19–2. They were just offered and admitted. [262]

Q. (By Mr. L. S. LYON.) Now, this K. T. & O. trap, where did the oil come in?

A. Here is the oil inlet here at 3 (indicating).

Q. And where did it pass?

A. Passed down here (indicating).

Q. Passed down sleeve 14, did it?

A. 13—14, yes, you are correct.

Q. Then it struck this baffle-plate 17.

A. Yes, sir.

Q. And it was showered over onto wall 2.

A. Yes, sir

Q. And passed onto wall 2 down to liquid, did it?

A. Yes, sir.

(Testimony of David G. Lorraine.)

Q. (By the COURT.) That is this patent here, isn't it? That is this device here.

A. No, there is a little difference; there is a larger sleeve here. For instance, this here pipe is 3 inches and the sleeve is 6 inches and then we have this large deflected plate.

Q. I see the pipe is at the bottom.

A. There is a deflector up at the end of the pipe.

Q. (By Mr. L. S. LYON.) Where did the gas separate from the oil? At what point during the progress of the oil in this trap, this K. T. & O. trap?

A. The gas goes up through this here slot and the gas comes off the oil here, holds the oil until the gas was cut off with this partition.

Q. I mean at what points?

A. Some gas came off here, you see.

Q. In the sleeve? A. Yes, sir.

Q. Some came off while it was going down the wall?

A. Well, it did not have much of a chance because the oil was always built up higher in this than it was on that side. This [263] acts the same as a dam in a river.

Q. You say that the oil stayed up on here?

A. Yes, sir.

Q. Up over the baffle-plate?

A. At times, yes; sometimes right—

Q. How often, about a third of the time the trap was working?

A. I watched for it 30 days and the oil was always up at least around here on this side (indicating).

(Testimony of David G. Lorraine.)

Q. If the oil was up above the baffle-plate the oil would not shower on the wall, would it?

A. Well, it would; it was on the wall—with the oil right there that wouldn't stop it (indicating).

Q. If it was up on here it surely would not shower onto the wall below the baffle-plate, would it?

A. I should say it would, yes.

Q. And run down the wall?

A. Why, it would have to run down through here or go over the top. It would have to; if this is its course, it would have to go down the wall in a big volume.

Q. Some of the time in the operation of this trap was the oil level down below?

A. When we would leave the trap out it would go to the bottom.

Q. Wouldn't this automatic valve keep this level, at a constant level?

A. That is not an automatic valve. That is just the oil outlet; the automatic valve would keep it level.

Q. (By the COURT.) Where is the oil outlet?

A. Right here at 31 (indicating).

Q. (By Mr. L. S. LYON.) Wouldn't that seek a level course across there?

A. Now, while the oil was flowing through there. If you stopped the oil from going in there this would come to a level, [264] but this partition is just the same as a dam in the river, to keep one side higher than the other. This oil is always higher

(Testimony of David G. Lorraine.)

on this partition on this side than it is on that side (indicating).

Q. If the oil is right up here at the top of this thing here, don't you just let it run right into that (indicating)?

A. Well, you would never get this to settle.

Q. Why not?

A. Because it would stir it up and shoot it around there; you want to stop that turbulent action.

Q. If it came there in a clump do you think it would stir it up any more than if it just passed along over the bottom there, if the oil was up there (indicating). A. I think so.

Q. How much more?

A. That is the object in deflecting the oil, to stop the turbulent action in there, to keep it—

Q. Now, you observed this running for 30 days, you say. A. Yes, sir.

Q. Where at?

A. Well, it was at 16D, I think, and I would not say the number of the well, but I think it was 82, what they call the Pacific Oil Company now.

Q. Where?

A. It was about three miles from Taft; I would not say—

Q. How could you tell that this oil was up above this baffle-plate?

A. Well, I was very anxious to find out just how that trap was working, because it was a new model, and one day I cut the oil off and there was froth and

(Testimony of David G. Lorraine.)
foam here, and then I had gage cocks to tell just where the oil was.

Q. Where were those gage cocks on the device?

A. In the tank. [265]

Q. Well, whereabouts; point them out on the device. How many of them were there, and where were they? A. There was two of them.

Q. Where were they?

A. I cannot show them on this section. I can show it on the Patent Office drawing which is something similar and show you about where they were. I would not want to say the distance, but they were up above the center portion of the trap. They were up above this oil wall.

Q. Well, in your patent drawing, when you came to draw your patent up, you showed the level of the oil as straight across, didn't you?

A. Yes, but—

Q. You have described the operation of it as being straight across, haven't you?

A. That was before this trap was put in action here, and before I tested it out on the one well.

Q. The patent was applied for before this K. T. & O. trap was sent out, do you say?

A. The patent? No, it was not applied for, but the patent attorney had the specific papers.

Q. When did you sign it, before or after the K. T. & O.? A. I believe it was after.

Q. Yet, you left the oil level straight across there.

A. Yes, sir.

(Testimony of David G. Lorraine.)

Q. And described the showering of the oil and running down this wall? A. Yes, sir.

Q. Now, how big was this baffle? How much area was there on this baffle of that K. T. & O. trap?

A. The area of the plate here?

Q. Yes, how big was it? What were the dimensions of it?

A. We have an area there of about 18 inches.

Q. That was as big as it was altogether? [266]

A. Not an inch over or under; it is a 6-inch baffle.

Q. Now then, when you came to making this Tonner trap, why did you change the dimensions of that spreader or baffle?

A. To stop the stirring action. I thought by putting this type of baffle in it it would prevent the stirring action.

Q. Why?

A. Why, I thought I would have space enough here for any well, but I haven't got it.

Q. You haven't got enough space?

A. Not for it to gravitate down, no.

Q. Does it hit over on the edge, the wall of the chamber?

A. Why, out here, when you put this on the Tonner well, this filled right up with oil, as I said yesterday, and it ran over the top.

Q. How often did it do that? How long did this trap operate on the Tonner lease, this trap here?

A. It is still in operation.

# United States
# Circuit Court of Appeals
## For the Ninth Circuit.

# Transcript of Record.

**(IN TWO VOLUMES.)**

DAVID G. LORRAINE,

Appellant,

vs.

FRANCIS M. TOWNSEND, MILON J. TRUMBLE and ALFRED J. GUTZLER, Doing Business Under the Firm Name of TRUMBLE GAS TRAP COMPANY,

Appellees.

**VOLUME II.**
**(Pages 321 to 578, Inclusive.)**

Upon Appeal from the United States District Court for the Southern District of California, Southern Division.

FILED
DEC 28 1922
F. D. MONCKTON,
CLERK.

Filmer Bros. Co. Print, 330 Jackson St., S. F., Cal.

# United States

# Circuit Court of Appeals

## For the Ninth Circuit.

# Transcript of Record.

**(IN TWO VOLUMES.)**

DAVID G. LORRAINE,

Appellant,

vs.

FRANCIS M. TOWNSEND, MILON J. TRUMBLE and ALFRED J. GUTZLER, Doing Business Under the Firm Name of TRUMBLE GAS TRAP COMPANY,

Appellees.

**VOLUME II.**

**(Pages 321 to 578, Inclusive.)**

Upon Appeal from the United States District Court for the Southern District of California, Southern Division.

Filmer Bros. Co. Print, 330 Jackson St., S. F., Cal.

(Testimony of David G. Lorraine.)

Q. And it still won't work?

A. Still works enough so we have to put a weight which you can see in the picture here, an extra weight, in order to raise the float on account of this here baffle here; we have to add an extra weight which is set up there.

Q. And the device always runs with the oil above the baffle-plate, does it?

A. There is always froth and foam up there; I would not want to say oil.

Q. There is the level of the oil in the Tonner device as it normally runs?

A. Why, I should say right in between these two gage cocks here, under normal conditions.

Q. That is just where the float would maintain it, isn't it? A. Yes, sir. [267]

Q. Just normal, across where the float would have it. A. Yes, on this side here (indicating).

Q. Where is it on this side (indicating)?

A. I got one pipe as I said yesterday, that is right in there, and that goes back 5 inches, and at most times you can get the froth and foam right out of that pipe.

Q. That doesn't necessarily show oil running over this surface and coming down?

A. No, but it would if we put it in at the top.

Q. You haven't got it in at the top?

A. No, sir.

Q. Where is that level in this diameter in the Tonner device, if you know?

A. Well, you could not tell exactly where that

(Testimony of David G. Lorraine.)

level is because there is changes on the wall on this side; this side is where the petition changes.

Q. Now, didn't you on July 18, 1921, file a reissue application of your patents? A. Yes, sir.

Q. And wasn't that for the purpose of correcting any errors in the earlier patent?

A. Well, we could not correct any errors in the drawings; we could not change the drawings.

Q. You did change the drawings, didn't you?

A. No, sir; we put the drawings in the same way.

Q. Now, then, I will ask you the question: Isn't there a port running up in back of that 2 there in the original patent? Isn't there a port indicated running up there and opening between that baffle-plate and that tube in the original drawing?

A. There should not be.

Q. You can see, can't you; there is one there, isn't there? A. Yes, sir.

Q. That has been blacked out in the reissue here, the drawing has been changed.

A. That is just an error; that is all that is. [268]

Q. There was a change in the drawing, then wasn't there?

A. That is just an error; that is no fault of mine that that was changed.

Q. You have described the oil level as just the same on July 18, 1921, as you had when you applied for your reissue?

A. Well, we could not change the wording as long as the drawing showed that was the difference.

Q. Did you say, then, in your reissue "one of the

(Testimony of David G. Lorraine.)

features of the present invention resides in means for very careful adjusting the level of the outlet member 31 as by swinging it about its axis in the member 32"—here, this paragraph (handing document to the witness)? A. Yes, sir.

Q. In which you describe very carefully that this level is maintained across.

A. On this side of the partition; not that side. This side was meant, controlling the float, because the float is not in that chamber (indicating).

Q. Well, now, you took this baffle out in your subsequent devices, didn't you? A. Yes, sir.

Q. But you put the baffle in up until the time this suit was filed, didn't you? A. No, sir.

Q. Had you made any devices prior to that time of the filing of this suit in which the baffle wasn't in any of these here?

A. Like this here? (Indicating.)

Q. How many devices had you made?

A. We had five of these in.

Q. Five of them like this?

A. And we took four of them out, as I told you.

Q. When did you take them out?

A. We took them out immediately after we saw how the [269] thing worked in there.

Q. Wasn't that after this suit was filed?

A. No, sir.

Q. Are you sure of that?

A. Well, now one of them might have stayed in the shop and I would not say as to that.

(Testimony of David G. Lorraine.)

Q. Where were those devices put in? Where were they installed?

A. Well, the General Petroleum, I think, got either three or four, and the Standard Oil Company one.

Q. When were those delivered to them?

A. I could not say, unless I got the record.

Q. You could not say whether they were delivered before or after the suit was filed?

A. I think they were delivered afterwards. I would not say for sure.

Q. They were delivered after the suit was filed?

A. I believe so.

Q. And they had the baffles in after the suit was filed, did they?

A. No, they didn't have that baffle in.

Q. They had a baffle in, didn't they?

A. Had nothing but this in the top (indicating).

Q. When did you take the baffles out of this device for the first time?

A. Well, I came back to the hop right from the well and either told Mr. Burrough at that time I was going to take the baffle out, either take it out or lower it.

Q. Either take it out or lower it? A. Yes, sir.

Q. What were you going to lower it for?

A. So it would not hold the froth so it would run over the top of that sheet, and so I would not have to use that extra weight. [270]

Q. Well, if you lowered it you would get it further down into the oil, wouldn't you?

(Testimony of David G. Lorraine.)

A. I think I would, yes, most assuredly.

Q. Those that you sent out were just like this one on the Tonner lease except they had this bottom floating spreader off?

A. That wasn't in there, no.

Q. Did it have that two-way jet there?

A. Yes, sir.

Q. How did that work? What was the progress of the oil in the trap with that two-way jet with the spreader out of it?

A. Well, it was much better.

Q. Where did the oil go when it came out of the inlet pipe?

A. It is not made just exactly like that spreader there, but it came down this way on each side, a 45-degree angle from that place here. This was brought down there and stopped at only just a short ways.

Q. It had it on there a little ways, though?

A. Well, perhaps five-eighths of an inch, so you could not help it if you put it out there, even the side of the pipe threw very little down there (indicating).

Q. Where would the oil go when it came out of that nozzle? A. Why, it generally went down.

Q. Would it come over and hit against this sand baffle, this interior sheet?

A. Some of the froth might hit it, yes.

Q. Did some of it hit on the outside, I mean on the edge of the chamber? A. This sheet?

(Testimony of David G. Lorraine.)

Q. Some of the oil struck the ring of the chamber?

A. I would not say it would, no, not here; no place in there. The only place it would go would be two sources, come out in two streams. [271]

Q. Would that strike the edge or not?

A. Well, I do not think it hardly could; some of it might have, because I have never seen that thing in action.

Q. From that it would at least boil down into the liquid, down in the bottom part of the trap?

A. Well, there is two large openings here, 7 inches on each side here, for that to flow out of on each side.

Q. You think it would drop down and break into the liquid below and not touch any wall on the way down?

A. It would be impossible to get that oil below that vertical partition without going to this wall.

Q. It did go down the wall?

A. It may have gone down in a solid volume or a thin volume and there would be no way in the world of telling.

Q. How thick was it when it went down the wall?

A. There would be the distance from that sheet to this sheet.

Q. Why didn't you leave that off and just let it plunge right down in there?

A. For the simple reason no man knows anything about baffling oil and gas, either, allowing his incoming gas and oil to stir up his water and sand

(Testimony of David G. Lorraine.)

with his oil, he would never get a separation of his water and his sand, if he kept it stirred up. The object of this invention is to get the oil into the trap and get it into a settled condition.

Q. And the way you do that is to spray it on to one of these other surfaces and let it flow down, is that right?

A. You do not have to let it flow down.

Q. Well, it runs down, then.

A. Yes, runs down.

Q. Now, then, did this work all right?

A. Certainly.

Q. How many of these did you make?

A. We made—Mr. Lacey has the record there, I think about 36. [272]

Q. Why did you then decide to change from that construction and pit in this elbow, this belled elbow? A. The belled elbow?

Q. Yes.

A. Like we did on this recent sale to the General Petroleum, we made several tests and found out it was a very good baffle.

Q. That is this one here?

A. This one right there (indicating).

Mr. F. S. LYON.—Exhibit 1.

Q. (By Mr. L. S. LYON.) You found this belled elbow was a better baffle than that two-way nozzle?

A. I consider it is, yes.

Q. Why? A. Because it is a little cheaper.

Q. Why is it any cheaper?

(Testimony of David G. Lorraine.)

A. Because there is a difference in the price and then here is the casting that goes onto that pipe and more machine work.

Q. Any other reason?

A. No. There is no other reason; it has a larger,—little larger opening there on the end.

Q. Spreads the oil better, doesn't it?

A. Spreads it?

Q. Yes.

A. The object is not to spread it; the object is to get the oil in there without stirring it up.

Q. Well, but it does that by spreading it around better, doesn't it?

A. No, there is no intent there to spread that oil. All the object in that is not to get a nozzle force into it.

Q. But how do you avoid getting the nozzle force by this belled elbow?

A. That is made just the opposite from what a fireman's hose nozzle is made. [273]

Q. So it comes out in a sheet form, doesn't it?

A. No, not sheet. For instance, the hose nozzle you use to shoot water a great distance, you reduce it down so it will be shooting that water a great distance, while, if you wanted to stop that water from going a great distance, you would put it in that bell form, wouldn't you?

Q. Have you actually watched this thing work so you want to testify how it actually works and what the shape of the flow of the liquid is as it

(Testimony of David G. Lorraine.)

comes out of that elbow? A. From a well, yes.

Q. How does it work?

A. Why, it flares right out in one big—

Q. Are you willing to say that the oil coming through there would not, due to its velocity, tend to climb up on this wall and spread out on that wall and come out like that (indicating)?

A. It might if it was dead oil and molasses, but I do not think it would from an oil well.

Q. Are you willing to testify it would not, if there was gas in the oil?

A. I am willing to testify like I have tested it, where I put it onto an oil well, that same nozzle, and you can see the action that comes out in one big spray.

Q. Why do you make the elbow into a bell?

A. It is not made into a bell.

Q. Why didn't you leave it perfectly round? Why did you cut off that edge, flatten it out, the edge adjacent to this sand screen?

A. The others are simply machined off.

Q. They were machined off on the other side, too, weren't they? A. On the other side?

Q. Why is this cut off, a cut edge adjacent to the sand screen if all you were trying to do was machine off the others?

A. You mean that flat place? [274]

Q. Yes.

A. I never left any instructions to put a flat place on it there. They must have cut that off in order to get it in there.

(Testimony of David G. Lorraine.)

Q. Why couldn't you—it is a fact, isn't it, that this incoming pipe projects over quite a little ways from the edge of the trap; isn't that true?

A. Yes, sir.

Q. And then the elbow is stuck on so it comes right approximately at this sand screen; isn't that so? A. That is not necessary.

Q. Well, it is true, isn't it?

A. Well, doesn't that close in?

Q. You could just as well cut this off here and cut this out in the middle, couldn't you?

A. It would be just as good, yes.

Q. But you did not do it?

A. No, it would not weld in there as easy; you could not get at it to weld it.

Q. Why not? You could weld it just as easy if you would move it over six inches, couldn't you?

A. Oh, no, there is no room to weld it.

Q. How much room is there?

A. There is only about 12 inches between this space here, this whole space here is only 12 inches.

Q. You put them approximately adjacent to this baffle here, or this wall here?

A. Well, we are not particular, just so we get it in there, if we can get it equidistant.

Q. For one thing, why do you point this on an angle? A. To keep it from stirring up the oil.

Q. How does that do that?

A. It does not shoot down there.

Q. Where does it go?

(Testimony of David G. Lorraine.)

A. It goes off to one side in place of stirring up the [275] bottom.

Q. Then where does it go as it starts off to one side. A. What, the oil?

Q. Yes.

A. Why, the oil, when it builds up a static head there, the weight of the oil forces the other oil down around the lower part of the partition.

Q. Does it drop into the oil without touching that wall?

A. I should say 90 per cent of the oil does when it comes from your well.

Q. Why don't you turn it down and let it go down and hit?

A. For the simple reason you do not want that velocity and force to stir up your oil.

Q. You testified yesterday didn't you, that what you were trying to do in your trap was to give this gas as much distance that it has to travel as possible? A. Yes, sir.

Q. So that is why, in this construction, you make it go up here, over here, and down there; is that true?

A. Well, it goes back here and just the opposite (indicating).

Q. It goes back up here, and through this pipe down here, and back up. A. Yes, sir.

Q. You want the gas to travel just as far as it can in your trap? A. Yes, sir.

Q. I think you said, didn't you, that you want

(Testimony of David G. Lorraine.)

to get the oil down into the receiver just as quickly as you can.

A. Want to get it into a settled state just as quick as you can; the oil will also travel as far from the oil intake, so it will free itself from water and sand and other materials.

Q. Just explain what you mean by that last part.

A. Well, it is right here explained right here in this [276] drawing (indicating). This oil travels further by being run in at the most remote part of the top from the most remote part of the side here. You see this oil travels all the way around here and while it is settling down here it must travel further to that oil outlet, whereas if you let that oil right in here at the top you would be taking gas, oil and sand right into your oil outlet.

Q. Didn't you say yesterday that you wanted to get the oil into this body of oil just as quickly as you could? A. Settled as quickly.

Q. Just as quickly as you could without moving any further than you had to? A. Yes, sir.

Q. You did not want it to move any further than you had to, do you?

A. Not from the point where it comes in at to where it strikes the oil.

Q. Now then, why can't you take and put this nozzle right in here so that it don't have to move any further and then you have got that much more distance for your gas to go?

A. If that is a true state of affairs, in your trap you would be getting your nozzle force too close

(Testimony of David G. Lorraine.)

to your oil, your heavy oil. I consider that this here, of many models I have made, this is the best arrangement; that is my answer to that.

Q. Don't you want a space here in which the gas can get off? A. Not necessarily.

Q. Where would the gas get off if you sprayed it right in there?

A. If you put your nozzle in there you would be blowing some of your froth and some of your foam up around through this gas channel here.

Q. Why not cut this off and let it come around here (indicating)?

A. That would not work good.

Q. Why not?

A. Because that was tried. That would all froth up, foam up and form a cushion there. [277]

Q. What you are after is to put this stuff in,—put the oil and the gas in at the top of the trap and let it pass down quietly to the oil, submerged oil below the volume of oil below; is that right?

A. The idea of this construction, of that construction there, is this—I told you about six times, is to get the oil into the trap and the gas without stirring up the lower portion of the settled oil sands and water. That is the idea of that construction.

Q. And wouldn't the further it dropped, the more speed it would get in going, the harder it would hit this volume of oil below?

A. This oil is not dropping; this oil is gravitating down here.

(Testimony of David G. Lorraine.)

Q. How about this one (indicating)?

A. That is the gas in the chambers.

Q. I mean this form of device with this elbow on, does the oil drop down into the volume of oil below, or what does it do?

A. It simply goes down of its own weight.

Q. The further it has to drop, the harder it hits, doesn't it? It keeps going faster all the time in there.

A. This oil here forces the oil down around this partition.

Q. Just answer my question. If you projected that oil two feet above this body would it be moving slower when it hit than if you projected it four feet, wouldn't it?

A. Yes, but you do not understand the construction of this. This is applied on the side, built up with oil, and you must have greater weight on your oil on this side, and naturally a higher level of oil to force the oil down around the other side of the partition.

Q. Now, you are willing to state that you have observed one of this kind of traps, 1922, with this belled elbow on it, and that the oil stands higher in it in normal operations on the wall between the sand screen and the wall than it does on the flat side of the [278] sand screen?

A. When it is on an oil well, yes.

Q. Normal operation of the well? A. Yes, sir.

Q. How much higher?

(Testimony of David G. Lorraine.)

A. It depends on several different things. How much foam is in the oil, and gas, how much gas is contained in the oil and regarding the flow of the oil.

Q. Does all of the oil that comes out of an oil well, has it any foam? You have been *taken* about foam on the oil; has all of it foam?

A. Every trap we have had on a flowing well, that is on a flowing well, these flowing wells have foam, yes.

Q. How much foam?

A. Well, that is a pretty hard question.

Q. All oil is not in the form of a foam, is it?

A. If there is any gas in the oil there is foam on the oil.

Q. Do you maintain in your trap with, say a 4-inch inlet as you have put here, that all of your inlet is filled with a froth, or half of it filled with regular body of oil and the other half with a froth?

A. For instance, I think this will answer your question: Out here at the Wilbur well of the Shell Oil Company they had a well that was producing about ten million feet of gas, and 50 barrels of oil was coming out of that gas in 24 hours. Now, that was nothing but like a fog that is all turned out into atmosphere; it looked like a fog, that is all.

Q. Now, how many wells are there that are in that condition; what is the percentage of them? I want just to show that he is taking an extreme. Now, on this particular Shell well you are talking

(Testimony of David G. Lorraine.)
about you put 165 pounds pressure on each one of your traps; isn't there four traps and each one of them operated under 165 pounds pressure on the trap? [279]

A. Well, I couldn't say I ever saw that pressure on those traps.

Q. They are operated under considerable pressure and that you know, don't you *know?*

A. I do not think they are, no.

Q. What do you know about it?

A. I have not been out there to see lately.

Q. Now, I think you said, didn't you, that this baffle plate was 2 inches over the edge of the shell in the Tonner well, in the Tonner trap?

A. Yes, sir.

Q. How far across is this baffle-place in that Tonner trap?

A. Well, as I told you, I did not have the drawings, but I think it is about 11 inches.

Q. Doesn't it go halfway across the trap?

A. This way?

Q. I mean this thing here: How far does that go, clear across the trap? A. 31½ inches.

Q. Then there is an aperture, 2 inch aperture, and it is 31 inches around and 2 inches across, is it?

A. I believe it is.

Q. What is the size of the oil inlet that goes into that trap? A. Three inches.

Q. You maintain that you can bring in, in a 3-inch inlet, enough material to clog an outlet that is 2 inches across and 31½ inches around?

(Testimony of David G. Lorraine.)

A. Yes, sir.

Q. You say that the product of that Shell well, when it escapes at atmospheric pressure, looks like a fog? A. Yes, sir.

Q. What does it look like after it goes through your trap?

A. Well, the Southern Counties Gas Companies declared it [280] was the cleanest separated oil they ever saw.

Q. Have you seen it? A. No, sir.

Q. You have not. Do you know what your gas traps are connected with down at that Shell, those four at the Shell plant, the Shell property?

A. Connected with?

Q. Yes, where does the gas line run to?

A. The southern counties gas line, that is, it did; I would not say now.

Q. What was the pressure on that line?

A. About 125 pounds, I should judge.

Q. Then, it would take over 125 pounds pressure in the trap to force the gas from the trap into the line of the gas company, wouldn't it?

A. I should think it would, yes.

Q. Isn't it true that you have to put a pressure in a gas trap in order to get a dry gas?

A. Yes, sir.

Q. And the gas companies want dry gas, don't they, in their lines?

A. Yes, and they want clean gas.

Q. That means dry gas, doesn't it?

A. Yes, sir.

(Testimony of David G. Lorraine.)

Q. Now, in the device shown in your patent which we have, figure 4 here, where is the expansion chamber?

A. The expansion chamber? I would not say that there is any expansion chamber in that trap.

Q. Don't you have to have an expansion chamber for gas to get out of the oil?

A. You have a clogging chamber. Mr. Trumble is the only man with the specifications that called for an expansion chamber.

Q. You know the one I mean, the one in which the gas passes from the oil.

A. Well, if there is an expansion chamber in this trap it must be the largest in diameter. It could not be the smallest [281] chamber; that must be the expansion chamber right there (indicating).

Q. Don't you have to have a chamber in which the oil comes out and spreads out and the gas takes off from the oil?

A. You would not call the oil chamber the expansion chamber.

Q. I do not care what you call it; I am asking you where, in your trap, figure 4 of your patent, is the chamber in which the gas expands and separates from the oil.

A. Well, we have never pointed this out as an expanding chamber, but that would be an expansion chamber, the one in the trap, because it is larger, and that is where the gas has room to expand. If you are going to expand something you must have space there for it to expand, larger space.

(Testimony of David G. Lorraine.)

Q. You mean, then, the top of the trap above the volume of oil? A. The top of the trap, yes.

Q. That sand screen or baffle that you have in your trap, No. 19, is an open communication at the top, isn't it? A. Yes, sir.

Q. Both sides communicate. A. Yes, sir.

Q. Then, couldn't you properly say that the whole trap was one chamber with a barrier in the middle to divide it?

A. No, any more than we can say that two rooms that have a partition are the same room.

Q. Well, if that partition did not go to the top of the room it would be one open chamber, as far as the air was concerned.

A. As far as the air—I don't know as it would; the wind might blow in one and not the other.

Q. (By the COURT.) Doesn't this operate as an expansion chamber here by reason of the gas going over here and passing up here (indicating)?

A. I would say so, yes, sir.

Q. (By Mr. L. S. LYON.) And to a certain extent the oil would [282] expand as it went up here, the gas would expand as it goes in here.

A. It would have to get to a larger place to expand.

Q. Well, it comes up from here, and gets up here and expands on this side of the baffle, and, of course, some on that side. (Indicating.)

A. You have to keep moving here. There would be your expansion chamber, the large chamber would be the expansion chamber.

(Testimony of David G. Lorraine.)

Q. It could expand partially in a smaller chamber and then expand some more in a larger one, couldn't it?

The COURT.—I think it is very apparent how it is moved.

Q. (By Mr. L. S. LYON.) Now then, how does the oil get into this trap? It comes in at the top, doesn't it? A. Comes in at one side, yes.

Q. At the top?

A. Well, not at the extreme top, no.

Q. Now, if there is pressure maintained in your trap, in those instances where there is pressure, the pressure is uniform throughout the trap, isn't it?

A. Through the whole trap, yes, I should say it would be.

Q. Now, you have a means of withdrawing the oil, haven't you? A. Yes, sir.

Q. And that consists in this port 31, marked in this figure 4? A. Yes, sir.

Q. Now, substantially that same apparatus is employed in your 1922 trap that you sold to the General Petroleum Company?

A. With this here (indicating)?

Q. Yes. A. No, that is not made the same.

Q. What is the difference?

A. The difference is it comes straight down here and hasn't that control on it. [283]

Q. What comes straight down?

A. This here nipple is standing in a vertical position in place of as you see it there.

(Testimony of David G. Lorraine.)

Q. You have a nipple in there which takes the oil out practically the same place.

A. Yes, sir, out here, you see (indicating).

Q. In this 1922 trap where would you say that the gas separates from the oil, what point in the trap?

A. Well, I would say it was separating from the time it left this nozzle until it went out of the top of the trap.

Q. Well, I mean at what point does the gas part from the oil? Of course, it is still going through all the time it gets out, but at what point on the trap does the gas leave the oil?

A. Your oil intake here, the oil comes in here and the oil goes below, some of it. This is still cleaning your oil as the gas travels right up to the very top.

Q. Now, then, during the entire progress of the oil down until it strikes this mass below, there is gas separating from it, isn't there? A. Yes, sir.

Q. There is still some gas in the oil after it gets to the bottom of the chamber?

A. That is where we get most of our gas from after the oil gets into the settled state.

Q. From the time the oil leaves the elbow until the oil is turned out of your separator there is gas coming out of it.

A. If there is nothing but gas or fog coming out of here,—of course there is some oil that leaves the gas right here (indicating) at the nozzle, but if there is heavy oil here there is no gas to speak of leaving that oil until it gets down into a settled state.

The COURT.—Is that all of this witness?

Mr. *S. F.* LYON.—One other question for cross-examination.

Mr. L. S. LYON.—We may want to ask him a question after the demonstration, if your Honor please. [284]

The COURT.—When do you want to go out?

(Discussion *in re* demonstration of the trap at the plant of the Trumble Manufacturing Company, West Alhambra.)

The COURT.—Well, we will meet at 2 o'clock.

(Whereupon a recess was taken until 2 o'clock P. M. of the same day, at the plant of the Trumble Manufacturing Company, for a practical demonstration of the gas trap.) [285]

Los Angeles, Cal., Friday, March 24, 1922,
2:30 o'clock P. M.

(Demonstration at gas trap at the Trumble Manufacturing Company, West Alhambra, California.)

Mr. HARRIS.—This is the so-called 1922 model which was purchased by the General Petroleum Company from the Lorraine Oil and Gas Separating Company, in regard to which testimony has been given. The holes in the side, this square hole at the left—at the right of the oil inlet was cut in my presence, and at that time there were no openings in the drum anywhere except the natural openings. The other two openings have been cut since. On the oil inlet pipe inside of the trap is an elbow which has been flared out to a sharp edge and which has been cut off so that it goes over against the partition. This elbow is at an angle—

The COURT.—You do not describe that exactly when you say it is cut off against the partition. It is simply a part of the baffle.

Mr. HARRIS.—On the outside, your Honor, it is cut off; the other side over next to the partition is machined off. The end of the elbow next to the partition is machined off so as to bring it up against the partition. The elbow is threaded on the inlet pipe and there is a space of probably an inch and a half between the round portion of the shell of the trap and the inlet end of the elbow, the elbow not being fully screwed up on the pipe.

The COURT.—The fact is, the elbow rests against the partition.

Mr. HARRIS.—The elbow rests against the partition and is machined off so it can rest against the partition. The gas outlet from the chamber is on the extreme right, near the top, that is the pipe that goes across to the chamber on the other side. There is a chamber on the other side. Now by looking in that [286] top your Honor can see the tops of both chambers so that we could not see anything until we cut it open. We did not know what was in there, in either of these chambers, until we cut it open.

The COURT.—This chamber over here is the elbow in the partition so it extends over next to the wall of the trap.

Mr. HARRIS.—And there, your Honor, is the

gas outlet pipe that comes down out of that further chamber, on the further side.

The COURT.—This is the gas outlet?

Mr. HARRIS.—This is the gas outlet pipe. In other words, the gas goes through this pipe which we see down here, through from this chamber on this side to the chamber on the far side.

The COURT.—Then it goes down and comes up again.

Mr. HARRIS.—Then it goes down around the partition and comes up.

The COURT.—Now, there is a pipe in the inside coming in from the right. What is the office of that?

Mr. HARRIS.—A pipe coming on from the outside?

The COURT.—Yes.

Mr. HARRIS.—I do not know what pipe your Honor refers to. Just a moment until I get up there. Mr. Lorraine says that is the oil discharge line you see coming in there.

The COURT.—That is the oil discharge line? There is a fixture that seems to be detached.

Mr. HARRIS.—The floats are taken down, the two floats are lying here on the ground.

The COURT.—That fixture operates with the floats.

Mr. HARRIS.—That is a part of the floats.

The COURT.—What is the office of the fixture extending to the outside?

Mr. HARRIS.—Those are the two floats at the bottom. One is the gas valve on the left hand side, as you face them, and on the right hand side is the oil valve. The gas valve of this [287] pipe and the oil valve in the other pipe on the other side. Now the bottom of this trap is filled with oil and we have placed this elbow on the outside of the oil inlet and have connected the pipes so that oil can be taken out of the bottom of the trap and put back into the trap through the normal hole in the pipe on the trap. We have also provided means for putting gas into this oil before it goes through the pumps so we can apply a mixture of gas and oil in there through this inlet pipe.

The COURT.—Could you put oil through without the gas?

Mr. HARRIS.—We can put oil through with or without gas.

Mr. BAGG.—Mr. Lorraine, have you any further explanations you want to make about this?

Mr. LORRAINE.—Yes, please. I would like to examine this pipe here. (Indicating.)

Mr. BAGG.—Any further explanation?

Mr. LORRAINE.—I would like to state first that this is not a fair test, although we are willing to go through with this test.

Mr. BAGG.—In what respect would you say it is not?

Mr. LORRAINE.—In the first place, the oil level, when this trap is in operation, is as high as that float I am holding, which you see right there, and there is a dam formed here in the incoming oil. In the next place, this pipe is either a 3-inch or 4-inch and it would not shoot that up in nozzle force, up at the top here as this—

The COURT.—You claim this ought to be the same size as this (indicating)?

A. To here, yes. This would have a tendency to throw your force over here. This is built right up here with froth and foam, as they come from an oil well, and the object of that is to scrub that froth and foam off of the top, and that is a scrubbing chamber over in the other side of the separator.

The COURT.—What do you mean by "scrubbing"?

Mr. LORRAINE.—It takes the crude oil out of the gas, what [288] would remain coming over here, when the froth goes through there, it would clean it as it turns, whip out like cracking a whip.

The COURT.—You say the froth goes through that hole?

Mr. LORRAINE.—Yes, sir, that is the object of that.

The COURT.—That is the gas outlet?

Mr. LORRAINE.—Yes, sir, that is the cleaning chamber.

The COURT.—Well, the froth, is that anything more than gas?

Mr. LORRAINE.—Oil and gas mixed.

The COURT.—That is oil and gas mixed?

Mr. LORRAINE.—Yes, sir.

The COURT.—That goes through there and what becomes of the froth?

Mr. LORRAINE.—That froth drops down into

other oil as it takes the turn around that other partition.

The COURT.—I see, between the two partitions.

Mr. LORRAINE.—Yes, sir, what whips around; it has got to go through to the other side and then drops down to the other side because this pipe runs through to the other chamber.

The COURT.—Then it drops down into the other chamber?

Mr. LORRAINE.—Yes, sir, as it goes below the other chamber it whips what crude oil is in there out, and then the lighter gas comes out the gas outlet. That is the object in that passageway, and that is the object of that scrubbing chamber, to clean this froth. This is all like soapsuds in here, which you would see if you went out there to a well with the oil at high temperature, that is, mixed with gas and oil. This chamber is filled right up here with froth and foam on the other wall, and that is the object of this, is to clean that lighter froth and foam over there in the other chamber. This is built right up with oil when it is operating, and discharges that oil out of that oil discharge valve there (indicating).

The COURT.—That is down below here?

Mr. LORRAINE.—Yes, that is the oil discharge valve on that side, on the exterior, and this is the gas discharge valve.

The COURT.—Over there (indicating)? [289]

Mr. LORRAINE.—Yes, sir.

The COURT.—Now, you say that the partition here does not extend low enough?

Mr. LORRAINE.—The partition is low enough. It is built up with oil when this is separating gas and oil. The oil in that outer chamber is as high as that float I am holding.

Mr. BAGG.—In what other respect would you say that this would not be a fair test? How about the pressure?

Mr. LORRAINE.—Well, of course, a well pump is far different from the pump that they have got on here, and the temperature of the oil would be different. I do not believe it is possible to mix gas and dead oil together like they are when they come from a well.

Mr. BAGG.—How about the water and the sand?

Mr. LORRAINE.—Well, of course, we could eliminate that because there is wells where I would be willing to concede that, where there is no sand in water and nothing but oil and gas.

Mr. BAGG.—Now, then, what, from your experience, as an oil and gas man, would be the effect of having these two openings, all of these openings closed and this practically airtight?

Mr. LORRAINE.—A pressure in here such as would build up with reference to the—

Mr. BAGG.—The effect it would have on the oil as it comes out of that nozzle into the face of this pressure.

Mr. LORRAINE.—Well, the effect would be entirely different. For instance, you could put enough pressure, hold back enough pressure with this trap as these valves are closed up with plugs, both valve outlets, to prevent the well from overflowing in there at all, providing this shell would stand it. It would be just the same as trying to flow oil against a concrete wall.

Mr. BAGG.—If there is somewhat less pressure in there than in here, in the shell, in the separator chamber, than the pressure in the gas line, or in the oil and gas line, as it comes from the oil well, what effect would that have upon the gas and oil as it [290] comes from the oil well in being discharged into this receiving chamber?

Mr. LORRAINE.—Why, it would start to expand the ends of this elbow here and fill the whole elbow.

Mr. BAGG.—What would be the form, then, in which it would come out? It would be in what kind of shape?

Mr. LORRAINE.—It would be froth and I would describe it as a ray from a lamp, made larger.

Mr. BAGG.—Well, how about the sprinkling can?

Mr. LORRAINE.—Yes, that would be a good illustration for it.

Mr. BAGG.—Now, that condition would not obtain in this separator and apparatus as it is presented here?

Mr. LORRAINE.—No, I hardly think it would; in fact, I know it would not.

The COURT.—Would the pressure inside, supposing these holes were closed, prevent the oil from flowing as readily as it would otherwise.

Mr. LORRAINE.—Yes, it would because you get practically the pressure within there. There is no question about that, but you have a pressure there.

The COURT.—Supposing you had a very heavy pressure, 250 pounds, for instance, would that prevent the oil from flowing?

A. As rapidly, yes; it would flow just the same.

Q. Could you have pressure enough in there so that the oil would not flow at all?

Mr. LORRAINE.—No, it always went to the bottom under all pressures that they put on gas traps and the lighter liquids, always have gone to the top, but not as rapidly.

Mr. F. S. LYON.—I will ask you this question, Mr. Lorraine: Now, then, you have seen how this trap is now connected. From your experience are you able to state where the oil will go on being ejected from this nozzle or—

Mr. LORRAINE.—Elbow where applied?

Mr. F. S. LYON.—Yes. [291]

Mr. LORRAINE.—You mean just as it leaves the elbow?

Mr. F. S. LYON.—Trace the course.

Mr. L. S. LYON.—What the passage of the oil will be when we turn it on. Now, if you can, tell us from your experience.

Mr. LORRAINE.—Well, I never saw this connected up with a reducer here. I can only give you my general idea of where the oil would flow, and it is according to how the oil would be pumped through there, but this oil, under ordinary conditions, would flow right into here, and froth and foam right in there (indicating).

Mr. F. S. LYON.—You know how it is connected. You have said you never saw the inside of your own trap, but you know how it would flow. Look at this thing and tell us how it will flow in here.

Mr. LORRAINE.—Did I say I never saw the inside of my own trap?

Mr. L. S. LYON.—In operation. Suppose you tell us how this will work now from your experience as a gas and oil man.

Mr. LORRAINE.—It depends on how much pump pressure.

Mr. L. S. LYON.—The more pump pressure the more of the oil will be forced on to the wall of the baffle, and onto the wall of the shell; is that correct?

Mr. LORRAINE.—The more it will be directed over this way, yes.

Mr. L. S. LYON.—And the more pressure that you have in coming as against the pressure that you have inside, the more of it will come onto the inside of this shell; is that correct?

Mr. LORRAINE.—The more pressure, yes.

Mr. L. S. LYON.—The more pressure there is on the incoming gas the more your oil will run down this surface here (indicating)?

Mr. LORRAINE.—Yes, sir.

Mr. L. S. LYON.—And the shell?

Mr. LORRAINE.—Yes, sir. You see this surface is over more than the partition or the shell? [292]

Mr. L. S. LYON.—I am speaking, in my question, about the inside of the shell, of the trap itself.

The COURT.—I see.

Mr. L. S. LYON.—If I understand Mr. Lorraine's statement, the more pressure there is on the incoming stream of oil and gas the more of the oil will be thrown on to the inner face of the shell of the trap than is distinguished from when there is pressure here going on to the surface of the baffle-plate on the other wall.

Mr. LORRAINE.—I understood your question, Mr. Lyon, as connected to here, at this installation.

Mr. L. S. LYON.—Yes.

Mr. LORRAINE.—Yes, but this trap runs this way, runs full of oil and your oil level is up here, ordinarily, and this chamber is full of froth and foam.

Mr. L. S. LYON.—Your outlet is down where you see it there?

Mr. LORRAINE.—But that doesn't make any difference with the oil level. We can prove that where they are operating.

Mr. L. S. LYON.—If we put 10 pounds pressure on this thing, now, describe to the court where the oil is going to strike. Is it going to float down in the middle or strike on one of these two walls.

Mr. LORRAINE.—Well, I can't tell the Court.

Mr. L. S. LYON.—Why not?

Mr. LORRAINE.—Because you can't put 10 pounds pressure on it.

Mr. L. S. LYON.—If we do where will it hit?

Mr. LORRAINE.—Well, it is operating with pressure on here.

Mr. L. S. LYON.—We have those holes open there, the way it is now. You say you can tell without looking in the gas trap where these things are going to happen. Look at this, will you, please and tell us what it is going to do.

Mr. LORRAINE.—That oil directed from this elbow will hit right in that oil (indicating).

Mr. L. S. LYON.—Will it land on that sand screen over there or will it not? [293]

Mr. LORRAINE.—It would not be deflected on the sand screen.

Mr. L. S. LYON.—Will it float down on that screen or not?

Mr. LORRAINE.—The whole thing will be filled with oil.

Mr. L. S. LYON.—We are going to operate it now, we are going to turn the pump on and let some oil go through there. Can you tell us where it will go?

Mr. LORRAINE.—Oh, on this line.

Mr. L. S. LYON.—Right here, here it is up here now.

Mr. LORRAINE.—There is no question but what you can drive that oil against that wall.

Mr. L. S. LYON.—Will it go on there?

Mr. LORRAINE.—Certainly it will go on.

Mr. L. S. LYON.—Which one will it go on?

Mr. LORRAINE.—With 10 pounds of pressure it would hit this wall (indicating).

Mr. L. S. LYON.—With five pounds pressure which one will it go on?

Mr. LORRAINE.—Well, that is just a matter of guesswork with this reducer here.

Mr. L. S. LYON.—You mean the inner portion of the shell itself?

Mr. LORRAINE.—Yes, sir. It will direct itself right out here.

Mr. L. S. LYON.—If we put 10 pounds pressure on this, if we put 10 pounds pressure on this thing right in here, the way it is connected you are willing to state that the oil will not go on that baffle over there, but will come over on the outside edge of the shell?

Mr. LORRAINE.—No, I would not state that.

Mr. L. S. LYON.—Which will it do?

Mr. LORRAINE.—That will depend on just how this nipple goes up here. If this nipple reached up here enough so that would break the oil through this top surface it would have a tendency to go on that wall. It would depend, too, on how the end of that pipe is cut; it would deflect it in several different ways. [294]

Mr. BAGG.—How about the amount of gas in the oil? Would that have any effect on it?

Mr. LORRAINE.—Well, as I said, Mr. Bagg, it would be a very difficult matter to state.

Mr. F. S. LYON.—Any other question, your Honor, of this man?

Mr. BAGG.—I would like to call the Court's attention to the fact that the oil level in this is away below the partition.

The COURT.—That is very apparent.

Mr. F. S. LYON.—Will you state, Mr. Trumble, how far this pipe extends in here?

Mr. TRUMBLE.—Just a standard thread, inch and a half pipe, screwed into the bushings. We can take it out in a very few minutes.

Mr. L. S. LYON.—I just want Mr. Lorraine to explain. He states he can tell these things.

Mr. LORRAINE.—It would be impossible to tell where the oil would strike with this installation. Anything I have not practiced with—anything I have tested out I am willing to make statements on and I am willing to prove through an actual demonstration.

Mr. L. S. LYON.—You state you can't tell where the oil will go on this before it is turned on.

Mr. LORRAINE.—Because I have never, your Honor, had this out in this manner.

Mr. HARRIS.—This trap is made of five-sixteenths inch steel, very heavy rivets in the top and heavy trap on the incoming oil pipe welded solidly into the shell where it goes through.

The COURT.—What point do you make about that, Mr. Harris?

Mr. HARRIS.—Simply that the trap is made so it will stand heavy pressure. It is not a trap that operates without pressure or they would not use this heavy steel, heavy rivets or heavy top.

Mr. F. S. LYON.—Where was this oil line that you say it always stays? How high up?

Mr. LORRAINE.—This oil level? [295]

Mr. F. S. LYON.—Yes.

Mr. LORRAINE.—Right over close. Depends upon how fast the oil is coming in as to where it stays, right around that flat iron housing on the other side, not on this side. This side is always high.

Mr. F. S. LYON.—How much difference in the level according to your theory will there be in the level of the oil on this, what we will call the inlet side and on the other, the gas chamber baffle side.

Mr. LORRAINE.—That depends on the density of the oil, how much gas there is in the oil and how fast the oil is coming in. On this side there is a variation in the level of the oil, but on the other side there is no variation to speak of.

Mr. F. S. LYON.—The level of the oil on the other side, you say, always remains below the end of the baffle on that side, must it not?

A. Below, no, sir; above.

The COURT.—He means the baffle on the other side.

Mr. F. S. LYON.—The short baffle, the submerged oil?

Mr. LORRAINE.—Oh, on the outer side, yes, sir.

Mr. F. S. LYON.—The level of the oil must always be below that in order to let it work.

Mr. LORRAINE.—No, it can work without it, but it is supposed to be below; of course, better below.

Mr. F. S. LYON.—If we filled this up so it is coming up there with 5 pounds of pressure upon it, that speed, how many barrels a day would be going through the pipe, as near as you can tell?

Mr. LORRAINE.—I will just tell you how difficult that question is to answer. Our orifices, this size, there is some wells flowing with a hundred pounds pressure, perhaps 600 barrels a day; No. 1 is flowing through less than an inch opening with 1800 barrels a day; at Tonner lease No. 3 I believe at the present it is flowing through an inch beam and only 300 barrels a day. It depends on the density of the oil, how much can be pumped through [296] that line and also the temperature upon it. No human being can answer your question directly.

Q. Will you be able to tell us about what amount is coming out if it was kept flowing at that rate?

A. I could not. For instance, you could use a piece of glass and it looked like oil and you might guess at that. It depends upon the stickiness, the thickness of this oil, how fast it is travelling; you can only guess.

Mr. F. S. LYON.—If you saw it coming up there can you tell us?

Mr. LORRAINE.—Absolutely not, and no one can tell you by looking at it.

Mr. HARRIS.—It would make no difference if you took the big pipe out and put this in (indicating)?

Mr. LORRAINE.—It would not make any difference with this demonstration; it is not connected like it.

Mr. HARRIS.—Mr. Lorraine says this small pipe going into the elbow on the outside of the trap make no difference in the flow of the oil.

Mr. LORRAINE.—If you had this discharging gas and oil you would have this discharging the gas and oil right in froth and foam.

Mr. HARRIS.—What I want to know is whether our man shall change this pipe connection for you or not.

The COURT.—I would not go to that trouble, no. You can describe the action of the oil as it entered without the gas into this machine which is now demonstrated.

Mr. LORRAINE.—Well, I should say under this demonstration this throwed oil upon the wall of the partition in a very heavy stream, a stream almost equal to the area of a small two inch pipe that leads up to the four-inch pipe. At no time could I see where there was any volume of oil thrown upon the wall even with that demonstration, with an open volume, that I consider a volume of oil. [297]

The COURT.—There was oil thrown over here, though?

Mr. LORRAINE.—On the wall, yes, your Honor.

The COURT.—On the outer wall?

Mr. LORRAINE.—Yes, sir; I admitted that on the vertical partition it was thrown on the wall.

The COURT.—Now, describe the operation as you see it when so applied.

Mr. LORRAINE.—As I see it when so applied here the stream seemed to me to widen out to a certain extent although there was very little gas in the oil. I must say, comparing that gas with the gas from oil wells, if that is all the gas that was necessary to save from oil that comes from wells, why, I do not think we would need any gas traps.

The COURT.—How did the gas take to the side of the partition and how to the side of the wall,—I mean how did the oil take to the side of the partition and how to the side of the wall with the gas appliance?

Mr. LORRAINE.—Well, it seemed to flow to me in a thicker stream, more loosely.

The COURT.—What proportion of it ran down the partition and what proportion down the wall, as you observed it?

Mr. LORRAINE.—I should say it was nearly equally divided.

The COURT.—What did you observe as to the foam on there?

Mr. LORRAINE.—As to the foam on the bottom of the trap, why, any oil pumped into a receptacle with an open top, if any force, if it flows any distance, will create a foam or froth regardless of any gas mixing in with it.

The COURT.—Is that the foam that comes with the gas from the well that you see on top of the oil?

Mr. LORRAINE.—I would not say so, no.

The COURT.—Is it similar?

Mr. LORRAINE.—That is a foam that is created by the churning action on the oil as it hits the bottom, the same as the froth [298] or foam that is created with nothing but water that flows a great distance.

The COURT.—How does that appear with the froth or foam as the oil comes from the well?

Mr. LORRAINE.—Well, I should say it would be a very poor comparison, as the oil is mixed so evenly with the gas in so many instances.

The COURT.—So that this is not a good demonstration, then, as to that?

Mr. LORRAINE.—No, I do not think it is.

Mr. BAGG.—I would like to ask him one or two more questions:

Q. You noticed both when the oil was being brought up here or pumped up there and when it was mixed with gas what did you notice with reference to the splashing effect, if any, and did all of the oil go down either the partition or the side walls of the receptacle? In other words, did it go down in a shower, did you notice, or observe whether this oil went in a shower both against the partition and the side walls of the receptacle and down to the center?

Mr. LORRAINE.—Well, I would say that some of that oil hits the wall and some bounced back over at times free from the wall. It was hitting right there where the partition connects on to the shell, right in the crevice like when it was pumped up there and did not bounce over there.

Mr. BAGG.—That is all.

Mr. F. S. LYON.—About how many barrels a day do you think the oil was coming out of that nozzle when the demonstration was made? Just as near as you can. I am not asking for an accurate estimate?

Mr. LORRAINE.—Well, it will only be a guess.

Mr. L. S. LYON.—Just give us your best guess.

Mr. LORRAINE.—I guess about 350 barrels in 24 hours.

Mr. F. S. LYON.—Can we step around to the back of the trap? [299]

The COURT.—Yes.

Mr. BAGG.—I would like to ask the witness one question:

Q. Have you observed that any portion of the oil, as it came out of that receptacle, fell directly down to the bottom without striking the side walls of either the partition or the side wall of the receptacle?

Mr. LORRAINE.—I did not see right down into the bottom, Mr. Baggs. From where I was standing I could not see the bottom.

Mr. L. S. LYON.—Mr. Harris, will you point out the gas and oil outlet valves of this trap and their adjustments on these two valves?

Mr. HARRIS.—As we stand facing with these valves, these two valves, one on the left-hand side and one on the right-hand side, the one on the right-hand side is cut through—

The COURT.—The left-hand side—

Mr. HARRIS.—The left-hand side is connected through a vertical pipe with the gas outlet in the top of the trap. The one on the right-hand side is connected through a shaft pipe with the oil outlet. Two valves are operated on separate shafts, in other words, the shaft that goes into the right-hand valve and the shaft that goes into the left-hand valve are separate, so that either valve can be working independent of the other. Attached to these two operating shafts are two levers which terminate in a slotted connection, each of the valves having its own slotted connection. When the lever on the oil valve is in its extreme upper position, as I understand it, the oil valve is connected.

Mr. LORRAINE.—Yes, sir; that is correct.

Mr. HARRIS.—When the lever on the gas valve

is in extreme upper position, as I understand it, the gas outlet valve is open.

Mr. LORRAINE.—That is it.

Mr. HARRIS.—In the operation of the trap which I saw at the Tonner lease the oil valve was operating somewhere near the middle of its range, and I assume that this trap in its operation [300] will operate somewhere the middle of its range, depending on the flow of gas confined in the trap.

Mr. LORRAINE.—Why, the adjustment on these valves here and the valves on the Tonner lease are entirely different. You can't change the position of the valves on the Tonner lease when once set.

Mr. HARRIS.—These can be set.

Mr. LORRAINE.—These can be set.

Mr. L. S. LYON.—During the time of the test on the Tonner lease did you, Mr. Lorraine, tell Mr. McLean and Mr. Burroughs of the General Petroleum Company that by regulating that gas valve a pressure could be maintained in the tank without putting another back pressure valve on the line?

Mr. LORRAINE.—The General Petroleum Company?

Mr. L. S. LYON.—Did you tell these men that?

Mr. LORRAINE.—At Tonner 3?

Mr. L. S. LYON.—No. Did you tell them that or make such a statement to them?

Mr. LORRAINE.—On that well?

Mr. L. S. LYON.—I don't know whether it was on that well, but did you make such a statement to McLean and Burroughs that by regulating that gas valve there you could keep a pressure in the tank without having a back pressure valve on the line?

Mr. LORRAINE.—But you can't maintain—

Mr. L. S. LYON.—Did you make the statement?

Mr. LORRAINE.—No, I did not, not like you are making it.

Mr. L. S. LYON.—What did you say to that effect?

Mr. LORRAINE.—I said you could close it in, but as far as maintaining a pressure on it, I did not make that statement, and you can't do it with those valves. If you will take the plugs out I will show you this myself. You can't maintain a pressure unless there was a big volume of gas, or volume of oil maintaining that pressure. In other words, you have to get another machine to sustain the pressure. [301]

Mr. L. S. LYON.—What did you say to McLean and Burroughs?

Mr. LORRAINE.—That is what I said.

Mr. L. S. LYON.—What did you say to them? Just tell us, please.

Mr. LORRAINE.—As near as I can remember, I said, you can regulate either the gas outlet or the oil outlet; you could close it in.

Mr. L. S. LYON.—Did you say a pressure could be made in the trap?

Mr. LORRAINE.—I did not say anything about pressure, no, sir; nothing about pressure; I never guaranteed it to any operator.

The COURT.—Mr. Harris, go ahead.

Mr. HARRIS.—By means of this connection it is

impossible to adjust the relative positions of these valves in various relationships so that the amount of opening in the oil valve and the relative amount of opening in the gas valve can be adjusted in various combinations and to get various conditions within the trap.

Mr. LORRAINE.—That is correct.

Mr. HARRIS.—As I hold it now, as I understand it, the gas valve is entirely closed?

Mr. LORRAINE.—Yes, that is correct.

Mr. HARRIS.—And, as I am holding the right-hand lever, it is about the middle of its stroke, and the left-hand lever which is on the gas valve is at the bottom of the stroke and so arranged both could be put through and apparently together in that manner, under these conditions, the oil valve will be partly open when the gas valve is entirely closed.

Mr. LORRAINE.—That is correct.

The COURT.—Will you explain more fully about the float?

Mr. LORRAINE.—If we could put that mechanism on there we could easily make it more clear, your Honor, and I think it would be fairer to both sides.

Mr. F. S. LYON.—Your Honor understands the oil line comes out this end and the gas line out this?

The COURT.—Yes, I understand that.

Mr. L. S. LYON.—You saw the flow of oil going through the [302] trap on demonstration. How many barrels a day would you say were going down there from your observation?

Mr. HARRIS.—I should estimate at least a thousand barrels a day of twenty-four hours.

Mr. L. S. LYON.—Mr. Paine, from your observation of the same thing, how many barrels of oil would you say you saw going through the trap per day on this experiment?

Mr. BAGG.—I think we will object to that and ask it be stricken out for the reason it is rebuttal testimony; and can be introduced at the proper time and for the further reason neither of these two gentlemen have qualified themselves to testify as to the amount of oil flowing or anything of the kind.

The COURT.—We will hear that on rebuttal, so you needn't take it now.

Mr. F. S. LYON.—We offer, then, all of these facilities that are here for measurements, and so forth, and that is the only object of bringing it up at this time.

The COURT.—All right.

Mr. F. S. LYON.—Your traps are shipped with the regular equipment and a 300-pound pressure gage, are they not?

Mr. LORRAINE.—As a rule they are.

Mr. L. S. LYON.—This is the gage that you ship with your traps to measure the pressure.

Mr. BAGG.—He shipped that because he wasn't able to get anything else, probably.

The COURT.—I think we can take care of that when we get in town.

Mr. L. S. LYON.—We just wanted it to show that we had opened the equipment that came with this trap and found this 300-pound pressure gage, that is all, your Honor.

The COURT.—Very well.

Mr. BAGG.—Explain to the Court how that works inside (indicating).

The COURT.—Now, you may explain that.

Mr. LORRAINE.—In the interior this rock shaft —(remark [303] inaudible)—this rock shaft operates with the counter-weight arm; I had it here in position and that link connects the valve levers, and when oil goes down in the trap that seals this valve absolutely tight. This valve here impinges on and locks, loops one sleeve around another sleeve, and the same action takes place here in a downwardly motion, when the oil goes up in the interior of the trap, and that would seal that tight, that sleeve impinging itself on and wrapping itself around, and that would flow all the oil into the oil line. When the trap is in operation this lever is in that position and you can see here in the valve, by looking out there, there is an opening in both valves.

Mr. F. S. LYON.—Now, that float arrangement, does your Honor understand that it is an ordinary float?

The COURT.—I suppose the float is to regulate the discharge of oil and gas.

Mr. BAGG.—I would like to call the Court's attention to the size of these two pipes, both the gas pipes and the oil pipe, as being practically the same size as the inlet.

Mr. HARRIS.—All 4-inch pipes.

The COURT.—The amount of the discharge of the gas would be regulated by what?

Mr. BAGG.—By the pressure to the opening.

The COURT.—You say there are two openings to the right?

Mr. BAGG.—Yes, your Honor.

The COURT.—But the supply of gas can be regulated more by pressure than the discharge of the oil?

Mr. BAGG.—I think not. Mr. Lorraine could answer that.

Mr. LORRAINE.—By holding your discharge to the gas openings 100 barrels of oil during that period, the gas line would be open much larger than the oil.

The COURT.—Oh, I see.

Mr. LORRAINE.—So that these work in synchronism, and when one valve—when one valve moves towards open position the other valve toward closing position and the orifice is regulated in the valve according to the volume of gas in the oil that is [304] flowing into the trap and takes care of that automatically. That is not a counter-weight valve or back pressure valve.

The COURT.—Did you say it regulated according to the oil as it is flowing into the trap?

Mr. LORRAINE.—Yes, sir.

The COURT.—As the oil is flowing in the separation is taking place?

Mr. LORRAINE.—Yes, sir.

The COURT.—Doesn't it depend somewhat on

the separation the amount of oil that is running out, or gas that is running out?

Mr. LORRAINE.—Well, I think I can make that plain to you: When there is a large volume of oil flowing into the trap the float rises higher and opens the oil valve just a little more, and it closes the gas valve to a certain extent and moves it towards the closing position, so that permits that big volume of oil to escape. It comes in equal division, that is, what you might call equal division.

The COURT.—The object of the trap is to separate the gas from the oil?

Mr. LORRAINE.—Yes, sir.

The COURT.—Now, then, if you get a good separation, absolute separation, are the valves controlled by the amount of oil flowing in?

Mr. LORRAINE.—That is controlled by the amount of oil flowing into the trap and also controlled by the—

Mr. BAGG.—For instance, just as an illustration: If there was twice as much oil coming in in volume as there was gas, that is, the proportion of the mixture as it comes into the oil from the oil well, as we will say the volume is twice as much oil as there is gas, then the oil line would be opened, its aperture would be opened twice as wide as the gas line in the same proportion. Then, if they came in half and half, then the arm would stand so that the oil opening and the gas opening would be the same. [305]

The COURT.—I think I understand that.

Mr. BAGG.—Yes.

The COURT.—But here is a broad statement. These valves are regulated by the amount of oil and gas that flows inside, I understand that, because the purpose of the tank is a separation of the oil from the gas, and if there were more oil in there than gas, why, of course, the oil valve would be open larger, and if there were more gas than oil it would operate the other way.

Mr. BAGG.—But the oil level would determine that.

Mr. LORRAINE.—We have the same orifice in each valve here that we have in the pipe. You see, when this is in this horizontal position, that gives you the same opening out of each pipe.

Mr. L. S. LYON.—I just want to ask Mr. Lorraine one question: When the oil valve is in this position so that its lever is in the central position, can the gas valve be regulated so that it is completely closed in your device?

Mr. LORRAINE.—No, because the oil level would change that position and immediately open up your gas valve.

Mr. L. S. LYON.—But can it be regulated now as this oil valve is in this position, so that the gas valve is completely closed?

Mr. LORRAINE.—It can't without the other bolt in there; I just put this one bolt in there.

Mr. L. S. LYON.—You could take the bolts out and put it into position, couldn't you?

Mr. LORRAINE.—You could, yes, but you would not have a gas and oil separator, because you would not have anything open at one end of it. That is the object of having that around there just the right length so it can close it completely.

Mr. L. S. LYON.—What is the reason for having two arms instead of one?

Mr. LORRAINE.—What is the object?

Mr. L. S. LYON.—Yes. [306]

Mr. LORRAINE.—Well, it is to adjust the valve.

Mr. L. S. LYON.—In your patent you only show one arm, don't you?

Mr. LORRAINE.—That is all right. There is another application on this; that wouldn't make any difference.

Mr. L. S. LYON.—The object of having two arms is so you can adjust one with relation to the other.

Mr. LORRAINE.—Well, the principal use of this is so you can raise the oil level. For instance, if you see your oil is getting too much gas, there is too much wastage of gas going out in the stock tank, you can raise or lower your oil level. You put both of these valves down, then you lower your oil level. You raise them up like that and you raise your oil level.

Mr. F. S. LYON.—We would like to have Mr. Harris show the court, if the Court will permit, how these two levers can be adjusted so that a back pressure can be maintained on the gas line in this trap by means of this lever and still have the trap operate as a gas and oil separator.

The COURT.—He can explain that.

Mr. F. S. LYON.—What the purpose of this link proposition is here.

Mr. HARRIS.—By this long lever which is applied to the two short levers the slot is near the central position, and it is so adjusted that in that position there is a certain opening of the oil valve on the right-hand side. It is possible by loosening these bolts to entirely close the gas valve so that the gas valve acts to hold a pressure on the trap. Now, if it was so adjusted and the oil level falls a little bit, the gas valve will be slightly opened and the gas will escape. In other words, it will act as a reducing valve to hold pressure on the interior of the trap.

Mr. L. S. LYON.—And it is so adjusted the trap will operate for the separation of gas and oil continuously?

Mr. HARRIS.—Operate continuously with any pressure in it. In other words, by adjusting these bolts in these slotted arms, which is adjusted to the gas valve which appears on the left-hand [307] side as a reducing valve for the purpose of building up a gas pressure in the top of the trap, you maintain a pressure on the trap?

Mr. F. S. LYON.—What do you consider the reason for those slotted arms on the device, Mr. Harris?

Mr. HARRIS.—For the purpose of adjusting the gas pressure in the interior of the trap.

The COURT.—As you have stated?

Mr. HARRIS.—As I have stated.

Mr. F. S. LYON.—Is there any other purpose you can see or you can understand for making such slots?

Mr. HARRIS.—I cannot see any other purpose for it at all.

The COURT.—Do you think of any other purpose, Mr. Lorraine?

Mr. LORRAINE.—I certainly do. The oil is of many different gravities. In other words, of many different weights. On the top of it it is always lighter when it comes into a gas trap than it is on the bottom of it. You must get an oil that will at least float a float, to operate this machine, and if you see your oil going through with your gas, with this adjustment feature here you can lower your oil level so your froth and foam will be still down low enough so it won't go out with that gas. If you should see gas going with your oil into the stock tank and you want to stop it, why, you can raise your oil level and carry a higher column of oil to prevent that, and that is the object of that, although I am willing to admit that you can choke the flow with this, but you can't keep a pressure on that trap with it; it has absolutely nothing to do, and I can prove it by a trap on an oil well. The minute you do that your pressure is going through that opening you have there, so it will not sustain or maintain a pressure upon the trap.

Mr. BAGG.—In other words, you can set this adjustment here any way you want and with a pressure on your tank as shown by the pressure gage, then you can take that pressure, or you can open your gas outlet and then it will immediately adjust itself and show no pressure. [308]

Mr. LORRAINE.—No, it would not be necessary

to do that. If you shut the flow of oil off, or the flow of gas, the pressure would release them instantly, right there, those valves. There would be no pressure held there whatsoever.

Mr. F. S. LYON.—Mr. Bagg, anything further?

Mr. L. S. LYON.—We will ask the clerk to take that pressure valve that came with the device.

Mr. BAGG.—All witnesses in the case will take notice that the Court is now adjourned until Monday morning at 10 o'clock.

The COURT.—Is that satisfactory to you?

Mr. F. S. LYON.—Entirely.

(Whereupon an adjournment was taken until Monday, March 27, 1922, at 10 o'clock A. M.) [309]

[Endorsed]: In the District Court of the United States for the Southern District of California, Southern Division. (Before Hon. Charles E. Wolverton, Judge.) Francis M. Townsend et al., Plaintiffs, vs. Davis *S.* Lorraine, Defendant. No. E-113—Equity. Reporter's Transcript of Testimony and Proceedings on Trial. Vol. III. Los Angeles, California, March 24, 1922. Filed Apr. 7, 1922. Chas. N. Williams, Clerk. By R. S. Zimmerman, Deputy Clerk. Reported by G. J. Kennelly, Doyle & St. Maurice, Shorthand Reporters and Notaries, Suite 507, Bankitaly International Building, Los Angeles, California, Main 2896. [310]

# INDEX.

[311]

Los Angeles, Calif., Monday, March 27, 1922, 10 A. M.

The COURT.—You may proceed.

Mr. BAGG.—If your Honor please, to-day was the day we agreed to file our answer to the supplemental bill of complaint. I have the answer prepared and have served a copy upon counsel for the other side, but I have not had an opportunity to look it over to see if there is any possible correction to be made, and would like, if your Honor please, to have until this afternoon to file it.

The COURT.—All right.

Mr. F. S. LYON.—Plaintiff offers in evidence this section of the Lorraine trap which was used in the demonstration Friday afternoon. We have sectioned it out and brought in the portion of the baffle-plate, wall elbow, delibery pipe and so forth which your Honor saw.

Mr. BAGG.—If your Honor please, it seems to me that would be a part of their evidence in chief.

Mr. F. S. LYON.—It is part of the demonstration.

(Testimony of David G. Lorraine.)

Mr. BAGG.—And this would not be the proper time to offer it, until they have rested, because the defendant is now having its turn.

The COURT.—I think you had better reserve it for rebuttal.

Mr. F. S. LYON.—All right, your Honor, I thought we had better clear it up right at this point in the record. [312]

**Testimony of David G. Lorraine, for Defendant (Recalled—Cross-examination).**

DAVID G. LORRAINE, recalled for further examination.

Cross-examination (Resumed).

(By Mr. L. S. LYON.)

Q. I show you a copy of the "Oil Weekly" dated February 25, 1922, and a cut in the lower left-hand side. Do you recognize that cut (handing same to witness)? A. Yes, sir.

Q. What is it a cut of?

A. Why, it is an advertisement of my separator.

Q. Is that a cut of a photograph of an actual device? A. Why, I believe it is; yes, sir.

Q. Do you know where it was taken?

A. Thompson No. 4 of the General Petroleum Company's property.

Q. When, if you know?

A. In March or April, 1921. Now, I wouldn't say just which of those months.

Q. Were you present when the photograph was taken? A. No, sir.

(Testimony of David G. Lorraine.)

Q. Was it taken subject to your direction?

A. Well, yes.

Q. Was that photograph as it appears in the publication handed you correctly show the manner in which that trap was connected and installed?

A. Yes, I believe it does, as near as I can remember.

Q. Now it is a fact, is it not, that in this photograph the gas outlet is shown as provided with a valve which I will mark "X" (marking)?

A. Yes, sir.

Mr. L. S. LYON.—The cut is offered in evidence as Plaintiff's Exhibit No. 20.

The COURT.—Any objection? [313]

Mr. BAGG.—No objection.

The COURT.—It will be received.

Q. (By Mr. L. S. LYON.) What is it in the action of a gas trap that has the effect of regulating or controlling the flow of an oil well?

A. Well, there are several different features that might be considered in a gas trap to regulate the flow.

Q. Well, in your gas trap.

A. Well, to allow the oil to flow out evenly.

Q. Why?

A. Because it is positive and certain in action.

Q. Well, does your gas trap necessarily require the oil coming out of the oil well to flow any evener than if it just came out of a pipe and didn't go into your gas trap at all?

A. Why, I would say it would, yes.

(Testimony of David G. Lorraine.)

Q. Why?

A. Why, on account of the two valves working in synchronism.

Q. What effect does that have?

A. It maintains an oil level in the trap in the same place.

Q. What effect does the oil level in the trap have on the flow of the oil from the trap?

A. Well, for instance, if you had a trap that only had a pipe and no oil discharge valve to control that oil outlet your oil would go from the bottom of your trap to the upper portion of your trap, and it would cause a clogging or jamming action on your well.

Q. But that would not be the effect if you just permitted the oil to escape out into the atmosphere —if you just had the oil coming out of the flow nipple and running out of the pipe into a sump hole? What difference does your trap make over such a condition as that?

A. Well, it has been proven it makes a great deal of difference.

Q. Well, what is that due to? [314]

A. It is due to just what I have tried to explain to you. If you have a separator or gas trap that will allow the oil to build up into the trap your oil will not escape as fast through the same size orifice as your gas will and consequently it throws the oil and gas into both lines.

Q. Yes, but I am speaking about the oil as it comes from the well. Now when the oil comes from

(Testimony of David G. Lorraine.)
the well and drops out of here what is there in your trap that makes that oil flow steadily and evenly so as to control the production of the well? What is there that differs from merely running that out into the atmosphere? What difference does this trap have on the flow from this pipe than if this pipe was just running out into an open tank?

A. The difference is that there is no stirring action in this gas and oil separator, no churning action.

Q. Well, would there necessarily be any churning action if you just had a reservoir and a pipe running over and running out of the reservoir?

A. Yes, sir.

Q. Why?

A. Well, for instance, you can take a bottle of this oil as it passes from an oil well at a certain temperature, and if you stir it or shake it it will blow the cork out of the bottle. But this action does not take place in this gas and oil separator.

Q. Now, is it not a fact that the only reason that a gas trap regulates the flow of the oil from the inlet pipe so as to control the flow of the well is due to the fact that there is a pressure in the gas trap which will maintain and require the oil to come out in somewhat nearly an even manner?

A. No, we haven't found it so. We have operated these traps in a vacuum and atmospheric pressure, which is, as a rule, called zero pressure.

Mr. BAGG.—I would like to have him answer the question.

(Testimony of David G. Lorraine.)

The COURT.—Very well; answer the question in full. [315]

A. And we have found that these separators will cause the well to flow steadily.

Q. (By Mr. L. S. LYON.) Now, if the pressure in the tank is merely atmospheric pressure is it not a fact that so far as the oil flowing from the well is concerned, just as much oil would flow, and just as freely, as if you didn't have a gas trap on the line at all.

A. Well, with this here separator—

Q. Well, just answer that question.

(Last question read.)

A. No.

Q. Why not?

A. Because we have proved it to be a fact that the well will flow better through the trap.

Q. Tell us why.

A. All right. This prevents—If your well goes to flowing heads it prevents your oil from flowing up into your trap and in place of throwing that oil into your gas line it keeps your oil in the same well. It will not allow your well to flow in heads. It cushions that head and stretches that flow out.

Q. Well, how does it do that if there is no pressure at all on your gas inlet? How does it make any difference about what is down here? The oil will come out of this pipe just as unrestricted, won't it?

A. Why, for a short period it builds up a certain

(Testimony of David G. Lorraine.)

pressure—a very slight pressure—to force that oil from the separator.

Q. Does that pressure make any difference?

A. I contend it causes the well to flow steady.

Q. Now you state that you maintain a certain level in your gas and oil trap. To maintain a constant level you have to take out of your trap as much oil as comes in, do you not? A. Yes, sir.

Q. And it has to flow at just the same amount in an hour as is coming into the trap? [316]

A. Yes, sir.

Q. Now you haven't yet made it clear to me why, if this pipe is connected with an oil well, and there is no pressure on this pipe at all, the oil will not run out of there just exactly the same as it would run into an open chamber. Will you explain that?

A. Well, there are some oil wells that flow like artesian water wells, but very few of them. As a rule they are inclined to flow in heads.

Q. All right. Go ahead and explain.

A. Well, a big volume of oil will be discharged into the trap when flowing under those conditions, and then a large volume of gas.

Q. Now what has that to do with the gas trap regulating the flow from the well? Does the gas trap regulate the flow from the well?

A. Well, in place of it allowing this here oil to come in this separator, this large gusher, it cushions the flow and holds the oil level at the same place. There is a slight variation, of course, in

(Testimony of David G. Lorraine.)

the oil level, but it is very light. It moves your gas valves toward the closing position and holds your oil level within five or six inches.

Q. Now, is it not a fact that a back pressure in an oil-receiving line due to the fact that you maintain a pressure in your gas trap is what regulates the flow from the well?

A. No, I wouldn't say so.

Q. Didn't you so state in your patent, Plaintiff's Exhibit No. 3, beginning at line 92 of page 3 of your specification as follows: "It has been found from practical experience in the operation of this apparatus that there is an increase in the production of oil from the same wells, because of the back pressure in the will line"? Did you state that?

A. Yes, sir; some wells. Very few of them.

Q. Now where does the oil go from this Tonner installation, [317] Defendant's Exhibit "D" from your trap (handing paper to witness)?

A. It goes into either one of these here stock tanks.

Q. It goes in at the top? A. Yes, sir.

Q. The discharge, then, from the oil outlet from your trap into the stock tank is higher than the oil level in your trap, is it not?

A. I believe it is, yes.

Q. How much higher?

A. Well, I couldn't say. It depends upon the ground level here as compared with the ground level of the tank.

(Testimony of David G. Lorraine.)

Q. Now, here is the oil outlet line, is it not, in the Tonner installation. A. Yes, sir.

Q. How much higher is this discharge into the storage tank than the oil outlet in your trap? You give us your best estimate of it.

A. Oh, perhaps three feet; maybe four. I wouldn't be able to estimate that exactly.

Q. Well, say four feet. Now, what was it that caused this oil to flow up this four feet into your trap on the Tonner lease?

A. Well, it might be possible that this oil level is four feet higher than this valve.

Q. It might be? A. Yes.

Q. And what was it, as a matter of fact, that caused this oil to flow up into these tanks on the Tonner lease? Oil will not flow up hill, will it, unless there is a pressure on it?

A. Well, as Mr. Rae stated, when there was 55 pounds pressure on that trap there was ample pressure there to raise the oil over a hundred feet.

Q. Well, then, it is a fact, is it not, that there was a pressure in this tank, in your trap on this Tonner No. 3, [318] sufficient to cause the oil to flow up and discharge into those higher storage tanks?

A. There was a pressure of some kind. There might have been a pressure of the oil—the weight of the oil.

Q. Would the weight of the oil cause it to flow up hill?

A. No, it will not, but it will reach its level.

(Testimony of David G. Lorraine.)

Q. Now, is it not a fact that the discharge point into those storage tanks is higher than the oil level in your trap on Tonner No. 3?

A. I believe it is.

Q. And to have the oil flow into those storage tanks some pressure must have been maintained in that trap?

A. If the oil discharge in the tank is higher than the oil level in the trap it would take some pressure to put it there.

Q. Now, what was the area on the oil outlet valve on your Tonner No. 3 trap when the valve was completely open? A. About fourteen inches.

Q. And what was the area when it was about half open? A. Three inches.

Q. In the normal operation of your trap where would that be—about half open or—

A. You would have approximately the same opening as your three inch line, which is about seven inches.

Q. How much pressure would it require to pass 1200 barrels of oil a day through that three inch oil valve opening and up into this tank, in your opinion?

A. That would depend upon the gravity of the oil and how free it would be from gas,—

Q. Well, you know what oil it was that was coming from that Tonner well. You saw it. Now about how much, in your opinion, pressure would it take to do what I have stated?

(Testimony of David G. Lorraine.)

A. Well, it would take less than a pound on that trap. [319]

Q. You would say a pound pressure?

A. Perhaps.

Q. To run 1200 barrels of oil through a three-inch opening?

A. Yes, sir; about that. It would depend upon the gravity of the oil.

Q. Well, what limits would you say? Would a half pound do it?

A. Well, I hardly think it would.

Q. You state that this gauge in this photograph, Defendant's Exhibit "D," shows the needle down at normal. What size gauge was that? What was the type of the gauge used?

A. An American pressure gauge, I believe.

Q. 300-pound gauge? A. I think so, yes.

Q. Was it like this gauge here that was delivered with this trap that you saw last Friday (exhibiting)?

A. No, it is a smaller gauge than that.

Q. How much smaller?

A. Oh, about two-thirds that size.

Q. It takes a 300-pound gauge? A. Yes.

Q. And the first division is 20 that is marked, is it? A. Yes, sir.

Q. Now how much pressure do you think it takes to register on a gauge like this one I have in my hand, to move the needle?

A. Well, none of those gauges are accurate.

(Testimony of David G. Lorraine.)

Q. How much pressure do you think it takes to move the needle on this gauge I have in my hand?

A. On a high pressure gauge of that kind it would take all the way from two to six pounds as a rule.

Q. To start moving the needle? A. Yes.

Q. Did that Tonner No. 3 gauge have a shut-off like this one I have in my hand? [320]

A. Yes, sir.

Q. Now, referring back to this Tico trap that you say you saw in 1905, where was that?

A. At Humble, Texas.

Q. And what was the name of the well on which it was used? A. I couldn't say that.

Q. What was the name of the company that owned the well?

A. I think it was the Producers' Oil Company, but I am not positive.

Q. Well, do you know?

A. No, I am not positive.

Q. Was Mr. Tico present when that was installed?

A. No, sir, not to my knowledge.

Q. What is Mr. Tico's full name?

A. I couldn't tell you. I don't know.

Q. Did you ever meet him? A. No, sir.

Q. Now, did you know at the time that trap was installed in 1905 that it was named the Tico trap.

A. I was told afterwards that that was the model.

Q. Now before testifying in this case you read and examined this Government pamphlets, Defendant's Exhibit "C," did you not? A. Yes, sir.

(Testimony of David G. Lorraine.)

Q. Who was present when you saw this Tico trap installed in 1905?

A. Well, there were several men making the installation. The only man I knew was Mr. Frank Smith.

Q. And what was his position?

A. Well, I think he was a driller. I met him in Houston, Texas.

Q. Do you know where he is now?

A. No, I couldn't say.

Q. When did you last see him? [321]

A. About three years ago.

Q. Where did you see him? A. In Taft.

Q. Do you think he is in Taft now?

A. I couldn't say.

Q. He was a driller, you say? A. Yes, sir.

Q. And who else was there that you know?

A. Why, I couldn't call any names.

Q. How long did it take him to install that trap there, that Tico trap?

A. Well, I couldn't tell you that.

Q. How long were you there watching it being put up? A. About a day, I believe.

Q. Were you invited there by somebody in the company? A. No, I was not.

Q. What were you doing

A. Why, I was looking for work in the oil field.

Q. Now if you wanted to go to that trap to-day how would you get there? Tell us.

A. You would go to Houston, Texas, and I think there is the T. P. railroad that runs out there about

(Testimony of David G. Lorraine.)
14 miles. You get off at the station and turn to the right and you go about a quarter of a mile, to the best of my memory.

Q. That is as nearly as you can describe where it was. A. Yes, sir.

Q. And you cannot remember the name of the company that it was installed for?

A. I didn't know that, but—

Q. Have you been back there since 1905 to see whether the trap is there or not? A. No, sir.

Q. How old are you, Mr. Lorraine?

A. Forty-five. [322]

Q. Now, this pamphlet, Defendant's Exhibit "C," you have testified that this was printed under the auspices of the Bureau of Mines. Your knowledge of that you gained from reading the pamphlet and seeing what is written on it? Isn't that a fact?

A. I have not read the pamphlet through, no.

Q. Well, how do you know it was printed under the auspices of the Bureau of Mines.

Mr. BAGG.—If the Court please, I don't think the witness testified that it was printed by the Bureau of Mines, any more than that the Court read it and I read it in presenting it to him. I don't think the witness testified to that effect. It is immaterial.

Mr. F. S. LYON.—Is it understood, then, that the witness did not so testify?

Mr. BAGG.—I don't think he did.

Mr. F. S. LYON.—You are willing to stipulate, so far as the record is concerned,—

(Testimony of David G. Lorraine.)

Mr. BAGG.—I will not stipulate anything with reference to it, but it is my recollection that he did not testify as to the source from which it came. It shows upon its face. I probably read it to him. I don't know why it is important and it seems like it is encumbering the record unnecessarily.

The COURT.—It wouldn't help the introduction of it anyway.

Mr. BAGG.—No.

Mr. F. S. LYON.—One of our objections to that is that it is self-serving and not the best evidence, your Honor. If it could be an official document it doesn't prove itself. Our original objection was as to its competency. Now on the cross-examination of this witness, he having been so far the only witness in regard to the document, we want to show very clearly that he has no knowledge of any such original document, so that there can be no question hereafter in regard to the fact that all that is before the Court is simply this paper, Exhibit "C," itself. [323]

The COURT.—I suppose that will be admitted.

Mr. BAGG.—Yes. The only authenticity we know anything about is what it shows on its face and the fact that it came from the Government.

Mr. F. S. LYON.—Well, we don't stipulate that it came from the Government. Our objection is absolutely an objection to its competency.

The COURT.—Very well.

Mr. BAGG.—Well, we reserve the right to prove where we got it.

(Testimony of David G. Lorraine.)

Q. (By Mr. L. S. LYON.) Now you have testified, I believe, that the traps are made for you by the Lacy Manufacturing Company of this city. Is that company contributing in any manner, by payment or otherwise, to the defense of this case?

A. No, sir.

Q. What is the character of the relationship between you and the Lacy Manufacturing Company?

A. None at all.

Mr. BAGG.—I object to that. There are allegations in the complaint with reference to relationship they are not a party to the suit.

Q. (By the COURT.) Well, you have the Lacy Manufacturing Company build this trap?

A. Yes, sir.

Q. And that is the relation you have with them, that you employ them to do that?

A. That is correct.

The COURT.—I think that is sufficient.

Mr. L. S. LYON.—That is all. [324]

Redirect Examination.

(By Mr. BAGG.)

Q. When did you *first the* interior of the Trumble trap?

A. Well, I never saw the whole of the interior of the Trumble trap at any one time, because it was never made so that any person*al*—as Mr. Paine stated here—could see the interior of it. But the first Trumble trap I saw with the manhole place removed was in March, 1921.

(Testimony of David G. Lorraine.)

Q. Now, you have testified earlier in the day to the fact that when an oil well flowed in heads the sudden inrush of the oil and gas from the well produced a pressure in the trap, or in your trap. Is that correct? A. Yes, sir.

Q. I will ask you to state if you know whether or not that is peculiar to your trap. A. Yes, sir.

Q. I will ask you to state whether or not it is fact that all gas traps connected to wells flowing in heads will have more or less pressure built up in the trap during the time this well is flowing in a head.

Mr. L. S. LYON.—We object to that, your Honor. In the first place it is grossly leading, and in the second place it is the same question which counsel has just asked the witness and the witness answered just the other way. He asked him if that action was peculiar to his trap and he said yes. Now counsel says, Well, is it not true in every trap—just trying to contradict his own witness. And it is leading and calls for a conclusion of the witness as to whether it is true in every trap or not.

The COURT.—The objection is overruled.

Mr. L. S. LYON.—Exception.

A. Any trap, regardless of how it is constructed, if it has an inlet and two outlets, which it must have, if there is a sudden head of oil or gas comes into that trap with the two [325] outlets it will build up a certain pressure; but the oil level is not

(Testimony of David G. Lorraine.)

checked as it is in our trap; the oil level runs higher.

Q. (By the COURT.) Would that cause a back pressure of the oil coming from the well?

A. Yes, sir.

Q. You have been talking about that somewhat, and I think are of the opinion that the back pressure was not a good thing for the oil well.

A. Yes, sir. And that is the oil builds up too high in the trap it will put more of a back pressure on the well and you have to carry a greater pressure.

Mr. BAGG.—That is all.

Recross-examination.

(By Mr. LYON.)

Q. Now, you stated the first time you ever saw a Trumble trap was in March, 1921.

A. The first time I ever saw a Trumble trap with the manhole removed so that I could see inside of it.

Q. You knew, as a matter of fact, how they were constructed before that—maybe not in detail, but as to the principle of operation of the trap?

A. There never was any detail drawing in advertisements, or description, disclosing the Trumble trap, to my knowledge.

Mr. L. S. LYON.—Now read the question and I will ask that it be answered.

(Last question read.)

A. I couldn't say that I did.

(Testimony of David G. Lorraine.)

Q. Well, what did you know about the Trumble trap before March, 1921. Had you ever seen one at all? A. Yes, sir.

Q. When did you first see one?

A. I think it was in 1916 or 1917. [326]

Q. How many Trumble traps do you think you saw between the time you first saw one and March, 1921? A. Oh, perhaps about 150.

Q. Now, when was it that you took the valve to Mr. Rae, the representative of the Trumble Company and tried to interest him in putting the valve on the Trumble trap?

A. I don't know as I tried to interest him in putting the valve on the Trumble trap. What I did was to try to interest him in my trap with my valve.

Q. Did you have a trap at that time?

A. I had a drawing of one; yes, sir.

Q. Did you show it to Mr. Rae?

A. Why, I am not positive; but I showed it to him later on.

Q. You didn't show him anything except the valve, did you, at that time?

A. I believe I did. I only had the valve model with me; but I had drawings with me. But I wouldn't like to say for sure, at that first meeting, that I showed Mr. Rae the drawing of the trap.

Q. And you would not like to say, would you, that you even said you had a trap, outside of having a valve for a trap? A. Yes, sir.

Q. What did you say to him?

(Testimony of David G. Lorraine.)

A. I explained the trap.

Q. Tell us just what you said to Mr Rae.

A. Well, I explained how my trap worked—

Q. Well, tell us what you said.

A. I told him that it had two valves that worked in synchronism and a vertical partition in the trap that held the oil on one side and the float worked in settled oil, but I couldn't tell you the exact words that I used.

Q. And you told him all of that, did you?

A. Yes, sir.

Q. And you showed him a model of the valve? [327]

A. Just of the valve.

Q. Now what was it you wanted Mr. Rae to do?

A. Why, I wanted him to take me and make me acquainted with the Trumble Gas Trap Company.

Q. For what reason.

A. So that I *could the* gas trap business over to them and make some dealings with them.

Q. Is it not a fact that what you said you wanted him to do was to interest them in putting your valves on their trap? A. No, sir.

Q. Did you say nothing to that effect at all?

A. No, sir.

Q. At no time had you ever made such a statement to Mr. Rae? A. No, sir.

Q. Now, when was it that you first showed those valves and had this discussion with Mr. Rae?

Mr. BAGG.—If your Honor please, I don't see the purpose of any of this testimony. Counsel

(Testimony of David G. Lorraine.)

started in to cross-examine him on when he first saw the Trumble trap, and I don't see what any conversation that he may have had with reference to valves with Mr. Rae has to do with this matter at all or how it is proper cross-examination, and I therefore object to it as unnecessarily encumbering the record.

The COURT.—What is the purpose of it?

Mr. F. S. LYON.—The redirect examination was in regard to when you first saw—or, in other words, knew of—the interior construction of the Trumble gas trap. Now we have already shown on the recross that he brought this valve there. He has testified that the valve was for the purpose of his own trap. We are now laying a foundation for impeachment. If he denies that he knew of the construction at that time we will show by Mr. Rae that the whole conversation was with regard to the Trumble trap and this was simply a case at this time of his presenting this outside [328] doublevalve and an attempt to get Mr. Rae interested in that and having the Trumble Company put it on its traps.

Mr. BAGG.—Well, suppose that that is the purpose of it, your Honor, you can only impeach a witness on some material matter, and that is not material to this case, about any conversation that he may have had with this proposed witness Rae with reference to the construction of some valve to go on his trap. That has no connection whatever with his having seen or known the interior

(Testimony of David G. Lorraine.)

or the working mechanism of the Trumble trap, and he can only impeach the witness on a material matter.

Mr. F. S. LYON.—To make the question of impeachment more direct, we will, in this connection, ask the witness if at the time of the presentation to Mr. Rae of this model of the valve he did not discuss with Mr. Rae the interior construction of the Trumble gas trap.

Mr. L. S. LYON.—And how your valves could be very easily put on and worked with the Trumble trap.

A. I had already installed—

Mr. BAGG.—Just a moment.

The COURT.—Just a minute.

Mr. BAGG.—We object to that as immaterial and not tending to prove or disprove any of the issues of this case. It doesn't make any difference what their conversation was with reference to the matter. The question I asked him was when he first saw the interior of a Trumble trap.

The COURT.—They are trying to prove now that he had prior knowledge of the interior of the trap.

Mr. L. S. LYON.—Yes.

The COURT.—I will allow the question in that form.

Mr. BAGG.—All right.

The COURT.—You can answer that shortly, without taking up much time.

A. I have already installed a trap with those

(Testimony of David G. Lorraine.)

valves on it with the Southern Pacific Company near Taft.

Q. (By the COURT.) Well, what have you to say about obtaining [329] knowledge from this patent as to the interior of the trap?

A. I hadn't obtained any knowledge from the interior of the Trumble trap. When I exhibited this valve model of my trap I had already built one and had it installed and found that it operated very successfully.

Q. (By Mr. L. S. LYON.) Was that the K. T. & O. trap you have told us about?

A. Yes, sir.

Q. Now, when was it you first had this discussion with Mr. Rae?

A. It was after I had installed the trap, about—

Q. Well, what was the date?

A. I couldn't say the exact date.

Q. What is the nearest date you can give?

A. Why, I should think it was in—I know the year; it was in 1920.

Q. At what time in 1920?

A. It was in April, because I had just come from the trap. In the middle part of April.

Q. Now, you didn't discuss with Mr. Rae at all how your valves could be fitted on a Trumble trap?

A. No, and I didn't see how they could.

Mr. BAGG.—Now, if your Honor please,—

The COURT.—Well, that is getting beyond the question here.

Mr. L. S. LYON.—That is all.

Mr. BAGG.—That is all. Now, if your *Honor, we* have, in the first place, the certified copy of the letters patent of the United States to George L. McIntosh, Assignor, granted March 11, 1913. We would like to have this marked Defendant's Exhibit "E."

Mr. F. S. LYON.—This is not pleaded in the answer, and I suppose the offer is limited to show the state of the art.

Mr. BAGG.—Yes, sir. That is all we want.

The COURT.—Yes. [330]

Mr. BAGG.—Now, for the same purpose we wish to introduce a certified copy of the letters patent to Walter Anderson Taylor, granted April 29, 1890, and would like to have it marked Defendant's "F."

For the same purpose we wish to introduce in evidence certified copy of letters patent of the United States to Arthur W. Barker, granted July 13, 1909, and I will ask that it be marked Defendant's "G."

For the same purpose we have the certified copy of letters patent of the United States, Ustes Vivian Bray, granted January 16, 1912, and ask that it marked Defendant's Exhibit "H."

We also have the letters patent of the United States, a certified copy, to Augustus Stiger Cooper, granted March 20, 1906, and ask that it be marked Defendant's Exhibit "I."

We have the certified copy of letters patent of the United States granted to Albert T. Newman,

Assignor, June 4, 1907, and ask that it be marked Defendant's Exhibit "J."

Now, if your Honor please, we have a witness here whom we want to let go, and would like to put him on the stand now so that he may go back to this work, and I will ask Mr. Johnson to take the stand if that is agreeable.

**Testimony of Walter P. Johnson, for Defendant.**

WALTER P. JOHNSON, a witness called on behalf of the defendant, having been first duly sworn, testified as follows:

Direct Examination.

(By Mr. BAGG.)

Q. Please state your name, age, place of residence, occupation?

A. Walter P. Johnson; age 20; residence 824 E. 29th Street, this city; occupation Machinist.

Q. What company are you employed by?

A. The Lacy Manufacturing Company.

Q. Are you engaged in any part of the building of the [331] Lorraine gas and oil separators?

A. How do you mean that?

Q. Well, do you have anything to do with the building or construction of them down there in the shop? A. I do.

Q. Were you employed by the Lacy Manufacturing Company at the time the oil and gas separator which was sold to the General Petroleum Company—March 17, 1922? A. Yes, sir.

(Testimony of Walter P. Johnson.)

Q. Did you have anything to do with the installation of the partition in the trap? A. I did.

Q. I will ask you to state, if you know, in the first place whether any part of the outer edge of the outlet of the elbow that is installed in that trap was machined or ground off.

A. It was ground off by hand.

Q. I will ask you to state if you know why that was ground off.

A. Because of a mistake in the setting. It was necessary to grind off the end so that the baffle-plate would seat down into place.

Q. (By the COURT.) You say that was a mistake in seating?

A. In seating. It was a mistake in seating.

Q. (By Mr. BAGG.) I will ask you to step down, if you will, and examine this part of the trap which leans there against the wall, and, if you can, explain to the court what you mean by that answer.

(Witness leaves stand and discusses exhibit.)

A. You will notice this is very close there now.

The COURT.—This is the same trap we had before?

Mr. BAGG.—Yes.

The COURT.—I examined that very thoroughly.

A. That was just a mistake in this nipple. This part is assembled first, and is supported from the outside threads, and [332] it was—the nipple seated or was let through a little too far before it was welded in, and then it was necessary to grind off the end to let this plate down. These plates

(Testimony of Walter P. Johnson.)

are made to size and have to go down over a certain size, and we filled it up and kept it from touching here on the side.

Q. I will ask you to state whether you are familiar with the construction of the other traps of this type that have been put out by the Lacy Manufacturing Company for Mr. Lorraine, the defendant in this case.

A. I have seen different types manufactured since that time.

Q. I will ask you to state whether, if you know, any other of the elbows that have been installed in any other traps have that face or edge cut off as you have described in this particular trap.

Mr. F. S. LYON.—We object to that as leading and suggestive. If the witness knows these things he ought to be able to describe them without the words being put in his mouth.

The COURT.—I will hear the answer. I think the question is a little leading, but I think he has already answered it, really.

(Last question read.)

A. No; there has been none since that manufactured to my knowledge.

Q. Do you know whether any were done that way before this trap?

A. I say, not to my knowledge.

The COURT.—Who constructed this model.

Mr. BAGG.—I think this man worked on that.

The COURT.—This model here.

Mr. BAGG.—Oh, no.

(Testimony of Walter P. Johnson.)

Mr. F. S. LYON.—No, we made that model.

Mr. BAGG.—That is all. [333]

Cross-examination.

(By Mr. F. S. LYON.)

Q. How much did you have to grind off this elbow in this trap which is here in the courtroom in order to make it fit?

A. I didn't measure it.

Q. You didn't do it?

A. I didn't measure the amount that I ground off.

Q. Well, did you do the grinding?

A. I did it myself.

Q. What is your best approximation of the amount that you ground off?

A. Well, to the best of my recollection, about three-eighths of an inch.

Q. Then the mistake that was made in setting, as you say, this unit in before welding it in place was by sliding it in three-eighths inch too far; is that it?

A. No, it was slid a good deal more than three-eighths too far.

Q. Then how about grinding it off three-eighths of an inch and no more to remedy the difficulty.

A. Well, all that was necessary was for the plate to set down, and three-eighths inch is sufficient to allow it to clear.

Q. Now, are you positive that no others of those elbows that went into any of these traps were

(Testimony of Walter P. Johnson.)

ground off in a similar manner. Do you know that of your own knowledge?

A. Well, I wouldn't—I will swear to you that any that I have worked on didn't have.

Q. Well, you didn't put them all in? A. No.

Q. Did you have working drawings for that construction of the trap? A. I don't know. [334]

Q. Were there any drawings for that trap there in the shop that you saw? A. Yes.

Q. Under whose instructions did you work in that shop on the work that you did on the Lorraine traps?

A. Under the foreman's instructions.

Q. What is his name? A. Ludkey.

Q. Then you made these assemblies under his direction and supervision, did you?

A. Well, it was to be under my supervision—

Q. Answer the question. Did you make them under Mr. Ludkey as the foreman's supervision and direction? A. Yes, sir.

Q. You don't know to what extent he had detailed drawings, do you? A. I do not.

Q. Did you ever see any detail drawings for this trap there? A. Yes.

Q. Well, what drawings did you see there?

A. I couldn't say.

Q. Did you see any drawings for the elbow or nipple or its assembly?

A. I couldn't say as to that.

Q. You don't remember, is that it?

A. No, I do not.

(Testimony of Walter P. Johnson.)

Mr. F. S. LYON.—That is all. [335]

Redirect Examination.

(By Mr. BAGG.)

Q. Mr. Johnson, do you know how far it is customary to place the edge of that elbow from that partition?

Mr. L. S. LYON.—we object to the witness being asked how far it is customary. He might be asked how far he put in any that he made.

Q. (By Mr. BAGG.) Well, if at all.

A. We measure them to see if the baffle will seat down.

Q. Then there was no particular specified distance that that was to be placed from the baffle-plate?

Mr. F. S. LYON.—We object to that as leading and suggestive. The witness has stated what he did. He has not had the drawings. He said that they adjusted them so that the baffle will seat down in place. Now I think that is as far as he can go.

The COURT.—He may answer.

(Last question read.)

A. I wouldn't say that there was any specified distance. There may be. I didn't have it.

Q. (By the COURT.) Did you construct that according to drawings? Did you work according to drawings in your shop? A. Yes, sir.

Q. Now did you construct that contrivance according to drawings? A. No, sir.

(Testimony of Luther L. Mack.)

Q. You just simply had instructions how to do it? A. Yes, sir.

Mr. BAGG.—That is all. [336]

**Testimony of Luther L. Mack, for Defendant.**

LUTHER L. MACK, a witness called on behalf of the defendant, having been first duly sworn, testified as follows:

Direct Examination.

(By Mr. BAGG.)

Q. Please state your name, age, place of residence and occupation.

A. Luther Mack, age, 42; residence, South Pasadena; occupation, Patent Attorney.

Q. How long have you been engaged in the business or profession of patent attorney?

A. About 12 years.

Q. Prior to your going into the business of patent attorney what other business were you engaged in?

A. Mechanical engineer.

Q. And before that what?

A. Well, I was a mechanical draftsman for a number of years before that.

Q. And you are still a draftsman as well as patent attorney? A. Yes, sir, I still am.

Q. I will ask you to state, Mr. Mack, if you have examined the file wrapper in connection with the Milon J. Trumble patent, a certified copy of which has been introduced in evidence and marked Plaintiff's Exhibit No. 2. A. Yes, I have.

(Testimony of Luther L. Mack.)

Mr. F. S. LYON.—I object to that as immaterial and not the proper subject of expert testimony. The file-wrapper can only be used in any case to show the prosecution of the Trumble application in the Patent Office, and on its face it shows what it is relevant for.

The COURT.—I don't know what is coming. I will allow [337] the question so far.

Mr. BAGG.—I was just asking him if he had examined it, that is all, your Honor, right now.

Q. I will ask you to state if you have examined the certified copy of the patent issued to Albert T. Newman and marked Defendant's Exhibit "J" which I hand you (handing same to witness)?

A. Yes, sir; I have.

Q. I will ask you to explain to the Court from the drawings and specifications as set out in that patent how that instrument or device described therein operates.

Mr. F. S. LYON.—We object to that on the ground that it is incompetent, no foundation laid, the witness not have qualified to answer the question. He has simply stated that he is a mechanical engineer and not that he knows anything about gas traps or their purpose, function or effect. He is merely a patent attorney and is a scribner or preparer of specifications for patents.

Q. (By the COURT.) I suppose you have had experience in obtaining patents, making drafts and supervising drafting, and so forth?

A. Yes, sir, for 12 years.

(Testimony of Luther L. Mack.)

Q. And have obtained patents?

A. Yes, sir; many of them.

The COURT.—I will overrule the objection.

Q. (By Mr. BAGG.) Proceed now and explain to the Court just how that patent works.

A. In the Newman patent there are two cylinders or receptacles marked "a" and "b" which are similar or nearly similar in size and arrangement, the lower part of which is adapted to receive oil and the upper part forms a chamber for the reception of gasses and oil. The oil and gas are admitted to the two cylinders through pipes "g" and "h" which project downwardly to the tops and are divided into a plurality of nozzles or jets marked "g-2" and "h-2" in the patent. The interior of the receptacles "a" and "b" have deflectors marked "K-prime" in the [338] cylinder "a" and "L-prime" in cylinder "b," against which the force of the oil is projected, causing the oil to be broken up, and so as not to disturb the volume of oil below the deflectors, the oil that is discharged through the bottom there feeding "cc" and outwardly and through a pipe "e." There is a float in one of the cylinders "b" not shown in cylinder "a," marked "m," which automatically regulates a valve "o" for discharging the oil therefrom. Of course this is from a superficial examination of this patent as described in the application for patent.

Q. Now, Mr. Mack, I will ask you to state if you know where the gas would come off, or where

(Testimony of Luther L. Mack.)

would the gas come, after it struck this baffle-plate or deflector?

A. The oil and gas come in together through the pipes "g" and "h," and the gas is released into the upper chamber or expansion chamber by the cylinders "a" and "B" and then is discharged therefrom through the next "a–3" and "b–3" and pipes "jj" which lead to a common pipe "j–2."

Q. Now after the oil strikes this deflector would any oil or gas come off from below the deflector?

Mr. F. S. LYON.—We object to that as incompetent, the witness not having qualified to answer the question. He has not shown that he knows anything about gas separators or that he has had any experience in oil; and, further than that, that inasmuch as these patents are offered in evidence solely for the purpose of showing the state of the art they are good for that purpose only so far as they show on their face themselves, and there is a written description connected with them which describes the operation and all.

The COURT.—I will let him answer the question, because the man is an expert, and his testimony may throw some light on the question we are trying to determine.

Mr. F. S. LYON.—It is not shown that he knows, actually, anything about what would be the operation in an oil well, with [339] oil and gas mixed, or that he ever had anything to do with oil wells or gas. He may be a mechanical engineer, and he may be a patent solicitor, but I doubt very much if every

(Testimony of Luther L. Mack.)

patent attorney or every mechanical engineer knows the action of this intermingled natural gas and oil coming out of oil wells. We have a description here in this particular device of the patent, second paragraph. It says: "Operation of the device is as follows," and it is described right there, and I don't think it is competent to contradict or change that description at all, what was given to the world by this patent. In other words, this is a part of the art. It is what was shown in the drawings and what was described there.

The COURT.—I suppose the Court will be governed by what is shown there, but, as I have remarked before, this man is an expert and is here for the purpose of elucidating that patent, and I will overrule the objection. You may answer.

(Last question read.)

A. Well, the force of the jet of oil and gas which comes through the pipes "g" and "h" would be downwardly, and if there were pressure in those pipes it would be projected very forcibly against this deflector "K-prime" and the hood or the deflector "L-prime" in cylinder "b." Naturally, the gas would then be liberated from the oil and would rise in expansion chambers and be discharged through the neck "a-3" while the oil would go down over the inclined surfaces of the deflector "K-prime" beyond the wall and then downwardly into the oil body below. The oil would strike and then subdivide in such way that—

The COURT.—What is the date of that patent?

(Testimony of Luther L. Mack.)

Mr. F. S. LYON.—1907.

Mr. L. S. LYON.—We make the further objection to this patent that this is not in the art to which this invention pertains. It is not the separating of oil and gas,—it is for a water well. It starts, at the top of it here, "water and gas separator." It has nothing to do with the oil industry at all, and does not belong to this art, and it is introduced solely to show the state [340] of the art and does not show what is the state of this art at all.

Mr. BAGG.—If your Honor please, it is an instrument for the purpose of separating gas from water, which is very similar, and we take it it would be equivalent, because of the fact that the water comes from these artesian wells impregnated with this gas and it becomes necessary to separate the gas from the water, and it is the same kind of gas, as I understand, that comes with the oil, the only difference being that in the oil and gas separator it separates it from the oil and in this case it separates the gas from the water, but the action is the same.

Q. (By the COURT.) Could that be used for the separation of oil and gas? A. Yes, sir.

Q. You are of that opinion, are you?

A. I am of that opinion; for the reason that, if my memory is correct, this is one of the citations mentioned in the file-wrapper. I am not positive about that.

Q. In the file-wrapper of the patent suit?

(Testimony of Luther L. Mack.)

A. Of the patent in suit; yes, sir.

Mr. L. S. LYON.—We make the further objection, your Honor, that is the defendant is estopped in this case claiming that the action of water is the same as oil. The Court ruled out testimony in regard to experiments when the defense stated that the action of water was not a fair test or representation of what oil would do in one of these devices.

The COURT.—Well, this patent has already been introduced without objection.

Mr. BAGG.—Now, if your Honor please, that is the only one we have with reference to water anyway; the rest of them are oil and gas separators.

The COURT.—Very well. You had better take up another patent. [341]

Mr. BAGG.—Yes.

Q. Now, Mr. Mack, I hand you Defendant's Exhibit "E," which is a certified copy of the letters patent issued to George L. McIntosh, assignor, and ask you to describe its action or the action of the device described therein when it comes to separating oil and gas.

Mr. F. S. LYON—If you Honor please, may we understand that our objection to the testimony of this witness, as to its foundation, and so forth, in regard to any attempt to explain these devices, is in, the one your Honor overruled a moment ago, and our exception?

The COURT.—Yes.

Mr. F. S. LYON.—On the ground that he has not

shown any familiarity with the action of oil and gas and has had no experience in the oil business whatever.

The COURT.—Very well; the objection will be overruled.

Mr. F. S. LYON.—And that will go to all these inquiries on the same line?

The COURT.—Yes.

Q. (By Mr. BAGG.) Now explain to the Court how that is—

A. This device has a receptacle marked "4." The device is called a mechanism for separating gas from liquids, with an inlet marked "3" at the bottom, substantially in the center of the receptacle, and projecting upwardly thereinto. The inlet has two or three deflectors on its upper end, against which this volume of liquid and gas projects when they overflow from the upwardly-projecting end, and the volume of oil collects in the lower part of the receptacle while the gas collects in the upper part. There is an outlet for the gas, marked "9," which has a hooded opening "10" above the level of the oil through which the gas is discharged from the expansion chamber, and the oil is taken off through an outlet "7" on the side.

Q. I will ask you to further examine the upper portion of this trap or device and explain to the Court the construction [342] of this upper portion, this hood here.

A. Well, the receptacle is composed of two parts, one of which telescopes into the other, the lower

(Testimony of Luther L. Mack.)
part being the larger in diameter, and the top of the upper part is crowned.

Q. I will ask you to state what the effect, if any, of this sliding or this telescoping of this upper portion into the lower portion would be upon the pressure in the expansion chamber, as you call it, and particularly in the gas-collecting chamber?

A. Well, I couldn't answer that question, I believe. I don't understand that those upper and lower portions are relatively movable. Of course if the upper portion "8" were movable with respect to the lower portion, a pressure which might be created in the upper portion would cause the elevation of the one with respect to the other.

Q. And what would be the effect as to maintaining a pressure in that receptacle?

A. Well, of course the weight of the upper portion would to a certain extent serve to maintain a pressure.

Q. A constant pressure?

A. A constant pressure.

Q. (By the COURT.) Is that air-tight between the two compartments?

A. I don't understand it to be. But that would be the effect in that case.

Q. Has that patent been used with success?

A. I don't know.

Q. (By Mr. BAGG.) I hand you herewith certified copy of letters patent of the United States dated January 16, 1912, and issued to Eustice Vivian Bray, Assignor, and ask you to describe the

(Testimony of Luther L. Mack.)

workings of that to the Court (handing paper to witness).

A. This device is for the purpose of separating gas from liquids and includes a receptacle having a conical lower portion "1" which is adapted to receive and hold a volume of oil or [343] liquid, and an upper portion "2" and a conical top "3." The oil and gas, or the liquid and gas, is admitted into the center of the top portion "3" through the pipe "16" and is projected downwardly into the lower portion "1." The bottom of the tank has a valve "5" that seats in an outlet "4" and is controlled by the level of the liquid through the agency of a float marked "7." The upper portion "1" of the receptacle has a plurality—three being shown—of imperforate cones, one above the other. The liquid collects in the bottom after having been projected on to and percolating down through these openings in the cone, and the gas rises into the top and is discharged through a pipe marked "17."

Q. Does that run down the wall as it passes through those cones?

A. In my opinion a quantity of that oil would run down the walls of the receptacle at least from a point immediately below the lowermost cone, only provided the apertures which are marked in the cone are so remotely spaced from the wall that it might by gravity drip off into the bottom. But, as shown in Figure 3, the apertures are very close to the side of the wall, and in my opinion at least

(Testimony of Luther L. Mack.)

a film of the liquid would run down underneath the surface of the lower cone and thence downwardly over the outer walls into the liquid. There is an overflow pipe marked "20" which connects with the side of the receptacle and projects downwardly below the normal level of oil for removing any surplus amount of oil in case the float "8" does not perform its work properly.

Q. Now, I will ask you to state whether or not this float "8" is so constructed as to maintain an oil level above the intake part of the oil outlet.

A. The float is so arranged as to at all times, under normal and proper operation, keep the mouth "22" of the overflow outlet submerged.

Q. I hand you herewith Defendant's Exhibit "I" which purports to be a certified copy of the letters of the United States [344] to August Steiger Cooper, dated March 20, 1906, and ask you to explain its operation to the Court (handing same to witness).

A. This is a device for the separation of oil, water and sand as done under that patent, and includes an elongated cylinder "A" which is arranged to receive oil, water and sand in the bottom of the receptacle, and has an inlet marked "b" in the form of a pipe which is arranged tangentially with respect to circumferences of the cylinder. This pipe has a continuation in the form of a blade, marked "A," so that the force of the oil and gas and other elements may be directed into the tank in a line tangent to the circumference of

(Testimony of Luther L. Mack.)

the tank and will continue. The oil will then be discharged on to the inner surface of the tank in sheet form and will be continued in a circle around the inner side, from whence it will drop downwardly into the oil below over the wall of the tank. The gas is discharged from the upper chamber of the tank above the level of the oil through a pipe marked "F," and the lower end has a draincock or sediment-cock marked "A-3" through which the sediment may be drawn off. The oil is taken out through a submerged outlet pipe "C" with a goose-neck on the inner end so that when that outlet is submerged and as the oil reaches a certain level it will drain off. It will nominally maintain a certain level of oil in there, and there is a valve marked "C" which is controlled by a float for regulating the discharge of oil.

Q. Now, as the oil level rises in the bottom of this receptacle, what effect does that have upon the float and the valve operated by that?

A. Well, the float will rise and the valve will open so as to discharge the surplus amount of oil.

Q. And what takes place when the oil gets below the predetermined point?

A. The reverse operation of the float will be effected and the valve closed.

Q. What, if any, effect has this peculiar arrangement at [345] the mouth of the inlet pipe with reference to the effect that it has upon the incoming oil and gas as it strikes the walls of the expansion chamber?

(Testimony of Luther L. Mack.)

A. Well, it would spread the volume of oil out over the inner surface of the tank in a thin sheet or film, and then, of course, by gravity the oil so spread would gravitate down into the body of oil below.

Q. And where would the gas separate from the oil in this particular device?

A. It would separate during the time this was spread out into a sheet. That is the purpose, as stated in the patent, for spreading it, so that the gas would be liberated from the oil.

Q. Then it would come from within this film of oil as it flows down the side wall of the receptacle?

A. Yes, sir.

Q. Is there a valve in connection with the gas outlet? A. There is no valve shown.

Q. I hand you Defendant's Exhibit "G," which is a certified copy of letters patent issued by the United States to Arthur W. Barker, granted July 13, 1909, and ask you to explain to the Court the operation of this particular trap (handing same to witness).

A. This device has a—

The COURT.—What is that called?

The WITNESS.—Natural gas separator for separating gas and oil.

A. (Continuing.) —the device includes a cylinder "1" to which oil is supplied through pipes marked "4," "5" and "6," a valve "11" being interposed between sections "4" and "5." The portion "6" of the inlet pipe is directed upwardly

(Testimony of Luther L. Mack.)

into the tank and has a spreader "7" on the top so that the force of the oil which is discharged from pipe "6" will be spread outwardly from the top and the oil broken up so as to liberate the gas and collect it in the upper portion of the cylinder. The oil is [346] collected in the body of the cylinder and is taken off through an overflow pipe marked "9" having a vent "10." Gas is taken off through a vertical pipe marked "12" having duplex nozzle "13" on the top, so that as the gas collects in the upper portion of the chamber it will be discharged through the portion "13" downwardly and direct to any suitable point. There is a drain pipe "3" with a valve on it at the intake tank, also for draining off the sediment or the oil when it gets below the pipe "9."

Q. (By the COURT.) That doesn't show the oil settling clear across the tank.

A. No, it doesn't, but there would be.

Mr. BAGG.—That is all.

Cross-examination.

(By Mr. F. S. LYON.)

Q. Did you ever see any of the traps in these prior patents to which you have referred in your direct examination in operation? A. No, sir,

Q. Do you know whether any of them were actually used? A. No, sir.

Q. Referring now, first, to the Cooper patent No. 815,407, Defendant's Exhibit "I," are you familiar

(Testimony of Luther L. Mack.)
with the principle upon which the well-known Cyclone Dust Collector is operated and based?

A. No, sir.

Q. Do you know what would be the effect and result upon the shell of this Cooper device of the projection of the blast of intermingled oil, sand and gas against it in the manner of the tangential blast referred to in this patent?

A. You ask me if I know. I wouldn't say that I know. I would say I believe.

Q. You have had no experience with it?

A. I believe it is as I have stated. [347]

Q. You have had no experience with it?

A. No.

Q. Are you able from any experience to tell us whether such blast of sand, oil and gas would cut through the side of such container and render the device inoperative? A. Not with oil and gas; no.

Q. You don't know? A. No. I do not.

Q. Then you agree that the operation of this device would be as described in the specification of the patent, and that is that: "Gas, oil, water, silt and sand are forced from near the bottom of the well by the compressed gas and pass through the pipe B into the cylinder A. Entering the cylinder under pressure tangentially to its circumference, the material thereby given a rotary motion, which separates the constituents conformably to their specific gravities. The gas rises and passes from the cylinder through pipe F to the compressor. The sand seeks the outer circle and drops down to the

(Testimony of Luther L. Mack.)

bottom, as is indicated by "S." You agree that that would be the operation, do you?

A. No, not necessarily.

Q. Are you able, from any experience you have had, to state what effect the scrubbing action of sand in the mode of operation specified by Mr. Cooper would mean?

A. Any matter, any element or any substance projected into a receptacle like that or given a motion like that would naturally seek the outer surface, theoretically. It doesn't always do it. Now, with oil and sand I am not able to state how the thing would work. If you would ask me how a sheet of water would be thrust around the inside of that, I would say unquestionably it would be so, because I have done it in experimenting.

Q. I ask you that question in regard to the well-known Cyclone Dust Collector, because that is the device used for separating air from shavings and sawdust, and so forth, in lumber-mills and all kinds of dust in factories from the air—the [348] same tangential whirling device. Now, I have also asked you the question whether you knew what would be the effect of the sand in an oil well being whirled about and thrown against the inner periphery of the shell of the trap. Can you state any knowledge in that regard?

A. I cannot state any knowledge; no, sir.

Q. You are then unable to state whether or not the reason why this Cooper trap never went into

(Testimony of Luther L. Mack.)

use was because it would immediately wear out due to the attrition of the sand on the metal, are you?

Mr. BAGG.—Now, we object to any inquiry with reference to that because of the fact that the Cooper trap "has not gone into use." There is no evidence here to show it has not gone into use. It is a patented article and I don't think it makes any difference whether it went into use or not.

The COURT.—Strike out the assumption that it has not gone into use and ask your question without it.

Q. (By Mr. F. S. LYON.) Well, answer the question without the assumption that it has not gone into use, then.

A. I would like the question read as it now stands.

(Question read.)

Mr. F. S. LYON.—I will restate the question.

Q. You are unable to state from any practical knowledge, then, whether or not the grinding action of the sand delivered with the incoming oil and gas would erode and destroy the shell of the trap if constructed and operated as disclosed and described in this Cooper patent, Defendant's Exhibit "I," are you?

A. No; I am not unable to state it. Of course there is an abrasive action on any parts that would be exposed to sand or any last, under pressure—no question about that—and a gradual wearing away. But how long it would take it to wear away the cylinder for that purpose would depend upon

(Testimony of Luther L. Mack.)

its thickness, hardness and temperature, and so forth.

Q. You are not familiar, are you, with any of the working [349] conditions in oil wells due to the sand in the oil?

A. Why, no, I can't state any positive knowledge of conditions at an oil well or of the operation of any of these devices from actual experience. It is theory.

Mr. BAGG.—If your Honor please, I think I will object to this line of examination for the reason that it is immaterial and does not tend to prove or disprove any of the issues of this case. Whether or not the Cooper trap went into operation has no bearing upon the state of the art. The mere fact that it was once patented and described and the disclosures made in the Patent Office would be sufficient to put all the world on notice with reference to its method of operation and whether it operated or not has nothing to do with this case, and we are not trying to determine the validity or operativeness of the Cooper patent. It simply would make no difference whether it was operative or not. Whether it operated or not wouldn't make any difference. We are not trying to set it aside. But I think this line of testimony is unnecessarily encumbering the record and is irrelevant and immaterial.

The COURT.—Well, this witness has elucidated these patents, and this may go as his knowledge in the premises.

(Testimony of William A. Trout.)

Mr. BAGG.—Yes; but so far as the use of the Cooper patent is concerned, or whether it ever operated or not, that wouldn't cut any figure.

The COURT.—I will allow the question.

(A recess was thereupon taken until two o'clock P. M.) [350]

AFTERNOON SESSION—2 o'clock.

The COURT.—You may proceed, Gentlemen.

Mr. BAGG.—If your Honor please, we have a witness here from out of town, and who holds a very important position, and it is necessary for him to remain away from his business as short a time as possible, and counsel have agreed that we may put him on out of order at this time.

The COURT.—Very well.

**Testimony of William A. Trout, for Defendant.**

WILLIAM A. TROUT, a witness called on behalf of defendant, having been first duly sworn, testified as follows:

Direct Examination.

(By Mr. BAGG.)

Q. Please state your name.

A. William A. Trout.

Q. Where do you reside?

A. No. 1917 Catalina Street, Long Beach, California.

Q. What is your business?

A. Master mechanic for the Shell Oil Company.

(Testimony of William A. Trout.)

Q. How long have you been engaged in the oil industry?

A. Fourteen years the 8th of next month.

Q. During that time have you had any experience with the building or installing of oil and gas traps or separators? A. I have.

Q. To what extent have you had experience along that line?

A. Well, I have worked on and remodeled one or two traps and have built one or two of my own invention.

Q. Do you remember the occasion of installing a trap up at Coalinga for the company you were employed by along in the year 1914?

A. I do. [351]

Q. I will ask you to state when you installed that trap, the exact date.

Mr. F. S. LYON.—May I ask the purpose of the examination?

Mr. BAGG.—I just want to fix the date prior to the time of your invention.

Mr. F. S. LYON.—This is not supposed to be a prior device, is it?

Mr. BAGG.—Yes, sir.

Mr. F. S. LYON.—We object, then, on the ground that it is not admissible under the pleadings.

Mr. BAGG.—It is for the purpose of showing the state of the prior art.

The COURT.—And only for that purpose?

Mr. BAGG.—And only for that purpose.

(Testimony of William A. Trout.)

The COURT.—Very well.

Q. (By Mr. BAGG.) Can you fix the exact date, or nearly the date?

A. The first trap was built and installed in the last two weeks in April, 1914, prior to the 2d of May, and in operation.

Q. I will ask you if you can make a drawing or sketch of that trap as it was installed at that time. A. Absolutely.

Mr. BAGG.—I have a sketch that the gentleman made in my office, but if counsel on the other side objects to my allowing the witness to use it I will have him make a sketch in the presence of the court (handing paper to counsel).

Mr. F. S. LYON.—I think you had better have the witness make his sketch. I don't think that one is a full sketch of it anyway. He may use it, or part of it, if he made it; I don't know.

Mr. BAGG.—Well, I will have him make a drawing of the trap itself.

Q. I wish you would make a drawing or sketch of that trap for the benefit of the Court. [352]

(Witness draws sketch.)

A. I think that is the essential or main points (handing paper to Court).

Q. Now, I will ask you to explain to the Court how this trap operates as you have described to the Court.

A. The gas and oil are brought in through this passage here opening into the side of the trap.

(Testimony of William A. Trout.)

Q. Better put a letter there for that so that the reporter can get it.

Mr. F. S. LYON.—Start with the numeral 2.

A. Two. (Marking.) And it is propelled around the shell of the trap by centrifugal force. Underneath this inner sleeve, whose head extends three-quarters of the way around, the gas going around with the oil rises through the opening at the top between the outer shell and the inner sleeve. Shall I mark that—

Mr. F. S. LYON.—Three.

A. (Marking.) The oil being propelled around the outer shell by centrifugal force, of its own weight falls to the bottom, carrying sand with it, and in turn, passes out through the valve in the bottom.

Q. (By the COURT.) All of the oil?

A. All of the oil and sand. The gas, rising through the opening in the top as described, passes upward through a series of baffle-plates and out through a top connection to a back-pressure valve.

Mr. F. S. LYON.—Mark that "4."

A. (Marking.) The fluid level in the trap—shall I mark that "5"?

Mr. F. S. LYON.—Yes.

A. (Marking.) —is maintained by a float which is connected by a series of levers to a rod—

Mr. F. S. LYON.—Six.

A. (Marking.) Which, in turn, connects the bottom valve to a sliding piston in the top of the trap.

(Testimony of William A. Trout.)

Will it be necessary to go into the purposes of those— [353]

Mr. BAGG.—Yes; if you will.

A. The purpose of this connection and the cylinder and sliding piston in the top are to balance the valve so that the opening and closing of the same will not be affected by varying pressures. The rod, piston and levers are supported by a spring connected thereto on the top of the trap, and which supports the weight or which maintains tension sufficient to balance the weight of the valve, rods, levers and piston.

Q. Now, Mr. Trout, in this sketch which you have here I believe you stated that the oil came in at one side and was sent in or came in at a kind of tangent to the line or side of the outer wall of the receptacle and was whirled around by the centrifugal force, and the oil gradually flowed down the side-walls of the receptacle to the bottom, and the gas, where did that separate from the oil?

A. During its revolution around the side-walls of the trap.

Q. Inside of this—

A. Yes. The oil, being heavier, followed the outer wall by centrifugal force while the lighter gas went to the center and separated.

Q. Now, you say that trap was installed prior to May 2, 1914? A. Yes, sir.

Q. Do you know when the trap was built?

A. The trap was built after the 15th of April, 1914.

(Testimony of William A. Trout.)

Q. Where was that trap installed?

A. On Well No. 17, 2, on what at that time was known as the W. K. Oil Company, now the Shell Company of California.

Q. That was in California here?

A. In California, in the Coalinga field.

Q. Is that trap in operation yet?

A. That I couldn't say. They have been shifted around considerably, and whether that particular one is in operation or not I don't know. That is, the first one. But there have [354] been a number of them in operation.

Q. Since that time? A. Yes, sir.

Q. You installed that trap, did you?

A. I did, sir.

Q. I will ask you to state to what extent that trap was successfully operated.

A. Well, it was not a huge success on that well, owing to the fact that the well was flowing by large heads and it didn't have capacity enough to handle the sudden rushes. Its size was not great enough. We found afterwards in some places by installing two of them in a battery they would operate on that kind of a well.

Q. And did you install it on another well, or was it removed to some other well?

A. It was moved around the lease, yes. It was moved to two or three other wells.

Q. And did it operate successfully on them?

A. Yes.

(Testimony of William A. Trout.)

Q. What was the capacity of this particular well you speak of?

A. Why, it flowed—the greatest flow I remember of was 2,300 barrels a day, but it came in large heads. It would lay off for hours and *the* come in a big head, and what the rate was while it was flowing in heads we had no way of measuring.

Q. Then the only objection there was to this trap was the fact that it didn't have sufficient capacity to handle this flow of oil?

Mr. LYON.—That is objected to as leading and suggestive and incompetent.

Mr. BAGG.—Take the witness. [355]

Cross-examination.

(By Mr. F. S. LYON.)

Q. You say, Mr. Trout, you are master mechanic of the Shell Oil Company of California?

A. At Long Beach.

Q. At Long Beach? A. Yes, sir.

Q. And who do you work under there?

A. Mr. William McDuffy.

Q. You are acquainted with Mr. Lorraine, defendant in this case? A. I am.

Q. You have on your property, of the Shell Oil Company of California at Long Beach Lorraine traps, have you? A. We have.

Q. Did you know they are involved in this litigation? A. I have heard so.

Q. What pressure are those Lorraine *trap* operated under on your wells at Long Beach?

(Testimony of William A. Trout.)

Mr. BAGG.—We object to that; that is not cross-examination.

The COURT.—What is that question, what pressure?

Mr. LYON.—Yes.

The COURT.—I think that is outside of the examination in chief.

Q. (By Mr. LYON.) Where was this trap built to which you refer and of which you have made the sketch to-day?

A. It was built at the oil fields, at the Shell Company shops, or at the time it was built the California Oil Fields Limited shops.

Q. Did you build it yourself? A. We did.

Q. What was its size?

A. The dimensions of it?

Q. Yes. [356]

A. Three feet in diameter; the body was four feet long and the runs 32 inches.

Q. And you have correctly, according to your best recollection indicated the oil level that would be kept upon this trap in the sketch you have produced?

A. That varied according to conditions. It was adjustable. The oil level was adjustable.

Q. Answer the question: Is that the level you kept on it as you used it as you have indicated there?

A. Roughly so, yes; somewheres near there.

Q. Well, within what limit?

A. Oh, within a couple of feet.

(Testimony of William A. Trout.)

Q. When did you come to California?

A. 1897.

Q. And from where? A. Oregon.

Q. With what company were you in 1897 in California? A. I was not with any company.

Q. What did you do in 1897?

A. In 1897 I was going to school.

Q. And when did you finish school.

A. About 1903.

Q. And after school what did you do?

A. Went to work.

Q. For whom first?

A. Worked for John H. Blakeway in San Francisco.

Q. When did you start to work for Blakely?

A. Oh, let's see; in the fall of 1902, some time.

Q. And what time in the fall of 1902?

A. That I don't just exactly recall.

Q. What do you mean by the fall, what time of year? A. The latter three months of the year.

Q. Are you unable to tell us within three months of when you went to work for him? [357]

A. Yes. I couldn't say any closer than that.

Q. What did you do for Blakely?

A. Went to work as an apprentice.

Q. At what? A. Sheet metal work.

Q. How long did you continue with him?

A. About a year, as nearly as I can remember.

Q. You say about a year. Was it a year or nine months or longer?

A. That I cannot say exactly.

(Testimony of William A. Trout.)

Q. What time of the year did you leave him?

A. That I don't remember.

Q. You left in 1903?

A. Somewheres about that.

Q. Or was it 1904 that you left him?

A. I wouldn't be positive of the date.

Q. Who did you next work for?

A. Worked for the Tay Pipe Company.

Q. When did you go to work for the Tay Pipe Company?

A. Immediately when I left Blakeway.

Q. And where? A. San Francisco.

Q. What did you do for them?

A. Sheet metal work.

Q. How long were you with them?

A. Oh, a matter of three or four months.

Q. Was it three or four months?

A. I couldn't say positively.

Q. You left at the end of that time? A. Yes.

Q. Where did you go to work next?

A. I cannot tell you the name of the next shop I have worked at; I have forgotten.

Q. You don't know how long you worked there? [358]

A. Yes. I worked four or five months or some such a matter.

Q. For whom did you work after that?

A. Worked for John Lawson & Sons.

Q. That was in your next employment?

A. Yes.

(Testimony of William A. Trout.)

Q. When was it you went to work for John Lawson & Sons? A. I don't recall the year.

Q. You don't recall the year? A. No.

Q. How long did you work for them?

A. A year or some such a matter.

Q. Was it one year or two?

A. I think a year, or possibly less.

Q. Was it more or less than a year?

A. That I couldn't say.

Q. Who did you work for next?

A. Taylor & Pritchard.

Q. Where? A. San Francisco.

Q. When did you go to work for them?

A. Immediately after I left Lawson's.

Q. Are you unable to give the date?

A. I couldn't give you the dates, no.

Q. Nor the year? A. Huh?

Q. Nor the year?

A. What year? I couldn't say.

Q. How long did you work for Taylor & Pritchard?

A. I worked for them about three or four months.

Q. Doing what, sheet metal work there?

A. Yes.

Q. After you left Taylor & Pritchard what did you next do?

A. Went to work for the Pacific Blower & Heating Company.

Q. At San Francisco? [359] A. Yes.

(Testimony of William A. Trout.)

Q. What year did you go to work for that company? A. I think the spring of 1905.

Q. You say the spring of 1905. Are you able to fix the date any nearer than that?

A. No. That is about as near as I could fix it.

Q. Did you ever set any gas trap? Did you build the one you say this sketch is like?

A. Yes, sir.

Q. What trap? A. McLaughlin trap.

Q. Any other?

A. The one we call the Daubenspeck trap.

Q. Any other? A. No.

Q. What were you doing in the year 1911?

A. In 1911 I was working for the California Oil Fields, Limited; I was foreman of their tank department.

Q. (By Mr. BAGG.) You say you were foreman of their tank department? A. Yes, sir.

Q. (By Mr. LYON.) Have you any record with you showing the date upon which this alleged first trap was made? A. I have a note in my book.

Q. A note what? A. A note in my note-book.

Q. Have you any original drawings of anything to show its construction? A. I have not.

Q. The trap you have illustrated in this sketch that you have made, has it any name?

A. Well, it has been called the Durward gas trap.

Q. Was it ever patented?

A. It was not. [360]

Q. Where did it get the name of Durward?

A. Mr. Durward was superintendent of the prop-

(Testimony of William A. Trout.)

erty soon after we started building them and he had drawings made at one time and attached his name thereto.

*A.* As a matter of fact, when were the drawings made?

A. I think the drawings were made in 1915 some time.

Q. Where are those drawings?

A. That I couldn't say; there are some of them in existence at the Shell Company's property at Coalinga, I believe.

Q. Where is Mr. Durward?

A. He is still there.

Q. As a matter of fact, Mr. Trout, on that property prior to building this trap you had a Trumble gas trap, didn't you? A. No, sir, I did not.

Q. Had not seen one around in that field at all?

A. Never did.

Q. How long after that was it you first had a Trumble gas trap?

A. It was some months afterwards; after we had set up our trap, experimented with it considerably and taken it off the well and the first Trumble trap was put on the same well.

Q. And put in the place of this trap, wasn't it?

A. Exactly.

Mr. LYON.—That is all.

Redirect Examination.

(By Mr. BAGG.)

Q. The Trumble trap, what kind of luck did you have with that?

(Testimony of William A. Trout.)

A. Well, it sanded up the same as the other trap had.

Q. You had the same trouble that you did with the other one?

Mr. LYON.—We object to that as leading and suggestive.

The COURT.—Did you have trouble with that trap? [361]

A. Not a great deal. We had some trouble. The first time we put it on it sanded up. The trap, though, stayed there for quite a long while in operation.

Q. (By Mr. BAGG.) Did this trap ever go out by reason of the sand striking against the side walls of the receiving chamber?

Mr. LYON.—We object to that as leading and suggestive and not redirect examination.

The COURT.—He may answer the question.

Q. (By Mr. BAGG.) Did it?

A. Which trap do you refer to?

Q. This trap you have drawn there.

A. No, sir.

Mr. BAGG.—That is all.

Mr. LYON.—That is all. I will ask that the witness remain in attendance.

The COURT.—Very well. You will remain here.

Mr. BAGG.—If your Honor please, would like to introduce this drawing which the witness has made.

The COURT.—Very well.

The CLERK.—Exhibit "K." [362]

**Testimony of L. L. Mack, for Defendant (Recalled —Cross-examination).**

L. L. MACK, recalled.

Cross-examination.

(By Mr. F. S. LYON.)

Q. Referring, Mr. Mack, to Defendant's Exhibit "G," the Barker patent No. 927476. Where did you say the oil and gas entered from the well into this trap?

A. Through the pipes 4, 5 and 6. Thst is the sections 4, 5 and 6 and at the opposite section.

Q. Then the action would be to simply allow the oil to drop down into a body which lies at the bottom of the trap, would it?

A. No. I wouldn't say that. There is a sprayer on top marked 7 here which would spray the oil out latterly from the nozzle there or outlet so that it would spray out and fall down in a shower if at all, which probably would have deflected it towards the sides of the wall probably similar to the trap down the side there and similar to a drop down by gravity into the body into the interior.

Q. And the entrained gas if it was escaping from such oil as the oil fell down would be apt to pass out from the body or descending oil?

A. There would be no body of descending oil. There would be a spray and the gas would be liberated from that spray as it was broken up. That is the object of part 7, to break up from a volume into a stream or spray so as to liberate the gas from the oil.

(Testimony of L. L. Mack.)

Q. With this heavy crude oil do you believe such a sprayer would efficiently break the oil into a spray?

A. I believe it would have a tendency to; I believe it would do it more or less, depending to its extent, depending on the weight of the oil and its gravity and a few other things.

Q. Would an 18 or 20 gravity crude oil which is forced [363] against what you have called sprayer 7 break the oil into a spray?

A. I believe it would; yes.

Q. That is your idea?

A. That is my opinion. I cannot specifically state as to any particular gravity of oil, but that is the best of my opinion, that it would.

Q. The action of this spray in this Barker patent would be to a greater extent then in your opinion comparable to the disclosure of the gas and oil in this patent of McIntosh, Defendant's Exhibit "E," would it?

A. It would be somewhat similar except I believe it would break up in that case, in the case of the device of Barker to an even greater extent than it would here under certain conditions. The object of the character of this outlet being partially disclosed is the same as the outlet of Barker except Barker has a top over the outlet which would tend to spray it more towards the sides.

Q. Referring to the Brea patent, 1,014,943, Defendant's Exhibit "H," did you say it was your un-

(Testimony of L. L. Mack.)

derstanding that the members, 1, 11, 12 and 13 were imperforate holes?

A. They are perforated cones, yes.

Q. I think also—

A. Well, I perhaps use that wrongly. I meant to say a perforated cone, at least they had perforations in them. I think I did say imperforate when I should not have said it.

Q. As a matter of fact, the description of this device as contained in the patent is a cylindrical casing which is perforated with a series of brackets forming, a number of said brackets at different levels a support and are screens 11, 12 and 13, the latter screen having three openings shown within the lower spray, is that correct?

A. Well, I cannot tell much of the size of the openings.

Q. They are described as screens instead of imperforated cones? [364]

A. They are described as screens, but they are perforated cones.

Q. Also screens?

A. They are perforated cones. I wouldn't say they were screens at all.

Q. From a practical standpoint you are unable to state how long it would take such screens to fill up with the sand that is contained in the oil, are you?

A. No, I would not be able to say. That would depend entirely on the size of those openings, the perforations, and the number of them, and the relative area of the inlet pipe and the total area of all

(Testimony of L. L. Mack.)

the perforations, as to how long one would fill up or bank up on the top of the cone.

Q. Have you any personal knowledge as to whether the device was ever built or used in accordance with this Brea patent? A. I do not.

Mr. BAGG.—We object to that as incompetent, irrelevant and immaterial; it does not tend to prove—

The COURT.—I think that is immaterial.

Mr. LYON.—Testing his knowledge in regard to each one of those.

The COURT.—What is that?

Mr. LYON.—The question is material whether or not he knows of any of them having been used, goes to the question of how much he knows about the device, if he never had seen it.

The COURT.—You may proceed then.

Mr. LYON.—I want to show, if he knows, what he has seen of this device.

Mr. BAGG.—All he has ever qualified to being is a patent draftsman and having had a large experience in drawing specifications and claims and drawing and making patent drawings that he could understand that operation of them. That is all.

The COURT.—If he knew their use it would be some evidence. That would go to his knowledge. [365]

Mr. LYON.—That is all.

**Testimony of A. A. Wharff, for Defendant.**

A. A. WHARFF, a witness called on behalf of the defendant, being first duly sworn, testified as follows:

The CLERK.—State your name, please.

A. A. A. Wharff.

Direct Examination.

(By Mr. BAGG.)

Q. Where do you live, Mr. Wharff?

A. Route 3, Anaheim.

Q. What is your age? A. Fifty-two.

Q. What business are you engaged in at the present time? A. Superintendent of production.

Q. For what oil company?

A. Merchants Oil Company.

Q. Where is the Merchants Oil Company located?

A. Richfield.

Q. How long have you been employed by the Merchants Oil Company?

A. Since the first of the month.

Q. Since the first of March? A. March.

Q. Where did you reside before you came to California?

A. Near Santa Maria—or before I came to California—in Illinois.

Q. How long have you been engaged in the oil industry? A. Since 1892.

Q. During that time what experience if any have you had with oil wells and oil production and oil and gas separators?

(Testimony of A. A. Wharff.)

A. Well, I have always been with the production end of the business, producing oil. [366]

Q. That is ever since 1892? A. Yes, sir.

Q. Prior to November 14th, 1914, had you ever superintended or been connected with the construction of any oil and gas separator?

Mr. F. S. LYON.—We object to that as leading and suggestive. All testimony, your Honor, in regard to oil use which is merely verbal is to be scrutinized very carefully and it certainly is very necessary that the critical date should not be put before the witness.

The COURT.—This is merely preliminary.

Mr. BAGG.—Yes, sir, that is all it is.

The COURT.—I will allow the question.

Mr. LYON.—You do not intend by this witness to prove any prior use or attempt to, anything that is prior? Because it certainly is absolutely essential that the proof as to dates, if he relies on anything that is prior, he should not allowed to say to the witness that it is prior to a certain date.

Mr. BAGG.—I am simply asking the witness the time prior to the date of their patent and then I will fix the date.

Mr. LYON.—That is exactly the ground of our objection, dates prior to our dates.

The COURT.—If the witness can give his date independently, let him do it.

Mr. BAGG.—Will you read the question, Mr. Stenographer?

(Question read.)

(Testimony of A. A. Wharff.)

A. Yes, we used them.

Q. (By Mr. BAGG.) Just answer the question yes or no. A. Yes, sir.

Q. Now then, will you state to the Court as nearly as you can, giving the dates as nearly as you can and the description of any oil and gas trap or separator that you were connected with or had reason to observe or was instrumental in constructing.

Mr. LYON.—That is purely for the purpose of showing the state of the art. [367]

Mr. BAGG.—Yes, sir.

A. We built some traps in Ohio.

Q. In what year?

A. I think something like 1893 or 1894.

Q. Now, I will ask you to describe to the Court if you can the character of those traps or their method of construction. If you wish to, you can draw it.

(Witness drawing.) The oil entering here.

The COURT.—Speak up louder.

A. The oil entering the pipe here and carrying that line on around.

Q. (By Mr. BAGG.) Mark that A or 1.

Mr. F. S. LYON.—He has marked it "inlet."

Q. (By Mr. BAGG.) All right.

A. The oil would enter here and pass down, well, anywhere from there to there, the four feet, some places we would use them probably four and maybe a little longer, maybe four feet, and throw it against the side of the trap. That trap we built there out of 9 or 10 inch casing—I disremember which—and

(Testimony of A. A. Wharff.)

the oil would flow against the side of the trap, sometimes we had on a 45 degree turn here, and sometimes an ell, which would be at right angles, and then the gas would pass out here. We used the gas in the boilers and there was some pressure on the gas lines all the time. The oil outlet was from the bottom of the trap here clear out and then out, in order to carry this flow of oil, making a level somewheres near level here in the trap, of the oil.

The COURT.—You used no float whatever in that trap?

A. No float whatever in that trap, no, sir.

Q. (By Mr. BAGG.) This trap was constructed you say, about when?

A. In 1893 or 1894, something like that.

Q. Now, suppose you had more than one well that you wanted to connect with that trap, I will ask you to state to the [368] Court how you did that.

Mr. F. S. LYON.—I object to that on the ground it is hypothetical.

The COURT.—How can that be material?

Mr. BAGG.—It would simply show that each well was constructed with an inlet just exactly like that one and sometimes they had three of those sprays, one in each well.

The COURT.—You may answer the question.

Q. (By Mr. BAGG.) Go on and answer the question.

A. By using a cap of three or four openings on the top—they were mostly pumping wells, and we could

(Testimony of A. A. Wharff.)

pump around and we aimed to put the oil from each well separate.

The COURT.—You mean each well would be thrown in there?

A. Yes, it would be the same condition as the first one, with the exception—

Q. (By Mr. BAGG.) Will you indicate how that would be constructed with another inlet with another well just on that drawing?

A. Well, yes, I could. The cap on this head was screwed on here, that opens say three—we often had them bored as many as we wished, as many as we could get connected in a ten-inch sap and they would be put just the same as that.

Mr. BAGG.—Now, if your Honor please, we would like to introduce this drawing which the witness has made and have it marked Defendant's Exhibit.

The CLERK.—"L."

Q. (By Mr. BAGG.) What oil companies have you been employed by?

A. Well, I was with the Caldwell Oil Company near Maxburg, Ohio, something like five years—the exact date I couldn't tell you—and I think after I left them I worked for Murphy, Shaw & Duffy, it was in the same field near Maxburg, Ohio; their offices were in Washington; from there I went to Jettys, Cullen & Dryer, near Lowell, Ohio, and I was with them a number of years. The property was sold to another company and I was with them and they transferred me to Illinois—I think it was June, [369] 1906. That was Bruner-Stewart &

(Testimony of A. A. Wharff.)

Company, Bruner, Brown & Hoag, and Jennings Brothers & Company under one head.

Q. How long were you with them?

A. I think right around four years, and maybe something like five years, possibly—the exact length of time I couldn't tell you.

Q. Since coming to California what oil companies have you been connected with?

A. I worked I think seven days for the Starey Oil Company, up near Schooley, in Ventura County, that was in 1910, November I think, and from there I went to the Santa Maria fields with the Union Oil Company.

Q. How long were you with them?

A. Ten years, seven months and one day last June, the 30th.

Mr. BAGG.—That is all. Take the witness.

Cross-examination

(By Mr. F. S. LYON.)

Q. Mr. Wharff, I judge according to your recollection that this early trap you think was in 1903 or 1904 was made out of 9 inch casing?

A. 9 or 10 inch.

Q. In other words, by casing you mean the pipe that is used in the oil fields? A. Yes, sir.

Q. It was simply made of a piece of pipe in which entered the oil and gas?

A. Yes. Sometimes we put up two and three joints of it alongside the rig.

Mr. LYON.—That is all.

Mr. BAGG.—That is all. [370]

**Testimony of W. H. Swoap, for Defendant.**

Mr. W. H. SWOAP, a witness called in behalf of the defendant, being first duly sworn, testified as follows:

The CLERK.—What is your name?

A. W. H. Swoap.

Direct Examination.

(By Mr. BAGG.)

Q. State your age and residence.

A. My age is 54, my residence, 135 South Walnut Street, Brea, California.

Q. What is your occupation?

A. I am what is called as an oil well warker; I have worked in practically all branches of the business with the exception of refining; I never was in the refining department.

Q. How long have you been connected with the oil industry? A. Since 1887.

Q. During the time you have been engaged in the oil industry, have you had occasion to examine the construction of any oil and gas separators?

A. Yes; a few, not many.

Q. When was that?

Mr. F. S. LYON.—That is for the purpose of showing a prior state of the art only?

Mr. BAGG.—Yes.

A. The first one was in 1897, in Pennsylvania.

Q. (By Mr. BAGG.) Can you draw a sketch of the arrangement?

(Testimony of W. H. Swoap.)

A. I am no draftsman, but I will make a sketch that will be imperfect.

Q. I wish you would draw a sketch as nearly as you can of the arrangement. [371]

A. (Witness drawing.) This was a drum built out of a boiler iron; it was constructed by the company that owned the lease; I don't know as it was ever patented or any attempt made to ever get a patent on it. The oil came in here like this, it was thrown against the side of the wall, ran down on the side of the steel, as far as I know anything about it, because it was impossible to see in there at the time the oil was going in; there was a manhole plate right in on this side of the top. The gas came up and left here; there was a safety valve on the top with a weight on to make a certain pressure on there and also to keep it from blowing up, because the wells that were in there were flowing wells and they were small, approximately five-barrel wells of paraffine base oil, very light oil, and these wells were flowing wells and had to be shut. What I mean by that, they were shut off the greater portion of the 24 hours, open in the morning and left flowing for four or five hours and then shut back and not let flow until the next day, and when the well was opened it exerted a great pressure coming in there and the gas left through this point here and there was a back pressure valve on the line on the ground. The only reason that these traps were installed at that time was to preserve the gas for use, because the extraction of gasoline from fuel oil at that time was only just in its infancy and the gas was not being used for that

(Testimony of W. H. Swoap.)

purpose in that section of the country at all, it was simply to preserve it and carry it out to the boilers to use for fuel.

Mr. BAGG.—I would like to introduce this drawing in evidence.

A. I did not explain how the oil come out there.

Q. Very well.

A. The outlet for the oil was in the bottom. There was no automatic control below there, it was simply what we called a stop cock. The capacity of the tank was enough.to hold the production of the well for the entire day and in the evening before we left for home we would go and open this valve and the pressure that was on the tank would force the oil into the stock tank which was over here and as soon as the oil was out we could close that valve. [372]

Mr. BAGG.—I would like to introduce this in evidence and mark it Defendant's Exhibit "M."

Q. What company are you connected with at the present time?

A. General Petroleum Corporation.

Q. Where? A. The Tonner tract.

Q. Do you know anything about this oil, or Lorraine sep*at*or that has been installed on tract No. 3 on the Tonner Tract?

A. You mean well No. 3?

Q. Well No. 3 at the Tonner tract?

A. Yes, sir. I was taking care of that well when

it was installed there and I am taking care of it still.

The COURT.—What separator was that, Lorraine?

Mr. BAGG.—That is the Lorraine separator installed on the Tonner tract, well No. 3. I wish you would state to the Court the history of this trap so fas as the pressure on the same goes.

A. Well, the gas from the well was being taken to an absorption plant to absorb gasoline out of it there, from a three-inch line. This same line conducted the gas from the other three wells, the whole four wells through the one line went through it. At the absorption plant it is necessary to maintain certain pressures to put it through the absorbers. The absorbers are high standpipes from 20 to 24 inches, I should judge, throughout, and the gas comes in at the bottom and goes out through the top and then it comes down through it and there is a series of trays to break the oil up as it comes down and that absorbs the gasoline out of the gas and then the oil goes through a compressor plant that puts it to a pressure of 120 pounds, something like that, to deliver it in the Southern Counties gas mains for fuel. They need to have two compressors at the absorption plant to transfer this gas from the absorbers it is necessary to have pressure enough that it will go that route [373] and they were carrying about 45 pounds. This particular well, No. 3, didn't have as much gas according to the amount of oil as some of the other wells handled and putting on that much

(Testimony of W. H. Swoap.)
pressure on the well tended to retard the flow so much that it stopped flowing for two or three nights, so the Los Angeles foreman notified the superintendent of the absorption plant that he would have to fix it and reduce the flow therefrom or they would turn the gas out into the atmosphere. The fact of the matter is the well stopped flowing and when I was on the job I turned the gas out and let it go for several hours in order to get it started again. So in order to overcome that they laid a four-inch line from the absorption plant and put it into a main of their own line, a four-inch line in diamaeter to reduce the pressure and they put it through one of the absorbers on which there were five, put it into one of them alone and let the gas escape immediately into the air and not absorb it instead of putting it down into the pressure tank they let it directly into the air and that is the way it is done now, and as soon as that was done that lowered the pressure on the discharge gas line so that there is not any pressure shown on the gauge. There is no doubt a slight pressure but the gauge does not record it.

Q. That slight pressure is due to what?

A. That slight pressure is due to the line friction, I would say.

Mr. BAGG.—That is all.

Cross-examination.

(By Mr. F. S. LYON.)

Q. Then if I understand your testimony correctly, Mr. Swoap, the gas pressure on oil wells,

(Testimony of W. H. Swoap.)

we may say, after they have been producing a while ordinarily fades away, doesn't it?

A. Well, some do and some does not. Now, the No. 1 Tonner well is making almost as much gas today as it was the day it came in. This No. 3 well, however, did fade some. [374]

Q. And that is true generally of all oil wells according to the particular circumstances. They maintain their pressure but the pressure does fade away and you have to eventually go to pumping to get the oil, is that so? A. Yes, sir.

Q. You are using a gas trap you could not use a greater pressure of gas upon it than you have got anyway, I mean pressure from the well?

A. No; unless you had some means of pumping it through.

**Testimony of William G. Lacy, for Defendant (Recalled).**

WILLIAM G. LACY, recalled.

Direct Examination.

(By Mr. BAGG.)

Q. What is your name?

A. William G. Lacy.

Q. And are you connected with the Lacy Manufacturing Company?

A. I am employed by them; yes.

Q. Have you a letter from Mr. F. M. Townsend dated June 14, 1921?

A. I have here, I believe.

(Testimony of William G. Lacy.)

Q. Is this letter which I hand you the original letter which you received or the Lacy Manufacturing Company received from Mr. F. M. Townsend?

Mr. F. S. LYON.—We are willing to admit that. I don't know that this witness knows it. If he shows us the letter. (Examining same.) That is Mr. Townsend's original letter on that date, that we will admit.

Mr. BAGG.—We wish to have this marked Defendant's Exhibit "N."

Mr. F. S. LYON.—That is the reply to it.

Q. (By Mr. BAGG.) No. I wish you would examine this paper which I hand you and state what it is.

A. That is the reply to Mr. Townsend's letter of June 14. [375]

Q. (By Mr. L. S. LYON.) Did you sign this letter?

A. No; that was signed by William Lacy, President, of the Lacy Manufacturing Company.

Q. I thought that your name—that you said your name was William Lacy?

A. I said my name was William G. Lacy.

Q. You appeared in response to a subpoena for William Lacy, President of the Company, did you not?

A. The subpoena was served upon me and I responded to it.

Q. (By Mr. BAGG.) Are you acquainted with the signature of—

Mr. F. S. LYON.—We are willing to stipulate

(Testimony of William G. Lacy.)

that is Mr. William Lacy's signature, as President of the Lacy Manufacturing Company. I am willing to save you the proof on that.

Mr. BAGG.—All right. Now, we wish to introduce this in evidence and have it marked as Defendant's Exhibit.

The CLERK.—Defendant's Exhibit "O."

Mr. F. S. LYON.—And it will be stipulated in regard to the other letter of the Lacy Manufacturing Company also in evidence.

Q. (By Mr. BAGG.) After you received the letter which was introduced the other day by the plaintiff in this case under date of December 20, 1920, what, if anything, did you or your company do with reference to investigating the Trumble oil and gas separator?

Mr. LYON.—We object to that until the witness has shown he has some personal knowledge.

Mr. BAGG.—I don't know that this Mr. Lacy did anything, and I would like to have the foundation laid.

Q. (By Mr. BAGG.) If you know what, if anything, was done.

A. I don't know that the company investigated—

Q. Did you take part in any of those investigations?

A. I was present when those investigations were made, I recollect.

Q. At the company's office? [376]

A. Lacy Manufacturing Company's office.

(Testimony of William G. Lacy.)

Q. Read the question, Mr. Stenographer.

(Question read.)

Q. Answer the question; what did you do?

A. We secured an opinion from a patent attorney by the name of Keene, a written opinion.

Q. I will ask you to examine this letter which I hand you, and state whether or not that is the original letter which you received from Mr. Keene?

Mr. F. S. LYON.—That is objected to as immaterial.

The COURT.—Well, it might be.

Mr. BAGG.—We would like to show they acted in good faith and on the advice of counsel.

(Discussion.)

Mr. F. S. LYON.—Now, the advice of counsel would not be competent, and cannot be introduced here at all.

Mr. BAGG.—I think it would be most competent, because it would show with the advice of counsel there that the Lorraine patent did infringe on the Townsend patent, and it would be a strong building up of their case.

The COURT.—But his evidence would not be competent.

Mr. BAGG.—And we will ask to have introduced the written opinion sent by him.

The COURT.—That could not be evidence of the fact and it could not be evidence of the advice itself.

(Discussion.)

The COURT.—I think the letter can go in for that one purpose, and that is to show that Mr. Lacy exercised due care in the construction of this patent to see that he was not infringing.

Mr. BAGG.—Yes, sir.

The COURT.—It can be put in for that purpose only.

Mr. BAGG.—That is our purpose.

Mr. LYON.—May I see the letter? (Examining same.) I am willing to permit you to use this to show that you acted under it [377] but I object to it more, your Honor, because I don't know the alleged writer of it and I don't know if he is even a lawyer. As a matter of fact, I never have heard of him, and consequently know nothing as to whether this is the advice of an attorney at law. There should be some proof of it, and not allow it to go in, that they acted on that advice. If they want to prove merely that it was the advice of a lawyer—

The COURT.—The witness has already stated he took the advice of a lawyer, and I presume the man that wrote this letter is a lawyer.

Mr. BAGG.—Under the laws of this state it would be extremely dangerous for him to have a letter-head of a lawyer and he not being entitled to practice in this state.

Mr. F. S. LYON.—I notice this doesn't say "Attorney at Law" on it, and while—

The COURT.—Who was the party?

Mr. F. S. LYON.—Dr. W. P. Keene, I never heard of him myself before. I wouldn't make this objection if I didn't think the man was not admitted to practice.

(Testimony of William G. Lacy.)

The COURT.—And not a member of the bar of this state?

Mr. F. S. LYON.—Not that I know of, and that is the reason I make this objection.

Q. You don't know whether he is a member of the bar or not, do you? A. I couldn't say.

Mr. F. S. LYON.—I know there are men here who are practicing and have the words "Law Offices" and so forth on their doors, and yet they are not members of the bar. Now, I don't want to go any further with that proposition than that, except we know they are giving advice—

The COURT.—Do you know about this, Mr. Bagg?

Mr. BAGG.—No. This happened before my connection with this case, and all I know is that we have such a letter, and I understood Mr. Keene was a patent attorney and reasonably well known. [378] That is all I know of him. I never met the gentleman myself.

The COURT.—Well, we will let it go in on the line I have suggested.

(Defendant's Exhibit "P.")

Mr. BAGG.—That is all.

Cross-examination.

(By Mr. LYON.)

Q. Mr. Lacy, is this a true reproduction of one of the Lorraine gas traps manufactured by your company showing the exterior valve mechanism and arm (exhibiting)?

(Testimony of William G. Lacy.)

A. Yes, I believe that was made from a photograph of one of the earlier traps.

Mr. F. S. LYON.—We will ask that it be received in evidence as Plaintiff's Exhibit 21, simply to show your Honor what we attempted to illustrate to you last Friday.

The COURT.—Very well.

(Plaintiff's Exhibit No. 21.)

Mr. F. S. LYON.—It only shows that one link arm there, and it shows the inlet at the top, at this particular well. We do not assert that this is the one that is in the courtroom.

Mr. BAGG.—I will ask Mr. Prout to take the stand. [379]

**Testimony of George H. Prout, for Defendant.**

GEORGE H. PROUT, called as a witness on behalf of the defendant, having been duly sworn, testified as follows:

Direct Examination.

(By Mr. BAGG.)

Q. Please state your name, age and residence.

A. Thirty-seven years of age; residence, No. 2655 North Sitchel, Los Angeles.

Q. What business are you engaged in?

A. Machinist.

Q. Where are you employed?

A. The Lacy Machine Company.

Q. What connection have you with the manufacture or construction of Lorraine gas and oil separators?

(Testimony of George H. Prout.)

A. I assembled practically all the gas separators that have been made by the Lacy Manufacturing Company in the last year.

Q. Are you familiar, then, with the interior mechanism and construction of these traps?

A. Yes, sir.

Q. I will ask you to state if there has been any other oil and gas separator put out by the Lacy Manufacturing Company for Mr. Lorraine which is like the model part of which stands up there against the wall, and the whole of which was originally sold to the General Petroleum Company on the 17th day of March this year.

Mr. F. S. LYON.—That is objected to as leading and suggestive, and as calling for a conclusion of the witness, and not the best evidence. The witness can detail the construction of it, but—

Mr. BAGG.—I am asking him whether he is familiar with that.

Mr. F. S. LYON.—No; this question is whether there was any [380] other one like this one.

(Question read.)

Mr. BAGG.—I will change that question to read:

Q. Has the Lacy Manufacturing Company turned out any other traps of the same model as the trap sold to the General Petroleum Company on March 17th, 1922? A. Yes.

Mr. LYON.—Same objection to that. I think if counsel wishes to he could ask the witness wherein any of these differed or was similar to this trap and let him state.

(Testimony of George H. Prout.)

The COURT.—I think I will allow the question. What was the answer? A. Yes, sir.

Q. (By Mr. BAGG.) Have you examined this part of the trap? A. No, sir.

Q. I wish you would examine the same, and particularly the elbow on the interior of the trap leading from the oil and gas inlet.

A. (Examining.) I never built that trap.

Q. Now, I will ask you to state wherein the other traps that you have superintended the construction of, and of this particular model, differ from this trap which you have just examined?

A. The ones that I have superintended and constructed, I always had clearance between the L—I had the L set equally spaced between the shell and the baffle-plate, and I see that L is set away over to the baffle-plate.

Q. Did you observe the edge of this elbow in this particular trap just now? A. Yes, sir.

Q. I will ask you to state wherein that differs from the elbow that you have put into these various traps you have testified to.

A. I see that one there has been chipped off or ground off the back to go up against the baffle-plate. [381]

Q. I will ask you to state if you know whether or not that is the only trap put out in that shape.

A. Yes, sir.

Mr. F. S. LYON.—We object to that.

The COURT.—I don't know whether he knows

(Testimony of George H. Prout.)

or not. He said he had nothing to do with the construction of this trap.

Mr. BAGG.—He says he didn't put this one out, but I asked him if he knows whether there were any traps put out.

(Question read.)

A. That is the only one to my knowledge, and I didn't know that was out.

Mr. BAGG.—That is all.

Cross-examination.

(By Mr. F. S. LYON.)

Q. Mr. Prout, what is your position with the Lacy Manufacturing Company?

A. Machinist.

Q. Do you work under a foreman?

A. Yes, sir.

Q. And a superintendent

A. I am not a superintendent.

Q. I say, and there is a superintendent of your department? A. Yes, sir.

Q. You say you have worked on the assembly of practically all the Lorraine traps made at the Lacy Manufacturing Company?

A. That is in the last year.

Q. They have working drawings for those traps, have they?

A. No, sir. I worked there nine months before I ever saw a drawing of—

Q. You worked under the instructions of your foreman? A. Yes, sir.

(Testimony of George H. Prout.)

Q. And you don't know what drawings he has, then? A. No. [382]

Q. You say you worked nine months before you saw a drawing of it. Then you have seen one recently? A. Yes, sir; one assembled drawing.

Mr. F. S. LYON.—That is all.

Q. (By Mr. BAGG.) Do you know, Mr. Prout, how much space there is between the baffle-plate and the edge of this elbow in the traps you have been putting out?

A. Half an inch. The reason I know that is that the L measures 9½ inches over all, and between the shell and the baffle is 10½ inches, and when I set the pipe to be welded I measured down just to 10 inches to the edge of my L and have my welder stick out to the baffle.

Mr. BAGGS.—That is all.

(Five minutes recess.) [383]

**Testimony of Robert W. Smith, for Defendant.**

ROBERT W. SMITH, a witness called in behalf of the defendant, having been first duly sworn, testified as follows:

Direct Examination.

(By Mr. BAGG.)

Q. Please state your name, age, residence and occupation or profession.

A. I am 27; I live at No. 910½ South Hope Street, Los Angeles, and I am a patent solicitor.

(Testimony of Robert W. Smith.)

Q. How long have you been a patent solicitor or been engaged in the business of soliciting patents?

A. Since 1912.

Q. About ten years? A. Yes.

Q. What firms have you been associated or connected with in that line of work?

A. Back in Washington in 1912 when I started in that business I was first with Vernon E. Hodges; then I was with Munn & Company; then I went to Cincinnati in 1916 with A. F. Herbsleb. That was in 1916. After the war I went back to Washington and was with Hubert E. Peck, and after I came to Los Angeles I was with Hazard & Miller, and then opened my own office.

Q. You are now engaged in a private practice of your own? A. I have my own office.

Q. During that time have you had any occasion to prepare drawings for various devices that have been patented?

A. I made some of my own drawings.

Q. Do you prepare specifications. A. Yes.

Q. For patents?

A. That had been the major part of my business throughout that entire period.

Q. Have you ever prepared a specification or an application [384] for a Mr. Lorraine?

Mr. F. S. LYON.—We object to that as incompetent, irrelevant and immaterial for any purpose in this case. If the object is to prove any application of Mr. Lorraine the proper procedure is to produce a certified copy of it from the Patent Office,

(Testimony of Robert W. Smith.)

and there is no application that is material in evidence.

Mr. BAGG.—This is just simply preliminary, your Honor, to show that he is familiar with this class of work, that is all.

The COURT.—You might leave that part of it out, then, as to whether he has prepared a specification or application from Mr. Lorraine.

Mr. BAGG.—Very well.

Q. Have you had occasion to prepare any specifications for any instrument or devices connected with the separation of other treatment of oil and gas as they come from oil and gas wells? A. I have.

Q. I hand you Defendant's Exhibit "E," which is a certified copy of the patent granted by the United States to George L. McIntosh, Assignor, under date of March 11, 1913, and will ask you to describe as nearly as you can the action of that device (handing paper to witness).

Mr. F. S. LYON.—If your Honor please, we object on the ground that it is incompetent, no foundation laid, the witness not having qualified to answer the question; and on the further ground that the documents speak for themselves. If they are part of the prior art they must be sufficiently plain, clear and exact to enable one skilled in the art to make and use a device embodying the invention. If your Honor has any question as to the competency of the witness in this connection, I would like to examine him as to his qualifications before he gives his testimony.

(Testimony of Robert W. Smith.)

Mr. *L. F.* LYON.—We would like to make the further objection that these are the same patents that have been gone over by the other expert witness, and, as I understand the rule in this district in regard to experting prior art patents, the Court [385] limits the parties to one expert to describe the same patent.

The COURT.—Is that the rule in this court?

Mr. BAGG.—I am not familiar with that rule, your Honor. I think it is purely within the discretion of the Court.

Mr. L. S. LYON.—I think it is, but in all the cases I have been in the Court has refused to sit and listen to the testimony of two experts for the same patent, to see what is in the same patent.

The COURT.—Is this all the experts you have now?

Mr. BAGG.—Yes, that is all.

The COURT.—I will hear him.

Mr. F. S. LYON.—What about the cross-examination with regard to qualifications, your Honor?

The COURT.—Do you want to cross-examine him as to whether he is an expert or not?

Mr. F. S. LYON.—Yes.

The COURT.—All right.

Mr. BAGG.—No objection.

Q. (By Mr. LYON.) You are not admitted as an attorney at law, are you? A. In California?

Q. Yes.

A. No, I don't practice in the state courts.

Q. Are you a graduate engineer from any school?

A. No, sir.

Q. In other words, all the mechanical training you have is such as you have picked up while working in the patent attorneys' offices, and you simply have been admitted to practice before the Patent Office as a drawer and amender of scribner of application for patents; is that correct?

A. I am a member of the bar of the District of Columbia, of the Federal Courts in the District of Columbia. I have not yet been admitted in California. As to my education, as I say I am not a graduate of a higher school technically, but my [386] technical education is confined to high school education in technical lines.

Q. Did you ever have any practical experience in any department of oil production or oil well drilling or operation? A. Have I had—

Q. Yes, any actual experience in any of those lines. A. No, sir.

Q. (By the COURT.) Do you claim to be a patent expert? A. Yes, sir.

Mr. F. S. LYON.—That is all.

The COURT.—Very well. Go ahead.

Q. (By Mr. BAGG.) Proceed and answer the question.

A. This patent granted in 1913 has an outer receptacle, an oil and gas inlet—

Q. (By the COURT.) Well, all I want to know is the operation of the oil as it enters the trap.

A. The oil and gas outlet has baffle-plates arranged beneath it so that the oil and gas as it en-

(Testimony of Robert W. Smith.)

ters will fall down on these baffle-plates and the action of the baffle-plates will be to shower the oil and gas outwardly so that the oil will drop to the bottom of the receptacle, some of it dropping directly and some of it being showered down the wall of the receptacle, the gas rising to the top and passing out through the gas outlet at the top.

The COURT.—Is that all?

Mr. BAGG.—That is all I want with this witness, on this particular patent.

The COURT.—What cross-examination have you?

Mr. F. S. LYON.—None on that at the present time.

Mr. BAGG.—Well, we will introduce them all at this time, if your Honor please.

Q. I hand you Defendant's Exhibit "J," which is a certified copy of the patent granted to Albert T. Newman, and ask you to explain to the Court the action of that.

A. In this there are two receptacles, the oil and gas [387] entering through this tee head so that they go into the two receptacles, in one case falling upon the baffle "K-prime" and in the other case upon the baffle "L-prime." The oil and gas will then be diverted to the side walls of the reception chambers so that the oil will fall to the bottom of the receptacle, the gas rising, and in the course of that oil and gas falling downwardly the oil and gas will pass off the baffles and flow down-

(Testimony of Robert W. Smith.)

wardly in a thin body on the side-walls.

Q. How about the operation on this side?

A. In this side the oil and gas will flow onto the baffle "L-prime" and will be diverted to the side-wall and flow downwardly in a thin body.

Q. (By Mr. F. S. LYON.) And is there any reference whatever in this specification of the Newman patent to oil? A. It is—

Q. Answer the question briefly, yes or no.

A. What is the question?

(Question read.)

A. I can't say whether the word "oil" is used in there or not.

The COURT.—This is a water and gas separator, is it not?

Mr. BAGG.—Yes. That is the patent.

Mr. F. S. LYON.—A water and gas separator.

Q. (By Mr. BAGG.) Now, I hand you Defendant's Exhibit "G," which is a certified copy of the letters patent issued by the United States to Arthur W. Barker, dated July 13, 1909, and ask you to explain the action of that device.

A. In this case the incoming flow is through the pipe "6" and will strike the baffle-plate "7," which is arranged above the pipe "6" so that the constituents of the flow will be thrown outwardly after striking that baffle, some of them dropping directly and some of them passing down the side-walls of that receptacle giving the gas a chance to pass upwardly and then outwardly through the

(Testimony of Robert W. Smith.)

pipe "12," and the heavier constituents to [388] drop to the bottom.

Q. (By the COURT.) This is the intake?

A. This is the intake.

Q. And that?

A. That is a baffle, "7," above the end of the pipe.

Q. And it will simply strike against the baffle and then be showered out?

A. Showered out in all directions.

The COURT.—Very well; I understand that.

Q. (By Mr. BAGG.) I now hand you Defendant's Exhibit "H," which is a certified copy of patent issued by the United States to Eustice Vivian Bray, of Coalinga, California, and ask you to explain to the court the action of this device.

A. In this case there are a plurality of cones arranged in the upper part of the receptacle. Those cones are perforated. The oil and gas inlet dropping onto those cones, the flow passes through the openings in the cone so that the oil is collected in the bottom of your receptacle and the gas rises to the top.

Q. (By the COURT.) There is where the separation takes place, in the cone?

A. The gas is in this portion of the receptacle here and must pass up and outwardly through the outlet at "17."

Q. The separation of the oil from the gas takes place in the cone? A. Yes.

Q. (By Mr. F. S. LYON.) You never saw device like this Bray patent, did you?

(Testimony of Robert W. Smith.)

A. No, I never have.

Q. And you have never had any experience which would enable you to state whether or not such a screen as the screens 11, 12 and 13 would be operative or whether they would immediately clog up with the sand from the well?

Mr. BAGG.—We object to that as immaterial. It doesn't make [389] any difference whether it is operative or not.

The COURT.—What is your opinion as to that?

A. My opinion depends entirely upon the proportion of sand in the oil, the gravity of the oil, and those factors, the constituents of the flow, to cover completely, whether or not those perforations would clog or not.

Q. It depends upon the size of the perforations too, does it not?

A. That is another factor, the size of the perforations.

Q. (By Mr. BAGG.) Now I will ask you to explain if there is any float action in that. If so, explain that to the Court.

A. In this reference there is a float action controlling the oil outlet.

Q. And what is the effect of that float action with reference to the opening of the oil outlet?

A. The opening of the oil outlet is controlled by the float.

Q. With reference to the oil level?

A. The float is arranged to maintain the oil outlet

(Testimony of Robert W. Smith.)

submerged below the oil level. By a submergence of the oil outlet.

Q. (By Mr. F. S. LYON.) The purpose of the pipe "20" and trap is to permit the oil to flow out if the sand is sufficient to stick this valve "5" in the float, is it not?

A. The purpose of that float is to maintain—

Q. No, the purpose of that pipe and these connections.

A. The purpose of that pipe and connection is for the oil outlet.

Q. Under what conditions?

A. During the operation of the trap as a separator of oil and gas.

Q. Is that the normal operation of the trap, to allow the oil to go out the pipe "20" and the trap?

A. The point of it is, that if an excess of sand accumulates in the bottom of the trap the float will still operate.

Mr. F. S. LYON.—All right; that is all.

Q. (By Mr. BAGG.) I hand you Defendant's Exhibit "I," which is a certified copy of the patent of the United States to August Steiger Cooper, dated March 20, 1906, and ask you to explain the operation of that to the Court.

A. In this case the oil inlet receptacle is arranged with a reduced opening so that the oil and gas as it enters is thrown [390] against a baffle-plate "a-prime," and the plate being arranged at a tangent to the receptacle, the incoming oil flow will be thrown around the side-walls of that receptacle to drop to

(Testimony of Robert W. Smith.)

the bottom. The gas then rises from within that film of oil and gas, the oil dropping to the bottom and being controlled by a valve which is connected to the float.

Q. Is there any oil pressure-maintaining means upon the trap, do you know?

A. The pressure-maintaining means will be the pressure which is maintained throughout the entire system. That is, if we have an oil and gas inlet coming from the well at the well pressure and have a large receptacle and an outlet pipe, which is relatively small, so that the same pressure is maintained through your oil and gas pipe.

Q. If this pipe discharged into the open air without any pressure as shown here there would be no pressure, would there?

A. That would depend upon the relation of that outlet pipe to the intake pipe and to the size of your receptacle.

Q. But there would be no means for maintaining the pressure other than the possible restriction of the area of the pipe conducting the gas away.

(No response.)

Q. Now, have you said that the oil will be delivered in a film by this Cooper construction?

A. The oil and gas entering here will be diverted by this baffle to a point adjacent to the side-wall of the receptacle and pass down that side-wall in an enveloping body. The thickness of that will of course depend upon the velocity and amount of flow.

Q. I think you used the term "film" on direct, did

(Testimony of Robert W. Smith.)

you not? What was your idea of a film in that expression?

A. I don't remember that I used the word "film." I would like to know if I did.

(Reporter reads from previous answer, p. ——, as follows: [391] "The gas then rises from within that film of oil and gas, the oil dropping to the bottom and being controlled by a valve which is connected to the float.")

A. By the use of the word "film" I mean that enveloping body of oil and gas passing down the side-wall.

Q. (By Mr. LYON.) Then what would you give as your idea of the thickness of the film of oil flowing down a gas-trap? You have used the term.

A. To my mind a film of oil is a body of oil practically of such a thinness that no further thinness can be obtained and still maintain a body of that oil.

Q. And you think that would be the result of this Cooper construction, that it would be so thin that it would be practically to the vanishing point, do you?

A. There would be that film on the wall. There would be part of that flow which would not pass down that wall.

Q. In other words, there might be a little spray of oil on the wall, in this Cooper device; is that it?

A. No; that is not my idea, that there might be a spray on there. Some of it would be directed on that wall to form a film, but it wouldn't be confined to throwing that film on the wall. There would be

(Testimony of Robert W. Smith.)

other portions that would drop directly through the trap.

Q. Have you any idea of which would be the greater amount or what proportion would drop direct, with this Cooper device, or what percentage would be filmed?

A. That question would depend upon the various factors, that is, the rate of flow, the constituents of the flow, and the gravity and viscosity of the oil.

Q. Now, what would be the effect, if you know, if the oil was directly discharged against the wall of the shell in an ordinary oil well producing oil and gas and the usual quantities of sand? How long would such a shell as that last, with the incoming oil, gas and sand directed against the inner space of the shell? [392] Have you any idea?

Mr. BAGG.—I object to that, your Honor, I don't think it is material, or that it cuts any figure in this case at all.

The COURT.—I think it is rather speculative.

Mr. F. S. LYON.—It is speculative, so far as this witness is concerned, I will admit, because he doesn't know anything except about writing a specification.

Mr. BAGG.—We object to that line of argument, if the Court please.

The COURT.—Yes.

Mr. F. S. LYON.—Your Honor is satisfied that this witness cannot answer it without knowing whether—

The COURT.—I would not make any comments upon the witness at the present time.

(Testimony of Robert W. Smith.)

Q. (By Mr. F. S. LYON.) You never saw any device like this Cooper patent, did you?

A. No, I have not.

Q. And have had no experience, as I understand it, with and of these gas traps?

A. You mean any of these in the references?

Q. Yes. Not the patents but the devices.

A. You mean in these patents in these references.

Q. Yes.

A. No, I have never seen any of those.

Mr. F. S. LYON.—That is all.

The COURT.—Have you any further witnesses?

Mr. BAGG.—I think we are through, your Honor, but I would like to hold the matter open until to-morrow morning.

Mr. F. S. LYON.—We have some other testimony, your Honor.

The COURT.—We will adjourn now until 10 o'clock to-morrow morning.

(An adjournment was thereupon taken until Tuesday, March 28th, 1922, at 10 o'clock A. M.) [393]

[Endorsed]: Original. In the District Court of the United States for the Southern District of California, Southern Division. (Before Hon. Charles E. Wolverton, Judge.) Francis M. Townsend et al., Plaintiffs, vs. Davis *S.* Lorraine, Defendant. No. E-113—Eq. Reporter's Transcript of Testimony and Proceedings. Vol. IV, Los Angeles, California, March 27, 1922. Filed Apr. 7, 1922. Chas. N. Williams, Clerk. By R. S. Zimmerman,

Deputy Clerk. Reported by J. P. Doyle, J. J. Petermichal. Doyle & St. Maurice, Shorthand Reporters and Notaries, Suite 507 Bankitaly International Building, Los Angeles, California, Main 2896. [394]

INDEX.

Plaintiff's Witnesses in Rebuttal:



[395]

Los Angeles, California, Tuesday, March 28, 1922, 10 A. M.

The COURT.—You may proceed in Townsend vs. Lorraine.

Mr. BAGG.—If your Honor please, we would like to file our amended answer. We have served a copy on the other side.

The COURT.—Very well.

Mr. BAGG.—If your Honor please, the other day we announced that we would make an effort to get the patent of the Tico trap. Since that time we have investigated the matter and have found that the Tico trap itself has never been patented, and consequently we cannot produce a certified copy of that patent.

Mr. F. S. LYON.—Now we disagree with counsel as to his statement of the facts. On May 9, 1916, the patent 1,182,873 that I referred to was granted to the Titusville Iron Company of Titusville, Pennsylvania, for practically the device as shown in the drawings of Defendant's Exhibit "C."

The COURT.—What is the date of the patent and who is the patentee?

Mr. F. S. LYON.—Charles E. Fisher; May 9, 1916.

The COURT.—Is that the same patent you refer to?

Mr. BAGG.—No, it is not, if your Honor please. That patent to Charles E. Fisher is for a valve mechanism and has no connection whatever with the patent in suit. However there is a description of the trap, but that description is only given for the purpose of showing the workings of this baffle. The trap itself has never been patented so far as we have been able to discover.

With that statement the defendant rests. [396]

## Testimony of W. C. Rae, for Plaintiffs (Recalled—In Rebuttal).

W. C. RAE, recalled as a witness on behalf of plaintiff, in rebuttal, having been first duly sworn, testified as follows:

Direct Examination.

(By Mr. L. S. LYON.)

Q. Mr. Rae, do you remember the first time you ever had a conversation with the defendant in this case, Mr. Lorraine, regarding his asserted invention of some apparatus in connection with that gas trap?

A. Yes, sir.

(Testimony of W. C. Rae.)

Q. Will you tell us what that conversation was, when it was and how it arose, as nearly as you can remember?

Mr. BAGG.—Now, if your Honor please, we object to that as incompetent, irrelevant and immaterial. In the first place, so far as disclosed here, it is not rebuttal testimony, and if it is intended, as I presume it is, for the purpose of impeaching some testimony Mr. Lorraine has given, we object to it for the reason that it would have a tendency to contradict an immaterial part of his testimony, and for the further reason that no foundation was laid for any such testimony in rebuttal, because neither the time nor the place nor those present was fixed in the examination.

Mr. L. S. LYON.—May it please the Court, on defendant's case defendant's counsel has asked the defendant when he had first seen a Trumble trap, and he stated that he had never seen one, the inside workings of one, until after such and such a date. We then, on cross-examination, asked the witness if he had not discussed the Trumble trap with this witness, and he said no, that he had merely gone to him and shown him his own trap, the Lorraine trap, on that date, which was earlier than the date on which Lorraine said he first saw and understood the workings of the Trumble trap. Now we propose to impeach by this witness that testimony given on cross-examination. [397] And your Honor will remember that I asked the witness particularly to relate the conversation, and if he did not state to

(Testimony of W. C. Rae.)

Mr. Rae that the only thing he had was a valve, and that he wanted to put the valve on the Trumble trap, and he stated that he knew how the Trumble trap worked; and all of this was denied by the witness, the conversation that he had was denied, and we now want to impeach that testimony by Mr. Rae, and Mr. Rae will tell us what conversation was had and what was said by Mr. Lorraine.

The COURT.—As I remember it, the words were not put in the witness' mouth, and the question asked if he did not say so and so, but, as I recall now, the witness was asked if he did not have a conversation on that subject with the witness Rae, and he denied that. I think you can go ahead and ascertain whether that conversation was had. I think your impeachment may go that far at least

Mr. BAGG.—The objection we raise, your Honor, is that it is an immaterial matter, and you cannot, of course, impeach a witness on an immaterial matter, and it will only go to encumber the record as far as that is concerned.

The COURT.—Well, I think that it goes further than that; it goes to the question of whether the witness then knew of the patent and its operation.

Mr. BAGG.—I think, your Honor, the question we asked the witness was what time it was that he first saw the interior or understood the mechanism of the Trumble trap. I may be mistaken, but I do not think there is any testimony about the fact that Mr. Lorraine had seen the Trumble trap, possibly, casually, as any other observer would see it, before

(Testimony of W. C. Rae.)

this, but he knew nothing about the workings of the trap before the time he testified.

Mr. L. S. LYON.—That is what Mr. Lorraine stated. Now [398] we expect to show by this witness that Mr. Lorraine disclosed to him and stated all the facts as to how the Trumble trap worked and showed the witness how his valve could be put on the Trumble trap, and this was before he had any Lorraine trap at all.

The COURT.—I will hear the testimony.

Mr. BAGG.—Exception.

Q. (By Mr. L. S. LYON.) State what that conversation was and what was said and the circumstances.

The COURT.—You may state whether you had such a conversation first.

A. I had a conversation with Mr. Lorraine, and he came to my office. He was sent up by a friend of mine, Mr. Barnes, of the National Supply Company. He showed me the drawings of a gas trap valve.

Mr. BAGG.—We object to that unless he fixes the time and place.

A. (Continuing.) The place was at my office, No. 916 Higgins Building, Los Angeles, about the end of 1919 or the first or second month of 1920. There were several visits he made to me. I don't recall the date, but it started at the end of 1919, as I recall it. I was going back and forward to the east every sixty days, and he came up to see me nearly every time I came home. The only thing he showed

(Testimony of W. C. Rae.)

me was the valve. He says, "I have examined a number of traps, and I think the Trumble trap is the best in the field so far"; that "I have a valve that I would like to sell your Company." I put up the proposition to our Mr. Trumble, who was experimenting with the valve at that time, and he says, "I am not interested." Then the thing was practically dropped for several months.

Mr. BAGG.—We object to any conversation that took place between him and Mr. Trumble. [399]

The COURT.—Yes.

A. (Continuing.) It was absolutely necessary for me to put it up to the man who was working on the valves for my company. I am merely the sales representative, remember.

The COURT.—But your conversation with Mr. Trumble is not—

A. Well, that is all right.

Q. (By Mr. L. S. LYON.) Now, when was it, if ever, that Lorraine first showed or explained to you that he had a trap, with relation to this conversation?

A. Six or eight months later.

Mr. BAGG.—Now we object to that because that has no bearing on this case, and the testimony of this witness does not show that Mr. Lorraine ever explained to him that he knew the interior working of the Trumble trap or the principle.

The COURT.—I think you had better confine yourself to that question.

Q. (By Mr. L. S. LYON.) Will you state anything further that you remember concerning the conversation you had with Mr. Lorraine, that early conversation, with respect to what he understood or whether he understood the construction or operation of the Trumble trap?

A. He didn't state anything about the knowledge he had of the Trumble trap workings at that time.

Q. What did he say?

A. He merely said he had a valve that would work with our trap and would like to sell it to us.

Mr. L. S. LYON.—That is all.

Mr. BAGG.—Now, if your Honor please, we move to strike out all of the testimony of this witness for the reason that it does not tend to impeach any of the evidence given by the defendant in an immaterial matter.

Q. (By the COURT.) Did he say to you at that time that [400] he was acquainted with the workings of the Trumble trap?

A. I didn't go into details

The COURT.—That is all he said?

A. That is all he was talking about.

Mr. L. S. LYON.—The evidence is, your Honor, by the defendant himself, that he had a conversation with this witness, so it does not come under the hearsay rule anyhow; and in the second place I asked the defendant yesterday that very question, didn't he state that the Trumble trap was the best trap, and that he had a valve that he wanted to put on it, and he said no, he did not.

(Testimony of Paul Paine.)

The COURT.—I will let the statement of the witness stand for what it is worth.

Mr. BAGG.—Exception.

The COURT.—Any cross-examination?

Mr. BAGG.—No questions. [401]

**Testimony of Paul Paine, for Plaintiffs (Recalled—In Rebuttal).**

PAUL PAINE, recalled as a witness on behalf of the plaintiff, in rebuttal, testified as follows:

Direct Examination.

(By Mr. F. S. LYON.)

Q. Mr. Paine, you are the same Mr. Paine who testified on direct examination? A. Yes.

Q. You heard the testimony of the defendant Lorraine with reference to the maintenance of an oil level in the Lorraine trap, did you? A. Yes.

Q. Based upon your knowledge and experience, what have you to say as to the maintenance of the oil level in the Lorraine trap?

Mr. BAGG.—Now we object to that as incompetent, irrelevant and immaterial. There is no dispute about the oil level in the Lorraine trap.

Mr. F. S. LYON.—Well, we are going to dispute it.

Mr. BAGG.—And I do not see how this has any bearing in rebuttal of any evidence that Mr. Lorraine may have given. In the first place, if he is testifying as an expert witness counsel should put to him the question that will enable him to make an

(Testimony of Paul Paine.)

answer and lay the proper foundation for his answer in the question. The mere fact that he heard what this witness testified I cannot see has any bearing on the case at all.

The COURT.—Well, there has been considerable testimony about the oil level, but whether or not it referred to the Lorraine trap is another question. It seems to me the principal controversy came up about the oil approaching a level.

Mr. F. S. LYON.—I will interrupt the proceedings just long enough to offer in evidence this portion of the 1922 Lorraine trap sold March 17, 1922, to the General Petroleum [402] Company by the defendant, being a portion of the trap that was demonstrated last Friday to the Court in the presence of counsel and the parties, as Plaintiff's Exhibit No. 22.

Mr. BAGG.—We object to it because, in the first place, it has not been properly identified, and in the next place it is not proper rebuttal.

The COURT.—Well, the Court saw that trap, and I am pretty well prepared to say, myself, that is part of the trap.

Mr. BAGG.—That is true, but we do not think this particular part of it has been properly identified. Of course I take it your Honor will probably recognize that as the trap you saw, but we think for the purpose of the record, it has not been properly identified, and in the next place, it is not proper rebuttal testimony.

The COURT.—Have you got the witness here?

(Testimony of Paul Paine.)

Mr. F. S. LYON.—Yes. I will ask that it be marked for identification now and proceed with this witness.

(Plaintiff's Exhibit No. 22.)

(Question read.)

Mr. BAGG.—We wish to insist upon the objection, your Honor, that it is not proper rebuttal testimony.

The COURT.—There has been considerable testimony about the manner in which the oil and gas as it came in would foam up and be some time in settling so as to get the level of the oil itself.

Mr. BAGG.—Yes, sir; he is asking about the oil level, and that might mean the oil level in the receiving chamber or in the settling chamber. I take it that the oil level he refers to is the oil level in the settling chamber which manipulates those floats.

Mr. F. S. LYON.—My question was all-comprehensive, as I did not wish to lead the witness.

The COURT.—Well, go ahead. I will hear it.

A. The oil, since it is a fluid, must come to a common [403] level in all portions of this trap so long as the chambers are inter-communicating as they are in this trap. Now it is possible that if the well were to make a very violent surge of oil in large quantity, just momentarily in the receiving chamber the level of the fluid would be higher, but that would obtain for a very limited period of time, and then the fluid must come, in response to physical laws, to a common height throughout the entire trap.

(Testimony of Paul Paine.)

Q. (By Mr. F. S. LYON.) If I understand Mr. Lorraine's testimony correctly, he stated that on this side (indicating) of the baffle-plate at which the oil is let into the trap the oil is always higher than it is on the other side of that baffle-plate. What have you to say in that regard?

A. I do not consider that to be the case, except in those instances of very violent surges, which would be, as I have said, only momentary.

Q. Reference has been made in the testimony to wet gas and dry gas. Please explain to us the difference between these two.

A. Natural gas as it occurs is found in two major classes: the first is in those wells which produce gas only, and such gas occurs frequently at very high pressure in very large volume and is the major source of supply for industrial and domestic consumption; and the second form of occurrence of gas is that which occurs along with the oil from oil wells. The oil itself consists of a great number of different oils which go to make up the crude oil as it occurs, and it is the function of the refiner to separate these different divisions into classes which have a commercial market. It was observed early in the history of the oil industry that that [404] gas which occurs along with the oil is particularly impregnated in many instances with vapors of gasoline, which are the most easily evaporated from crude oil, and in that manner the natural gas came to have a division into the two classes of dry gas, which is that gas that occurs in wells which produce

(Testimony of Paul Paine.)

gas only, and wet gas, which is that gas which comes along with the oil and contains some gasoline vapors. The division is not a sharp one, because sometimes the so-called dry gas has a very small quantity of gasoline vapors in it. During the period from 1905 to 1910 the attention of the oil industry became directed particularly to the value of this small quantity of gasoline vapor contained in the gas occurring along with the oil from oil wells, and that has led to the so-called casing-head gasoline industry and the recovery from the gas of those vapors and the obtaining of an additional amount of gasoline, which is equal to about fifteen per cent of the total quantity of gasoline produced in the United States.

Q. (By the COURT.) Can dry gas be reduced to oil?

A. No, sir. The dry gas is used for fuel in oil field operations, or is carried to where there is a market and a demand. That is the major source of supply, natural gas, for cities and industrial plants. The dry gas continues as a gas always.

Q. There is no way of reducing that to an oil or liquid? A. No; that continues as gas.

Q. (By Mr. F. S. LYON.) It is what is known as a fixed gas? A. Yes, sir.

Q. If I understand the statement correctly, Mr. Lorraine has stated that the maintenance of a pressure on the gas trap throws the gasoline vapors back into the oil. Is [405] that correct?

A. That cannot be, because of the physical laws involved. After this gasoline vapor has gone over

(Testimony of Paul Paine.)

from the form of liquid in the oil into a form of vapor contained along with the gas, then the pressure and the temperature factors necessary to recondense that into a liquid again are so great that they are beyond the conditions which obtain around any gas traps ordinarily in use in the oil fields. The function of the pressure on a gas trap is presented, in that it prevents, through the holding of a pressure, the evaporation of gasoline into that vaporous condition.

Q. In other words, preventing the gasoline from vaporizing and mixing with the fixed gas.

A. Then when the gasoline in the oil, or a portion of it, has been vaporized and has become mixed with those gases which continue as gases and are not condensable, then in order to recondense those vapors pressures and temperatures must be reached which are practically never present in these gas traps.

Q. I believe you explained that in your former testimony in referring to the fact that ten pounds or less pressure with certain gases would hold them in a liquid form where, if mixed with other gases and absorbed in them, it would be necessary to run as high as 150 to 300 pounds pressure to liquify that gas which could be held as a liquid at ten pounds or less.

A. Yes. The principle involved is that known as the principle of partial pressures. The usual operating pressure at plants at which the gasoline vapors are recondensed into the liquid form ranges around two hundred and fifty pounds.

(Testimony of Paul Paine.)

Q. If I remember correctly, Mr. Lorraine said that in his trap he could separate the oil according to gravity. [406] Is such a fractionation of the oil as it comes from a well possible as it comes from one of the gas traps?

Mr. BAGG.—Now, your Honor, I do not think Mr. Lorraine made that statement. In the next place, I don't think it has any bearing on this case. As to separating the oils with reference to their specific gravity, that is the general principle on which all these gas and oil separators are based. As they come from the oil and gas well they separate into the various ingredients that compose the crude oil, the sand dropping to the bottom, the water next, and the oil next, and the gas on top. I don't think there is any dispute about that, and I don't think Mr. Lorraine ever intended to make any distinction as the attorney for the other side has attempted to put in the mouth of the witness.

(Last question read.)

Mr. F. S. LYON.—In view of the objection I will reframe the question.

Q. How is oil fractionated to take off of the oil different gravities or boiling points?

Mr. BAGG.—Now, if your Honor please, we object to that as unnecessarily encumbering the record. It is getting the record full of a lot of matter that is highly edifying, no doubt, and adds considerable to our knowledge, but at the same time I do not see where it cuts any figure in this particular case, with reference to the separation and refine-

(Testimony of Paul Paine.)

ment of oils, because I do not think anybody ever contended for a moment that oil and gas separators were intended for the purpose of refining oil or separating them according to their various degrees of density. I am sure there is nothing in the evidence that shows any such contention on our part.

Mr. F. S. LYON.—We will argue what Mr. Lorraine's testimony [407] was, but he used the term that the oil was separated according to gravity, and I just want the Court to understand that no separation of the oil into different gravities of oil is possible in this gas trap.

Mr. BAGG.—We admit that we do not make any pretention to refining the oil or separating it into its various constituents.

Mr. F. S. LYON.—And that nothing of the kind is possible in a gas trap.

Mr. BAGG.—We will not admit that.

The COURT.—I understand the witness made some reference to the oil settling according to gravity.

Mr. BAGG.—Yes, we admit that. There is no question about it.

The COURT.—I don't remember anything that he might have said about the separation of the oil according to gravity; but you can ask that question. I will say to counsel that these matters of the separation of oil and all those things are entirely new to me, and I don't know what bearing they are going to have hereafter when the argument is reached, and

(Testimony of Paul Paine.)
I have allowed a wider latitude on that account than I otherwise would.

Mr. BAGG.—In view of that fact, your Honor, we wish the Court to understand that we are not trying for one moment to shut off any information the Court would like to have, and if any of our objections come to that point we will waive them.

The COURT.—I will hear the answer.

A. I have never known a gas trap to separate the oil into its constituents with respect to their specific gravity, and I cannot conceive it as being possible.

Q. (By Mr. F. S. LYON.) Why not?

A. Because the oil is so intimately mixed that even when it is stored for a long period in tanks there is [408] only a very minute difference in the gravity between the weight of the oil at the top of the tank and the weight of the oil at the bottom of the tank, and even then there is nowhere near any line of demarcation between the different fractions.

Q. Now, very briefly explain to the Court what is necessary and how the crude oil is separated into its various boiling points and gravities.

Mr. BAGG.—Now, if your Honor please, we would like to have the record show that we are not objecting to this because of the fact that it is for the general information of the court.

The COURT.—Yes; explain that briefly.

A. The oils that go to make up crude oil evaporate at different temperatures. The oil is then heated, and that oil which evaporates at the lower tempera-

(Testimony of Paul Paine.)

ture comes off first and is condensed, and that comprises gasoline; then that portion of the oil which evaporates at a slightly higher temperature is taken off through raising the temperature.

Q. (By the COURT.) And what do they call that?

A. And that, out here in California, is called distillate; and then the oils which come off at still higher temperatures go over into the fractions of coal oil, kerosenes and stove-oil, and the progressively heavier oils, and that is the fundamental process involved in the refining of oil, is to heat it and take off—

The COURT.—I think that is enough on that.

Mr. F. S. LYON.—That is all I want.

Mr. BAGG.—Now we also wish the record to show, your Honor, that we admit all of this as being true.

The COURT.—Very well.

Q. (By Mr. F. S. LYON.) When one of these gas traps is used [409] and the oil is to be moved from the trap to a tank or a point higher than the trap is it possible to so discharge the oil without a pressure in the trap?

Mr. BAGG.—Now, if your Honor please, we object to that because it is not proper rebuttal, and because Mr. Lorraine testified positively that that was the fact. The record will show that he admits that.

Mr. F. S. LYON.—You admit that it is necessary,

(Testimony of Paul Paine.)

to maintain a pressure in the trap to discharge the oil?

The COURT.—I think that is self-evident. I don't see how it could be otherwise.

Mr. BAGG.—No, sir; it could not be.

Mr. F. S. LYON.—All right.

Q. You are familiar with the gas lines from these gas traps to gasoline absorption plants, are you?

A. I have installed a good many of them.

Q. What have you to say in regard to whether such lines contain valves in them or not?

A. So far as I have ever observed, they—

Mr. BAGG.—Wait a minute. That is objected to as incompetent, irrelevant and immaterial unless he proves it with reference to the Lorraine trap. We certainly do not think that as to all these traps it makes any difference.

Mr. F. S. LYON.—I think we can show a general construction, your Honor, and general use in regard to all these absorption plants, that they always have in the gas lines the valves we have been referring to here, and that they are necessary elements.

Q. (By the COURT.) Is it necessary to have such valves? I will hear the answer in that form.

A. Gas lines from gas traps to absorption plants invariably have—

Q. Well, is it necessary? A. Yes, sir. [410]

Q. (By Mr. F. S. LYON.) Would a separation device or gas trap in which the oil and gas or products from the well are delivered into the trap, with

(Testimony of Paul Paine.)

the end of the delivery pipe directed against the inside of the trap, be commercially practicable?

A. No, it would not.

Mr. BAGG.—Wait a minute. We object to that as incompetent, irrelevant and immaterial. I don't see the purpose of it.

Mr. F. S. LYON.—It shows the prior art, your Honor. For instance, just leave this pipe here (indicating) and let it discharge right against that baffle-wall in there. A direct discharge against the wall is what I have inquired about.

Mr. BAGG.—Whether it is practicable or not, your Honor, doesn't cut any figure in this case whatever.

The COURT.—I think that is going outside of the given instruments that have been introduced here in evidence. None has been introduced here which gives that operation, that I know of; the one that comes nearest to it is where it is simply thrown up and it takes different directions as it strikes the top of the—

Mr. F. S. LYON.—Yes; and there is one of the witnesses who testified to one in which it was a vorticle arrangement. I am coming to that in the next question.

The COURT.—Where it goes right around the trap?

Mr. F. S. LYON.—Yes.

The COURT.—Well, you may ask him about that.

Q. (By Mr. F. S. LYON.) Would, in your opinion, a device in which the intermingled oil, gas and

(Testimony of Paul Paine.)

sand, or whatever was delivered from the well, was delivered against the inner periphery of a cylindrical drum or trap—

The COURT.—You better get that patent and refer to it.

Q. (By Mr. F. S. LYON, Continuing.) —like Defendant's Exhibit [411] "I," Cooper Patent No. 815, 407 (handing same to witness), be commercially operative?

Mr. BAGG.—Now we object to that as incompetent, irrelevant and immaterial. This exhibit was merely introduced for the purpose of showing the state of the art, and whether or not a trap or device that was patented at the time this one was patented was a workable arrangement is immaterial.

The COURT.—I think that calls for a conclusion. You can only determine that by actual experience with the trap.

Q. (By Mr. F. S. LYON.) Well, what would the action of the sand delivered from an oil well producing gas and oil be in a trap constructed like this Cooper patent?

Mr. BAGG.—We object to that for the same reason.

The COURT.—I think I will hear it.

A. The action would have a wearing effect on the inside of the trap through the sand-blast of the sand coming along with the oil.

The COURT.—Yes; I think that was admitted to be true.

Mr. BAGG.—Yes, sir.

(Testimony of Paul Paine.)

Q. (By Mr. F. S. LYON.) And is there or is there not such sand-blast with these oil-and-gas-producing wells? A. Yes; with many of them.

Q. To what extent has that been found a serious factor with gas traps?

Mr. BAGG.—Now, if your Honor please, we object to that. The fact that it wears out is no evidence that it is impracticable. Though it may wear out every twenty-four hours, that doesn't show that it is not an operative device, and that is the purpose, as I stated a few minutes ago, of introducing this, to show the state of the art, and if that worked for an hour it is an operative device. The mere fact that the plaintiff comes in here and presents [412] a trap that operates longer than that or is an improvement upon that does not give him any right to claim that because this patent might be inoperative according to his expert's testimony, that would not make it show the state of the art, as we contend in this case.

The COURT.—Utility is one of the elements of a patent, is it not?

Mr. BAGG.—Yes, sir; utility is one, but it doesn't make any difference about the degree of utility. If that would operate for five or ten minutes, or an hour, it would be an operative device.

Mr. F. S. LYON.—In the Webster loom case, your Honor, the patent was sustained solely on the ground that the Webster loom produced—now, my figures are subject to correction from the fact that I am speaking entirely from memory—it was either

(Testimony of Paul Paine.)

40 or 60 yards, where the old looms produced either 20 or 30—in other words, double the amount, approximately, and the quality of increased utility is patentable in that sense; but I am going further to show by this witness, by the next question, that a gas trap that would not stand up twenty-four hours would be, from a commercial standpoint, practically without any utility whatever, and I will ask the witness to answer my last question, if that is correct.

Mr. BAGG.—Well, we object to that.

The COURT.—I think you have illustrated this trap sufficiently for me. I understand the principles of the trap, and I think I understand the action of the sand or what effect the action of sand driven on to that would have. I think that has been illustrated sufficiently.

Mr. BAGG.—I was going to say, if your Honor please, that, as I understand counsel, he has misconceived our idea in this matter. We are not trying to vitiate or void the patent of Mr. Trumble; we are simply showing [413] the state of the art. Now, if we were using this as an anticipation then his argument might have some bearing because of the fact that this patent we were introducing would be impracticable according to this theory. Of course there is evidence here to show that his statement as he makes it is not true, that is, that it would not wear out immediately; but we are not trying to void his patent.

Mr. F. S. LYON.—You may take the witness.

(Testimony of Paul Paine.)

Cross-examination.

(By Mr. BAGG.)

Q. Mr. Paine, if the oil and gas as they come from the oil well were so mixed and churned up that it resembled a broth or foam, and the well was flowing either constantly or in heads, there would be, while the well was flowing, in the Lorraine trap, as you understand its operation, a higher level on the side of the partition into which the oil and gas in this foamy condition comes than on the opposite side or in the settling portion of the trap where the oil has already been separated from the gas and the gas has passed off, would there not?

A. There would be,—only in case this froth or foam were so attenuated that it had the characteristics of a gas instead of a fluid. But if it behaves as a fluid then it must come to a common level, and the well, on the occasion of a very heavy surge of oil, might cause the fluid level to be slightly higher upon this receiving chamber for an instant, but only for an instant.

Q. Now, let us take this trap here (model), and take this baffle-plate for the purpose of this present question. We will suppose that the oil level as it stands in the trap has reached, we will say, halfway or midway between the bottom and the top of the trap so that there is what is known as an oil seal around the bottom of the [414] trap. Now, if a large portion of oil should come in here, either by reason of the well flowing in heads or for any other reason, would it not be possible, with this gas com-

(Testimony of Paul Paine.)

ing in from the oil and gas well—or this oil and gas—in a foamy condition, this oil down here being in a settled condition, as long as the oil was flowing wouldn't this level in here be considerably higher than the level in here (indicating)?

A. Since there is a large open communication at the bottom the surface of the oil must reach a common level on both sides.

Q. Yes. That is, after the oil has ceased flowing there it is piling up in here, is it not?

A. Well, as I have said, that piling up influence I would only look to be momentary after the well had stopped and then started flowing at a very violent rate; but that condition would only obtain for a very short period because the level of the fluid on both sides of that compartment must be the same.

Q. That would be true if the specific gravity of the oil in the lower chamber was exactly the same as that in the smaller chamber, would it not? In other words, these two chambers here, with this communication underneath, and being tight here, would have a tendency to balance in proportion to their respective weights, wouldn't they?

A. Well, I am sorry, but that statement is not intelligible to me.

Q. Well, now, suppose, for instance, we will say this was soapsuds on this side and this was pure water on this side, and you were pouring in soapsuds and water in here (indicating on model). Now, these would range themselves in proportion to their respective specific gravities, would they not?

(Testimony of Paul Paine.)

A. Yes.

Q. This would be higher on this side because of the lower specific gravity, and this would be lower on this side—I should say higher specific gravity, I presume—at any rate [415] lighter—than it would on this side, would it not?

A. Yes. The relative weights of those two fluids would affect the—

Q. The oil level. A. The behaviour.

Q. Yes.

A. But where it is oil on both sides I cannot conceive of the level being at different heights.

Q. Well, now, oil after the gas has been taken off from it is of a lower specific gravity or heavier than the oil as it comes from the oil well impregnated and surcharged with gas, is it not?

A. Yes; after the gas has left it, oil is, of course heavier than the mixture of the oil and gas.

Q. Then if the oil and gas as it comes from the well is in a foamy or frothy condition, thoroughly surcharged with gas, that comes into the smaller chamber in the Lorraine trap, and while in that chamber the gas is taken off, then the oil in the bottom of the chamber would be heavier than the oil in this side, would it not?

A. That depends upon how fast the gas separates from the oil. The mixture of fluid coming into the trap is lighter than an equal volume of oil after the gas has been removed from it.

Q. And these would range themselves like a balance in a scale, would they not, according to their

(Testimony of Paul Paine.)

relative weights? This on this side would have to be as heavy in proportion as that on the other side?

A. If your liquid there is so foamy and frothy that it behaves as a gas instead of a fluid, then of course its behaviour must be in response to the law of gases—it will spread out in order to occupy all of the space containing it instead of—

Q. This would be lighter over on this side in proportion to the gas content, would it not? [416]

A. Well, I would like to have that question clearer. I don't want to evade.

Q. Well, as a matter of fact, if this oil well produces gas and oil, as it comes from the oil and gas well this oil and gas is lighter than the oil itself after the gas is taken out, is it not?

A. Undoubtedly.

Q. Now, you spoke a little while ago about the pressure on the inside or interior part of an oil and gas separator not producing or having any effect upon the oil and gas while in the separator. Is that correct?

A. No. I wouldn't say that the pressure has no effect. I think the pressure has a very strong influence.

Q. As a matter of fact, then, if the oil and gas comes into an oil and gas separator and there is pressure in that separator, regardless of how it comes, it would have a tendency to cause the oil to hold—some writers say dissolve—the lighter hydrocarbon, such as gasoline, and only the dry gas would come off. Isn't that correct?

(Testimony of Paul Paine.)

A. Its tendency would be to prevent the evaporation of some of those lighter gases.

Q. Yes; just to hold it in; while if it did come off then it would require some other process to put it back in?

A. Yes. That is because of the fact that the capacity of the gas to carry gasoline in the form of a vapor decreases as the pressure is raised.

Q. In other words, when the gasoline onces comes off of the oil it is like a prisoner escaping from jail —you have got to go and catch him again. Isn't that correct?

A. If it comes off of the oil and is mixed with those gases which are always gases, then it is more difficult [417] to recondense it.

Q. And you couldn't do that in a gas trap?

A. Not normally, under any of the ordinary operations of gas traps.

Q. Now these gas-valves you speak of on gas lines are put on for the purpose of reducing the flow or producing pressure in the absorption plants, are they not?

A. Well, they are put on for a variety of purposes. One valve which should be put on all gas lines of that type—and I will say are put on everywhere in good practice—is a pop valve, which, when the pressure reaches a certain height, will release and allow the gas to escape. That is for the protection of the machinery, just as a pop valve is put on a boiler. Then other are valves put in the line so that the flow of gas may be entirely closed or re-

(Testimony of Paul Paine.)

stricted. Then there are also regulating valves put on, and which I think you have in mind, that is, valves which regulate the pressure of the gas on the absorption plant side of the valve. These are put on there for the purpose that if the pressure is varying on the gas trap—and accordingly on the gas line—it may be desired to have the gas delivered to the absorption plant at a uniform pressure. The working of those valves is to maintain that uniform pressure on the discharge side or absorption plant side of the valve.

Q. And that is usually put on near the absorption plant, is it?

A. The regulating valve is usually put on near the absorption plant. The pop valve is more commonly put on up near the trap.

Q. The oil requires a pop valve on all traps, does it not, for safety? A. I don't know that it does.

Q. Well, they all have them on, don't they? [418]

A. Oh, yes. The interests of the company require that.

Q. You don't know whether the law requires the safety valve or not?

A. The law requires it on boilers, but I don't know specifically as to—

Q. Well, these pop valves you speak of let the gas off into the atmosphere anyway, don't they?

A. Usually, when the gas reaches a certain degree of pressure.

Q. Yes, but they don't have anything to do towards maintaining the pressure?

(Testimony of Paul Paine.)

A. Oh, yes; that is exactly what they do. They hold the pressure from going above a certain point, but of course they don't hold it down—

Q. But they don't maintain the pressure. The valve would be there and would not operate until the pressure got to a dangerous point, would it not?

A. Certainly. They don't maintain a uniform pressure.

Q. And they don't maintain any pressure? As a matter of fact when the pressure gets above that it allows the pressure to escape, does it not?

A. Well, it allows such gas to escape as is necessary to hold the pressure at the point for which the valve has been set.

Q. Yes, or to reduce the pressure to that point?

A. Yes; to bring so that that is the maximum pressure the gas would reach.

Q. So that it does not maintain the pressure. Its purpose is to reduce the pressure, is it not?

A. Well,—

Q. Well, suppose, now, as an illustration, that that valve was set for two hundred pounds. Now until two hundred pounds was reached in that gas trap that valve would [419] not operate at all, would it? A. No.

Q. It would be just like any other closed portion of the trap? A. Yes.

Q. Now, when that gas pressure, if it did, got above two hundred pounds, say it got up to four hundred pounds, then this trap would open and reduce the pressure would it not?

(Testimony of Paul Paine.)

A. That wouldn't be the action. When the gas reached two hundred pounds or very slightly over, of pressure, then it would allow sufficient gas to escape so that the pressure would continue at two hundred pounds.

Q. It would cut it down from above two hundred pounds to two hundred pounds, would it not—let it escape?

A. From very slightly over two hundred pounds.

Q. Yes. Well, whatever it was, whether it was little or much? A. Yes.

Mr. BAGG.—That is all.

Redirect Examination.

(By Mr. F. S. LYON.)

Q. I want to ask you one further question on direct, Mr. Paine: You were present at the demonstration of this Lorraine trap a portion of which has been marked for identification as Plaintiff's Exhibit No. 22, on last Friday, were you? A. Yes.

Q. Did you observe the amount of oil that was being pumped from that demonstration? A. Yes.

Q. Based upon your experience, are you able to tell us [420] approximately what the flow of oil was in that demonstration?

Mr. BAGG.—Now, we object to that as incompetent, irrelevant and immaterial.

The COURT.—I think that question was asked out there somewhere.

Mr. F. S. LYON.—I thought your Honor said we should ask it this morning. That is the reason I ask it.

(Testimony of Paul Paine.)

The COURT.—Well, then you may ask it.

A. The rate of flow I would judge to have a minimum of anywhere from 1,000 to 1,200 barrels per day, at that rate.

Q. (By the COURT.) Is that the rate at which it was flowing out there?

A. Yes, sir. At that rate, if it had continued, it would have flowed an amount of not less than somewhere from 1,000 to 1,200 barrels a day. And I might add that, for my own information, I checked up the capacities of centrifugal pumps which were pumping the oil and found that the amount given simply in the manufacturer's catalogue as the capacity of that pump was something in excess of that.

Q. Was that pump running to its capacity when it was operating there?

A. No, sir; I don't think it was. The maximum capacity of that pump was given as somewhere about 3,000 barrels a day and its minimum capacity in the range of 1,300 barrels a day. I don't think it was working anywhere near its maximum capacity.

Q. (By Mr. F. S. LYON.) Just for the information of the Court, based upon your experience with these flow nipples which control the flow from an oil well, will you illustrate how large a stream it takes for various productions per day, very briefly?

A. Well, this flow-plug is a restricted opening put in [421] the pipe coming from the well for the purpose of holding back some of the flow. I

(Testimony of Paul Paine.)

have had a well with a flow-plug in it of one inch in diameter produce 14,000 barrels a day.

Q. Did that have a high pressure?

A. Yes, sir; but I don't know how much it was. The well at Santa Fe Springs has been producing about 4,000 barrels a day with openings of about three-quarters of an inch, and it has had some six or seven hundred pounds pressure on it. A tremendously large quantity of oil can come through a small opening if there is enough punch back of it.

Q. Then it is a very small stream of oil, comparatively, that will amount to 100 barrels a day?

A. Oh, yes.

Mr. LYON.—That is all.

Recross-examination.

(By Mr. BAGG.)

Q. But it all depends on, first, the pressure, and next the viscosity of the oil, whether or not it is light or heavy. So then do you think that with the pressure on that line out there the other day, only ten pounds pressure, that that was sufficient to put a thousand barrels of oil through that small pipe?

A. Well, it was not the pressure that was putting the oil through the pipe—that is, it was not the ten pounds of gas pressure that was putting it through. There was a centrifugal pump driving it through.

Q. Well, you don't know what pressure was developed on that centrifugal pump, do you?

A. No.

Q. There was nothing stated in the evidence as to what that was.

(Testimony of Paul Paine.)

A. No. But we can compute—I haven't done it in detail, but one can compute from the action of a centrifugal [422] pump how much fluid is being pumped by means of it. I didn't do that; I simply referred to the manufacturer's catalogue of the capacities of the pumps, because I have used those tables in my actual designing and construction of plants.

Q. So then about all the testimony you can give in this matter is simply the statement that the amount of oil that would flow through that line or any other lines would depend altogether on the amount of pressure on the line behind the flowing oil and the character of the oil as to its viscosity?

A. And a great many other comparatively inconsequential items, such as size of line and things of that kind. But my observation of the flow was simply compared with the flow of similar streams of oil at wells where I have had an opportunity to measure how much oil the wells were producing.

Q. Now, that oil that was used out there in that experiment was dead oil which had been afterwards impregnated, that is, gas allowed to mingle with it, as it came up through that pipe, was it not?

A. Yes.

Q. And that oil out there was pretty high gravity, was it not?

A. I don't know the gravity. I would judge it to be about a medium gravity of oil, eighteen or nineteen gravity, probably, which would be considered low rather than high gravity.

(Testimony of Paul Paine.)

Q. Did you see the oil as it came from that elbow that day?

A. Yes, I went up and looked at it.

Q. Now, as a matter of fact, you observed how the oil acted after it came out of the elbow, didn't you?

A. Yes.

Q. Some of it did strike this baffle-plate and flow [423] down the side, didn't it (indicating)?

A. Yes.

Q. Some of it struck over here, didn't it?

A. Went over to the—

Q. And some of it had splashed out through here (indicating). The Court had to put an overcoat on in order to protect himself from it?

A. There was very small—a few drops.

Q. Came out through here? A. Yes.

Q. And a good deal of it dropped straight down into the pit, didn't it, in a kind of shower?

A. I looked down there and it seemed dark to me, and I didn't ascertain what proportion was dropping down there.

Q. But you would say there was some that went down there?

A. There was some that dropped.

Q. And just splashed around on that baffle-plate, didn't it? It came out and hit the baffle-plate and some of it struck over here and sort of splashed—a sort of splashing effect?

A. Some went down on the baffle-plate and some went over in the corner.

Mr. BAGG.—That is all. [424]

**Testimony of W. L. McLaine, for Plaintiffs (In Rebuttal).**

W. L. McLAINE, called as a witness on behalf of the plaintiff, in rebuttal, having been first duly sworn, testified as follows:

Direct Examination.

(By Mr. L. S. LYON.)

Q. Please state your name.

A. W. L. McLaine.

Q. You are connected with the General Petroleum Company, are you not? A. Yes, sir.

Q. What is your office with that company?

A. Director of Production.

Q. As Director of Production for the General Petroleum Company have you charge of Tonner No. 1 well of that company? A. Yes, sir.

Q. Do you know if there is a Lorraine gas and oil separator on that well? A. There is.

Q. Were you ever present when an adjustment was made of the gas outlet valve of that trap?

A. Yes.

Q. Will you tell us the reasons for making that adjustment and what adjustment was made?

A. Well, I wouldn't be able to determine as to the adjustments, on account of not knowing the mechanism on the inside of the trap, but I think both valves were regulated at the same time. The well was producing oil through a flow nozzle—

Mr. BAGG.—Now, if your Honor please, we object to any further testimony on the part of this

(Testimony of W. L. McLaine.)

witness. In the first place he does not appear to know why; and he testifies that both valves were adjusted, or two valves were adjusted, [425] and he has not distinguished whether that has anything to do with the gas outlet or not. I think it is immaterial and that it is unnecessarily encumbering the record.

Mr. L. S. LYON.—If you will give him time I think he will state what he knows and what he did. He said he had never seen the inside of the trap, not that he didn't know what had been done with the valves.

The COURT.—Proceed.

A. We had a line running from several wells, a gas line, into our absorption plant in which we were maintaining a certain pressure, and it was decided that—we wanted to see if we couldn't reduce the pressure from this particular well by taking the flow nozzle out of the discharge of the well and running the lead line from the well direct into the trap and adjusting the valves on the trap to see if we couldn't maintain a certain pressure on a regular pressure on the well in order not to throw too much pressure on the well which would restrict the flow of the oil, and I suggested to our superintendent of production that I thought by taking the flow nozzle out of the line we possibly might discharge the oil directly into the trap and regulate the valves to maintain the required pressure on the trap.

(Testimony of W. L. McLaine.)

Q. (By the COURT.) Do you mean the back pressure?

A. The back pressure, instead of using the flow nozzle.

Q. (By Mr. L. S. LYON.) And what did you do in that regard?

A. Well, there were two adjusting nuts on the oil valve and the gas valve that—

Q. Maybe you can use this Plaintiff's Exhibit 12 for the purpose of showing the Court what you refer to (handing the same to witness).

A. Each one of these valves has an arm with a slot cut across it in this manner by which the bolts can be loosened and this valve could be turned down and the other one up, or *vice versa;* you could set them at any particular angle [426] that you might require. But by lowering this valve on the oil side and raising the gas side it would open the oil valve and close the gas valve, and *vice versa.*

Q. (By the COURT.) These are the discharge valves. So we loosened the nuts on here and made two or three adjustments and watched the gage on the top of the trap to see if she would maintain the pressure we wanted, and when we got it to the pressure we desired we let it set and the trap held the pressure on the well that we were looking for.

Q. (By Mr. L. S. LYON.) After the adjustment what pressure, if you remember, was maintained in that trap?

A. I believe it was in the neighborhood of 28 pounds, as nearly as I can remember.

(Testimony of W. L. McLaine.)

Mr. L. S. LYON.—You may inquire, Mr. Bagg.

Mr. BAGG.—Now, if your Honor please, we move to strike out the testimony of this witness for the reason that it is not rebuttal. It does not rebut anything that has gone before.

The COURT.—I think it elucidates the situation. That manner of operating these slots was gone into while we were out at the trap on that day, and there was some testimony as to the effect of the back pressure on the well.

Mr. BAGG.—Very well.

Q. (By Mr. BAGG.) Now this was on the Tonner No. 1, was it not? A. No. 1; yes, sir.

Q. There was a back pressure on the gas line from that trap regardless of whether you moved the valves or not, was there not?

A. There was a slight pressure, yes.

Q. This was not on the Tonner No. 3?

A. On Tonner No. 1.

Mr. BAGG.—That is all. [427]

Mr. L. S. LYON.—That is all. [428]

## Testimony of M. J. Trumble, for Plaintiff (Recalled —In Rebuttal).

M. J. TRUMBLE, recalled on behalf of the plaintiff in rebuttal, testified as follows:

Direct Examination.

(By Mr. F. S. LYON.)

Q. Mr. Trumble, I show you a device here that has been marked Plaintiff's Exhibit 22 for Identification. What is it?

(Testimony of M. J. Trumble.)

A. That is a part of the Lorraine trap that was out in my shop.

Q. That is a part of the Lorraine trap that was demonstrated to the Court last Friday afternoon?

A. Yes, sir.

Q. And was purchased by the General Petroleum Company from the Lorraine Gas Trap Company?

A. Yes, sir.

Q. Have any changes been made in it, other than simply to cut off the section of the iron to remove it, from the condition in which it was shown on Friday afternoon? A. No, sir.

Mr. F. S. LYON.—We offer in evidence Plaintiff's Exhibit 22.

Mr. BAGG.—We offer the same objection, for the reason that it is not proper rebuttal testimony.

The COURT.—The objection is overruled.

Q. (By Mr. F. S. LYON.) I show you two devices and ask you if you know what they are (exhibiting same to witness). A. A pressure gage.

Q. Where did it come from?

A. It came in the tools or mechanics for the trap, in the box.

Q. And that box was opened for the purpose of making certain connections to demonstrate to the court these values on [429] the side of this trap?

A. Yes, sir.

Q. And this pressure gage was set out in this box—

Mr. BAGG.—If the Court please I would like to object to the leading form of the question.

(Testimony of M. J. Trumble.)

The COURT.—I think that was very leading.

Mr. F. S. LYON.—So do I, but I don't think there is any dispute about the facts.

The COURT.—Someone testified out there that that gage was not used but it was another gage of the same kind only smaller.

A. The same pressure, but not the same size dial. The box was opened up there at the time and this was taken out.

Q. That is the size of gage that is ordinarily used by this trap?

A. Three hundred pounds; yes, sir.

Mr. F. S. LYON.—We offer this as Plaintiff's Exhibit 23. That is all.

Mr. BAGG.—That is the same size pressure gage that you use on your trap, is it not?

A. On the high pressure traps.

Q. Down on the Tonner No. 3 you had the same kind of looking valve or gage, did you not?

A. A lower pressure.

Mr. BAGG.—No further questions.

Mr. F. S. LYON.—That is all. [430]

**Testimony of Ford W. Harris, for Plaintiff (Recalled—In Rebuttal).**

FORD W. HARRIS, recalled by the plaintiff in rebuttal, testified as follows:

Direct Examination.

(By Mr. F. S. LYON.)

Q. You are the same Ford W. Harris who has

(Testimony of Ford W. Harris.)

heretofore testified on behalf of the plaintiff in this case? A. I am.

Q. Have you examined and are you familiar with Defendant's Exhibits "E," "F," "G," "H," "I" and "J"? A. I am.

Q. Will you please take these respective patents and state briefly the construction and principles upon which the devices of these patents operate and compare the same with the plaintiff's patent, Plaintiff's Exhibit No. 1, and with the defendant's traps, Plaintiff's Exhibits 22, and 10? Be as brief as you can and get right down to the meat of each one as you go along so as not to take any more time than necessary.

A. Referring to Defendant's Exhibit "E," which is the McIntosh patent, this patent shows a pipe 9, through which the liquid is passed into the trap, this liquid being shown as passing up in a sort of fountain and flowing downwardly over some members which are secured on the outside of the pipe.

The COURT—Isn't that fastened up against an obstruction there?

A. There is no obstruction above the inlet pipe, unless it should strike against the top of the trap, which might occur if there was considerable pressure.

Q. It is the action, then, of a fountain?

A. Of a fountain, yes. It simply comes up through there, bubbles up and falls back. So far as the drawing shows, it does not strike the outer surface, or outer wall, of the condenser at all, nor

(Testimony of Ford W. Harris.)

is there any means for maintaining [431] pressure, nor are there any floats in there for maintaining the level of the oil in the trap.

Q. (By Mr. F. S. LYON.) The separation in that trap, then, of the gas, is what might be termed to be in a body of oil, is it?

Mr. BAGG.—Now this is an intelligent witness, and I suggest that—

Mr. F. S. LYON.—All right. Your Honor has heard his testimony.

A. Defendant's Exhibit "F," which is the Taylor patent, is a steam separator. It has nothing to do with gas traps that I can see. The steam enters through a pipe 3a and flows down over a series of baffles, presumably the steam contains some water, and the steam is taken out from the very bottom of the trap through a passage 17 put out through a pipe 5. There is a float in the bottom for draining out the condensed water from time to time as it collects in the trap. There is no means for maintaining pressure on this trap, and it is not intended to separate gas and oil, and I question seriously whether it would separate gas and oil.

Defendant's Exhibit "G," being the Barker patent, shows a chamber into which the natural gas is blown through a pipe 5 and at the end of the pipe 5 has a hood or cap 7 over it, evidently intended to keep the gas from striking the top. It acts as a deflector plate and tends to throw any liquid that may be carried in the gas downwardly.

(Testimony of Ford W. Harris.)

There is no means of spreading the oil—in fact it is not intended there should be any great quantity of oil in this trap, which is, naturally, a gas trap. There is no means of spreading the oil on the wall of the chamber, no means of maintaining the pressure in the chamber, and there is no float valve for maintaining the oil level in the chamber.

Defendant's Exhibit "H," being the Bray patent, is a patent that I am quite familiar with as it was the principal [432] patent cited by the Patent Office when this case was prosecuted before the office and we had some difficulty with the Patent Office in connection with this patent. The difficulty with the Patent Office arose from the great similarity in external appearance between this trap and the Trumble trap. It is a cylindrical trap with a cone top and cone bottom, has a float in it, and there was an appearance of similarity which, after some discussion, we convinced the Patent Office was not a real similarity. The oil is introduced and passes through a series of screens and comes into the bottom of the trap. There is no means of spreading it in a thin film nor of maintaining any pressure inside of the trap. It is simply a sort of screening proposition.

Defendant's Exhibit "I," being the Cooper patent, shows a trap for operating on the centrifugal principle. I have quite recently had some experience with centrifugal separators for internal combustion engines which work on a somewhat similar order. I don't think this drawing is a correct

(Testimony of Ford W. Harris.)

representation, as it shows the liquid on the bottom on a level and quiescent. If the combined liquid and gas enters this separator with any degree of force there would be produced inside of the separator a vortex similar to that which we all observe—

Q. Take a sheet of paper and illustrate to the Court what you mean.

A. A vortex similar to that which we observe when we pull the plug out of a bathtub. Of course this would be a much more vertical vortex, but nevertheless it would be a vortex. In the sketch which I am making the fluid entering through a pipe marked "1" tangential shell tends to rotate the whole mass of material inside the shell—the gas and oil or anything there may be in there—in the form of a vortex, and a typical form of vortex, due to centrifugal action, [433] would be something similar to that which I will show by a line which I will mark "2," the shaded portion below that line representing the oil. In other words, the whole mass of material inside the trap would tend to be churned up, and it would tend to cling to the outside of the trap, this tendency becoming less as the material got deeper in the trap, so that we would have a whirlpool in there which would tend to throw the sand and the water toward its outer surface and the oil in the middle. I believe the action of the sand on this trap would be very destructive, from my observation of the wearing of nipples and pipes in wells that carry considerable sand. If the

(Testimony of Ford W. Harris.)

velocity of the oil and liquid entering this trap is low there would be no such vortex action, or a very slight one, but in that case the oil would simply be thrown in and would flow down inside the trap. Operated according to the patent, however, this condition that I show in my sketch would be the one that would obtain.

Mr. F. S. LYON.—The sketch produced by the witness is offered in evidence as Plaintiff's Exhibit 4.

A. (Continuing.) This patent shows no means for maintaining pressure inside the trap.

The next patent, Defendant's Exhibit "J," being the Noon patent, shows a water separator—a separator for taking gas from water. The fluid containing the gas enters the tops of the two chambers through pipes which are provided with nozzles g2, and these nozzles are pointed directly downwardly so that the fluid is driven down with considerable force. To break the force and prevent the liquid from being violently injected into the fluid in the bottom of the trap baffle-plates, k′ and l′, are provided in the two chambers. The purpose of these baffle-plates, as I believe the patentee says, is that "the hoods k′ and l′ preventing the gas from boiling up the water accumulated in [434] the bottom of the tank, thus preventing much spray and protecting the float 'm' from the incoming rush of gas." It is my opinion that while some of this water striking on top of these hoods may reach the wall of the tank, this trap was not in-

(Testimony of Ford W. Harris.)

tended or designed or used to effect the separation in this manner, because if that had been desired these hoods would simply have been placed up directly under the inlet and the inlet itself would have been placed higher. I think the only purpose and the only function of them is to prevent this boiling up and to protect the float as specified by the patentee.

Mr. F. S. LYON.—You may inquire.

Cross-examination.

(By Mr. BAGG.)

Q. Mr. Harris, you state in examining these various patents that you draw your conclusions as to their operation somewhat upon the intent of these various devices. Is that correct?

A. I don't think I so stated.

Q. Now in this last patent, the Noon patent, you spoke about the intent of these baffles.

A. The purpose of the baffles.

Q. Well, didn't you use the word "intent"?

A. It is possible that I did.

Q. Now you are basing your testimony, then, in a large measure upon what you understand to be the intent of these baffles and not what the actual effect would be?

A. I think I am basing it on what I think would be the result if the trap were constructed in that manner.

Q. But you do base your conclusions in a measure on [435] the intent as described in the specifications?

(Testimony of Ford W. Harris.)

A. Well, I don't know just exactly what you mean by that. I base my conclusions on what I consider would result if the trap were constructed in that way and what the patentee says in the specification as to what would result.

Q. Now, let us take Defendant's Exhibit "J," which is the Newman patent. If the water and gas should come in in a large flow in any kind of a head, as it came down through these nozzles as you might call them, the separators in the oil line, and struck upon these baffle-plates, they would strike with a splashing effect, would they not?

A. On the side of the container, yes.

Q. Yes, and producing a splashing effect upon these side-walls? A. Yes.

Q. And oil would do the same thing?

A. Yes.

Q. It would splash over here against the side-walls and flow down the side? A. Yes.

Q. Now these baffle-plates are more or less rectangular in shape, are they not, as shown here in the drawing?

A. Baffle-plate k′ is rectangular because the Figure 4 shows it to be rectangular.

Q. Now that plate, then, if this was a cylindrical form, this outer shell, would have a space on each end of it which would be more or less vacant, would it not, that is, it would not be tied to anything that would deflect the coming in or going out of the oil or gas, would it? A. That is correct.

Q. So that there would be space in there for the

(Testimony of Ford W. Harris.)

gas to come up past the ends of these baffle-plates in there, wouldn't there? [436]

A. Yes, but it would have to come up through the mass of oil coming down.

Q. Now, if this struck this baffle-plate here the tendency would be just like that of water flowing on the roof of a house, would it not, it would flow down hill, and would it not flow over towards the sides of this receptacle? A. Yes.

Q. Then there would be a space up and down the ends of this baffle which would let the oil and gas come up that way, would there not?

A. I don't think so. I don't think it would allow any oil to come up there.

Q. Well, I mean gas. I beg pardon.

A. Some gas would come out—

Q. All the gas that was in the oil as it struck this plate and started down there, if it came out, would have to come out down in here (indicating), would it not? A. Yes.

Q. Now I will ask you to examine this Cooper patent, which is Defendant's Exhibit "I," and state if there is not a plate shown in this drawing across the wall or adjacent to the wall of the receptacle, against which this oil and sand is projected as it comes in. A. The plate a'.

Q. Yes; and that is evidently for the purpose of reinforcing this side-wall against wear, is it not?

A. Well, I don't know what the purpose of it is.

Q. Well, you would say it would, would you not?

A. It would help, certainly,—

(Testimony of Ford W. Harris.)

Q. Now, if that was made out of glazed material it would take it a long time to wear out, would it not? A. Yes.

Q. Now if the oil was coming into that trap in the same [437] manner and at the same speed that was illustrated at the test out at the plaintiff's plant the other day in the presence of the Court, flowing in practically the same size stream, there wouldn't be any vortex effect in that, would there?

A. Well, of course, it would depend upon the size of the tank.

Q. Well, if it was what you might call a reasonable, moderate size tank.

The COURT.—That plate is called a steel wearing plate.

Mr. BAGG.—Yes.

Q. A steel wearing plate would be, evidently, then, from your description, for the purpose of preventing wear on the side-walls of this receptacle?

A. Yes, sir. It is evident that the patentee understood that he would get wear there.

Q. Now I don't remember whether you answered my question with reference to that—probably the Court didn't hear you. If the oil was coming in through this Cooper trap at the same rate of speed that was illustrated out at the experiment the other day at the plant of the plaintiff there would not be any vortex such as you have described in that event, would there?

A. It depends entirely upon the size of the trap. You will have to define "reasonable size," Mr. Bagg.

(Testimony of Ford W. Harris.)

Q. Well, say it was three feet across.

A. The same diameter, approximately, as the one you had out there?

Q. Yes.

A. I don't know positively, but it would seem to me that that oil would come down in a very similar manner if it were arranged that way, but of course I don't consider that that is the way Cooper ever did it or that that is [438] the way that patent would work.

Q. Well, that is the way it is described anyway.

A. No, sir.

Q. Now as to "intended to work" you are coming back to the intention of the patentee; you are not testifying as to the structure as actually described?

A. I am coming back to his method of operation as described. He says it goes around by centrifugal force, and that it flows down in films as it goes down this trap.

Q. Yes, it would have to flow down some place, would it not? It couldn't keep going round and round.

A. It goes around long enough to form this vortex.

Q. But if it was coming in at the slow rate it was coming in at the other day it wouldn't have that vortex effect, would it?

A. If the trap were made of this size and according to these specifications and the oil was flowing in at a suitable rate I would say the oil would go down over the wall of the trap here, the same as in Mr.

(Testimony of Ford W. Harris.)

Lorraine's trap, but I don't think this patent was intended to or ever did work that way, this particular trap.

Q. Well, that is just your opinion?

A. Well, no; I am going by what it said in the patent. It says: "The material was thereby given a rotary motion which separates the constituents conformably to their specific gravity." Now if that goes in there fast enough to form a centrifugal motion automatically it goes in at a high speed.

Q. Yes, but that would depend altogether on the well to which it was attached, as to whether it flowed rapidly or produced largely or not, would it not?

A. I don't think so. I think it is the size and construction of the trap. [439]

Q. Well, I am speaking about a three foot trap such as the Lorraine patent. It would depend altogether on the character of the flow of that oil from that particular oil well, would it not?

A. As applied to a well producing say one thousand or fifteen hundred barrels a day?

Q. Approximately, yes.

A. Approximately the amount this well was flowing.

Q. Such as the witness has described.

A. Well, I think it might work very much the same way.

Q. Now all oil and gas wells do not produce a large quantity of sand, do they? A. No.

Q. Some of them don't produce hardly any sand, do they?

(Testimony of Ford W. Harris.)

A. In a great many of them the oil is very clean.

Q. So that it would be only in those cases where there was a large amount of oil and the flow was very violent that there would be any material wear upon the—

A. No, I wouldn't say that. I say it is wholly dependent upon the sand and the character of the sand.

Q. And the amount too, is it not?

A. And the amount.

Q. You have never had any experience in watching the interior workings of these traps, have you?

A. No, except what I had out there the other day.

Mr. BAGG.—That is all.

Mr. F. S. LYON.—That is all. Plaintiff rests.

Mr. BAGG.—Defendant rests.

(Discussion *re* time for arguments and filing of brief.)

Mr. F. S. LYON.—I believe that new act is not in effect yet, is it, in regard to Judges from outside districts signing orders and entering decrees and so forth? I think, Mr. Bagg, we had better have a stipulation that Judge Wolverton may sign decrees, hear petitions for rehearing, [440] etc., in this case outside of the district with the same effect as though it were done here.

The COURT.—And after the expiration of the designated time—

Mr. F. S. LYON.—Yes, with the same force and

effect as though duly designated and personally present within the district.

Mr. BAGG.—Yes.

The CLERK.—Will you file a written stipulation to that effect?

The COURT.—The Court will take a recess until two o'clock.

(A recess was thereupon taken until two o'clock P. M.) [441]

AFTERNOON SESSION—Two o'clock.

(Discussion *re* arguments.)

Mr. F. S. LYON.—Before starting the argument, your Honor, I wish to submit for ruling hereafter a motion I now make to strike from the record and exclude from consideration Defendant's Exhibit "C," upon each of the grounds stated in the objection to the exhibit, and upon the further ground that this printed copy, if it is a copy, is not admissible without certification; that it is a private document and not a public document, and that there is no proof of it.

Now in that connection I also call your Honor's attention to the fact that, so far as the so-called Tico trap is concerned, there is no assertion of date for the Tico trap in any manner in the publication; and the same publication shows the Trumble gas trap, and it has nothing in it to show which was prior in point of time; and I will submit in our brief our authorities such as we have been able to find which bear upon the subject, calling your Honor's attention particularly to the statute which is the statute of organization of the Bureau of Mines and to the fact that that statute does not in any manner particularly make evidence any copy such as this, and there is no provision in that statute which makes any of its investigations or anything of that kind evidence, and it would depend solely upon the applicability of the doctrine of public documents; and I call your Honor's attention in that connection to the fact that there is no proof here that there is an original document filed by an officer whose duty it was to compile this, or any of the other requirements, and to the fact that the procedure in that connection is to require proof in certified form of all those documents.

The COURT.—The Court will take the matter under advisement, to be determined later on in the case when it is finally submitted. [442]

Mr. F. S. LYON.—And whichever way the ruling is it will be understood we are reserving exceptions on behalf of the parties.

The COURT.—Yes.

(Final Argument.)

[Endorsed]: In the District Court of the United States for the Southern District of California. (Before Hon. Charles E. Wolverton, Judge.) Francis M. Townsend et al., Plaintiff, vs. David G. Lorraine, Defendant. No. E–113—Equity. Reporter's Transcript. Vol. V. Filed Apr. 7, 1922. Chas. N. Williams, Clerk. By R. S. Zimmerman, Deputy Clerk. Los Angeles, California, March 28, 1922. Reported by J. P. Doyle. Doyle & St. Maurice, Shorthand Reporters and Notaries Suite,

507 Bankitaly International Building, Los Angeles, California, Main 2896. [443]

---

In the District Court of the United States, Southern District of California, Southern Division.

No. E–113.

FRANCIS M. TOWNSEND, MILON J. TRUMBLE and ALFRED J. GUTZLER, Doing Business Under the Firm Name of TRUMBLE GAS TRAP CO.,

Plaintiffs,

vs.

DAVID G. LORRAINE,

Defendant.

**Opinion.**

FREDERICK S. LYON, LEONARD S. LYON and FRANK L. A. GRAHAM, for Plaintiffs.

CHARLES BAGG, for Defendant.

WOLVERTON, District Judge:

Complainants are the rightful owners of a patent on a crude oil and natural gas separator, No. 1,269,134, issued June 11, 1918, application for which was filed November 14, 1914, and claim that the same has been infringed by defendant's reissue patent of November 8, 1921, No. 15,220, and an apparatus recently constructed of somewhat similar design. The defendant does not question the validity of complainant's patent, but claims that he does not infringe, for two reasons: First, that complainants are estopped, by reason of the proceeding before the Patent Examiner, from claiming any other means than that which spreads the whole of the incoming oil, gas, water and sand upon the walls of an expansion chamber in a thin film; and, second, that in view of the state of the art at the date of complainants' patent, they are precluded from claiming any other form of structure than that set out and described in the drawings and specifications accompanying such application.

The claims of complainants' patent which it is thought have been infringed are 1, 2, 3, 4, and 13. The first claim reads: [444]

> "In an oil and gas separator, the combination of an expansion chamber arranged to receive oil and gas in its upper portion, means for spreading the oil over the wall of such chamber to flow downwardly thereover, gas take-off means arranged to take off gas from within the flowing film of oil, an oil collecting chamber below the expansion chamber, an oil outlet from said collecting chamber, and valve controlled means arranged to maintain a submergence of the oil outlet."

Claim 2 varies from claim 1, so far as it is necessary to indicate, in that it comprises "means within the chamber adapted and arranged to distribute the oil over the wall of the chamber in a downwardly flowing film, gas take-off means arranged to take gas from within the envelop of downwardly flowing oil, and means for maintaining gas pressure upon such oil.''

So of claim 3: ''The combination of an expansion chamber having a surface adapted to sustain a flow of oil thereover in a thin body, means for distributing oil on to such surface, pressure-maintaining means arranged and adapted to maintain a pressure on one side of the flowing oil;'' and of 4: ''Means for maintaining pressure within the chamber, * * * means within the chamber adapted to cause the oil to flow in a thin body for a distance to enable the gas contained and carried thereby to be given off while the oil is subjected to pressure.'' Claim 13 includes ''an imperforate spreader cone, having its apex pointing upwardly, located inside said chamber in such a manner as to spread a thin film of oil over the inner wall of said chamber.''

Turning to the file-wrapper showing the proceedings before the examiner, claim 1 as made in the application contains the element ''means for reducing the oil into a finally divided condition to reduce the tension on the gas contained therein.'' Claim 2, ''Oil dividing means arranged in the expanding chamber to reduce the oil to a thin film-like condition.'' Claim 3, ''gas freeing means consisting of means to reduce the oil to a thin film arranged within the expanding chamber;'' and claim 4, ''A cone arranged near the top of such chamber to receive the incoming oil and spread it over the wall of the chamber in a thin film-like form.'' [445]

When the application came to the examiner, claims 1, 2 and 3 were each rejected on the application on the patents of Barker and Bray, and 4 on patent of Bray. The action of the examiner induced the petitioner to add the following to his specifications:

> "It will be noted that the action upon the oil while flowing down the wall of the expansion chamber in a thin film under pressure permits the free, dry gas to readily escape therefrom, while the pressure exerted upon the oil surface backed by the wall of the chamber holds the lighter liquids, such as gasoline, in combination with the oil body, and I desire to be understood as pointing out and claiming this action as being of great benefit to the crude oil derived from the well on account of keeping the gasoline series in combination with the main body of oil."

Also to cancel claims 1, 2, and 3, and to insert claims 1, 2, 3, and 4 as now contained in the issued patent.

The examiner again rejected claims 1 to 4 inclusive, on patent of Bray, and 5 to 13, inclusive, were held not to patentably distinguish from Bray, and accordingly were rejected. In response to these objections, the applicant added claims 13 and 14 as now contained in the patent. As presented, the examiner again rejected claims 14 and 15, being claims 13 and 14 in patent, also claims 1 to 13, inelusive, as not to patentably distinguish from references of record. The applicant replied to the action of the examiner, stating, among other things, that the "applicant's invention consists of a containing vessel, an imperforate cone adapted to spread the whole body of the oil to the outer edge of the vessel, and means for taking off gas from the interior of the cone near the center of the vessel"; this to distinguish from the Bray patent. He says, further: "Moreover Bray does not take off his gas below his screens, and the claims of Trumble are quite specific in stating that the gas is taken off inside the cone."

The matter coming again before the examiner, on reconsideration, all the claims were allowed as contained in the patent. Claim 9 (original claim 8) was rejected as met by patent of Bray, and has been eliminated from the patent. [446]

A patentee, where he is required by the rulings of the Patent Office to modify and restrict his claims, to obviate anticipation by previous patents, is by the limitations he thus imposes upon such claims, and where the patent is for a combination of parts, his claims must be limited to a combination of all the elements which he has included in his claims as necessarily constituting that combination. Phoenix Caster Co. vs. Spiegel, 133, U. S. 360, 368; New York Asbestos Mfg. Co. vs. Ambler Asbestos A. C. C. Co., 103 Feb. 316. And it was said in Roemer vs. Peddie, 132 U. S. 313, 317:

> "When a patentee, on the rejection of his application, inserts in his specification, in consequence, limitations and restrictions for the purpose of obtaining his patent, he cannot, after

> he has obtained it, claim that it shall be construed as it would have been construed if such limitations and restrictions were not contained in it."

See, also, National Hollow Brake-Beam Co. vs. Inter-changeable Brake-Beam Co., 106 Fed. 693, 714, where the Court adds:

> "But this is the limit of the estoppel. One who acquiesces in the rejection of his claim because it is said to be anticipated by other patents or references is not thereby estopped from claiming and securing by an amended claim every known and useful improvement which he has invented that is not disclosed by those references."

Two thoughts were uppermost with the patentee in making the changes indicated: First, to avoid the objection with reference to Barker and Bray with means for reducing the oil into a finely divided condition; and, second, to confine the oil in its flow down a wall or surface with maintained pressure meanwhile. The theory of the patentee is obviously that, pressure being maintained, the dry gas will readily escape from a thin film or body of oil passing down and against a wall or other surface, without at the same time taking off the lighter liquids, such as gasoline, which will yet remain in the crude oil and add to its value.

The limitation and restriction which the patentee has imposed upon his patent must be gathered from his addition to his specifications and the claims which were finally approved by the examiner. He says in the added specifications that the free, dry gas readily escapes, while the pressure exerted upon the oil surface, backed by the wall of the chamber, holds the lighter liquids in the oil body. In his claims, however, he asserts a broader scope for his [447] invention, as in claim 3, which comprises "the combination of an expansion chamber having a surface adapted to sustain a flow of oil thereover in a thin body, means for distributing oil on to such surface, pressure-maintaining means arranged and adapted to maintain pressure on one side of the flowing oil." All this was approved and allowed by the examiner.

Construing the whole together, the added specifications and the claims, I am impressed that the patentee is not confined to means of causing the oil to flow down the outer wall of the chamber, but that his patent includes any means that will cause the oil to flow down any surface as well, such as a baffle-plate or inner partition or wall, which is reached after the emulsified oil enters the chamber. I think therefore, the patentee is not estopped by the proceedings before the Patent Office to insist upon the broader claims.

Now, as to the state of art: Defendant makes no question as to the validity of complainants' patent, and does not rely upon any anticipation for defeating it. The point he makes, therefore, as to what must be accomplished by a combination device would seem to be irrelevant. The question remains, Are the elements of complainants' device so restricted, in view of the state of the art, as to subject them to so narrow a construction as to limit them to the very means shown by the drawings and specifications in patent?

The complainants' patent contains a cone device near the top of the upper chamber, with its lower rim extending in its full circumference in close proximity with the wall of the chamber, with no gas take-off above the cone, so that the entire emulsified body flows down a wall; the gas to be taken off below the cone. Defendant's patent, referred to in counsel's brief as Model 1, has an inner partition set away from the wall on one side more than one-third the distance of the diameter of the chamber, and extending below the oil level. To this partition, at some distance from the top of the chamber, is attached a baffle-plate extending downward on an incline of perhaps 45 degrees, and to within an inch and a half or two inches from the wall for the entire segment [448] cut off by the partition. The oil inlet, consisting of a pipe, extends downward to within a short distance of a baffle-plate. The pipe has two openings, so that the stream of oil is divided and projected on the baffle-plate in two directions laterally. The device is provided with a gas take-off above the partition and one from underneath the baffle-plate; all to pass off eventually from the upper portion of the major chamber.

Model 2 contains a like partition to that described in Model 1. The oil inlet consists of a pipe extending into the side of the minor chamber, supplied with what is called a nipple, bell-shaped, to allow the oil to spread when discharged into the chamber.

The nipple is set at an angle with and extended within proximity of the inner wall, the effect of which is, when the oil is discharged into the chamber, to carry part of it down the inner partition wall, part down the outer wall, at and near the intersection of the inner with the outer wall, and part of it down by gravity without reaching either wall. The device is provided with a gas take-off above the nipple.

This sufficiently describes the models to make the application later. I may add further that the nipple in the model in evidence is machined off on one side to sit closely against the partition wall. Defendant says this was done through mistake in setting the nipple, the machine having allowed it to extend too far inwardly. If this is true, it only shows how easy it is to set the nipple in without discovery. But we are dealing with the model in evidence, which complainants say infringes their patent.

The patents introduced as showing the prior art are readily disposed of. Exhibits "E," the McIntosh patent, "F," the Taylor patent, "G," the Barker patent, and "J," the Newman patent, all inject the oil from the well in the form of a spray, having the effect to reduce it to a finely divided condition, and the gas is thus permitted to escape. None of them are provided with baffle-plates except Newman, but the oil does not reach them except as sprayed upon them, and I think none of these patents contain the element in combination of pressure within the chamber. All these patents are obviated in their evidentiary effect by the restrictions of [449] complainants' specifications and claims as made before the examiner. Exhibit "H," the Bray patent, is subject to the same criticism. The oil is there precipitated upon perforated cones, and only slightly, if at all, flows down the wall of the chamber or other surface. Exhibit "I," the Cooper patent, injects the crude oil tangentially to the wall of the chamber, and causes it to flow down the wall more or less, and the gas escapes upwardly and passes out in a take-off at the top of the chamber. But this patent in combination contains no element of pressure. It must be observed that we are dealing with a combination patent, and all the elements must be read with reference thereto.

Trout, in his testimony and by a drawing submitted by him, exhibits a trap having a chamber into which the oil is injected tangentially to the outer wall, but provided with a sleeve, which allows the gas to escape upward from a segment at the upper part of the sleeve, and baffle-plates are provided above, which can only have the effect to deflect the gas as it passes upwardly and thence outwardly by the gas outlet. This trap, as described, had the element of pressure. The testimony is not persuasive, however, as no drawings were presented, although, as witness testifies, some of them are in existence. The device was called the Durward trap, after the superintendent of the property, who constructed the trap. Durward himself was not called, although accessible. The trap was never patented, and fell into disuse.

In Parker vs. Stehler, 177 Feb. 210, it is laid down that:

> "The Courts have recognized the rule that oral testimony of witnesses speaking from memory only in respect to past transactions and old structures claimed to anticipate a patented device, physical evidence of which is not produced, is very unreliable, and that it must be so clear and satisfactory as to convince the Court beyond a reasonable doubt before it will be accepted as establishing anticipation."

See, also, Diamond Patent Co. vs. S. E. Carr Co., 217 Fed. 400, 402.

The testimony of Swoap and of Wharff is subject to the same criticism.

Respecting Exhibit "C," which is a pamphlet entitled "Traps for Saving Gas at Oil Wells," written by W. R. Hamilton, it was issued by the Bureau of Mines, at the Government Printing Office, in 1919, and treats largely of traps for separating gas from oil, and by [450] illustration presents many traps, from the most primitive to the more complex and modern; some having the element of pressure in combination. The document is not certified as official. The statements it contains in regard to the elements of the traps exhibited lack the solemnity of an oath, and the opposing party is deprived of the opportunity of cross-examination. I was inclined to admit this document as competent evidence at the trial, and did admit it subject to further consideration. I am now convinced, however, that the document is not admissible. If admissible, it could have no greater weight than the testimony of a witness speaking to the facts therein narrated, and could not be of compelling force, under the rule announced by the two authorities last above cited.

It is argued that the principle of subjecting oil to pressure, for the purpose of keeping lighter hydrocarbons in solution in the oil while the dry gas constituent separates from the body of the oil, is old, but this overlooks the theory of complainants that they have discovered a more efficient way of separating the gas from the oil, whereby a greater proportion of oil value is secured than had theretofore been derived by the use of any trap in existence or previously operated. Utility has been abundantly proven by the success achieved by complainants' device.

The defendant's trap, Model No. 1, infringes, in that the baffle-plate furnishes a surface down which the oil flows, with pressure against the oil, by which the gas escapes from the oil and passes out of the chamber by the take-off. So of defendant's device, Model 2, the oil is injected in part at least, against the partition, as well as against the chamber wall, so that it flows down thereon with pressure on the moving oil, from which the gas escapes. While part of the oil is reduced to a spray which falls by gravity to the settled fluid below, its action does not obviate the objectionable feature of a part flowing down the partition and a part down the wall. I am of the opinion also that defendant's trap will likewise infringe with the nipple constructed, as he claims it should be, according to drawings and specifications. [451]

I find that defendant's patent infringes claims 3 and 4 of complainants' patent, that his Model No. 2 infringes claims 1 to 4, inclusive, and that claim 13 is not infringed. Complainants wll have a decree accordingly, and the cause will be continued for an accounting.

[Endorsed]: No. E-113. U. S. District Court, Southern District of California. F. M. Townsend et al. vs. David G. Lorraine. Opinion of Judge Wolverton. Filed Sep. 11, 1922. Chas. N. Williams, Clerk.. By R. S. Zimmerman, Deputy Clerk. [452]

---

United States District Court, Southern District of California, Southern Division.

IN EQUITY—No. E-113.

FRANCIS M. TOWNSEND, MILON J. TRUMBLE and ALFRED J. GUTZLER, Doing Business Under the Firm Name of TRUMBLE GAS TRAP CO.,

Plaintiffs,

vs.

DAVID G. LORRAINE,

Defendant.

**Interlocutory Decree.**

This cause having heretofore come on regularly to be heard and tried in open court before United States District Judge Charles E. Wolverton, upon the proofs, documentary and oral, taken and submitted in the case and being of record therein, the plaintiffs being represented by Messrs. Frederick S. Lyon, Leonard S. Lyon and Frank L. A. Graham, and the defendant by Charles Bagg, Esq., and the case having been submitted on briefs to the Court for its consideration and decision, and the Court being now fully advised in the premises, and its opinion having been rendered and filed herein,

IT IS HEREBY ORDERED, ADJUDGED AND DECREED as follows:

1. That plaintiffs Francis M. Townsend, Milon J. Trumble and Alfred J. Gutzler, doing business under the firm name of Trumble Gas Trap Co., are the rightful owners of the United States letters patent No. 1,269,134, granted on June 11, 1918, to them, for a certain new and useful invention, to wit, a Crude Oil and Natural Gas Separator, and that the validity of said patent was not denied or put in issue by defendant in the above case; that said letters patent are good and valid in law, particularly as to [453] claims 1, 2, 3 and 4 thereof, which said claims are as follows:

> "1. In an oil and gas separator, the combination of an expansion chamber arranged to

receive oil and gas in its upper portion, means for spreading the oil over the wall of such chamber to flow downwardly thereover, gas take-off means arranged to take off gas from within the flowing film of oil, an oil collecting chamber below the expansion chamber, an oil outlet from said collecting chamber, and valve controlled means arranged to maintain a submergence of the oil outlet.

"2. In an oil and gas separator, the combination of an expansion chamber, inlet means arranged to permit the entrance of oil and gas into the chamber, means within the chamber adapted and arranged to distribute the oil over the wall of the chamber in a downwardly flowing film, gas take-off means arranged to take gas from within the envelop of downwardly flowing oil, means for maintaining gas pressure upon such oil.

"3. In an oil and gas separator, the combination of an expansion chamber having a surface adapted to sustain a flow of oil thereover in a thin body, means for distributing oil on to such surface, pressure-maintaining means arranged and adapted to maintain a pressure on one side of the flowing oil, withdrawing means arranged to take gas from the chamber, and means for withdrawing oil from the chamber.

"4. In an oil and gas separator, the combination of an expansion chamber, means for delivering oil and gas into the chamber, means for maintaining pressure within the chamber, means for drawing oil from the chamber, and means within the chamber adapted to cause the oil to flow in a thin body for a distance to enable the gas contained and carried thereby to be given off while the oil is subjected to pressure.'"

2. That defendant has infringed upon claims, 1, 2, 3 and 4 of plaintiff's patent No. 1,269,134, by making and causing to be made, and selling and causing to be sold, and causing to be used, apparatus embodying the invention patented in and by the said respective claims 1, 2, 3 and 4 of plaintiff's patent aforesaid, and that the apparatus made and sold by defendant referred to in this case as Defendant's Model No. 1, and described in reissued letters patent of the United States, No. 15,220 granted [454] November 8, 1921, to defendant, infringes upon said claims 3 and 4 of plaintiff's said patent, and that the apparatus made and sold by defendant referred to in this case as Defendant's Model No. 2, infringes upon said claims 1, 2, 3 and 4 of plaintiff's said patent both when made with the nipple machined off on one side to sit closely against the partition wall as illustrated in said Model No. 2, and when made without such machining or setting, as defendant claims such device was intended to be constructed; and that defendant has not infringed upon claim 13 of plaintiff's patent.

3. That defendant, his agents, servants, employees, confederates, attorneys and associates, and each and every of them, be and they are, and each of them is, hereby permanently enjoined and re-

strained from making, using, or selling, or causing to be made, used or sold, any device or apparatus embodying or containing the invention described and claimed in and by the said claims 1, 2, 3 and 4 of plaintiff's patent No. 1,269,134, and either or any of said claims, and from infringing upon and from contributing to the infringement of said claims, or either of them, and that a permanent Writ of Injunction issue out of and under the seal of this Court, commanding and enjoining said defendant, his agents, servants, employees, confederates, attorneys and associates, and each of them, as aforesaid.

4. That the plaintiffs have and recover of and from the said defendant David G. Lorraine, the profits which said defendant has realized and the damages which plaintiffs have sustained, from and by reason of the infringement aforesaid; and for the purpose of ascertaining and stating the amount of said profits and damages this cause is hereby referred to Honorable Charles C. Montgomery, Esq., as Special Master, *pro hac vice,* to ascertain, take, state and report to this Court an account of [455] all the profits received, realized or accrued by or to the defendant, and to assess all the damages suffered by the plaintiffs from and by reason of the infringement aforesaid, and that on said accounting the plaintiffs have the right to cause an examination of the respective agents, servants, employees, confederates, attorneys and associates, and each of them, *ore tenus,* and also be entitled to the production of the books, vouchers, documents

and records of the defendant, his agents, servants, employees, confederates, attorneys and associates, and each of them, in connection with the accounting, and that the said defendant, his agents, servants, employees, confederates, attorneys, and associates, and each of them, attend for such purpose before the Master from time to time as the master shall direct.

5. That the plaintiffs have and recover their costs and disbursements in this suit to be hereafter taxed, and that plaintiffs have the right to apply to the Court from time to time for such other and further relief as may be necessary and proper in the premises.

Costs taxed favor plff. at $43.70.

Dated: Los Angeles, California, September 26, 1922.

CHAS. E. WOLVERTON,
United States District Judge.

Approved as to form as provided in Rule 45.

——————————,
Attorney for Defendant.

Decree entered and recorded September 29, 1922.

CHAS. N. WILLIAMS,
Clerk.
By Louis J. Somers,
Deputy Clerk. [456]

[Endorsed]: No. E—113. United States District Court, Southern District of California, Southern Division. Francis M. Townsend et al., Plaintiffs, vs. David G. Lorraine, Defendant. In Equity. Interlocutory Decree. Filed September 29, 1922.

Chas. N. Williams, Clerk. By Louis J. Somers, Deputy. Lyon & Lyon, Frederick S. Lyon, Leonard S. Lyon, 312 Stock Exchange Building, Los Angeles, Cal., Attorneys for Plaintiffs. [457]

---

In the District Court of the United States, Southern District of California, Southern Division.

E-113—IN EQUITY.

FRANCIS M. TOWNSEND, MILON J. TRUMBLE and ALFRED J. GUTZLER, Doing Business Under the Firm Name of TRUMBLE GAS TRAP COMPANY,

Plaintiff,

vs.

DAVID G. LORRAINE,

Defendant.

**Petition for Appeal.**

To the Honorable BENJAMIN F. BLEDSOE, District Judge:

The above-named defendant feeling aggrieved by the decree rendered and entered in the above-entitled cause on the 29th day of September, A. D. 1922, does hereby appeal from said decree to the United States Circuit Court of Appeals for the Ninth Circuit, for the reason set forth in the assignment of errors filed herewith and he pays that this appeal be allowed and that citation be issued as provided by law, and that a transcript of the record, proceedings, and papers and documents upon which said decree was based, duly authenticated be sent to the United States Circuit Court of Appeals for the Ninth Circuit under the Rules of such court in such cases made and provided; and your petitioner further prays that the proper order relating to the required security to be required of him be made.

WESTALL & WALLACE.
By JOSEPH F. WESTALL,
Solicitors and Counsel for Defendant. [458]

---

In the District Court of the United States, Southern District of California, Southern Division.

E-113—IN EQUITY.

FRANCIS M. TOWNSEND, MILON J. TRUMBLE and ALFRED J. GUTZLER, Doing Business Under the Firm Name of TRUMBLE GAS TRAP COMPANY,

Plaintiff,

vs.

DAVID G. LORRAINE,

Defendant.

**Assignment of Errors.**

Now comes the defendant in the above-entitled cause and files the following assignment of errors upon which he will rely upon his prosecution of the appeal in the above-entitled cause, from the decree made by this honorable court on the 29th day of September, 1922.

I.

That the United States District Court for the Southern District of California erred in decreeing that the letters patent No. 1,269,134, granted June 11, 1918, to Milon J. Trumble, assignor of one-third to Francis M. Townsend and one-third to Alfred J. Gutzler, covered or were for a new and useful invention, or in any respect or at all involved invention in the subject matter of any of its claims, or that such subject matter had patentable utility over the prior art sufficient to sustain a finding of invention. [459]

II.

That the United States District Court for the Southern District of California erred in decreeing that the validity of said letters patent No. 1,269,134, granted June 11, 1918, to plaintiffs, was not denied or put in issue by defendant herein.

III.

That the United States District Court for the Southern District of California erred in decreeing that said letters patent No. 1,269,134, granted on June 11, 1918, to plaintiffs herein, are or were good and valid in law in any respect and particularly erred in finding that said letters patent were valid as to claims 1, 2, 3, and 4 thereof.

IV.

That the United States District Court for the Southern District of California erred in decreeing that defendant had infringed upon claims 1, 2, 3, and 4 of said letters patent No. 1,269,134, by making and causing to be made, and selling and causing to be sold, and causing to be used apparatus embodying the alleged invention patented in and by said respective claims 1, 2, 3, and 4 of said letters patent, and in finding and decreeing that any act of making or causing to be made, or selling or causing to be sold, or using or causing to be used, constituted infringement of said letters patent or any of the claims thereof.

V.

That the United States District Court for the Southern District of California erred in decreeing that certain apparatus made and sold by defendant referred to in this cause as Defendant's Model No. 1, and said to be described in reissued letters patent of the United States No. 15,220, granted November 8, 1921, to defendant, infringes upon said claims 3 and 4 of [460] said letters patent.

VI.

That the United States District Court for the Southern District of California, erred in finding and decreeing that certain apparatus made and sold by defendant referred to in this cause as Defendant's Model No. 2 infringes upon claims 1, 2, 3, and 4 of plaintiffs' said patent, or upon any of those claims, either when made with the nipple machined off on one side to sit closely against the partitioned wall as illustrated in said Model No. 2 or when made without such machining or setting, as defendant claims said device was intended to be constructed.

VII.

That the United States District Court for the

erred in decreeing that plaintiffs have and recover their costs and disbursements in said suit to be taxed.

XIV.

That the United States District Court for the Southern District of California, Southern Division, erred in failing to find and decree that said cause should be dismissed for want of equity at plaintiffs' cost, and that this defendant should have and recover his costs and disbursements herein expended.

XV.

That the United States District Court for the Southern District of California, Southern Division, erred in failing and refusing to find and decree that said letters patent No. 1,269,134, granted on June 11, 1918, to plaintiffs was not null and void in law for want of patentable invention over the prior art.

XVI.

That the United States District Court for the Southern District of California, Southern Division, erred in failing to find and decree that said letters patent No. 1,269,134, granted on June 11, 1918, contains nothing more than an aggregation of old and well known parts and features, and did not involve or contain a patentable combination.

XVII.

That the United States District Court for the Southern District of California, Southern Division, erred in failing and refusing to find and decree that any differences in construction or in association of means and devices described and claimed in said letters patent No. 1,269,134, were not sufficiently different from similar and identical devices of the prior art or different [463] in their operative effect to constitute the subject matter of said letters patent in suit of sufficient usefulness to sustain the grant of a valid patent.

XVIII.

That the United States District Court for the Southern District of California, Southern Division, erred in failing and refusing to find and decree that if said patent No. 1,269,134, granted June 11, 1918, and the claims thereof, were not construed with extreme strictness in the light of the specification and drawings of said letters patent and confined in interpretation to the exact, or substantially exact form of the means for spreading the oil in a thin film and taking off the gas, the claims of said letters patent could not be held valid in view of the prior art.

XIX.

That the United States District Court for the Southern District of California, Southern Division, erred in failing and refusing to find and decree that said letters patent, or any of the claims thereof, could not be given an interpretation broad enough to find infringement by making, using, or selling of any of the devices by defendant made, sold, or used without being anticipated and void in view of the prior art introduced in evidence.

XX.

That the United States District Court for the

Southern District of California, Southern Division, erred in finding that the patentees in said letters patent No. 1,269,134, granted June 11, 1918, are not confined to means causing the oil to flow down the outer wall of the chamber, but that their patent includes any means that will cause the oil to flow down any surface as well, such as a baffle-plate or inner partition of the wall, which is reached after the emulsified oil enters [464] the chamber, and in finding that said patentees are not estopped by the proceedings before the patent office to insist upon a broad interpretation of their claims.

XXI.

That the United States District Court for the Southern District of California, Southern Division, erred in finding, as to the showing of the prior art, that Exhibits "E," the McIntosh patent, "F," the Taylor patent, and "J," the Newman patent, all inject the oil from the well in the form of a spray having the effect to reduce it to finally divided condition, in that none of these patents spray in the oil from the well in a finally divided condition, but all flow the oil in in a manner like that of the patent in suit.

XXII.

That the United States District Court for the Southern District of California, Southern Division, erred in finding that Exhibit "E," McIntosh patent and Exhibit "F," the Taylor patent, are not provided with baffle-plates.

XXIII.

That the United States District Court for the

Southern District of California, Southern Division, erred in finding or "thinking" (as it is expressed in the opinion) that only one of the patents introduced in evidence contains the element of pressure in combination with other elements within the chamber, in that the fact is that all the patents and particularly the nearest reference repeatedly emphasized such element of pressure within the chamber.

XXIV.

That the United States District Court for the Southern District of California, Southern Division, erred in finding that all of the patents introduced in evidence to show the prior art, were obviated in their evidentiary effect by restrictions of [465] complainants' specification and claims as made before the Examiner, in that, said letters patent in suit contains no limitations or restrictions, nor does the language of the claims differentiate the device of the patent in suit from several of those referred to and discussed by the Court.

XXV.

That the United States District Court for the Southern District of California, Southern Division, erred in failing and refusing to decree that Exhibit "I," the Cooper patent, is a complete anticipation of the claims in suit and which were sustained by the Court, unless the language of said claims are held to apply strictly to the means shown and described in the specifications and drawings, and in finding that said Cooper patent is to be differentiated from the subject matter of the claims sustained by said District Court, because said Cooper patent in combination contains no element of pressure, when as a matter of fact, Cooper repeatedly and distinctly says that the element of pressure is incident to and part of the intended mode of operation of his device.

XXVI.

That the United States District Court for the Southern District of California, Southern Division, erred in finding that the testimony of Trout as to a certain prior art trap was not persuasive and was not sufficiently corroborated, and in finding that evidence to show the state of the art (as distinguished from anticipation) should establish such state of the art beyond a reasonable doubt.

XXVII.

That the United States District Court for the Southern District of California, Southern Division, erred in ruling Exhibit "C," not admissible as evidence, and not of great weight as evidence, and not as of compelling force, under the strict [466] rule as to the weight of evidence to establish an anticipation, said Exhibit "C" having been introduced in evidence to show the state of the art on the question of invention and construction of the claims in suit.

XXVIII.

That the United States District Court for the Southern District of California, Southern Division, erred in finding, or assuming that there is any evidence in the record by which utility of the subject matter of the claims in suit or any of them could be held to be "abundantly" or in any respect proven, in that there is no evidence comparing any alleged advantages of the structures of the prior art with that of the patent in suit, or setting forth or proving any facts relating to such prior art, oil and gas separators, by which such comparison could be made or upon which it could be predicated.

XXIX.

That the United States District Court for the Southern District of California, Southern Division, erred in finding that *that* the success achieved by plaintiffs' device is evidence of utility over the prior art, in that the prior art devices might have equal or greater advantages to that of the device of the patent in suit, so far as the evidence discloses.

XXX.

That the United States District Court for the Southern District of California, Southern Division, erred in finding or implying that if the oil strikes the chamber wall in any of defendant's devices and therefore flows to any degree down thereon with pressure on the moving oil, even though part of the oil is reduced to a spray which falls by gravity to the settled fluid below, that such action or mode of operation constitutes infringement in whole or in part on any of the claims of said letters patent No. 1,269,134. [467]

XXXI.

That the United States District Court for the Southern District of California, Southern Division, erred in finding that defendant's trap will infringe upon any of the claims of said letters patent in suit with the nipple constructed, as he claims it should be, according to the drawings and specification.

XXXII.

That the United States District Court for the Southern District of California, Southern Division, erred in finding that defendant's patent infringes claims 3 and 4 of complainants' patent, and likewise in finding that defendant's Model No. 2 infringes claims 1 to 4, inclusive, of said complainants' patent.

XXXIII.

That the United States District Court for the Southern District of California, Southern Division, erred in failing and refusing to find and decree that said letters patent No. 1,269,134 and each of the claims charged in this suit to be infringed, were and are null and void in view of the prior art and by reason of anticipation by the prior art as shown by the evidence herein.

WHEREFORE the appellant prays that said decree be reversed and that said District Court for the Southern District of California, Southern Division, be ordered to enter a decree reversing the decision of the lower court in said cause and dismissing said cause for want of equity at plaintiffs' costs.

WESTALL & WALLACE,
By JOSEPH F. WESTALL,
Attorneys for Appellant. [468]

[Endorsed]: No. E-113—In Equity. In the District Court of the United States in and for the Southern District of California, Southern Division. Francis M. Townsend et al., Complainants, vs. David G. Lorraine, Defendant. Petition for Appeal and Assignment of Errors. Filed Oct. 16, 1922. Chas. N. Williams, Clerk. By Edmund L. Smith, Deputy Clerk. Westall and Wallace, Attorneys at Law, Suite 516 Trust & Savings Bldg., Los Angeles, Phone 65683, Attorneys for Defendant. [469]

---

In the District Court of the United States, Southern District of California, Southern Division.

E-113—IN EQUITY.

FRANCIS M. TOWNSEND, MILON J. TRUMBLE and ALFRED J. GUTZLER, Doing Business Under the Firm Name of TRUMBLE GAS TRAP COMPANY,

Plaintiff,

vs.

DAVID G. LORRAINE,

Defendant.

**Order Allowing Appeal.**

On motion of Joseph F. Westall, Esq., of the firm of Westall and Wallace, solicitors and of counsel for defendant, it is hereby ordered that an appeal to the United States Circuit Court of Appeals for the Ninth Circuit from the decree heretofore filed and entered herein be, and the same is hereby allowed, and that a certified transcript of the record, testimony, exhibits, stipulations, and all proceedings be forthwith transmitted to said United States Circuit Court of Appeals for the Ninth Circuit. It is further ordered that the bond on appeal be fixed in the sum of $250.00.

Dated this 16th day of October, 1922.

BLEDSOE,

U. S. District Judge. [470]

[Endorsed]: No. E–113—Equity. In the District Court of the United States in and for the Southern District of California, Southern Division. Francis M. Townsend et al., Complainants, vs. David G. Lorraine, Defendant. Order Allowing Appeal. Filed Oct. 16, 1922. Chas. N. Williams, Clerk. By Edmund L. Smith, Deputy Clerk. Westall and Wallace, Attorneys at Law, Suite 516 Trust & Savings Bldg., Los Angeles, Phone 65683, Attorneys for Defendant. [471]

---

In the District Court of the United States, Southern District of California, Southern Division.

E–113—IN EQUITY.

FRANCIS M. TOWNSEND, MILON J. TRUMBLE and ALFRED J. GUTZLER, Doing Business Under the Firm Name of TRUMBLE GAS TRAP COMPANY,

Plaintiff,

vs.

DAVID G. LORRAINE,

Defendant.

**Bond on Appeal.**

KNOW ALL MEN BY THESE PRESENTS: That we, David G. Lorraine, as principal, and Fidelity and Deposit Company of Maryland, a corporation of the State of Maryland, and duly licensed to transact business in the State of California, having complied with the laws of the State of California and with the Statutes of the United States, and particularly with the Act of August 13, 1894, as amended by the Act of March 23, 1910, of the United States and with the rules of the above-entitled court, as surety, are jointly and severally held and firmly bound unto Francis M. Townsend, Milon J. Trumble and Alfred J. Gutzler, and to them jointly and severally, in the penal sum of Two Hundred and Fifty ($250.00) Dollars, to be paid to them and their respective executors, administrators, and assigns; to which payment, well and truly to be made, we bind ourselves and each of us, jointly and severally, and each of our heirs, executors, and administrators, by these presents.

Sealed with our seals and dated this 18th day of October, 1922. [472]

WHEREAS, the above-named David G. Lorraine has taken an appeal to the United States Circuit Court of Appeals for the Ninth Circuit to reverse the interlocutory decree made, rendered, and entered on the 29th day of September, 1922, in the District Court of the United States for the Southern District of California, Southern Division, in the above-entitled cause, granting a certain injunction against said defendant, as in said interlocutory decree set forth;

AND, WHEREAS, said District Court of the United States for the Southern District of California, Southern Division, has fixed the amount of defendant's bond on said appeal in the sum of Two Hundred and Fifty ($250.00) Dollars;

NOW, THEREFORE, the condition of this obligation is such that if the above-named defendant shall prosecute his said appeal, and any appeal allowed to be taken to the Supreme Court of the United States to effect, and answer all costs which may be adjudged against him, if he fails to make good said appeal, then this obligation shall be void; otherwise to remain in full force and effect.

DAVID G. LORRAINE,
Principal.

FIDELITY AND DEPOSIT COMPANY OF MARYLAND,

[Seal] By W. M. WALKER,
Attorney-in-Fact.

R. W. STEWART,
Agent.

The premium for this bond is $10/00 per annum. [473]

State of California,
County of Los Angeles,—ss.

On this 18th day of October, in the year one thousand nine hundred and twenty-two, before me, C. M. Evarts, a notary public in and for said County and State, residing therein, duly commissioned and sworn, personally appeared W. M. Walker, known to me to be the duly authorized attorney-in-fact of the Fidelity and Deposit Company of Maryland, and the said W. M. Walker, acknowledged to me that he subscribed the name of the Fidelity and Deposit Company of Maryland thereto as surety and his own name as attorney-in-fact.

IN WITNESS WHEREOF, I have hereunto set my hand and affixed my official seal the day and year in this certificate first above written.

[Seal] C. M. EVARTS,

Notary Public in and for the County of Los Angeles, State of California.

Examined and recommended for approval, as provided in Rule 29.

WESTALL and WALLACE,
By JOSEPH F. WESTALL,
Solicitors and of Counsel for Defendant.

I hereby approve the foregoing bond this 18th day of October, 1922.

BLEDSOE,
United States District Judge. [474]

[Endorsed]: No-113—Equity. In the District Court of the United States, in and for the Southern District of California, Southern Division. Townsend et al., Complainant, David G. Lorraine, Defendant. Bond on Appeal. Filed Oct. 18, 1922. Chas. N. Williams, Clerk. By R. S. Zimmerman, Deputy Clerk. Westall and Wallace, Attorneys at Law, Suite 516 Trust & Savings Bldg., Los Angeles, Phone 65683, Attorneys for Defendant. [475]

In the District Court of the United States, Southern District of California, Southern Division.

IN EQUITY—E-113.

FRANCIS M. TOWNSEND, MILON J. TRUMBLE and ALFRED J. GUTZLER, Doing Business Under the Firm Name of TRUMBLE GAS TRAP COMPANY,

Plaintiff,

vs.

DAVID G. LORRAINE,

Defendant.

**Stipulation Re Transcript of Record on Appeal and Exhibits.**

The above-named defendant having taken an appeal in this suit, to the United States Circuit Court of Appeals for the Ninth Circuit, from the interlocutory decree entered on the 29th day of September, 1922;

IT IS HEREBY STIPULATED AND AGREED SUBJECT TO THE APPROVAL OF THE COURT:

Both parties to this suit so desiring, the provisions of Equity Rules 75, and 76, except the second paragraph of Equity Rule 76, promulgated by the United States Supreme Court, applicable to appeals, are hereby waived; and that the testimony and proceedings in court on the trial of this cause be included in the Transcript of Record on Appeal by producing therein a true and correct copy of pages 1—544 of the Reporter's Transcript herein, as the properly prepared statement of the said proceedings and evidence on behalf of both parties under the provisions of Equity Rule 75, which the parties request be approved as such by the Court. [476]

SUBJECT TO THE APPROVAL OF THE COURT, IT IS FURTHER STIPULATED:

That such transcript on appeal shall further include a true and correct copy of the following papers, documents, orders, and proceedings entered and on file in the above-entitled cause:

( 1) Bill of complaint, filed January 3, 1921;

( 2) Answer, filed January 28, 1921;

( 3) Notice and motion to amend answer filed March 4, 1922;

( 4) Order entered March 13, 1922, continuing motion to amend answer to March 20, 1922;

( 5) Order entered March 20, 1922, continuing motion to amend answer to March 27, 1922;

( 6) Order of March 22, 1922, proceedings on final hearing and continuing to March 23, 1922, and allowing filing of supplemental bill and answer;

( 7) Order of March 23, 1922, proceedings on further hearing;

( 8) Supplemental bill of complaint, filed March 23, 1922;

( 9) Order entered March 24, 1922, proceedings in further hearing;

(10) Order entered March 27, 1922, proceedings in further hearing;

(11) Order entered March 28, 1922, proceedings in further hearing;

(12) Answer to supplemental bill, filed March 28, 1922;

(13) Transcript of proceedings and testimony on trial of this cause, as set forth in the attached stipulated transcript of such proceedings;

(14) Opinion of Court filed September 11, 1922;

(15) Interlocutory decree filed and entered September 29, 1922;

(16) Petition for appeal filed October 16, 1922;

(17) Assignment of errors filed October 16, 1922;

(18) Order allowing appeal entered October 16, 1922;

(19) Bond on appeal with endorsements, filed October 18, 1922;

(20) Citation with endorsements, filed October 20, 1922;

(21) This stipulation; [477]

(22) A certificate under seal stating the cost of the record and by whom paid;

(23) The names and addresses of parties to this appeal and their attorneys, Westall and Wallace (Joseph F. Westall and Ernest L. Wallace), 902 Trust and Savings Building, Los Angeles, California, solicitors and of counsel for Defendant-Appellant, David G. Lorraine, Los Angeles, California, and Frederick S. Lyon, Leonard S. Lyon, Stock Exchange Building, and Frank L. A. Graham, Higgins Building, Los Angeles, Cali-

fornia, solicitors and of counsel for plaintiff-appellees, Francis M. Townsend, Milon J. Trumble, and Alfred J. Gutzler, doing business under the firm name of Trumble Gas Trap Company, Los Angeles, California;

All of the above shall constitute, together with the book of exhibits hereinafter mentioned, the transcript of record of said cause on appeal, upon which record said appeal shall be heard and determined (except in so far as the immediately foregoing language may be qualified by the second paragraph of Equity Rule 76), which transcript, except said book of exhibits, shall be certified by the Clerk of this Court to the United States Circuit Court of Appeals for the Ninth Circuit.

IT IS FURTHER STIPULATED AND AGREED SUBJECT TO THE APPROVAL OF THE COURT:

That all exhibits filed by either party herewith shall be forthwith transmitted by the Clerk of this Court at the expense of defendant, to the Clerk of the United States Circuit Court of Appeals for the Ninth Circuit at San Francisco for use on said appeal, and that there shall be printed at the expense of defendant and under the supervision of the Clerk of the United States Circuit Court of Appeals for the Ninth Circuit, in a Book of Exhibits which shall form a part of the printed transcript of the record on appeal for use in said United States Circuit Court of Appeals for the Ninth Circuit on said appeal, copies of the following papers or documentary exhibits: [478]

Plaintiffs' Exhibit 1, Patent 1,269,134 to Milon J. Trumble, dated June 11, 1918.

Plaintiffs' Exhibit 2, File-wrapper and contents Trumble Patent in suit.

Plaintiffs' Exhibit 3, Copy of Patent 1,373,664.

Plaintiffs' Exhibit 4, Copy re-issued Patent No. 15,220.

Plaintiffs' Exhibit 6, Blue-print table of pressures, etc.

Plaintiffs' Exhibit 7, Paine's sketch of Stark Trap.

Plaintiffs' Exhibit $8^1$, $8^2$, $8^3$, $8^4$, Photographs.

Plaintiffs' Exhibit $9^1$, $9^2$, Sketches by Harris.

Plaintiffs' Exhibit 11, Drawing of Lorraine Trap.

Plaintiffs' Exhibit 12, Drawing produced by defendant.

Plaintiffs' Exhibit 13, Account sales of Lorraine Gas & Oil Automatic Separator.

Plaintiffs' Exhibit 14, Letter—Dec. 10, 1920, Townsend to Lacy Mfg. Co.

Plaintiffs' Exhibit 15, Letter—Dec. 13, 1920, Lacy Mfg. Co. to Townsend.

Plaintiffs' Exhibit 16, Blue-print drawing produced by Lorraine.

Plaintiffs' Exhibit 17, Sketch Drawing produced by Lorraine.

Plaintiffs' Exhibit 18, Three prints from drawings of Lorraine Separator.

Plaintiffs' Exhibit 20, Cut of Lorraine Trap, p. 46, Oil Weekly, Feb. 25, 1922.

Plaintiffs' Exhibit 21, Photo of Trap.
Plaintiffs' Exhibit 24, Harris sketch.
Plaintiffs' Exhibit 25, Enlargement of Drawing of Trumble Patent.
Plaintiffs' Exhibit 26, Enlargement of drawing of Trumble Patent. [479]
Defendant's Exhibit "B," Pencil Sketch.
Defendant's Exhibit "D," Photo of Lorraine Trap.
Defendant's Exhibit "E," Certified copy of patent 1,055,549 to George L. McIntosh.
Defendant's Exhibit "F," Certified copy of patent 426,880 to Walter Anderson Taylor.
Defendant's Exhibit "G," Certified copy of patent 927,476 to Arthur W. Barker.
Defendant's Exhibit "H," Certified copy of patent 1,014,943 to Eustace Vivian Bray.
Defendant's Exhibit "I," Certified copy of patent 815,407 to Augustus Steiger Cooper.
Defendant's Exhibit "J," Certified copy of patent 856,088 to Albert T. Newman.
Defendant's Exhibit "K," Pencil sketch of a trap.
Defendant's Exhibit "L," Pencil sketch of a trap.
Defendant's Exhibit "M," Pencil sketch of a trap.
Defendant's Exhibit "N," Letter—Townsend to Lacy Mfg. Co.
Defendant's Exhibit "O," Letter—Lacy Mfg. Co. to Townsend.
Defendant's Exhibit "P," Letter—Dr. W. P. Keene to Lacy Mfg. Co.

Said Book of exhibits shall be printed and copies thereof furnished counsel pursuant to the rules of said Circuit Court of Appeals for the Ninth Circuit and three complete printed copies of the transcript and book of exhibits shall be served and furnished to solicitors and counsel for plaintiffs and defendant respectively.

Dated this 20th day of October, 1922.

FREDERICK S. LYON,
LEONARD S. LYON,
F. L. A. GRAHAM,
Solicitors and of Counsel for Plaintiffs-Appellees.
WESTALL and WALLACE,
By JOSEPH F. WESTALL,
Solicitors and of Counsel for Defendant-Appellant.

It is so ordered this 20 day of October, 1922.

BLEDSOE,
U. S. District Judge. [480]

[Endorsed]: No. E–113—Equity. In the District Court of the United States in and for the Southern District of California, Southern Division. Francis M. Townsend et al., Complainants-Appellees, vs. David G. Lorraine, Defendant-Appellant. Stipulation and Order as to Transcript of Record on Appeal and Exhibits. Filed Oct. 20, 1922. Chas. N. Williams, Clerk. By Edmund L. Smith, Deputy Clerk. Westall and Wallace, Attorneys at Law, Suite 516 Trust & Savings Bldg., Los Angeles, Phone 65683, Attorneys for Defendant-Appellant. [481]

In the District Court of the United States, Southern District of California, Southern Division.

IN EQUITY—E-113.

FRANCIS M. TOWNSEND, MILON J. TRUMBLE and ALFRED J. GUTZLER, Doing Business Under the Firm Name of TRUMBLE GAS TRAP COMPANY,

Plaintiffs-Appellees,

vs.

DAVID G. LORRAINE,

Defendant-Appellant.

**Praecipe for Transcript** of **Record on Appeal, and Exhibits.**

To the Clerk of the Court:

Sir: Please prepare certified transcript of record on appeal in the above-entitled cause on appeal, and forward to the Clerk of the United States Circuit Court of Appeals for the Ninth Circuit all the exhibits filed by either party in said cause, at San Francisco, California, all pursuant to stipulation as to transcript of record on appeal and exhibits signed by solicitors and counsel for the parties to said cause on the 20th day of October, 1922, and ordered by United States District Judge Bledsoe of this Court, and filed in this cause in your office. Please furnish solicitors for defendant-appellant with an estimate of your charges and fees in connection with the above matters, and indicate the amount of deposit required by you in the premises.

Respectfully,

WESTALL and WALLACE,

By JOSEPH F. WESTALL,

Solicitors and of Counsel for Defendant-Appellant.

Dated: Los Angeles, California, October 20, 1922. [482]

[Endorsed]: No. E–113—Equity. In the District Court of the United States in and for the Southern District of California, Southern Division. Francis M. Townsend et al., Complainants, vs. David G. Lorraine, Defendant. Praecipe for Transcript of Record on Appeal, and Exhibits. Filed Oct. 20, 1922. Chas N. Williams, Clerk. By L. J. Cordes, Deputy Clerk. Westall and Wallace, Attorneys at Law, Suite 516 Trust & Savings Bldg., Los Angeles, Phone 65683, Attorneys for Defendant-Appellant. [483]

---

In the District Court of the United States, Southern District of California, Southern Division.

FRANCIS M. TOWNSEND, MILON J. TRUMBLE and ALFRED J. GUTZLER, Doing Business Under the Firm Name of TRUMBLE GAS TRAP COMPANY,

Plaintiff,

vs.

DAVID G. LORRAINE,

Defendant.

**Certificate of Clerk U. S. District Court to Transcript of Record.**

I, Chas. N. Williams, Clerk of the United States District Court for the Southern District of California, do hereby certify the foregoing volume containing 483 pages, numbered from 1 to 483, inclusive, to be the Transcript of Record on Appeal in the above-entitled cause, as printed by the Appellant and presented to me for comparison and certification, and that the same has been compared and corrected by me and contains a full, true and correct copy of the bill of complaint, answer, notice and motion to amend answer, and minute orders of March 13th, 1922, March 20th, 1922, March 22d, 1922, March 23d, 1922, supplemental bill of complaint, and minute orders of March 24th, 1922, March 27th, 1922, March 28th, 1922, answer to supplemental bill, transcript of proceedings and testimony, opinion of Court, interlocutory decree, petition for appeal, assignment of errors, order allowing appeal, bond on appeal, stipulation as to transcript of record on appeal and exhibits, praecipe and the original citation.

I DO FURTHER CERTIFY that the fees of the Clerk for comparing, correcting and certifying the foregoing record on writ of error amount to $145.55, and that said amount has been paid me by the appellant herein. [484]

IN TESTIMONY WHEREOF, I have hereunto set my hand and affixed the seal of the District Court of the United States of America, in and for the Southern District of California, Southern Division, this 15th day of November, in the year of our Lord one thousand nine hundred and twenty-two, and of our Independence the one hundred and forty-seventh.

[Seal] CHAS. N. WILLIAMS,

Clerk of the District Court of the United States of America, in and for the Southern District of California.

By R. S. Zimmerman,

Deputy. [485]

---

[Endorsed]: No. 3945. United States Circuit Court of Appeals for the Ninth Circuit. David G. Lorraine, Appellant, vs. Francis M. Townsend, Milon J. Trumble and Alfred J. Gutzler, Doing Business Under the Firm Name of Trumble Gas Trap Company, Appellees. Transcript of Record. Upon Appeal from the United States District Court for the Southern District of California, Southern Division.

Filed November 29, 1922.

F. D. MONCKTON,

Clerk of the United States Circuit Court of Appeals for the Ninth Circuit.

By Paul P. O'Brien,

Deputy Clerk.

In the United States Circuit Court of Appeals for the Ninth Circuit.

No ——.

FRANCIS M. TOWNSEND, MILON J. TRUMBLE and ALFRED J. GUTZLER, Doing Business Under the Firm Name of TRUMBLE GAS TRAP COMPANY,

Plaintiffs,

vs.

DAVID G. LORRAINE,

Defendant.

**Order Extending Time to and Including December 5, 1922, to File Record and Docket Cause.**

It appearing that the Clerk of the United States District Court for the Southern District of California is now preparing a transcript of the record for use in the above-entitled cause on appeal in accordance with stipulation and praecipe heretofore filed in said cause, but has not finished such transcript and will not be able to complete the same until after the return day of the citation herein,—

IT IS HEREBY ORDERED that defendant-appellant have to and including December 5th, 1922, within which to file the transcript on appeal and docket this cause on appeal in the United States Circuit Court of Appeals for the Ninth Circuit, and that the time of defendant-appellant to that end is hereby extended and enlarged.

Dated: Los Angeles, California, November 13, 1922.

BLEDSOE,
United States District Judge.

[Endorsed]: No. 3945. United States Circuit Court of Appeals for the Ninth Circuit. Francis M. Townsend, Milon J. Trumble and Alfred J. Gutzler, Doing Business Under the Firm Name of Trumble Gas Trap Company, Plaintiffs-Appellees, vs. David G. Lorraine, Defendant-Appellant. Order Extending Time to and Including December 5, 1922, to File Record and Docket Cause. Filed Nov. 14, 1922. F. D. Monckton, Clerk. Refiled Nov. 29, 1922. F. D. Monckton, Clerk.

# United States
# Circuit Court of Appeals
## For the Ninth Circuit.

3

DAVID G. LORRAINE,

Appellant,

vs.

FRANCIS M. TOWNSEND, MILON J. TRUMBLE and ALFRED J. GUTZLER, Doing Business Under the Firm Name of TRUMBLE GAS TRAP COMPANY,

Appellees.

## BOOK OF EXHIBITS.

Upon Appeal from the United States District Court for the Southern District of California, Southern Division.



Filmer Bros. Co. Print, 330 Jackson St., S. F., Cal.

# United States
# Circuit Court of Appeals
## For the Ninth Circuit.

---

DAVID G. LORRAINE,

Appellant,

vs.

FRANCIS M. TOWNSEND, MILON J. TRUMBLE and ALFRED J. GUTZLER, Doing Business Under the Firm Name of TRUMBLE GAS TRAP COMPANY,

Appellees.

---

## BOOK OF EXHIBITS.

---

Upon Appeal from the United States District Court for the Southern District of California, Southern Division.

---

Filmer Bros. Co. Print, 330 Jackson St., S. F., Cal.

# INDEX TO BOOK OF EXHIBITS.

Index. Page

**Plaintiff's Exhibit No. 1.**

No. 1269134.

THE UNITED STATES OF AMERICA.

To all to Whom These Presents Shall Come:

WHEREAS MILON J. TRUMBLE, of

Los Angeles, California,

has presented to the Commissioner of Patents a petition praying for the grant of Letters Patent for an alleged new and useful improvement in

CRUDE-PETROLEUM AND NATURAL-GAS SEPARATORS,

He having assigned one-third of his right, title, and interest in said improvement to Francis M. Townsend and one-third to Alfred J. Gutzler, both of Los Angeles, California, a description of which invention is contained in the specification of which a copy is hereunto annexed and made a part hereof, and has complied with the various requirements of Law in such cases made and provided, and

WHEREAS upon due examination made the said Claimant is adjudged to be justly entitled to a patent under the Law.

NOW THEREFORE these letters patent are to grant unto the said Milon J. Trumble, Francis M. Townsend, and Alfred J. Gutzler, their heirs or assigns for the term of seventeen years from the eleventh day of June, one thousand nine hundred and eighteen, the exclusive right to make, use and vend the said invention throughout the United States and the territories thereof.

IN TESTIMONY WHEREOF I have hereunto set my hand and caused the seal of the Patent Office to be affixed in the District of Columbia this eleventh day of June, in the year of our Lord one thousand nine hundred and eighteen, and of the Independence of the United States of America the one hundred and forty-second.

[Seal] F. W. H. CLAY,
Acting Commissioner of Patents.

M. J. TRUMBLE.
CRUDE PETROLEUM AND NATURAL GAS SEPARATOR.
APPLICATION FILED NOV. 14, 1914.

**1,269,134.** Patented June 11, 1918.

2 SHEETS—SHEET 2.

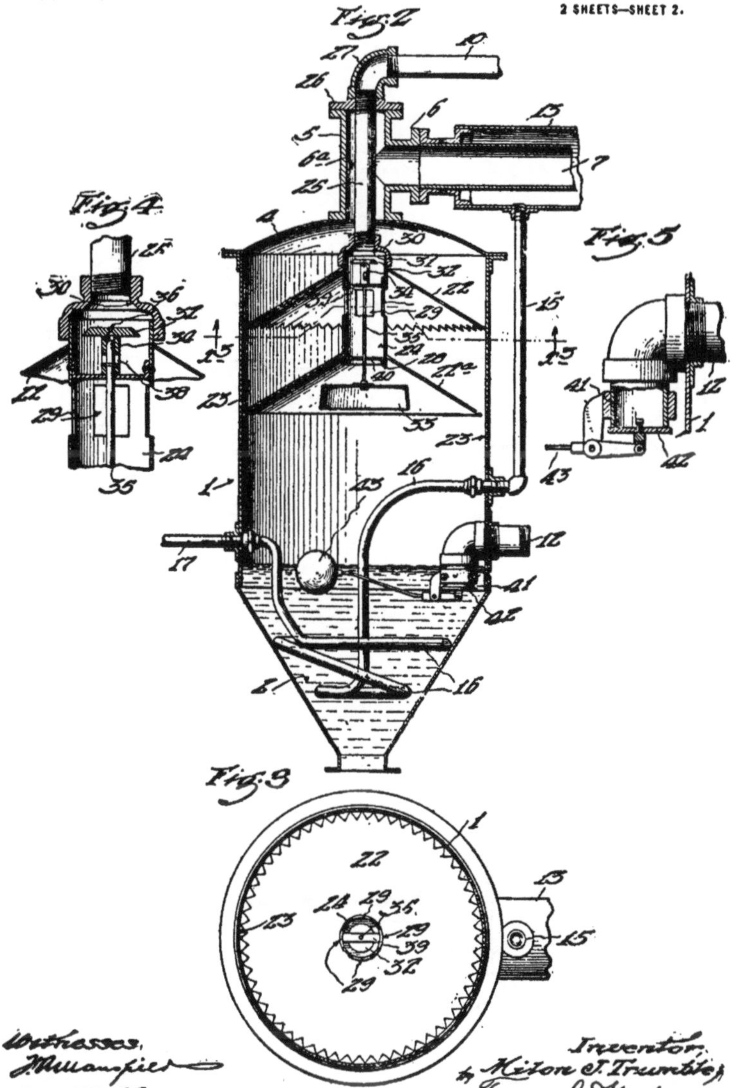


Witnesses:
[illegible]
Ford W. Harris

Inventor,
by Milton J. Trumble,
Townsend & Graham
his attys

M. J. TRUMBLE.

CRUDE PETROLEUM AND NATURAL GAS SEPARATOR.

APPLICATION FILED NOV. 14, 1914.

**1,269,134.** Patented June 11, 1918.

2 SHEETS—SHEET 1.

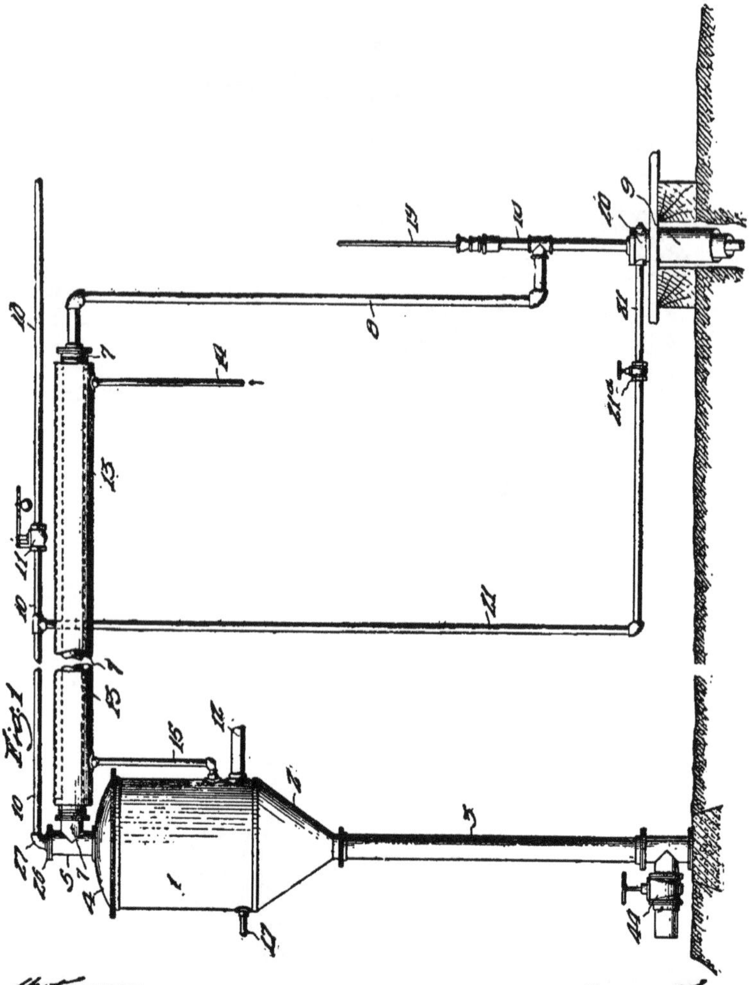

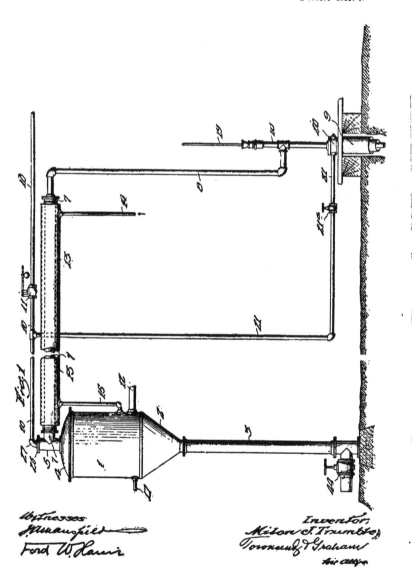

Witnesses
J. Mansfield
Ford W. Harris

Inventor,
Milon J. Trumble,
Townsend & Graham
his Attys.

# UNITED STATES PATENT OFFICE.

**MILON J. TRUMBLE, OF LOS ANGELES, CALIFORNIA, ASSIGNOR OF ONE-THIRD FRANCIS M. TOWNSEND AND ONE-THIRD TO ALFRED J. GUTZLER, BOTH OF LO ANGELES, CALIFORNIA.**

## CRUDE-PETROLEUM AND NATURAL-GAS SEPARATOR.

**1,269,134.** Specification of Letters Patent. **Patented June 11, 19**

Application filed November 14, 1914. Serial No. 872,110.

*To all whom it may concern:*

Be it known that I, Milon J. Trumble, a citizen of the United States, residing at Los Angeles, in the county of Los Angeles and State of California, have invented a new and useful Crude-Petroleum and Natural-Gas Separator, of which the following is a specification.

My invention is designed to effect a separation of natural gas from the crude oil when the same is first delivered from the well, and is also designed to accomplish the cleaning of the oil from free water and sand at the same time.

My invention is also adapted to overcome the unseating of the pump valve in an oil well when an excess of natural gas is present with the oil, thereby materially increasing the amount of oil produced from wells where high pressure of gas exists.

My invention is also adapted to reduce the velocity and equalize the delivery of oil from wells in which the pressure of gas causes the oil to flow in gushes or by heads.

My invention is also effective in maintaining the lighter gravity series of the crude oil in combination with the heavier series of the oil, thereby producing from oil wells a product of lighter gravity than where the oil and gas is permitted to separate on exposure to ordinary atmospheric conditions.

My invention consists in the arrangement and combinations of parts as hereinafter set forth.

The accompanying drawings illustrate my invention and one method of its application:

Figure 1 is an elevation showing my invention as applied to an oil well, the velocity reducing means and gas line being broken to contract the view.

Fig. 2 is a vertical section of the expansion chamber and fragment of velocity reducing means and oil heating means.

Fig. 3 is an inverted section of the expansion chamber taken on the line $x^3$—$x^3$ Fig. 3.

Fig. 4 is an enlarged vertical section of oil stopper for gas line.

Fig. 5 is an enlarged view of the oil outlet, the lower end of the outlet pipe and closure therefor being shown in section.

As shown in the drawings the expansion chamber 1 is provided with a tapering settling chamber 2 which in turn communica with and is supported by a stand pipe The top 4 of the chamber is centrally p forated and a T 5, having its stem 6 posed horizontally, is arranged to regi with the central perforation in the top Velocity reducing means, such as pipe communicates with the horizontal ste of the T 5 and delivers oil and gas i the interior of the chamber 1 through vertical member $6^a$ of the T. The veloc reducing means 7, as shown, consists o pipe many times greater in area than oil pipe 8 which delivers the oil from well 9.

A gas pipe 10 is arranged to lead the from the upper end of the T 5, and o nects with the gas collector within the pansion chamber, which will hereafter explained. A gas pressure regulating va 11 is provided in gas pipe 10 and cont the passage of gas from the expans chamber to any suitable receiver or dest tion.

Oil outlet means 12 is provided from chamber at a point intermediate the tling chamber of the expansion chamber the oil inlet means. When the charac of the oil demands, oil heating means be provided such as the steam jacket which surrounds the velocity reducer 7. steam pipe 14 is arranged to supply to such jacket. When desired a pipe takes steam from the jacket 13 and deliv the same to a coil 16 arranged in the tling chamber 2 to further heat the oil wi in the chamber. The exhaust from the 16 is effected through pipe 17 which pas through the wall of the chamber.

A pump tube 18 having a sucker rod 1 shown mounted on the casing head 20 o well 9. A gas pressure pipe 21 is sho as connecting the gas pipe 10 with the c ing head of the well to apply the press of the gas within the expansion chamber the top of the oil within the well 9 wh surrounds the pump tube 18. A valve may be provided in the pressure pipe 21 regulate the pressure within the well, such valve may be dispensed with, in wh event the pressure on the oil surroundi the pump tube and the oil in the oil pi will be equalized.

Oil dividing means are provided inte

of the expansion chamber, such as cones 22 and 22ª. Preferably these dividing cones are spaced apart, one below the other, as shown, with their peripheries nearly in contact with the wall of the chamber. The oil is delivered onto the upper surface of the cone 22 and spreads over the same evenly and is brought into contact with the wall 23 of the chamber and flows downwardly thereover until it is collected in the settling chamber. The cone 22ª is provided as a supplemental spreader to insure the even distribution of the oil over the wall of the chamber. As shown in the form herein illustrated, the cones 22 and 22ª are supported in the upper portion of the expansion chamber by means of gas collecting chamber 24 and pipe 25. Pipe 25 is screwed into flange 26 which in turn is seated upon and closes the upper end of the vertical member 6ª of the T 5, to which is connected the elbow 27 of the gas pipe 10. The lower end of the pipe 25 is screwed into a reducer 30, arranged with its larger end looking downward to receive the upper end of the gas collecting chamber 24 which is open at its lower end 28. The gas collecting chamber is also provided with gas admitting ports 29, 29 and 29 arranged between the cones 22 and 22ª which are protected thereby to prevent the entrance of oil into the gas line.

A valve seat 31 is provided in the reducer 30, and a valve stopper 32 is provided with means for seating the same when oil may rise within the chamber to a point where it may gain access to the gas line 10. As shown, this stopper 32 comprises a flat circular body having a beveled edge adapted to fit the valve seat in the reducer, and is provided with a perforated shank 34 through which the float rod 35, actuated by float 33, passes to open and close a vacuum breaker opening 36 in the stopper before actuating the stopper 32. A slot 38 is provided in the shank 34 which permits of the passage of a pin therethrough and through the float rod to limit the movement of such rod in breaking the vacuum to permit the unseating of the stopper.

The movement of the stopper, float rod, and float, away from the valve seat is limited by the end of shank 34 coming into contact with a bracket 39. A bracket 40 is arranged near the bottom of the gas collecting chamber to guide the float rod 35.

A trapped oil outlet from the chamber is provided which consists of a submerged pipe 41 provided with a closure 42 which is actuated by means of a float 43 which is so adjusted as to cause the pipe to be closed to the exit of oil before the level of oil reaches the open end of the pipe, to thereby prevent the escape of gas along with the outflowing oil from the chamber. Other forms of traps will readily suggest themselves, but the form herewith disclosed illustra function desired to be derived. A 44 is provided to permit sand and wat be drawn from stand pipe and set chamber when desired.

It is to be remembered that an impo feature of my invention is to provide whereby the oil is divided into a t broken body whereby the gas may r escape from its engagement therewit it is also desirable to maintain the oil divided form for a period of time suffi for all the gas to free itself from t It is also desirable to maintain the o quiescent condition for a short per time within the settling chamber to sand and water to settle out of the before the oil is drawn from the cha

In practice, oil is pumped from the by actuating the sucker rod 19, which f the oil through the pipe 8 to the ve reducer 7 from which it flows throug T 5 onto the top of the cone 22 by it is spread over the wall of the expa chamber and envelops the gas coll The oil may be heated while passing thr the velocity reducer by means of the jacket 13 up to or about the boiling of water. This heating will result in ening the surface tension upon any contained in the oil and allows the sa settle in the bottom of the chamber. settling of the water may be materiall sisted by the pressure of the gas wi chamber, and, if necessary, more heat be applied to the oil by means of the s coil 16.

In case oil is being taken from a w the flowing type, and is intermittent heads, the velocity reducer will afford for equalizing the flow of oil into th pansion chamber. In case a large a of oil should be delivered to the expa chamber, and the egress thereof shoul slower than the amount coming in, the 33 will actuate the stopper 32 and the gas outlet to prevent oil from ge into the gas line. The gas pressure then increase and thereby force the oil the chamber until the float will drop to open the gas line again. The pre regulating valve 11 in the gas line 10 trols the exit of gas from the expa chamber and thereby keeps a constant sure on the oil within the chamber.

In the case of a wet gas, or a gas rated with an oil of the lighter series, as gasolene, it is desirable to mainta high a pressure within the expansion c ber as may be practicable for the reason a large amount of such light series thereby be compressed and remain wi oil. In some instances this action ha sulted in producing an oil three p higher in gravity, according to the B scale, than had been produced by subjecting the oil and gas to the atmospheric action to permit the gas to escape. This increase in lightness of gravity brought the producer an increase of price amounting to fifteen cents per barrel.

The admission of gas under pressure to the casing head of a well to equalize the pressure of gas within the pump tube, prevents the oil from slipping in the pump barrel on account of the valve being unseated by the upwardly flowing gas. This action as produced by my invention increases the pumping efficiency and the production of the well, thereby making profitable producers of wells that have theretofore been unprofitable.

The oil flowing from the expansion chamber through outlet 12 may be conducted to any suitable tank or receptacle either by the pressure of the gas within the chamber or by gravity.

When desired, a safety valve may be provided for the chamber 1 to permit escape of gas from the expansion chamber in the event that valve 32 should become stuck or remain seated on its seat. In some cases the float 33 and valve 32 may be dispensed with.

It will be noted that the action upon the oil while flowing down the wall of the expansion chamber in a thin film under pressure permits the free, dry, gas to readily escape therefrom, while the pressure exerted upon the oil surface backed by the wall of the chamber holds the lighter liquids, such as gasolene, in combination with the oil body, and I desire to be understood as pointing out and claiming this action as being of great benefit to the crude oil derived from the well on account of keeping the gasolene series in combinaton with the main body of oil.

I claim as my invention:

1. In an oil and gas separator, the combination of an expansion chamber arranged to receive oil and gas in its upper portion, means for spreading the oil over the wall of such chamber to flow downwardly thereover, gas take-off means arranged to take off gas from within the flowing film of oil, an oil collecting chamber below the expansion chamber, an oil outlet from said collecting chamber, and valve controlled means arranged to maintain a submergence of the oil outlet.

2. In an oil and gas separator, the combination of an expansion chamber, inlet means arranged to permit the entrance of oil and gas into the chamber, means within the chamber adapted and arranged to distribute the oil over the wall of the chamber in a downwardly flowing film, gas take-off means arranged to take gas from within the envelop of downwardly flowing oil, and means for maintaining gas pressure up such oil.

3. In an oil and gas separator, the com nation of an expansion chamber havin surface adapted to sustain a flow of thereover in a thin body, means for d tributing oil onto such surface, press maintaining means arranged and adap to maintain a pressure on one side of flowing oil, withdrawing means arranged take gas from the chamber, and means withdrawing oil from the chamber.

4. In an oil and gas separator, the com nation of an expansion chamber, means delivering oil and gas into the chamb means for maintaining pressure within chamber, means for drawing oil from chamber, and means within the cham adapted to cause the oil to flow in a t body for a distance to enable the gas c tained and carried thereby to be given while the oil is subjected to pressure.

5. The combination in an oil and gas s arator of an expansion chamber havin settling chamber communicating therewi means for delivering oil and gas into upper portion of the expansion chamber cone arranged near the top of such cha ber to receive the incoming oil and spre it over the wall of the chamber in a t film like form, a gas collector arranged low the cone, gas outlet means from s collector, a float controlled valve arran to close the gas outlet means, means drawing oil from the separator above settling chamber, and means for discharg sand or water from said settling chamb

6. The combination in an oil and gas arator of an expansion chamber havin settling chamber communicating therewi means for delivering oil and gas to chamber, oil spreading and dividing me arranged to receive the oil within the cha ber and spread it in a thin film like fo over the wall of the chamber to flow the down, gas collecting and outlet means ranged below the oil spreading means a within the flowing film of oil, a float c trolled valve arranged to close the gas o let when the oil rises to a predetermi height within the expansion chamber, a fl controlled valved outlet from the separa whereby a predetermined depth of oil maintained in the settling chamber to s merge the oil outlet and prevent the pass of gas therethrough and means for discha ing sand or water from the settling chamb

7. In a gas and oil separator, the co bination of an expansion chamber, spreading means arranged within the cha ber, oil supplying means arranged to liver oil into such expansion chamber a onto the spreader means, gas collect means arranged within the chamber, outlet means communicating with the

collecting means, oil outlet means from the expansion chamber arranged to draw oil from a point above the bottom thereof, such oil supplying means comprising an oil pipe from a suitable oil supply and a velocity reducing means connected therewith, whereby the oil and gas are delivered to the expansion chamber in quiescent condition.

8. In an oil and gas separator, the combination of an expansion chamber, means within the chamber adapted to spread the oil in a thin film like form over the wall of the chamber, gas outlet means leading from the chamber, consisting of a gas pipe, pressure regulating means arranged to control the passage of gas from the chamber through such gas pipe, oil outlet means arranged to draw oil from the expanding chamber, means to maintain such oil outlet means submerged within the body of oil in the chamber, velocity reducing means arranged to receive oil and gas from a source of oil supply and deliver the same into the expansion chamber, and means for heating the oil within the velocity reducing device before its entry into the expansion chamber.

9. In an oil and gas separator, the combination of an expansion chamber having a centrally perforated top, a T arranged with one end of its cross to register with the central perforation in the top and having its stem disposed horizontally, a bushing on the upper end of the cross of the T, a gas pipe leading from such bushing, a gas pipe depending from such bushing through the cross of the T and into the expansion chamber, means for delivering oil to the chamber through the stem of the T, oil dividing means secured to the lower end of the last named gas pipe, and means for drawing oil from the chamber.

10. In an oil and gas separator, the combination of an expansion chamber having a centrally perforated top, a T arranged with one end of its cross to register with the central perforation in the top and having its stem disposed horizontally, a bushing on the upper end of the cross of the T, a gas pipe leading from such bushing, a gas pipe depending from such bushing through the cross of the T and into the expansion chamber, means for delivering oil to the chamber through the stem of the T, means for heating the oil during its passage to such chamber, oil dividing means secured to the lower end of the last named gas pipe, and means for drawing oil from the chamber.

11. In an oil and gas separator, the combination of an expansion chamber having a centrally perforated top, a T arranged with one end of its cross to register with the central perforation in the top and havi[ng its] stem disposed horizontally, a bushing [on the] upper end of the cross of the T, a gas [pipe] leading from such bushing, means for [regu]lating the flow of gas from such expa[nsion] chamber through such gas pipe to [main]tain pressure within such chamber, [a gas] pipe depending from such bushing th[rough] the cross of the T and into the expa[nsion] chamber, means for delivering oil t[o the] chamber through the stem of the [T, oil] dividing means secured to the lower e[nd of] the last named gas pipe, and mean[s for] drawing oil from the chamber.

12. In an oil and gas separator, the [com]bination of an expansion chamber, a se[ttling] chamber communicating therewith ad[apted] to receive the oil and retain the same [in a] quiet condition for a period of ti[me, a] stand pipe arranged to support the e[xpan]sion chamber and settling chamber a[nd to] receive sand and water from the se[ttling] chamber, a submerged and trapped [outlet] for the oil arranged to draw oil fro[m the] separator, means for introducing oil [into] the expansion chamber, means adapt[ed to] divide the oil within the chamber, [a gas] outlet from the chamber, and mean[s for] preventing oil entering such gas outlet.

13. In an oil and gas separator, a[n ex]pansion chamber; inlet means for feed[ing] foam composed of oil and gas throug[h the] central part of the top of said expa[nsion] chamber; an imperforate spreader cone, [hav]ing its apex pointing upwardly, locate[d in]side said chamber in such a manner [as to] spread a thin film of oil over the [inner] wall of said chamber, and means for t[aking] gas from the central portion of [said] chamber.

14. In an oil and gas separator, a[n ex]pansion chamber; inlet means for fe[eding] a foam composed of oil and gas thr[ough] the central part of the top of sai[d ex]pansion chamber; an imperforate spr[eader] cone, having its apex pointing upwa[rdly,] located inside said chamber in such a [man]ner as to spread a thin film of oil ove[r the] inner wall of said chamber, and a gas c[ollecting] pipe supporting said cone and pa[ssing] vertically upward through said inlet m[eans,] said pipe being arranged to take of[f gas] from the apex of said cone.

In testimony whereof, I have here[unto] set my hand at Los Angeles, California[, this] 9th day of November, 1914.

MILON J. TRUMB[LE.]

In presence of—
F. M. Townsend,
A. J. Gutzler.

[Endorsed]: E—113. Townsend et al. vs. Lorraine. Plffs. Exhibit 1. Filed March 22, 1922. Chas. N. Williams, Clerk.

No. 3945. United States Circuit Court of Appeals for the Ninth Circuit. Filed Jan. 26, 1923. F. D. Monckton, Clerk.

---

**Plaintiff's Exhibit No. 2.**

2—390.

UNITED STATES OF AMERICA,

DEPARTMENT OF THE INTERIOR,

UNITED STATES PATENT OFFICE

To All to Whom These Presents Shall Come, GREETING:

THIS IS TO CERTIFY that the annexed is a true copy from the Records of this Office of the File Wrapper, Contents and Drawings, in the matter of the

Letters Patent of

Milon J. Trumble, Assignor of One-third to Francis M. Townsend and One-third to Alfred J. Gutzler,

Number 1,269,134, Granted June 11, 1918,

for

Improvement in Crude Petroleum and Natural Gas Separator.

IN TESTIMONY WHEREOF I have hereunto set my hand and caused the seal of the Patent

Office to be affixed in the District of Columbia this 5th day of March, in the year of our Lord one thousand nine hundred and twenty-one and of the Independence of the United States of America the one hundred and forty-fifth.

[Seal] M. H. COULSTON,
Commissioner of Patents.

872110 , 1914

(EX'R'S BOO

, 191

PATENT No. 1269134

Name Milon J. Trumble,
Assor of 1/3 to Francis M. Townsend and 1/3 to Alfred J.
both of Los Angeles Cal.

, filed

of Los Angeles,

County of

State of California

Invention Crude Petroleum and Natural Gas Separator.

Division of App., No.

PARTS OF APPLICATION FILED.

| | ORIGINAL. | | RENEWED. |
|---|---|---|---|
| Petition | Nov. 14 | , 1914 | |
| Affidavit | " " | , 1914 | |
| Specification | " " | , 1914 | |
| Drawing 2 sheets | " " | , 1914 | |
| ~~Photo Copy~~ | | , 191 | |
| First Fee Cash $ 15, | Nov. 14 | , 1914 | |
| ~~" " Cert.~~ | | , 191 | |
| Appl. filed complete | Nov 14 | , 1914 | |
| Examined and Passed for Issue | Jan. 16, | 1918 | |
| J H. Lightfoot | Exr. Div. 25 | | Exr. Div. , |
| Notice of Allowance | Jan. 18 | , 1918 | , |
| | By Commissioner, | | By Commissione |
| Final Fee ~~Cash~~ Cert dated | May 8 | , 1918 | , |
| " " Cert. $20 | May 13 | , 1918 | . |
| Patented | JUN 11 1918 | | , |

Attorney ~~Townsend & Graham,~~ ~~Higgins Bld'g, Los Angeles, Calif.~~

sub

~~Associate~~ Attorney Graham and Harris
#933 Higgins B'ld'g. Los Angeles, Cal

(No. of Claims Allowed 14 ) Title as Allowed Crude Petroleum and Natural Gas Separat

Natural Gas Separator Cl. 183 - 105)

O.G. Claims 4,8 and 11

872110

LETTER-HEAD.

15 Received 872,110
Nov.
14
1914. M. C. G. November 9, 1914.
Chief Clerk U. S. Patent Office.

Hon. Commissioner of Patents,
Washington,
D. C.

Sir:

Enclosed herewith find application papers of Milon J. Trumble, for United States Letters Patent on CRUDE PETROLEUM AND NATURAL GAS SEPARATOR.

We enclose herewith Post Office Money Order, No. 324,757, dated November 9, 1914, for $15.00, which amount please apply as the filing fee for this application.

Kindly file, acknowledge receipt, and oblige.

Very respectfully,
TOWNSEND & GRAHAM.

C.
Encs.

Mail Room 872,110
Nov.
14,
1914 4605
U. S. Patent Office.

PETITION AND POWER OF ATTORNEY.

To the Hon. Commissioner of Patents:

Your petitioner, Milon J. Trumble, a citizen of

the United States, residing at Los Angeles, in the County of Los Angeles, and State of California, whose Post Office address is 1918 Santee Street, Los Angeles, Los Angeles County, California, prays that Letters Patent may be granted to him for the CRUDE PETROLEUM AND NATURAL GAS SEPARATOR, set forth in the annexed specification, and he hereby appoints the firm of

TOWNSEND & GRAHAM,

(the individual members of which firm are Francis M. Townsend and Frank L. A. Graham), of 1029 Higgins Building, Los Angeles, California, his attorneys, with full power of substitution and revocation, to prosecute this application, to make alterations and amendments therein, to receive the patent and to transact all business in the Patent Office connected therewith.

MILON J. TRUMBLE.

SPECIFICATION. 4606

To All Whom It May Concern:

Be it known that I, MILON J. TRUMBLE, a citizen of the United States, residing at Los Angeles, in the County of Los Angeles, and State of California, have invented a new and useful CRUDE-PETROLEUM AND NATURAL GAS SEPARATOR, of which the following is a specification:–

My invention is designed to effect a separation of natural gas from the crude oil when the same is first delivered from the well, and is also designed to accomplish the cleaning of the oil from free water and sand at the same time.

My invention is also adapted to overcome the unseating of the pump valve in an oil well when an excess of natural gas is present with the oil, thereby materially increasing the amount of oil produced from wells where high pressure of gas exists.

My invention is also adapted to reduce the velocity and equalize the delivery of oil from wells in which the pressure of gas causes the oil to flow in gushes or by heads.

My invention is also effective in maintaining the lighter gravity series of the crude oil in combination with the heavier series of the oil, thereby producing from oil wells a product of lighter gravity than where the oil and gas is permitted to separate on exposure to ordinary atmospheric conditions.

My invention consists in the arrangement and combinations of parts as hereinafter set forth.

4607

The accompanying drawings illustrate my invention and one method of its application:

Fig. 1 is an elevation showing my invention as applied to an oil well, the velocity reducing means and gas line being broken to contract the view.

per A

Fig. 2 is a vertical section of the expan~~ding~~sion chamber and fragment of velocity reducing means and oil heating means.

Fig. 3 is an inverted section of the expan~~ding~~sion chamber taken on line x3-x3 Fig. 3. per A

Fig. 4 is an enlarged vertical section of oil stopper for gas line.

Fig. 5 is an enlarged view of the oil outlet, the lower end of / the outlet pipe and closure therefor being shown in section. "

As shown in the drawings, the expan~~ding~~sion chamber 1 is provided with a tapering settling chamber 2 which in turn communicates with and is supported by a stand pipe 3. The top 4 of the chamber is centrally perforated and a T 5, having its stem 6 disposed horizontally, is arranged to register with the central perforation in the top 4. Velocity reducing means, such as pipe 7, communicates with the horizontal stem 6 of the T 5 and delivers oil and gas into the interior of the chamber 1 through the vertical member $6^a$ of the T. The velocity reducing means 7, as shown, consists of a pipe many times greater in area than the oil pipe 8 which delivers the oil from the well 9. " A

4608

A gas pipe 10 is arranged to lead the gas from the upper end of the T 5, and connects with the gas collector within the expan~~ding~~sion chamber, which will hereafter be explained. A gas pressure regulating valve 11 is provided in gas pipe 10 and controls the passage of gas from the expan~~ding~~sion chamber to any suitable receiver or destination. Oil out- " A let means 12 is provided from the chamber at a point intermediate the settling chamber of the A expan~~ding~~sion chamber and the oil inlet means. When the character of the oil demands, oil heating means may be provided such as the steam jacket 13 which surrounds the velocity reducer 7. A steam pipe 14 is arranged to supply steam to such jacket. When desired a pipe 15 takes steam from the jacket 13 and delivers the same to a coil 16 arranged in the settling chamber 2 to further heat the oil within the chamber. The exhaust from the coil 16 is effected through pipe 17 which passes through the wall of the chamber.

A pump tube 18 having a sucker rod 19 is shown mounted on the casing head 20 of a well 9. A gas pressure pipe 21 is shown as connecting the gas pipe 10 with the casing head of the well to apply
A the pressure of the gas within the expan~~ding~~sion chamber to the top of the oil within the well 9 which surrounds the pump tube 18. A valve 21[a] may be provided in the pressure pipe 21 to regulate the pressure within the well, or such valve may be dispensed with, in which event the pressure on the oil surrounding the pump tube and the oil in the oil pipe 8 will be equalized.

4609

Oil dividing means are provided interior of the
A expan~~ding~~sion chamber, such as cones 22 and 22[a]. Preferably these dividing cones are spaced apart, one below the other, as shown, with their peri-

pheries nearly in contact with the wall of the chamber. The oil is delivered on to the upper surface of the cone 22 and spreads over the same evenly and is brought into contact with the wall 23 of the chamber and flows downwardly thereover until it is collected in the settling chamber. The cone $22^a$ is provided as a supplemental spreader to insure the even distribution of the oil over the wall of the chamber. As shown in the form herein illustrated, the cones 22 and $22^a$ are supported in
A the upper portion of the expan~~ding~~sion chamber by means of gas collecting chamber 24 and pipe 25. Pipe 25 is screwed into flange 26 which in turn is seated upon and closes the upper end of the vertical member $6^a$ of the T 5, to which is connected the elbow 27 of the gas pipe 10. The lower end of the pipe 25 is screwed into a reducer 30, arranged with its larger end looking downward to receive the upper end of the gas collecting chamber 24 which is open at its lower end 28. The gas collecting chamber is also provided with gas admitting ports 29, 29 and 29 arranged between the cones 22 and $22^a$ which are protected thereby to prevent the entrance of oil into the gas line.

A valve seat 31 is provided in the reducer 30, and a valve stopper 32 is provided with means for seating the same when oil may rise within the chamber to a point where it may gain access to the gas line 10. As shown, this stopper 32 comprises a flat

4610

circular body having a beveled edge adapted to fit

the valve seat in the reducer, and is provided with a perforated shank 34 through which the float rod 35, actuated by float 33, passes to open and close a vacuum breaker opening 36 in the stopper before actuating the stopper 32. A slot 38 is provided in the shank 34 which permits of the passage of a pin therethrough and through the float rod to limit the movement of such rod in breaking the vacuum to permit the inseating of the stopper.

The movement of the stopper, float rod, and float, away from the valve seat is limited by the end of shank 34 coming into contact with a bracket 39. A bracket 40 is arranged near the bottom of the gas collecting chamber to guide the float rod 35.

A trapped oil outlet from the chamber is provided which consists of a submerged pipe 41 provided with a closure 42 which is actuated by means of a float 43 which is so adjusted as to cause the pipe to be closed to the exit of oil before the level of oil reaches the open end of the pipe, to thereby prevent the escape of gas along with the out flowing oil from the chamber. Other forms of traps will readily suggest themselves, but the form herewith disclosed illustrates the function desired to be derived. A valve 44 is provided to permit sand and water to be drawn from stand pipe and settling chamber when desired.

It is to be remembered that an important feature of my invention is to provide means whereby the oil is divided into a thin or broken body whereby the

4611

gas may readily escape from its engagement therewith and it is also desirable to maintain the oil in its divided form for a period of time sufficient for all the gas to free itself from the oil. It is also desirable to maintain the oil in a quiescent condition for a short period of time within the settling chamber to permit sand and water to settle out of the same before the oil is drawn from the chamber.

In practice, oil is pumped from the well by actuating the sucker rod 19, which forces the oil through the pipe 8 to a velocity reducer 7 from which it flows through the T 5 on to the top of the cone 22 by which it is spread over the wall of the expan~~ding~~sion per A chamber and envelops the gas collector. The oil may be heated while passing through the velocity reducer by means of the steam jacket 13 up to or about the boiling point of water. This heating will result in weakening the surface tension upon any water contained in the oil and allows the same to settle in the bottom of the chamber. The settling of the water may be materially assisted by the pressure of the gas within the chamber, and, if necessary, more heat may be applied to the oil by means of the steam coil 16.

In case oil is being taken from a well of the flowing type, and is intermittent in its heads, the velocity reducer will afford means for equalizing the flow of oil into the expan~~ding~~sion chamber. In case a " A large amount of oil should be delivered to the expan~~ding~~sion chamber, and the egress thereof should be " A slower than the amount coming in, the float 33 will actuate the stopper 32 and close the gas outlet to prevent oil from getting into the gas line. The gas pressure will then increase and thereby force the oil from the chamber until the float will drop down to open the gas line again. The pressure regulating valve 11 in the gas line 10 controls the exit of gas

A from the expan~~ding~~sion chamber and thereby keeps a constant pressure on the oil within the chamber.

In the case of a wet gas, or a gas saturated with an oil of the lighter series, such as gasoline, it is desirable to maintain as high a pressure within the

A expan~~ding~~sion chamber as may be practicable for the reason that a large amount of such light series will thereby be compressed and remain with the oil. In some instances this action has resulted in producing an oil three points higher in gravity, according to the Baume scale, than had been produced by subjecting the oil and gas to the atmospheric action to permit the gas to escape. This increase in lightness of gravity brought the producer an increase of price amounting to fifteen cents per barrel.

The admission of gas under pressure to the casing head of a well to equalize the pressure of gas within the pump tube, prevents the oil from slipping in the pump barrel on account of the valve being unseated by the upwardly flowing gas. This action as produced by my invention increases the pumping efficiency and the production of the well, thereby making profitable producers of wells than have heretofore been unprofitable.

4613

The oil flowing from the expan~~ding~~sion chamber through outlet 12 may be conducted to any suitable tank or receptacle either by the pressure of the gas within the chamber or by gravity.

When desired, a safety valve may be provided for the chamber 1 to permit escape of gas from the expan~~ding~~sion chamber in the event that valve 32 should become stuck or remain seated on its seat. In some cases the float 33 and valve 32 may be dispensed with.

I claim as my invention:- 4614

1. The combination in an oil and gas separator, of an expanding chamber having a settling bottom, means for delivering oil and gas to the upper portion of the chamber, means for reducing the oil into a finally divided condition to reduce the tension on the gas contained therein, gas outlet means from the chamber and an oil outlet from the chamber intermediate the settling bottom of the chamber and the gas outlet.

2. The combination in an oil and gas separator of an expanding chamber, a settling chamber communicating therewith, oil dividing means arranged in the expanding chamber to reduce the oil to a thin film like condition, means for delivering the oil to be freed from gas to the dividing means, a gas outlet from the chamber and a trapped oil outlet arranged to draw oil from the separator intermediate the settling chamber and the oil dividing means.

3. The combination in an oil and gas separator of an expanding chamber, a settling chamber communicating therewith arranged to receive the oil after being freed from gas, gas freeing means consisting of means to reduce the oil to a thin film arranged within the expanding chamber, a gas collector arranged within the oil dividing means, a gas outlet arranged to communicate

A[2] with the gas collector, and a trapped oil outlet arranged to draw oil from the separator at a point intermediate the settling chamber and the gas outlet.

5 ~~4~~. The combination in an oil and gas separator

of an expan~~ding~~sion chamber having a settling chamber per A communicating therewith, means for delivering oil and gas into the upper portion of the expan~~ding~~sion chamber, a cone arranged near the top of such chamber to receive the incoming oil and spread it over the wall of the chamber in a thin film like form, a gas collector arranged below the cone, gas outlet means from such collector, a float controlled valve arranged to close the gas outlet means, ~~and~~ means for drawing oil from the separator above the settling chamber, " A and means for discharging sand or water from said settling chamber. " A

6. ~~5~~. The combination in an oil and gas separator of an expan~~ding~~sion chamber having a settling " A chamber communicating therewith, means for delivering oil and gas to the chamber, oil spreading and dividing means arranged to receive the oil within the chamber and spread it in a thin film like form over the walls of the chamber to flow theredown, gas collecting and outlet means arranged below the oil spreading means and within the flowing film of oil, a float controlled valve arranged to close the gas outlet when the oil rises to a predetermined height within the expan~~ding~~sion chamber, ~~and~~ a float " A controlled valve outlet from the separator whereby a predetermined depth of oil is maintained in the settling chamber to submerge the oil outlet and prevent the passage of gas therethrough, and means " A

for discharging sand or water from the settling chamber.

7. ~~6~~. In a gas and oil separator, the combination of an expan~~ding~~sion chamber, oil spreading means arranged within the chamber, oil supplying means arranged to deliver oil into such expan~~ding~~sion chamber and on to the spreader means, gas collecting means

4616

arranged within the chamber, gas outlet means communicating with the gas collecting means, oil outlet means from the expan~~ding~~sion chamber arranged to draw oil from a point above the bottom thereof, such oil supplying means comprising an oil pipe from a suitable oil supply and a velocity reducing means connected therewith, whereby the oil and gas are delivered to the expan~~ding~~sion chamber in quiescent condition.

8. ~~7~~. In an oil and gas separator, the combination of an expan~~ding~~sion chamber, means within the chamber adapted to spread the oil in a thin film like form over the wall of the chamber, gas outlet means leading from the chamber, consisting of a gas pipe, pressure regulating means arranged to control the passage of gas from the chamber through such gas pipe, oil outlet means arranged to draw oil from the expanding chamber, means to maintain such oil outlet means submerged within the body of oil in the chamber, velocity reducing means arranged to re-

ceive oil and gas from a source of oil supply and deliver the same into the expan~~ding~~sion chamber, and means for heating the oil within the velocity reducing device before its entry into the expan~~ding~~sion chamber.

---

~~9. 8. In an oil and gas separator, the combination of an expan~~~~ding~~~~sion chamber, means within the chamber to divide the oil to reduce the surface tension of the oil upon the gas contained therein, an oil outlet from the bottom of the chamber, an oil inlet through the top of such chamber arranged to deliver oil on to the top of the oil dividing means, and gas outlet means comprising a pipe passing upwardly through the oil inlet supporting the oil dividing means within the expan~~~~ding~~~~sion chamber.~~

4617

9. ~~10~~. ~~9~~. In an oil and gas separator, the combination of an expan~~ding~~sion chamber having a centrally perforated top, a T arranged with one end of its cross to register with the central perforation in the top and having its stem disposed horizontally, a bushing on the upper end of the cross of the T, a gas pipe leading from such bushing, a gas pipe depending from such bushing through the cross of the T and into the expan~~ding~~sion chamber, means for delivering oil to the chamber through the stem of the T, oil dividing means secured to the lower end

of the last named gas pipe, and means for drawing oil from the chamber.

10. ~~11~~. ~~10~~. In an oil and gas separator, the combination of an expan~~ding~~sion chamber having a centrally perforated top, a T arranged with one end of its cross to register with the central perforation in the top and having its stem disposed horizontally, a bushing on the upper end of the cross of the T, a gas pipe leading from such bushing, a gas pipe depending from such bushing through the cross of the T and into the expan~~ding~~sion chamber, means for delivering oil to the chamber through the stem of the T, means for heating the oil during its passage to such chamber, oil dividing means secured to the lower end of the last named gas pipe, and means for drawing oil from the chamber.

a

11. ~~12~~. ~~11~~. In an oil and gas separator, the combination of an expan~~ding~~sion chamber having a centrally perforated top, a T arranged with one end of its cross to register with the central perforation in the top and having its stem disposed horizontally, a bushing on the upper end of the cross of the T, a gas pipe leading from such bushing, means

4618

for regulating the flow of gas from such expan~~ding~~sion chamber through such gas pipe to maintain pressure within such chamber, a gas pipe depending from such bushing through the cross of the T and into the expanding chamber, means for delivering oil to the chamber through the stem of the T, oil dividing means secured to the lower end of the last named gas pipe, and means for drawing oil from the chamber.

a
12. ~~13.~~ ~~12.~~ In an oil and gas separator, the combination of an expan~~ding~~sion chamber, a settling chamber communicating therewith adapted to receive the oil and retain the same in a quiet condition for a period of time, a stand pipe arranged to support the expan~~ding~~sion chamber and settling chamber and to receive sand and water from the settling chamber, a submerged and trapped outlet for the oil arranged to draw oil from the separator, means for introducing oil into the expan~~ding~~sion chamber, means adapted to divide the oil within the chamber, a gas outlet froh the chamber, and means for preventing oil entering such gas outlet.

per A
" A
" A

4619

IN TESTIMONY WHEREOF, I have hereunto set my hand at Los Angeles, California, this 9th day of November, 1914.

MILON J. TRUMBLE.

In presence of:

F. M. TOWNSEND.
A. J. GUTZLER.

OATH.

State of California,
County of Los Angeles,—ss.

Milon J. Trumble, the above-named petitioner, being duly sworn, deposes and says that he verily believes himself to be the original, first and sole inventor or discoverer of the CRUDE PETROLEUM AND NATURAL GAS SEPARATOR described and claimed in the annexed specification; that he does not know and does not believe that the same was ever known or used before his invention or discovery thereof; or patented or described in any printed publication in any country before his invention or discovery thereof, or more than two years prior to this application; or in public use or on sale in the United States for more than two years prior to this application, and that no application for patent on said invention has been filed by him or his legal representatives or assigns in any foreign country.

And said MILTON J. TRUMBLE states that he is a citizen of the United States, and resident of

Los Angeles, in the County of Los Angeles, and State of California.

MILON J. TRUMBLE.

Subscribed and sworn to before me this 9th day of November, 1914.

[Seal] FRED A. MANSFIELD,

Notary Public in and for the County of Los Angeles, State of California.

Div. 31, Room 169. Paper No. 2.

DEPARTMENT OF THE INTERIOR,

UNITED STATES PATENT OFFICE,

WASHINGTON.

Dec. 2, 1914.

U. S. PATENT OFFICE.

Dec. 4, 1914.

MAILED.

Townsend and Graham,
Higgins Building,
Los Angeles, Cal.

Please find below a communication from the EXAMINER in charge of the application of Milon J. Trumble, filed Nov. 14, 1914, Se. No. 872,110. Crode Petroleum and Natural Gas Separator.

THOMAS EWING,
Commissioner of Patents.

This application has been taken up for examination.

On page 2, line 14, after the word "of" insert the word *the.*

Throughout the description of the specification and the claims the ..rd expansion should be substituted for the word "expanding." The vessel or chamber, *per se,* does not expand.

Claims 1, 2, 3 are each rejected on patents 927,476, Barker, July 13, 1909, and 1,014,943, Bray, Jan. 16, 1912 (48—142).

Claim 4, is rejected on the patent above to Bray.

Claim 5 in addition to the patents above is rejected on patents 428,399, Moore, May 20, 1890, and 611,314, Cullinan, Sep. 27, 1898 (48—142).

Claim 8 is rejected on the patent above to Bray and the patent 426,880, Taylor, April 29, 1890. (48—142.)

ELY,
Examiner. Div. 31.

STOKES.

4620
SERIAL No. 872,110.

U. S. PATENT OFFICE,
RECEIVED A
MAR. 15, 1915.

DIVISION 31, PAPER No. 3.

Mail Room,
Mar.
15,
1915.
U. S. Patent Office.
Div. 31,
Room 169,
Milon J. Trumble,

Crude Petroleum and Natural Gas Separator,
Filed Nov. 14, 1914,
Serial No. 872,110.

Los Angeles, Calif., Mar. 10, 1915.

Hon. Commissioner of Patents,
Washington,
D. C.

Sir:

Examiner's letter of December 2d, 1914, together with the United States patents cited therein, has been carefully considered, and in response thereto we amend as follows:

Page 2, line 14, after the word "of" insert—the—

Chance the word "expanding" to be "expansion" in the following places— In lines 6, 9, and 16 of page 2. Lines 3, 6, 10 and 25 of page 3. Line 2 and 14 of page 4. Lines 14, 28 and 29 of page 6. Lines 7 and 13 of page 7. Lines 1 and 6 of page 8.

At the close of the specification, page 8, insert the following

---

—It will be noted that the action upon the oil while flowing down the wall of the expansion chamber in a thin film under pressure permits the free, dry, gas to readily escape therefrom, while the pres-

A′ Cls. 1–4.

4621

Ser. No. 872,110.

sure exerted upon the oil surface backed by the wall of the chamber holds the lighter liquids, such as gasoline, in combination with the oil body, and I desire to be understood as pointing out and claiming this action as being of great benefit to the crude oil derived from the well on account of keeping the gasoline series in combination with the main body of oil.

---

Cancel claims 1, 2 and 3, and insert the following:

---

1. In an oil and gas separator, the combination of an expansion chamber arranged to receive oil and gas in its upper portion, means for spreading the oil over the wall of such chamber to flow downwardly there over, gas take-off means arranged to take off gas from within the flowing film of oil, an oil collecting chamber below the expansion chamber, an oil outlet from said collecting chamber, and valve controlled means arranged to maintain a submergence of the oil outlet.

2. In an oil and gas separator, the combination of an expansion chamber, inlet means arranged to permit the entrance of oil and gas into the chamber, means within the chamber adapted and arranged to distribute the oil over the wall of the chamber in a downwardly flowing film, gas take-off means arranged to take gas from within the envelope of downwardly flowing oil, and means for maintaining gas pressure upon such oil.

4622

3. In an oil and gas separator, the combination of an expansion chamber having a surface adapted to sustain a flow of oil thereover in a thin body, means for distributing oil on to such surface, pressure maintaining means arranged and adapted to maintain a pressure on one side of the flowing oil,

withdrawing means arranged to take gas from the chamber, and means for withdrawing oil from the chamber.

4. In an oil and gas separator, the combination of an expansion chamber, means for delivering oil and gas into the chamber, means for maintaining pressure within the chamber, means for drawing oil from the chamber, and means within the chamber adapted to cause the oil to flow in a thin body for a distance to enable the gas contained and carried thereby to be given off while the oil is subjected to pressure.

---

Renumber claims 4, 5, 6, 7, 8, 9, 10, 11 and 12, as 5, 6, 7, 8, 9, 10, 11, 12 and 13, respectively.

We amend renumbered claim 5 as follows:

Line 2 of said claim, change "expanding" to—expansion—; same claim, line 10, cancel "and"—; last line, before the period, insert—above the settling chamber, and means for discharging sand or water from said setting chamber—.

4623

In renumbered claim 6, lines 2 and 11, change "expanding" to —expansion—; line 11, cancel "and"—; last line, before the period, insert—and means for discharging sand or water from the settling chamber—.

Renumbered claim 7, lines 2, 4, 7 and 12, change "expanding" to —expansion—.

Renumbered claim 8, lines 2, 12 and 14, change "expanding" to —expansion—.

Renumbered claim 9, lines 2 and 9, change "expanding" to expansion—.

Renumbered claim 10, lines 2 and 9, change "expanding" to —expansion.

Renumbered claim 11, lines 2 and 9, change "expanding" to—expansion—

Renumbered claim 12, lines 2 and 8, and 11, change "expanding" to —expansion—.

Renumbered claim 13, lines 2, 5 and 9, change "expanding" to —expansion—.

## REMARKS.

Renumbered claim 9 has not been amended for the reason that applicant's attorneys do not believe that it comes within the inventions shown by Bray or Taylor. While Bray has a central delivery for the oil through pipe 16, the gas outlet 17 takes the gas from the apex of the cylindrical cone 3 which closes the top of the chamber. 4624

In Taylor we find a steam outlet arranged below the baffles, but not passing upwardly through the inlet. The construction claimed by applicant is "gas outlet means comprising a pipe passing upwardly through the oil inlet supporting the oil dividing means within the expansion chamber." This construction enables applicant to provide a gas take-off within the envelope of oil which is flowing down the sides of the expansion chamber, and obviates the necessity of the gas passing upwardly through the descending film of oil as it passes from the cone 22 on to the sides of the chamber and yet permits the oil to be introduced through the same fitting which supports the hoods and gas take-off means.

Applicant's specification and claims are now written to define the peculiar arrangement of applicant's device, which is the means for causing the oil to be divided into a thin flowing body while giving off the natural gases carried in combination therewith, and applying pressure to the surface of the oil while undergoing this action. This results in maintaining all of the light, volatile liquids in combination with the oil and raising the gravity of the oil when delivered into the receiving tank.

In the Barker patent there are no means shown for expanding the oil in such a manner that it may flow in a thin film, but must fall from the head 8 either in a solid body, or if accompanied by suffi-

4625

cient pressure, in a sprayed or broken up condition.

The patent to Bray shows that the oil after being admitted into the expansion chamber is divided into drops or streams and flows downwardly in that condition.

In the patent to Cullinan the oil is delivered into the expanding chamber in a solid stream, while in the patent to Moore the oil is delivered through the pipe B at a point near the bottom of the chamber.

The patent to Taylor is not considered as being adaptable for the use of applicant's invention for the reason that no separation of oil and gas could be effected until both had passed the second baffle from the bottom, as a film of oil started on the upper baffle would flow downward to the lower edge and would fall over in a flowing mass on to the next succeeding plate, and any gas freed in the upper chamber would have to pass through such body of oil before it could escape into the chamber formed between the two upper baffles, and this same action would take place at each succeeding fall of oil from the edge of the baffle plates. This would cause an agitation of the oil which would have a tendency to carry off the lighter series of oil with the gas.

4626

Applicant has discovered that in order to maintain the lighter series of oils in combination with the heavier series when separating the gas, it is necessary to reduce the oil to a thin regularly flowing body which is not subjected to any breaking up action, and to permit the gas to escape therefrom without agitation.

All of the references cited would cause a breaking up of the flowing body of oil, or agitation thereof, and result in the carrying away of the light volatile oils with the gas.

In actual practice applicant has demonstrated that by the use of his separators the oil delivered therefrom has all of the light gasolines in permanent combination with the crude oil, such crude oil being from two to three degrees lighter, according to the Baume scale, than oil which had been passed through other forms of separators. Affidavits to this effect will be furnished if the Examiner would care to have the same on file in this case.

None of the references cited show the invention as claimed in applicant's amended claims, nor would any of the devices disclosed in the references be capable of performing the function of applicant's device.

Favorable action is requested.

Respectfully submitted,

TOWNSEND & GRAHAM,

Attys. for Trumble.

FMT—H.

Div. 31, Room 169. Paper No. 4.

DEPARTMENT OF THE INTERIOR

UNITED STATES PATENT OFFICE

WASHINGTON

April 9, 1915.

Townsend and Graham,
Higgins Building,
Los Angeles, Cal.

Please find below a communication from the EXAMINER in charge of the application of
M. J. Trumble, filed Nov. 14, 1914, Ser. No. 872,110,
Crude Petroleum and Natural Gas Separator.

THOMAS EWING,

Commissioner of Patents.

This application has been reconsidered in view of the amendment filed March 15, 1915.

Claims 1 to 4, presented by the above amendment, are each rejected on the patent of record to Bray 1,014,943.

Claims 5 to 13 inclusive are also held not to patentably distinguish from the patent of record to Bray and accordingly rejected.

In this patent to Bray, while the cones may be shown as perforated, these cones will nevertheless deliver the oil to the wall of the expansion chamber to flow downwardly thereover. This patent is also provided with means for automatically drawing off the heavy residue or deposit as well as means for allowing an overflow of the oil therefrom through the pipe 20. The gas is drawn off from the top of the chamber and whether this gas outlet is centrally located or eccentrically located is immaterial.

ELY,
Examiner, Div. 31.

STOKES.

Mail Room
Mar
30
1916
U. S. Patent Office

Serial No. 872,110

B

U. S. Patent Office,
Received
Mar 31, 1916
Division 31, Paper No. 5

Div. 31
Room 169,
M. J. Trumble,
Crude Petroleum and Natural Gas Separator,
Filed Nov. 14, 1914,
Ser. No. 872,110.

Los Angeles, Cal., Mar. 21, 1916.

Hon. Commissioner of Patents,
Washington,
D. C.

Sir:-

In response to Examiner's letter of April 9, 1915, we amend as follows:

Insert new claims:

---

13 —~~14.~~ In an oil and gas separator, an expansion chamber; inlet means for feeding a foam composed of oil and gas through the central

B′ part of the top of said expansion chamber; an imperforate spreader cone, having its apex pointing upwardly, located inside said chamber in such a manner as to spread a thin film of oil over the inner wall of said chamber, and means for taking gas from the central portion of said chamber.

14 ~~15.~~ In an oil and gas separator, an expansion chamber; inlet means for feeding a foam composed of oil and gas through the central part of the top of said expansion chamber; an imperforate spreader cone, having its apex pointing upwardly, located inside said chamber in such a manner as to spread a thin film of oil over the inner wall of said

Cls. 14—15.

chamber, and a gas outlet pipe supporting said cone and passing vertically upward through said inlet means, said pipe being arranged to take off gas from the apex of said cone.

REMARKS.

We cannot understand the Examiner's statement that the cones in the Bray patent will deliver oil to the wall of the expansion chamber. The cones Bray calls screens and they evidently make a tight closure with the walls of the chamber. An explanation is requested.

Very respectfully,

TOWNSEND & GRAHAM,

Attys. for Trumble.

FWH-H.

Div. 31, Room 169.

Paper No. 6.

DEPARTMENT OF THE INTERIOR.

UNITED STATES PATENT OFFICE.

WASHINGTON.

April 8, 1916.

U. S. PATENT OFFICE

Apr. 8 1916

Mailed

Townsend & Graham,
Higgins Bldg.,
Los Angeles, Cal.

Please find below a communication from the EXAMINER in charge of the application of
Milon J. Trumble, No. 872,110, filed Nov. 14, 1914,
Crude Petroleum and Natural Gas Separator.

THOMAS EWING,

Commissioner of Patents.

This case as amended Mar. 30, has been considered.

In reconsidering the case in view of applicant's remarks, it is not seen wherein the applicant is justified in saying that the screens make a tight closure with the walls of the chamber. In the patent the casing 2 is provided with a series of brackets 10 to support the screens. This would not look as if the screens make a tight closure with the walls of the chamber. Furthermore, it is said that the oil or liquid passes down over the screens and into the cone 1. This would imply that the heavy liquid reaches the casing and passes down its walls. There does not seem to be any authority for assuming that the liquid will do otherwise. While the screens in the patent are perforated, it is evident that these perforations are for the passage of the separated gases on their way to the outlet pipe 17, since the heavy liquid would be too thick to pass therethrough.

Claims 14, 15, presented by the above amendment are rejected on the patent to Bray.

Claims 1 to 13, inclusive, are held not to patentably distinguish from the references of record, and are again rejected.

ELY,
Examiner.

STOKES.

Mail Room Serial No. 872,110 Paper No. 7
Dec.
4
1916
U. S. Patent Office

PATENT OFFICE
DEC 6 1916
DIVISION XXV

Div. 31,
Room 169,
Crude Petroleum and Natural Gas Separator,
Filed Nov. 14th, 1914,
Serial No. 872,110,
MILON J. TRUMBLE.

Los Angeles, Cal., Nov. 24, 1916

Hon. Commissioner of Patents,
Washington,
D. C.

Sir:–

We have Examiner's letter of April 8th, 1916, and note the Examiner's theory of the operation of the invention of Bray. We cannot agree with this theory, and find nothing in the Bray patent to justify it.

Applicant's invention consists of a containing vessel, an imperforate cone adapted to spread the whole body of the oil to the outer edge of the vessel, and means for taking off gas from the interior of the cone near the center of the vessel.

In the Bray patent there is no means for spreading the oil, and even if the Examiner holds that screens 11, 12 and 13 spread a portion of the oil, he cannot avoid the conclusion that oil will pass through a screen and that some oil will not be spread.

Moreover, if we adopt the Examiner's viewpoint and hold that the openings in the screens are to let the gas through, it is even more evident that the screens of Bray are not equivalent to the imperforate cones of Trumble. Bray put holes in his screens because he had to, to make his ap-



paratus work, whether we consider the gas or the oil as passing through the holes.

Moreover, Bray does not take off his gas below his screens, and the claims of Trumble are quite specific in stating that the gas is taken off inside the cone.

In addition to these broad structural and functional differences, some of the claims recite specific structure which Bray certainly does not show. We think the Examiner is deceived by a certain similarity in appearance of unessential features of the Bray and Trumble inventions.

A reconsideration and allowance is requested. As we have apparently reached an issue with the Examiner, we would request that he act as fully upon the claims as he would upon appeal.

Very respectfully,
TOWNSEND & GRAHAM,
Attys. for Trumble.

FWH—H.

[On reverse side:] U. S. Patent Office, Serial No. Received Dec. 5 1916 Division 31, Paper No.

Div. 25, Room 315 Paper No. 8.

FFD/OIL

DEPARTMENT OF THE INTERIOR.

UNITED STATES PATENT OFFICE.

WASHINGTON.

Jan. 10, 1917.

U. S. PATENT OFFICE
JAN 10 1917
MAILED

Townsend & Graham,
Higgins Bldg.,
Los Angeles, Cal.

Please find below a communication from the EXAMINER in charge of the application of Milon J. Trumble, Ser. No. 872,110, filed Nov. 14, 1914, for Crude Petroleum and Natural Gas Separator.

THOMAS EWING,
Commissioner of Patents.

In response to applicant's letter of Nov. 24, 1916:

On reconsideration of this case all the claims except claim 9 (original claim 8) may be allowed.

Claim 9 is held as met by the patent to Bray, of record. Claim 8 is further objected to as being indistinct in the use of the expression "means within the chamber to divide the oil to reduce the surface tension of the oil upon the gas contained therein."

J. H. LIGHTFOOT,
Examiner, Div. 25.

OIL.

Mail Room Serial No. Paper No.
Feb.
6
1917
U. S. Patent Office

Serial No. 872110, Paper No. 9.
PATENT OFFICE
FEB. 8, 1917
DIVISION XXV

Div. 25,
Room 315,
Milon J. Trumble,
For Crude Petroleum and Natural Gas Separator,
Filed Nov. 14, 1912,
Serial No. 872,110.

Hon. Commissioner of Patents,
Washington,
D. C.

Sir:

We, the undersigned, having Power of Attorney in the above identified application giving us full power of substitution and revocation, hereby substitute for ourselves the firm of GRAHAM & HARRIS, Registration No. 10717, (the individual members of which firm are Frank L. A. Graham and Ford W. Harris), of 933 Higgins Building, Los Angeles, California, with full power of substitution and revocation, to prosecute said application, to make alterations and amendments therein, to receive the patent, and to transact all business in the Patent Office connected therewith.

Signed at Los Angeles, in the County of Los

Angeles, and State of California, this 24th day of January, 1917.

TOWNSEND & GRAHAM,

H.

Mail Room Serial No. 872,110 Paper No. 10
Dec.
20
1917
U. S. Patent Office

PATENT OFFICE
DEC 21 1917
DIVISION XXV

Div. 25,
Room 315,
Milon J. Trumble,
Crude Petroleum and Natural Gas Separator,

Los Angeles, Cal., December 11, 1917.

Hon. Commissioner of Patents,
Washington,
D. C.

Sir:

In response to Examiner's letter of January 10th, 1917, we amend as follows:

C Cancel claim 9.

Renumber the remaining claims.

REMARKS.

The above action being fully responsive, puts the case in condition for final allowance, which is respectfully requested.

Very respectfully,

GRAHAM & HARRIS,
Attys. for Trumble.

FWH—H.

IN REMITTING THE FINAL FEE GIVE THE SERIAL NUMBER AT THE HEAD OF THIS NOTICE.

Address Only Serial No. 872110
The Commissioner of Patents,
Washington, D. C.

2—181

AC.

DEPARTMENT OF THE INTERIOR.
UNITED STATES PATENT OFFICE.
WASHINGTON.

Jan. 18, 1918.

UNCERTIFIED CHECKS WILL NOT BE ACCEPTED.

Milon J. Trumble, Assor.

Sir: Your APPLICATION for a patent for an IMPROVEMENT in

Crude Petroleum & Natural Gas Separator,

filed Nov. 14, 1914, has been examined and ALLOWED.

The final fee, TWENTY DOLLARS, must be paid not later than SIX MONTHS from the date of this present notice of allowance. If the final fee be not paid within that period, the patent on this application will be withheld, unless renewed with an additional fee of $15, under the provisions of Section 4897, Revised Statutes.

The office delivers patents upon the day of their date, and on which their term begins to run. The printing, photolithographing, and engrossing of the several patent parts, preparatory to final signing and sealing, will require about four weeks, and such work will not be undertaken until after payment of the necessary fee.

When you send the final fee you will also send, DISTINCTLY AND PLAINLY WRITTEN, the received.

Final fees will NOT be received from other than the applicant, his assignee or attorney, or a party in interest as shown by the records of the Patent Office.

Respectfully,
JAMES T. NEWTON,
THOMAS EWING,
Commissioner of Patents.

GRAHAM & HARRIS,
933 Higgins Bldg.,
Los Angeles, Cal.

2—327.

---

CERTIFICATE OF DEPOSIT.
$20 Rec'd
Mar. 13, 1918, H
C. C. U. S. Pat. Office

MEMORANDUM.

of

FEE PAID AT UNITED STATES PATENT OFFICE.

(Be careful to give correct Serial No.)

Serial No. 872,110. Series of 1900.

Inventor : Milon J. Trumble.

Patent to be issued to Milon J. Trumble, Francis M. Townsend and Alfred J. Gutzler.

Name of invention, as allowed: Crude Petroleum and Natural Gas Separator.

Date of Payment: May 8, 1918.

Fee: Final Fee by Certificate of Deposit #7201.

Date of filing: November 14, 1914.

Date of circular of allowance: January 18, 1918.

The Commissioner of Patents will please apply the accompanying fee as indicated above.

GRAHAM & HARRIS,
Attorneys.

Send patent to
GRAHAM & HARRIS,
933 Higgins Bldg.,
Los Angeles, Cal.

M. J. TRUMBLE.
CRUDE PETROLEUM AND NATURAL GAS SEPARATOR.
APPLICATION FILED NOV. 14, 1914.

**1,269,134.** Patented June 11, 1918.
2 SHEETS—SHEET 1.

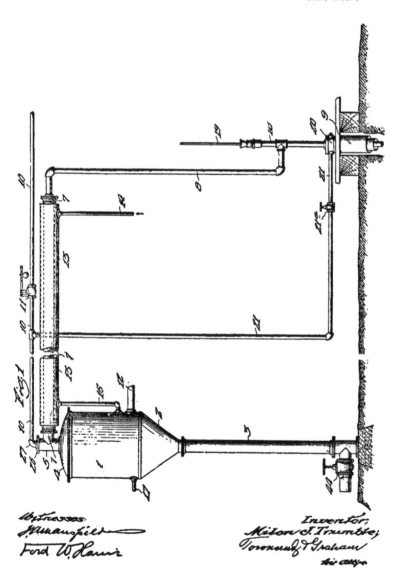

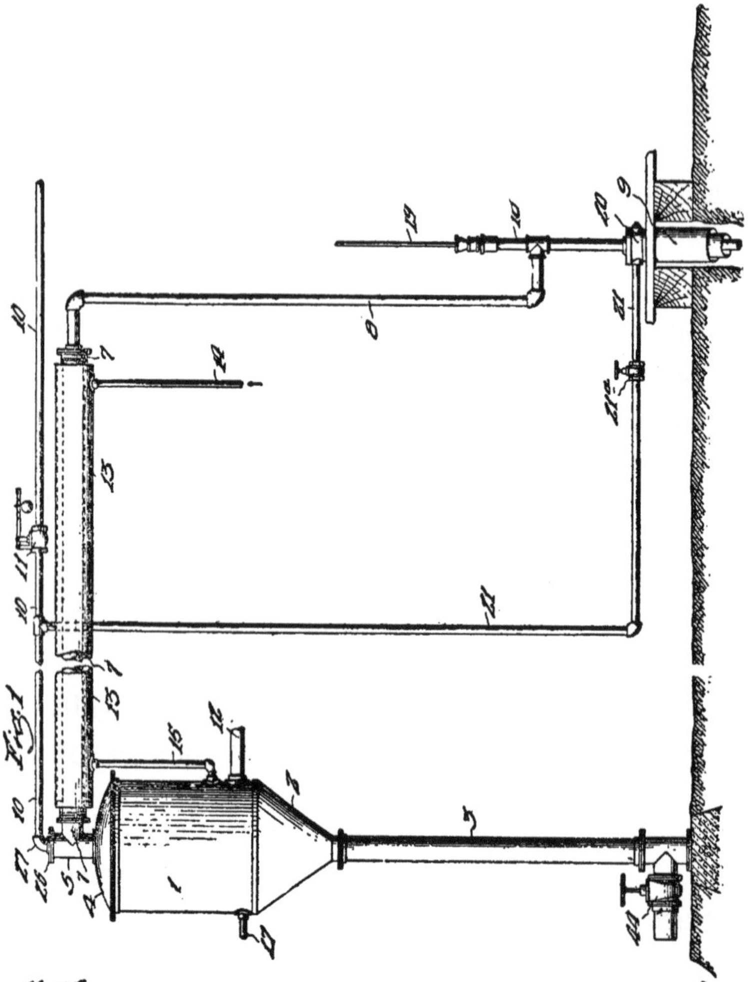

Witnesses:
J. Mansfield
Ford W. Harris

Inventor,
Milon J. Trumble,
Townsend & Graham
his attys.

M. J. TRUMBLE.

CRUDE PETROLEUM AND NATURAL GAS SEPARATOR.

APPLICATION FILED NOV. 14. 1914

**1,269,134.** Patented June 11, 1918.

2 SHEETS—SHEET 2.

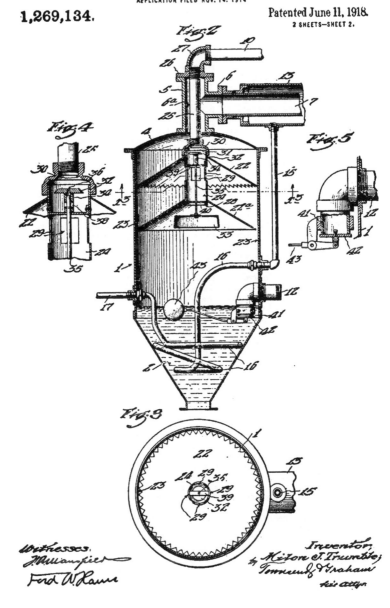

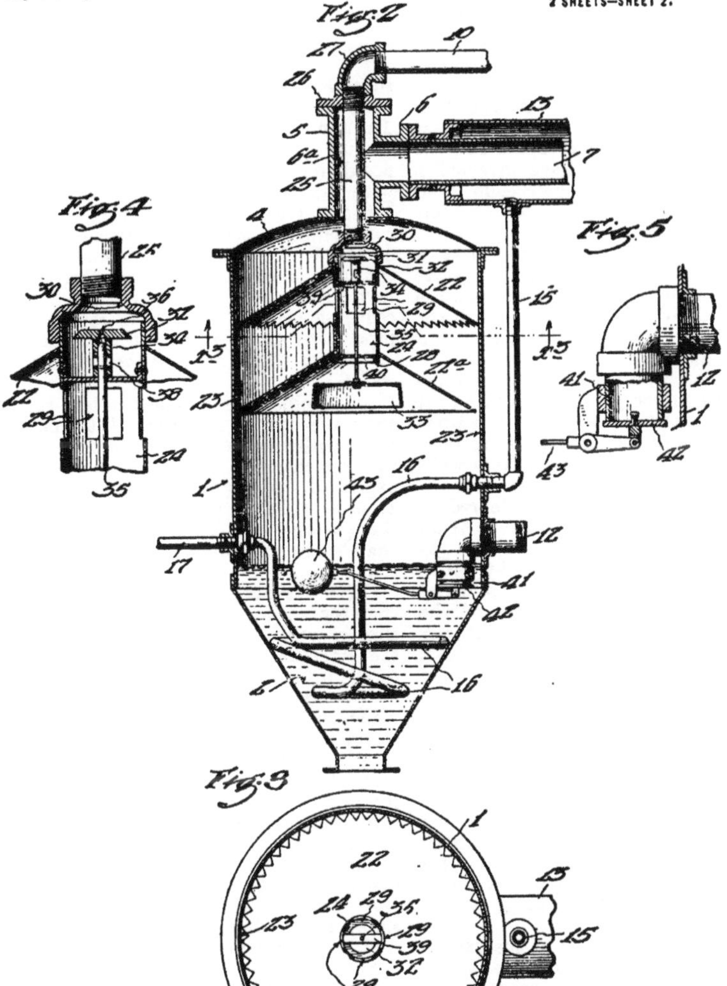


Witnesses.
J. Williamsfield
Ford W. Hawes

Inventor,
Milton J. Trumble;
by Townsend & Graham
his attys.

# UNITED STATES PATENT OFFICE.

ILON J. TRUMBLE, OF LOS ANGELES, CALIFORNIA, ASSIGNOR OF ONE-THIRD TO FRANCIS M. TOWNSEND AND ONE-THIRD TO ALFRED J. GUTZLER, BOTH OF LOS ANGELES, CALIFORNIA.

## CRUDE-PETROLEUM AND NATURAL-GAS SEPARATOR.

69,134. Specification of Letters Patent. Patented June 11, 1918.

Application filed November 14, 1914. Serial No. 872,110.

*ll whom it may concern:*

e it known that I, MILON J. TRUMBLE, itizen of the United States, residing at Angeles, in the county of Los Angeles State of California, have invented a and useful Crude-Petroleum and ural-Gas Separator, of which the following is a specification.

y invention is designed to effect a ıration of natural gas from the crude when the same is first delivered from the l, and is also designed to accomplish the ning of the oil from free water and l at the same time.

y invention is also adapted to overcome unseating of the pump valve in an oil l when an excess of natural gas is pres- with the oil, thereby materially increas- the amount of oil produced from wells re high pressure of gas exists.

y invention is also adapted to reduce velocity and equalize the delivery of oil wells in which the pressure of gas es the oil to flow in gushes or by heads.

y invention is also effective in main- ing the lighter gravity series of the e oil in combination with the heavier es of the oil, thereby producing from e s a product of lighter gravity than re the oil and gas is permitted to rate on exposure to ordinary atmos- ric conditions.

y invention consists in the arrangement c m na ns of parts as hereinafter set ho bi tio

he accompanying drawings illustrate my ntion and one method of its application:

igure 1 is an elevation showing my in- tion as applied to an oil well, the velocity cing means and gas line being broken ontract the view.

ig. 2 is a vertical section of the expan- chamber and fragment of velocity re- ing means and oil heating means.

ig. 3 is an inverted section of the ex- sion chamber taken on the line $x^3$—$x^3$ . 3.

ig. 4 is an enlarged vertical section of oil per for gas line.

ig. 5 is an enlarged view of the oil out- the lower end of the outlet pipe and therefor being shown in section.

hown in the drawings the expansion r 1 is provided with a tapering settling chamber 2 which in turn communicates with and is supported by a stand pipe 3. The top 4 of the chamber is centrally perforated and a T 5, having its stem 6 disposed horizontally, is arranged to register with the central perforation in the top 4. Velocity reducing means, such as pipe 7, communicates with the horizontal stem 6 of the T 5 and delivers oil and gas into the interior of the chamber 1 through the vertical member $6^a$ of the T. The velocity reducing means 7, as shown, consists of a pipe many times greater in area than the oil pipe 8 which delivers the oil from the well 9.

A gas pipe 10 is arranged to lead the gas from the upper end of the T 5, and connects with the gas collector within the expansion chamber, which will hereafter be explained. A gas pressure regulating valve 11 is provided in gas pipe 10 and controls the passage of gas from the expansion chamber to any suitable receiver or destination.

Oil outlet means 12 is provided from the chamber at a point intermediate the settling chamber of the expansion chamber and the oil inlet means. When the character of the oil demands, oil heating means may be provided such as the steam jacket 13 which surrounds the velocity reducer 7. A steam pipe 14 is arranged to supply steam to such jacket. When desired a pipe 15 takes steam from the jacket 13 and delivers the same to a coil 16 arranged in the settling chamber 2 to further heat the oil within the chamber. The exhaust from the coil 16 is effected through pipe 17 which passes through the wall of the chamber.

A pump tube 18 having a sucker rod 19 is shown mounted on the casing head 20 of a well 9. A gas pressure pipe 21 is shown as connecting the gas pipe 10 with the casing head of the well to apply the pressure of the gas within the expansion chamber to the top of the oil within the well 9 which surrounds the pump tube 18. A valve $21^a$ may be provided in the pressure pipe 21 to regulate the pressure within the well, or such valve may be dispensed with, in which event the pressure on the oil surrounding the pump tube and the oil in the oil pipe 8 will be equalized.

Oil dividing means are provided interior

of the expansion chamber, such as cones 22 and $22^a$. Preferably these dividing cones are spaced apart, one below the other, as shown, with their peripheries nearly in contact with the wall of the chamber. The oil is delivered onto the upper surface of the cone 22 and spreads over the same evenly and is brought into contact with the wall 23 of the chamber and flows downwardly thereover until it is collected in the settling chamber. The cone $22^a$ is provided as a supplemental spreader to insure the even distribution of the oil over the wall of the chamber. As shown in the form herein illustrated, the cones 22 and $22^a$ are supported in the upper portion of the expansion chamber by means of gas collecting chamber 24 and pipe 25. Pipe 25 is screwed into flange 26 which in turn is seated upon and closes the upper end of the vertical member $6^a$ of the T 5, to which is connected the elbow 27 of the gas pipe 10. The lower end of the pipe 25 is screwed into a reducer 30, arranged with its larger end looking downward to receive the upper end of the gas collecting chamber 24 which is open at its lower end 28. The gas collecting chamber is also provided with gas admitting ports 29, 29 and 29 arranged between the cones 22 and $22^a$ which are protected thereby to prevent the entrance of oil into the gas line.

A valve seat 31 is provided in the reducer 30, and a valve stopper 32 is provided with means for seating the same when oil may rise within the chamber to a point where it may gain access to the gas line 10. As shown, this stopper 32 comprises a flat circular body having a beveled edge adapted to fit the valve seat in the reducer, and is provided with a perforated shank 34 through which the float rod 35, actuated by float 33, passes to open and close a vacuum breaker opening 36 in the stopper before actuating the stopper 32. A slot 38 is provided in the shank 34 which permits of the passage of a pin therethrough and through the float rod to limit the movement of such rod in breaking the vacuum to permit the unseating of the stopper.

The movement of the stopper, float rod, and float, away from the valve seat is limited by the end of shank 34 coming into contact with a bracket 39. A bracket 40 is arranged near the bottom of the gas collecting chamber to guide the float rod 35.

A trapped oil outlet from the chamber is provided which consists of a submerged pipe 41 provided with a closure 42 which is actuated by means of a float 43 which is so adjusted as to cause the pipe to be closed to the exit of oil before the level of oil reaches the open end of the pipe, to thereby prevent the escape of gas along with the outflowing oil from the chamber. Other forms of traps will readily suggest themselves, but the form herewith disclosed illustra function desired to be derived. A 44 is provided to permit sand and wa be drawn from stand pipe and set chamber when desired.

It is to be remembered that an impo feature of my invention is to provide whereby the oil is divided into a t broken body whereby the gas may re escape from its engagement therewith it is also desirable to maintain the oil divided form for a period of time suf for all the gas to free itself from th It is also desirable to maintain the oil quiescent condition for a short perio time within the settling chamber to p sand and water to settle out of the before the oil is drawn from the char

In practice, oil is pumped from the by actuating the sucker rod 19, which f the oil through the pipe 8 to the vel reducer 7 from which it flows throug T 5 onto the top of the cone 22 by w it is spread over the wall of the expai chamber and envelops the gas col The oil may be heated while passing thr the velocity reducer by means of the s jacket 13 up to or about the boiling of water. This heating will result in ening the surface tension upon any contained in the oil and allows the sa settle in the bottom of the chamber. settling of the water may be materiall sisted by the pressure of the gas wi chamber, and, if necessary, more heat be applied to the oil by means of the coil 16.

In case oil is being taken from a w the flowing type, and is intermittent heads, the velocity reducer will afford for equalizing the flow of oil into th pansion chamber. In case a large am of oil should be delivered to the expa chamber, and the egress thereof shoul slower than the amount coming in, the 33 will actuate the stopper 32 and the gas outlet to prevent oil from ge into the gas line. The gas pressure then increase and thereby force the oil the chamber until the float will drop to open the gas line again. The pre regulating valve 11 in the gas line 10 trols the exit of gas from the expai chamber and thereby keeps a constant sure on the oil within the chamber.

In the case of a wet gas, or a gas rated with an oil of the lighter series, as gasolene, it is desirable to mainta high a pressure within the expansion c ber as may be practicable for the reason a large amount of such light series thereby be compressed and remain wit oil. In some instances this action ha sulted in producing an oil three p higher in gravity, according to the B scale, than had been produced by subjecting the oil and gas to the atmospheric action to permit the gas to escape. This increase in lightness of gravity brought the producer an increase of price amounting to fifteen cents per barrel.

The admission of gas under pressure to the casing head of a well to equalize the pressure of gas within the pump tube, prevents the oil from slipping in the pump barrel on account of the valve being unseated by the upwardly flowing gas. This action as produced by my invention increases the pumping efficiency and the production of the well, thereby making profitable producers of wells that have theretofore been unprofitable.

The oil flowing from the expansion chamber through outlet 12 may be conducted to any suitable tank or receptacle either by the pressure of the gas within the chamber or by gravity.

When desired, a safety valve may be provided for the chamber 1 to permit escape of gas from the expansion chamber in the event that valve 32 should become stuck or remain seated on its seat. In some cases the float 33 and valve 32 may be dispensed with.

It will be noted that the action upon the oil while flowing down the wall of the expansion chamber in a thin film under pressure permits the free, dry, gas to readily escape therefrom, while the pressure exerted upon the oil surface backed by the wall of the chamber holds the lighter liquids, such as gasolene, in combination with the oil body, and I desire to be understood as pointing out and claiming this action as being of great benefit to the crude oil derived from the well on account of keeping the gasolene series in combinaton with the main body of oil.

I claim as my invention:

1. In an oil and gas separator, the combination of an expansion chamber arranged to receive oil and gas in its upper portion, means for spreading the oil over the wall of such chamber to flow downwardly thereover, gas take-off means arranged to take off gas from within the flowing film of oil, an oil collecting chamber below the expansion chamber, an oil outlet from said collecting chamber, and valve controlled means arranged to maintain a submergence of the oil outlet.

2. In an oil and gas separator, the combination of an expansion chamber, inlet means arranged to permit the entrance of oil and gas into the chamber, means within the chamber adapted and arranged to distribute the oil over the wall of the chamber in a downwardly flowing film, gas take-off means arranged to take gas from within the envelop of downwardly flowing oil, and means for maintaining gas pressure u such oil.

3. In an oil and gas separator, the com nation of an expansion chamber havin surface adapted to sustain a flow of thereover in a thin body, means for tributing oil onto such surface, press maintaining means arranged and adap to maintain a pressure on one side of flowing oil, withdrawing means arranged take gas from the chamber, and means withdrawing oil from the chamber.

4. In an oil and gas separator, the com nation of an expansion chamber, means delivering oil and gas into the cham means for maintaining pressure within chamber, means for drawing oil from chamber, and means within the cham adapted to cause the oil to flow in a t body for a distance to enable the gas c tained and carried thereby to be given while the oil is subjected to pressure.

5. The combination in an oil and gas s arator of an expansion chamber havin settling chamber communicating therew means for delivering oil and gas into upper portion of the expansion chambe cone arranged near the top of such cha ber to receive the incoming oil and spr it over the wall of the chamber in a t film like form, a gas collector arranged low the cone, gas outlet means from s collector, a float controlled valve arran to close the gas outlet means, means drawing oil from the separator above settling chamber, and means for discharg sand or water from said settling chamb

6. The combination in an oil and gas arator of an expansion chamber havin settling chamber communicating therewi means for delivering oil and gas to chamber, oil spreading and dividing me arranged to receive the oil within the cha ber and spread it in a thin film like fo over the wall of the chamber to flow the down, gas collecting and outlet means ranged below the oil spreading means within the flowing film of oil, a float c trolled valve arranged to close the gas o let when the oil rises to a predetermi height within the expansion chamber, a fl controlled valved outlet from the separa whereby a predetermined depth of oil maintained in the settling chamber to s merge the oil outlet and prevent the pass of gas therethrough and means for discha ing sand or water from the settling chamb

7. In a gas and oil separator, the co bination of an expansion chamber, spreading means arranged within the cha ber, oil supplying means arranged to liver oil into such expansion chamber a onto the spreader means, gas collect means arranged within the chamber, outlet means communicating with the

collecting means, oil outlet means from the expansion chamber arranged to draw oil from a point above the bottom thereof, such oil supplying means comprising an oil pipe from a suitable oil supply and a velocity reducing means connected therewith, whereby the oil and gas are delivered to the expansion chamber in quiescent condition.

8. In an oil and gas separator, the combination of an expansion chamber, means within the chamber adapted to spread the oil in a thin film like form over the wall of the chamber, gas outlet means leading from the chamber, consisting of a gas pipe, pressure regulating means arranged to control the passage of gas from the chamber through such gas pipe, oil outlet means arranged to draw oil from the expanding chamber, means to maintain such oil outlet means submerged within the body of oil in the chamber, velocity reducing means arranged to receive oil and gas from a source of oil supply and deliver the same into the expansion chamber, and means for heating the oil within the velocity reducing device before its entry into the expansion chamber.

9. In an oil and gas separator, the combination of an expansion chamber having a centrally perforated top, a T arranged with one end of its cross to register with the central perforation in the top and having its stem disposed horizontally, a bushing on the upper end of the cross of the T, a gas pipe leading from such bushing, a gas pipe depending from such bushing through the cross of the T and into the expansion chamber, means for delivering oil to the chamber through the stem of the T, oil dividing means secured to the lower end of the last named gas pipe, and means for drawing oil from the chamber.

10. In an oil and gas separator, the combination of an expansion chamber having a centrally perforated top, a T arranged with one end of its cross to register with the central perforation in the top and having its stem disposed horizontally, a bushing on the upper end of the cross of the T, a gas pipe leading from such bushing, a gas pipe depending from such bushing through the cross of the T and into the expansion chamber, means for delivering oil to the chamber through the stem of the T, means for heating the oil during its passage to such chamber, oil dividing means secured to the lower end of the last named gas pipe, and means for drawing oil from the chamber.

11. In an oil and gas separator, the combination of an expansion chamber having a centrally perforated top, a T arranged with one end of its cross to register with the central perforation in the top and havi stem disposed horizontally, a bushing upper end of the cross of the T, a ga leading from such bushing, means for lating the flow of gas from such expa chamber through such gas pipe to tain pressure within such chamber, pipe depending from such bushing th the cross of the T and into the expa chamber, means for delivering oil t chamber through the stem of the dividing means secured to the lower e the last named gas pipe, and mea drawing oil from the chamber.

12. In an oil and gas separator, the bination of an expansion chamber, a se chamber communicating therewith ad to receive the oil and retain the same quiet condition for a period of ti stand pipe arranged to support the e sion chamber and settling chamber a receive sand and water from the se chamber, a submerged and trapped for the oil arranged to draw oil fro separator, means for introducing oil the expansion chamber, means adapt divide the oil within the chamber, outlet from the chamber, and mean preventing oil entering such gas outlet.

13. In an oil and gas separator, a pansion chamber; inlet means for feed foam composed of oil and gas throug central part of the top of said expa chamber; an imperforate spreader cone, ing its apex pointing upwardly, locate side said chamber in such a manner spread a thin film of oil over the wall of said chamber, and means for t gas from the central portion of chamber.

14. In an oil and gas separator, a pansion chamber; inlet means for fe a foam composed of oil and gas thr the central part of the top of sai pansion chamber; an imperforate spr cone, having its apex pointing upwa located inside said chamber in such a ner as to spread a thin film of oil ove inner wall of said chamber, and a gas c pipe supporting said cone and pa vertically upward through said inlet m said pipe being arranged to take o from the apex of said cone.

In testimony whereof, I have her set my hand at Los Angeles, California 9th day of November, 1914.

MILON J. TRUMB

In presence of—
F. M. TOWNSEND,
A. J. GUTZLER.

1914.

CONTENTS:

1. Application papers.
2. Rejection Dec. 2, 1914.
3. Amend't A. Mch. 15, 1915.
4. Rejection, April 9, 1915.
5. Amendment B, Mch. 30, 1916.
6. Rejection, April 8, 1916.
7. Letter to Office, Dec. 4, 1916.
8. Rej. Jan. 10, 1917.
9. Sub. P. Att'y Feb. 6, 1917.
10. Amend't C, Dec. 20, 1917.

183-105 25

105 Gas Sep. Serial No. 872,110 Div. 31 260
461

2 Sheets—Sheet 1

M. J. TRUMBLE.

CRUDE PETROLEUM AND NATURAL GAS SEPARATOR.

APPLICATION FILED NOV. 14, 1914.

**1,269,134.** Patented June 11, 1918.

2 SHEETS—SHEET 1.

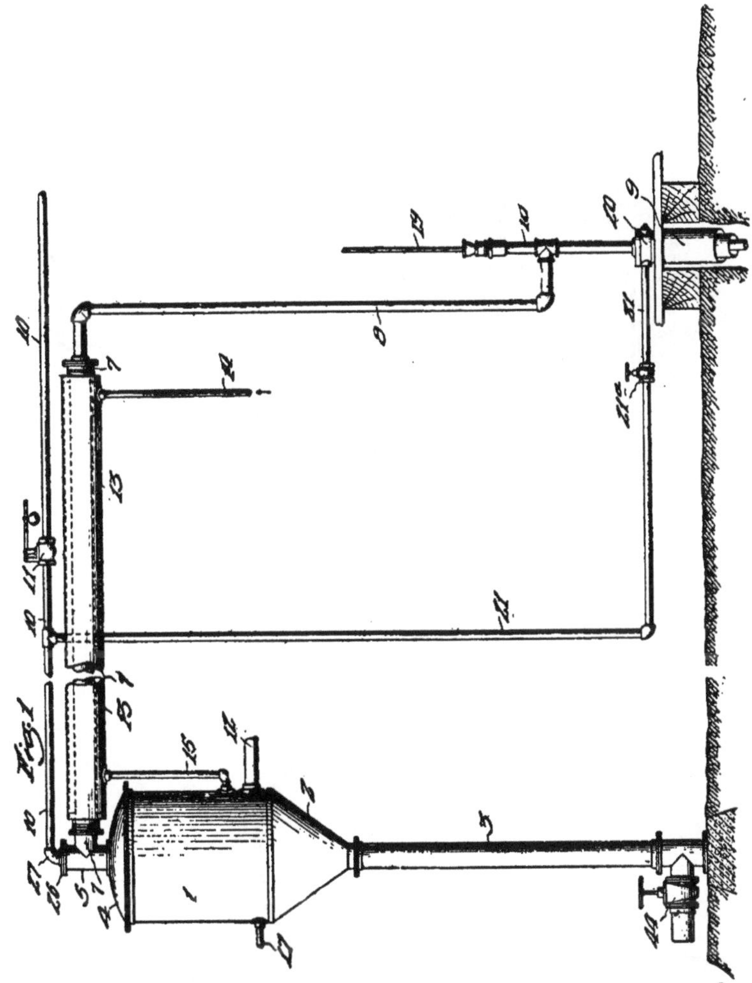


Witnesses:
[illegible]
Ford W. Harris

Inventor,
Milton J. Trumble,
Townsend & Graham
his attys.

105

25
Serial No. 872,110 Div. 31
2 Sheets—Sheet 2

M. J. TRUMBLE.
CRUDE PETROLEUM AND NATURAL GAS SEPARATOR.
APPLICATION FILED NOV. 14, 1914

**1,269,134.** Patented June 11, 1918.
2 SHEETS—SHEET 2.

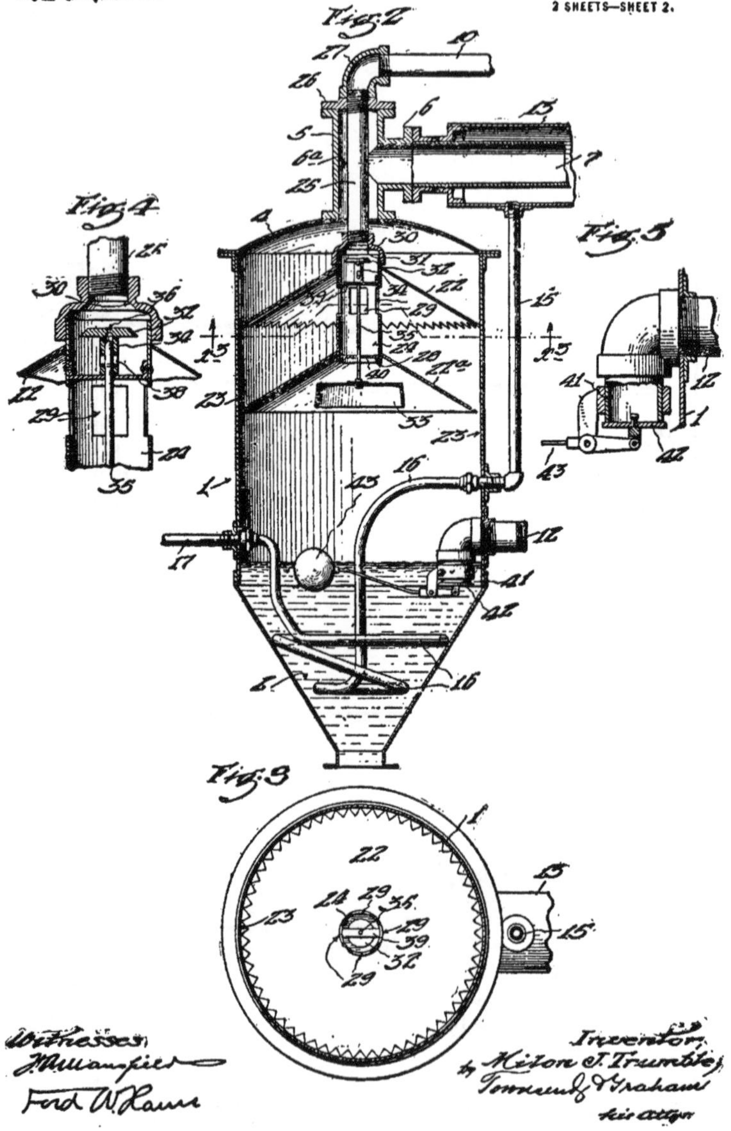

[Endorsed]: E—113—Eq. Townsend et al. vs. Lorraine. Plffs. Exhibit 2. Filed March 22, 1922. Chas N. Williams, Clerk.

No. 3945. United States Circuit Court of Appeals for the Ninth Circuit. Filed Jan. 26, 1923. F. D. Monckton, Clerk.

---

### Plaintiff's Exhibit No. 3.

[Endorsed]: E-113—Eq. Townsend vs. Lorraine. Plaintiff's Exhibit 3. Filed March 22, 1922. Chas. N. Williams, Clerk.

No. 3945. United States Circuit Court of Appeals for the Ninth Circuit. Filed Jan. 26, 1923. F. D. Monckton, Clerk.

D. G. LORRAINE.

GAS, OIL, AND SAND SEPARATOR.

APPLICATION FILED FEB. 5, 1920.

**1,373,664.** **Patented Apr. 5, 1921.**

2 SHEETS—SHEET 1.

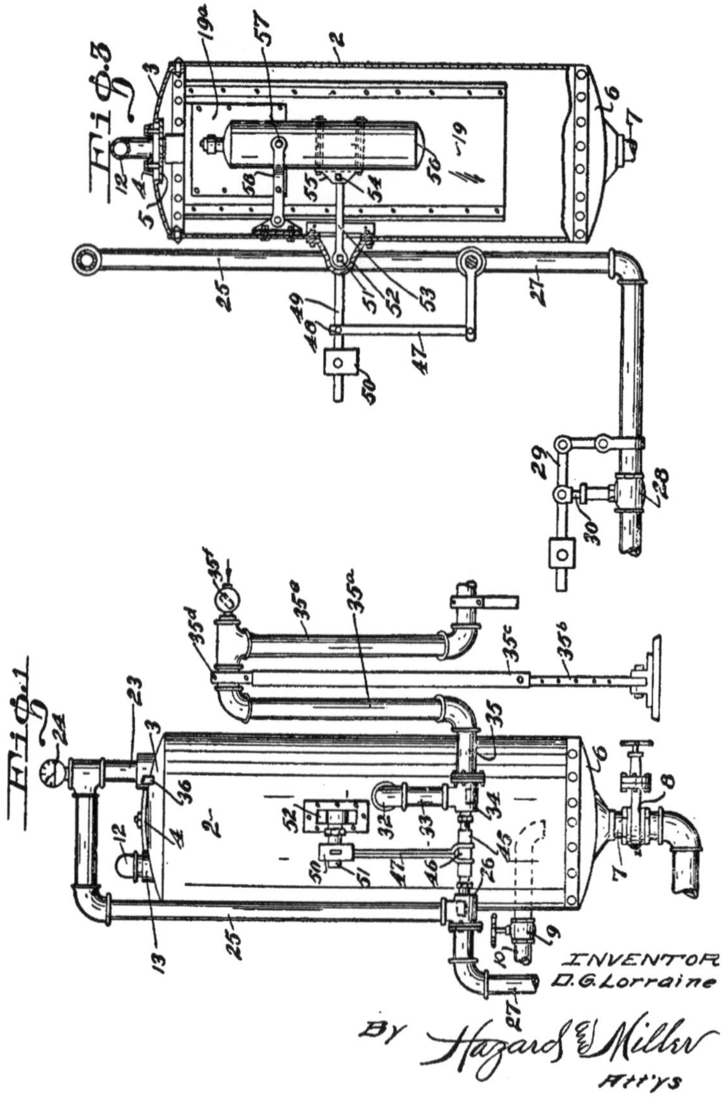

D. G. LORRAINE.
GAS, OIL, AND SAND SEPARATOR.
APPLICATION FILED FEB. 5, 1920.

1,373,664. Patented Apr. 5, 1921.
2 SHEETS—SHEET 2.

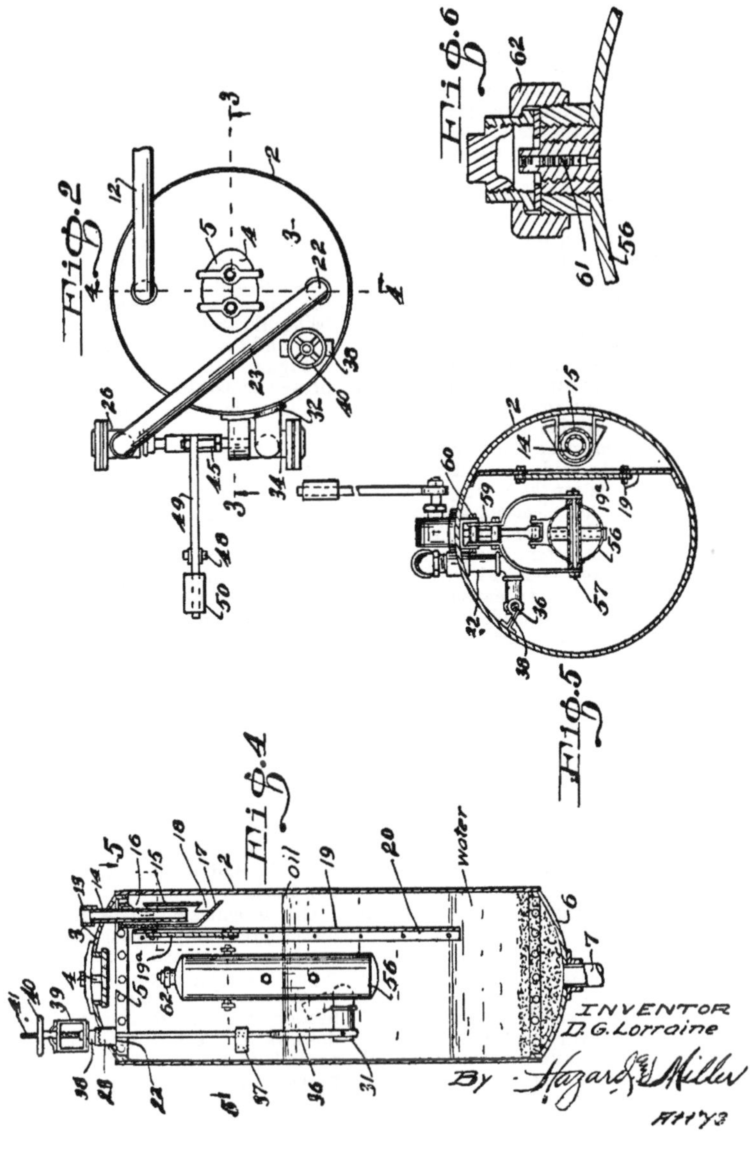

# UNITED STATES PATENT OFFICE.

DAVID G. LORRAINE, OF LOS ANGELES, CALIFORNIA.

## GAS, OIL, AND SAND SEPARATOR.

1,373,664. Specification of Letters Patent. Patented Apr. 5, 19'

Application filed February 5, 1920. Serial No. 356,438.

*To all whom it may concern:*

Be it known that I, David G. Lorraine, a citizen of the United States, residing at Los Angeles, in the county of Los Angeles and State of California, have invented new and useful Improvements in Gas, Oil, and Sand Separators, of which the following is a specification.

This invention relates to apparatus for the efficient separation of gas and oil and especially for the separation of these substances when same are recovered from oil wells, and an object of the invention is to provide an apparatus for the effectual recovery of the gaseous content of emulsified oils as well as for the separation of oils and gas when the same is not in the form of an emulsion. The invention has also for an object to provide an apparatus in which as few parts as practicable are arranged within a separating chamber and to provide for ready access to the said parts; and another object is to provide for the arrangement and organization of other elements wholly exterior to the separating chamber so that necessary adjustments, attention, and replacements may be given and had without the necessity of dismantling or materially interrupting the operation of a plant.

A further object of the invention is to provide means for maintaining a pressure continuously on the oil or emulsion as it is continuously admitted to the separating chamber so that the emulsion may be subjected to sufficient pressure to express the gaseous constituent of the emulsion from the mixture.

Another object of the invention is to provide effectual means for automatically controlling the discharge, separately, of the oil and gases from the apparatus and especially to provide a pneumatic device adapted to stand the pressures that may obtain within the separating chamber.

Another object of the invention is to provide hydrostatic means for controlling the discharge of oil and for preventing the loss of gas with the outwardly passing oil and to provide such a hydrostatic means readily variable to afford different heads of oil pressure to overcome the interior pressure during the operation of the apparatus.

Another object of the invention is to provide for the minute adjustment of the oil discharge mechanism.

Another object is to provide an orga zation in which the several parts are sim in construction, readily removable and newable, and to provide coördinate va means operative by a single connection w the said pneumatic control means.

A further object of the invention is provide a novel method for treating emulsions to facilitate the recovery o: t gaseous content of the emulsion.

With these and other objects in view invention consists of the method, the co bination and in details and arrangements the parts, an embodiment of the inventi being illustrated in the accompanying dra ings and described and claimed herein.

Figure 1 is a side elevation of the proved apparatus.

Fig. 2 is a top plan view of the ap ratus.

Fig. 3 is a section on line 3—3 of Fig. 2

Fig. 4 is a section on line 4—4 of Fig. 2

Fig. 5 is a section on line 5—5 of Fig. 4

Fig. 6 is a detail sectional view of upper portion of the pneumatic float.

The apparatus includes, preferably, upright cylindrical tank or receptacle 2 any suitable dimensions having a head that has a manhole 4 which is shown in F 2 as closable by a manhole plate 5. T lower end of the tank is provided with a b tom 6 having a central outlet pipe 7 which is provided a cut-off valve 8 whi when opened, will provide for the washi out of such material as sand or other su stance collecting in the bottom of the rec tacle 2 and which discharge may be fac tated by the introduction of water or ot fluid under pressure by the opening of valve 9 arranged in a pressure pipe 10 one side, and in the lower portion, of t receptacle 2.

Oil or emulsion to be treated is int duced, preferably, through the top 3 of t receptacle to which is connected a sup pipe 12 leading in through a connection on the top of the receptacle and which co nection is shown in Fig. 4 as including downwardly extending tube 14 havi around its lower end a sleeve or hood provided at its upper portion with a g outlet 16; the lower end of the sleeve 15 ha ing an inclined bottom 17 below the op mouth 18 formed in the side of the slee and which mouth is disposed toward the s jacent surface of the wall of the receptacle 2. The inlet sleeve 15 is arranged, as shown in Figs. 2 and 4, as in close juxtaposition to one side of the cylindrical receptacle; and extending vertically and transversely across this receptacle there is provided a partition wall 19 the flanges 20 of which may be riveted or otherwise secured as at 21 to the interior surface of the receptacle 2.

The lower end of this partition 19 terminates well above the bottom of the receptacle and the partition forms a confined vertical passageway down which the oil issuing from the mouth 18 of the inlet sleeve 15 is directed and is caused to pass beneath the lower edge of the partition 20 before the lighter materials, such as gas and oils, can rise to a predetermined variable height in the receptacle. This enables the sand that may be obtained in the oil or emulsion, coming from the supply pipe 12, to settle toward the bottom while the oil passes around the lower edge of the partition and ascends in the larger compartment formed on the opposite side of the partition.

The accumulating gases ascending to the top of the receptacle pass into an outlet 22 and into a gas pipe 23 in which may be provided a gage 24. The outwardly passing gas flows down a pipe 25 in the lower end of which is mounted a valve 26 controlling the flow from the receptacle 2. From the valve 26 is continued a delivery gas pipe 27 in which there is mounted a valve 28, Fig. 3, designed to hold a predetermined back pressure in the gas line and receptacle 2; the valve, in this case, being provided with a weight lever arm 29 connected to the stem 30 of the valve.

Oil is discharged from the receptacle through an adjustable outlet consisting, in this case, of an elbow 31 turnable about an axis in the connection 32 which is extended to the outside of the receptacle 2 and has a downturned branch 33 leading into an oil regulating valve 34 from which the oil passes into a delivery pipe 35.

One of the features of the present invention resides in means for very carefully adjusting the level of the outlet member 31 as by swinging it about its axis in the member 32; and to secure this refinement of adjustment the outlet member has attached to it a link or rod 36 extending up through a guy 37 and thence through a bearing 38, Fig. 2, to which is attached a yoke or other suitable member 39 to which is rotatively secured a handwheel 40 that engages the upper threaded end 41 of the rod 36. Therefore, by turning the handwheel 40 in one or the other direction it will lift or lower the rod 36 and consequently set the opening of the outlet member 31 at a desired position so as to enable the separation or outflow of the oil at or above a desired predetermined level with respect to the liquid conte the receptacle 2, as it is obvious tha oil will vary in clarity according tc specific gravity of its constituents an upper portion of a mass of oil is ther clearer than the lower portion, and to s the separation of this clearer or better ity the outlet member 31 is made adjust

A further feature of the present inve resides in the provision of a huoyan substantial means for automatically cor ling the opening and closing simultane of the gas outlet valve 26 and the oil o valve 34, and to that end these valve. shown as mounted on a common rock 45 on which is provided a lever 46 tc outer end of which is connected a lin the upper end of which is attached at 48 lever arm 49 on which may be adjust counterbalancing weight 50.

The lever arm 49 is secured to the tiguous end of a short rock shaft 51 mou in a box 52 attached to the side of th ceptacle 2 this box having diverging forming a mouth opening to the interi the receptacle 2, and in this mouth oscillates a lever 53 attached to the po of the shaft 51 extending into the bo. The swinging end of the lever 53 is otally connected at 54 to a bearing 55 is attached to the adjacent side of a p matic cylinder 56.

For the purpose of stabilizing and taining this pneumatic cylinder in a ver position it has connected to it at 57 the arms of a yoke 58 which has rearwardl tending parallel arms 59 these being oted on a bearing 60 arranged withir receptacle 2; the length of the lever d 58 and the lever 53 being substantially e and therefore holding the float 56 in ar right position and causing it to move vertical position at all times as determ by variation of the level of the oil ir receptacle 2.

As above mentioned there is mainta in the receptacle a gas pressure as d mined by the adjustment of the pre regulating valve 28, and it is desirab provide, therefore, a pneumatic float a which will not only be of a highly buo characteristic but also will be able to stand the exterior pressures applied wit crushing, and therefore, the float 56 is in the form of a hollow cylinder and h its upper end a check valve 61 that ma covered by a removable cap 62, and pneumatic float 56 may be charged compressed air or gas to any desired de or equal to the maximum at which pres would be maintained in the receptacle

The operation of the apparatus is sub. tially as follows:

The valve 28 having been set by its r lating means 29 to hold a given pressu the receptacle 2 then as oil and gas is supplied to the receptacle by the oil supply pipe 12 the latter will be directed down the small compartment through the inlet or feed sleeve 15 by which the oil is showered on to the adjacent portion of the receptacle wall whence it flows downwardly between the wall and the partition 19 any gases being liberated rising to the top of this compartment and passing over the upper end of the partition 19 and accumulating in the upper end of the receptacle 2.

The oil mass, or emulsion in some cases, passes beneath the lower end of the partition 19 and thus facilitates the deposit or separation of solids such as sand or other heavy substances, and the lighter portion of the oil with the gas and the emulsion rises into the larger compartment of the receptacle and passes upwardly to a level that is determined by the position of the pneumatic float 56. The gas will be compressed to a pressure substantially equal to that determined by the valve 28 and maintained there, and as the oil accumulates it will lift the float 56 and through means of the lever arm 53 rock the rock shaft 51 and through means of its connection to the valve shaft 45 rock the latter so as to open or increase the opening of the oil valve 34 and permit the discharge of oil through the outlet member 31 past the valve 34 and to the discharge 35. Thereupon the level of the oil will fall until in such action the float 56 in descending will again close or partially close the oil valve 34 and open or increase the opening of the gas valve 26. This alternate action of the automatic valve mechanism will continue so long as oil is supplied by the main oil pipe 13 to the apparatus.

From the above it will be seen that I have provided a method for separating or facilitating the separation of the gas and oil and separately discharging the same from emulsions; and furthermore have provided a method in which, by maintaining a predetermined pressure in the oil receptacle, the latter is subjected to pressures having the effect of expressing the gaseous content from emulsions, the gaseous constituent in the emulsion being driven from the denser liquids by the increase in the pressure on the oil within the receptacle 2. This, therefore, prevents the loss of the valuable gaseous constituent such as occurs in apparatus in which the oil passes immediately from a well or other source to an apparatus in which it is subject only to atmospheric pressure.

In the event of treating oils that are not emulsified the valve 34 may be omitted or may be fixed open and a static head or pressure may be secured by maintaining a column of oil in the discharge pipe 35 by providing the latter with a safety extension 35[a]. This extension is shown as in the form of a vertical partition arranged in the receptacle and terminating short of the top and bottom of the receptacle, and means arranged between the said partition and a contiguous wall of the receptacle for introducing the oil from the well whereby the oil is caused to flow downwardly and around the lower end of the partition and rise on the opposite side and the gas that freely escapes from the oil rising to the top of the receptacle and collecting above the upper end of the partition; said means including an inlet oil tube arranged in the upper part of the receptacle, said tube having a surrounding discharge sleeve with a deflecting bottom plate for discharging the oil toward the receptacle wall; said sleeve having upper and lower outlets for gas and oil respectively.

3. In an apparatus for separating oil and gas, a receptacle having separate discharge means for the oil and gas, each having an exterior valve actuated by a common crank, said valves being set for one to open as the other closes and float means within the receptacle actuated by the height of the liquid, operatively connected to said crank to actuate said valves.

4. In an apparatus for separating oil and gas, the combination of a receptacle having oil inlet means, separate oil and gas outlet means comprising pipes extending from the interior of said receptacle, stop cocks arranged in said pipes, one to open as the other closes and actuated by a common crank, float means within the receptacle operatively connected to said common crank adapted to actuate the valves with reference to the height of the liquid within the receptacle.

5. In an apparatus for separating gas and oil the combination of a receptacle, a float within the receptacle for actuating gas and oil outlet valves, said float comprising a sealed chamber, two supporting arms of equal lengths pivoted to said float at their proximal ends and pivoted to the wall of the receptacle at their distal ends to hold the chamber in substantially vertical alinement during movement thereof, the pivotal centers of said arms being off-set in vertical alinement at their distal ends, one of said arms having an extension thereon extending through the wall of the receptacle and being operatively connected to an actuating arm of said valves.

In testimony whereof I have signed my name to this specification.

DAVID G. LORRAINE.

---

## Plaintiff's Exhibit No. 4.

[Endorsed]: E-113—Eq. Townsend vs. Lorraine. Plaintiff's Exhibit 4. Filed March 22, 1922. Chas. N. Williams, Clerk.

No. 3945. United States Circuit Court of Appeals for the Ninth Circuit. Filed Jan. 26, 1923. F. D. Monckton, Clerk.

D. G. LORRAINE.

OIL, GAS, AND SAND SEPARATOR.

APPLICATION FILED JULY 18, 1921.

Reissued Nov. 8, 1921. **15,220.**

2 SHEETS—SHEET 1

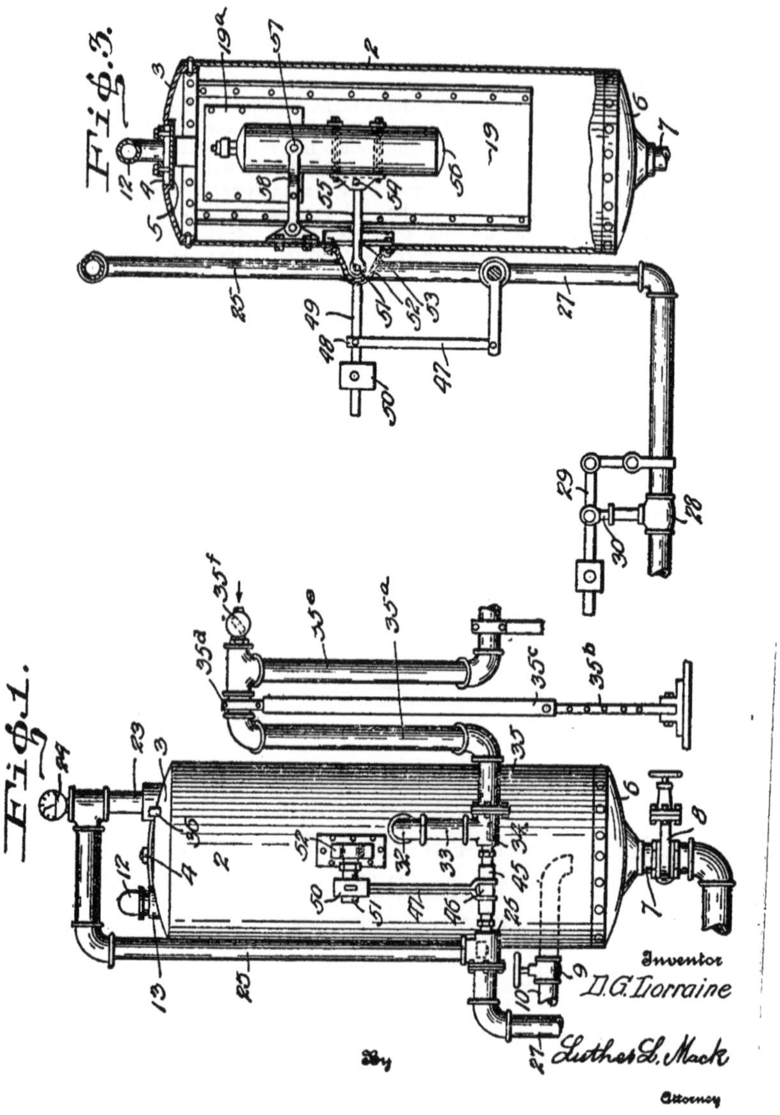

D. G. LORRAINE.
OIL, GAS, AND SAND SEPARATOR.
APPLICATION FILED JULY 18, 1921.

**Reissued Nov. 8, 1921.** **15,220**
3 SHEETS—SHEET 2.

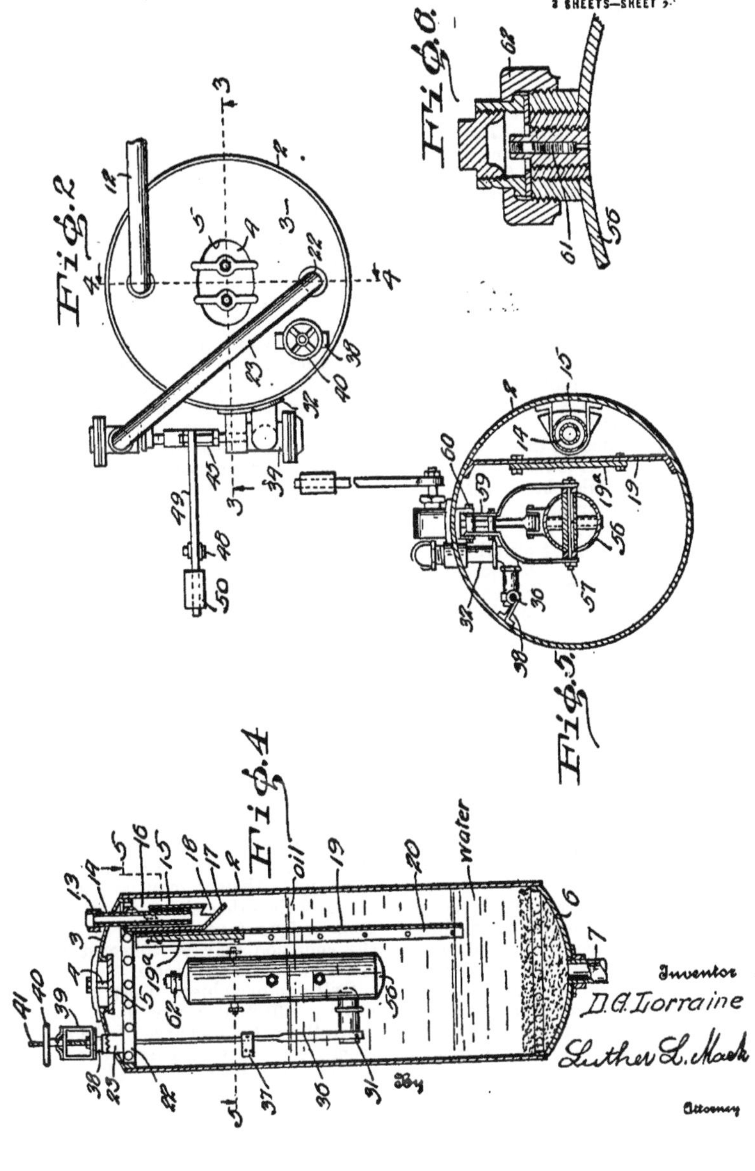

# UNITED STATES PATENT OFFICE.

DAVID G. LORRAINE, OF LOS ANGELES, CALIFORNIA.

## OIL, GAS, AND SAND SEPARATOR.

**15,220.** Specification of Reissued Letters Patent. **Reissued Nov. 8, 1921.**

Original No. 1,373,664, dated April 5, 1921, Serial No. 356,438, filed February 5, 1920. Application for reissue filed July 18, 1921. Serial No. 486,317.

*To all whom it may concern:*

Be it known that I, DAVID G. LORRAINE, a citizen of the United States, and a resident of Los Angeles, in the county of Los Angeles and State of California, have invented new and useful Improvements in Oil, Gas, and Sand Separators, of which the following is a specification.

This invention relates to and has for a primary object the provision of an apparatus for the efficient separation of gas and oil from oil wells, where the same is not in the form of emulsion, and does not contain sand and water and without pressure, and adaptable also for the purpose without modification, for maintaining a pressure whereby an emulsion may be broken up into its constituent elements.

It will be understood in the consideration of my invention that it is necessary and desirable to treat the oils obtained from the oil wells so as to thoroughly separate the gas and oil for different uses and so as to remove from the volume of oil all traces of water and sand which may be ejected from the well with the oil. To this end it is an object of my invention to provide an apparatus embodying a receptacle into which the volume of oil and accompanying substances is delivered from the well, for the purpose of separating the gas, water and sand therefrom. Said receptacle is preferably of substantial size and contents, so that a maximum volume of oil may be treated therein during the operation of the apparatus, and the separating operation facilitated.

Heretofore gas and oil separators have been so arranged as to provide a relatively shallow volume of oil, so that only a small volume may be treated at one operation, for the separation of the gas and other elements therefrom, and it is an object of my invention therefore to provide a receptacle of ample capacity, whereby a large volume and a substantially uniform level of oil may be maintained at a point above the mean elevation of the receptacle and in the major portion thereof, thereby rendering it possible to separate the gas from a maximum volume of oil. The upper and minor portion of the receptacle constitutes a gas collection or accumulation chamber from which the gas liberated from the oil is delivered by a suitable means to points remote from the separator for use.

It is a further object of my invention to sub-divide the interior of the gas and oil receptacle into two separate segmental chambers, by means of a vertical partition arranged therein and in communication at the top of the receptacle with the gas collection chamber, and at the bottom thereof with the oil receiving portion of the receptacle. The oil with its constituent elements is delivered from the well into the upper partition of the receiving chamber on one side of the said partition, which is relatively smaller than the other chamber of the receptacle, and the oil or emulsion rises to a higher level in the receiving chamber than the level of the volume of oil in the main chamber.

The arrangement of the said partition prevents the disturbance of the main volume of oil and permits the settling of the sand and water to the bottom of the tank, while the injection of the oil and its constituent elements from the well into the receiving chamber of the receptacle liberates a large volume of gas, as the oil enters the receptacle and thus the gas liberated collects in the upper portion of the receptacle. The heavier elements settle to the bottom of the tank and rise to and maintain a substantially uniform level in the main chamber thereof.

It is a further object of my invention to provide in an apparatus of the character described, means whereby the emulsified oil from oil wells may be injected by the natural pressure of the wells into and treated under pressure in the separating receptacle, for breaking up the emulsion into its constituent elements, so that the oil and valuable ingredients thereof may be withdrawn for use, and the other elements separated therefrom and whatever gas may be contained in the emulsion may be collected in the gas chamber of the receptacle.

Still another object of my invention is to provide effectual means for automatically controlling the discharge of the oil and gas from the apparatus separately and especially to provide a float device adapted to withstand the pressure that may remain within the separating receptacle and whereby the discharge of the gas and oil is effectually controlled.

Another object is to provide in an apparatus of the character mentioned, a receptacle having means capable of being employed for the separation of gas, oil, sand and water, and the accumulation of the gas in the upper portion of the receptacle without any pressure for the purpose of expelling the same for use, and also capable of being employed for demulsifying oils and reducing the same to their constituent elements, when the emulsion is supplied to the separating receptacle under pressure from the wells and whereby a substantial back pressure may be maintained in the gas discharge line for holding emulsion under a greater pressure within the receptacle than is necessary in the case of the separation of oil and gas in other forms than emulsion.

Another object of the invention is to provide hydrostatic means for separating the gas and oil and for preventing the loss of gas with the outwardly passing oil and to provide such a hydrostatic means readily variable to afford different heads of oil pressure to overcome the interior pressure during the operation of the apparatus, without overflowing the receptacle.

Another object of the invention is to provide for the minute adjustment of the oil discharge mechanism.

Another object is to provide an organization in which the several parts are simple in construction, readily removable and renewable, and to provide coördinate valve means operative by a single connection with the said pneumatic control means.

A further object of the invention is to provide a novel method for treating oil emulsions to break up and separate the ingredients of the emulsion.

With these and other objects in view the invention consists of the method, the combination and in details and arrangements of the parts, an embodiment of the invention being illustrated in the accompanying drawings and described and claimed herein.

Figure 1 is a side elevation of the improved apparatus.

Fig. 2 is a top plan view of the apparatus.

Fig. 3 is a section on line 3—3 of Fig. 2.

Fig. 4 is a section on line 4—4 of Fig. 2.

Fig. 5 is a section on line 5—5 of Fig. 4.

Fig. 6 is a detail sectional view of the upper portion of the pneumatic float.

The apparatus includes, preferably, an upright cylindrical tank or receptacle 2 of any suitable dimensions having a head 3 with a manhole 4 which is shown in Fig. 2 as closable by a manhole plate 5. The lower end of the tank is provided with a bottom 6 having a central outlet pipe 7 in which is provided a cut-off valve 8 which, when opened, will provide for the washing out of such material as sand or other substance collecting in the bottom of the receptacle 2 and which discharge may be facilitated by introduction of water or other fluid u pressure by the opening of a valve 9 ranged in a pressure pipe 10 at one and in the lower portion, or the receptac

Oil or emulsion to be treated is introdu preferably, through the top 3 of the re tacle to which is connected a supply pip leading in through a connection 13 on top of the receptacle and which connec is shown in Fig. 4 as including a d wardly extending tube 14 which may around its lower end a sleeve or hoo provided at its upper portion with a outlet 16; the lower end of the sleev may have an inclined bottom 17 below open mouth 18 formed in the side of sleeve and which mouth is disposed to the adjacent surface of the wall of receptacle 2.

The inlet sleeve 15 is arranged, as sh in Figs. 2 and 4, as in close juxtapositio one side of the cylindrical receptacle; extending vertically and transversely ac this receptacle there is provided a parti wall 19, the flanges 20 of which may be eted or otherwise secured, as at 21, to interior surface of the receptacle 2.

The lower end of this partition 19 terminate or have communication with opposite chamber well above the botto the receptacle and the partition forms a fined vertical passageway or chamber d which the oil issuing from the mouth 1 the inlet sleeve 15 is directed and is ca to pass beneath the lower edge of, or thro the partition 20 before the lighter mater such as gas and oils, can rise to a pred mined variable height in the recept This enables the sand that may be conta in the oil or emulsion, coming from the ply pipe 12, to settle toward the bot while the oil passes around the lower of or through parts in the partition, ascends in the larger compartment for on the opposite side of the partition. upper end of said partition may also minate short of the top of the receptacl have parts there in communication with gas chamber.

The accumulating gases ascending to top and minor portion of the receptacle into an outlet 22 and into a gas pipe 2 which may be provided a gage 24. The wardly passing gas flows down a pipe 2 the lower end of which is mounted a v 26 controlling the flow from the recept 2. From the valve 26 is continued a livery gas pipe 27 in which there is mo ed a valve 28, Fig. 3, designed to hol predetermined back pressure in the gas and receptacle 2; the valve, in this c being provided with a weight lever ar connected to the stem 30 of the valve.

Oil is discharged from the recep

through an adjustable outlet consisting, in this case, of an elbow 31 turnable about an axis in the connection 32 which is extended to the outside of the receptacle 2 and has a downturned branch 33 leading into an oil regulating valve 34 from which the oil passes into a delivery pipe 35.

One of the features of the present invention resides in means for very carefully adjusting the level of the outlet member 31 as by swinging it about its axis in the member 32, and to secure this refinement of adjustment the outlet member has attached to it a link or rod 36 extending up through a guy 37 and thence through a bearing 38, Fig. 2, to which is attached a yoke or other suitable member 39 to which is rotatively secured a handwheel 40 that engages the upper threaded end 41 of the rod 36 Therefore, by turning the handwheel 40 in one or the other direction it will lift or lower the rod 36 and consequently set the opening of the outlet member 31 at a desired position so as to enable the separation or outflow of the oil at or above a desired predetermined level with respect to the liquid content of the receptacle 2, as it is obvious that the oil will vary in clarity according to the specific gravity of its constituents and the upper portion of a mass of oil is therefore clearer than the lower portion, and to secure the separation of this clearer or better quality the outlet member 31 is made adjustable.

A further feature of the present invention resides in the provision of a buoyant yet substantial means for automatically controlling the opening and closing simultaneously of the gas outlet valve 26 and the oil outlet valve 34, and to that end these valves are shown as mounted on a common rock shaft 45 on which is provided a lever 46 to the outer end of which is connected a link 47 the upper end of which is attached at 48 to a lever arm 49 on which may be adjusted a counter-balancing weight 50.

The lever arm 49 is secured to the contiguous end of a short rock shaft 51 mounted in a box 52 attached to the side of the receptacle 2, this box having diverging walls forming a mouth opening to the interior of the receptacle 2, and in this mouth there oscillates a lever 53 attached to the portion of the shaft 51 extending into the box 52. The swinging end of the lever 53 is pivotally connected at 54 to a bearing 55 that is attached to the adjacent side of a pneumatic cylinder or float 56

For the purpose of stabilizing and maintaining the float cylinder in a vertical position it has connected to it at 57 the side arms of a yoke 58 which has rearwardly extending parallel arms 59 these being pivoted on a bearing 60 arranged within the receptacle 2, the length of the lever device 58 and the lever 53 being substantially equal and th fore holding the float 56 in an upright sition and causing it to move in a vert position at all times as determined by v ation of the level of the oil in the rec tacle 2.

When treating an emulsion as above m tioned, there is maintained in the recept a gas pressure as determined by the adj ment of the pressure regulating valve and it is desirable to provide, therefore pneumatic float as 56 which will not only of a highly buoyant characteristic but will be able to withstand the exterior p sures applied without crushing, and th fore, the float 56 is made in the form o hollow cylinder and has at its upper en check valve 61 that may be covered b removable cap 62, and the pneumatic 56 may be charged with compressed air gas to any desired degree or equal to maximum at which pressure would be ma tained in the receptacle 2.

The operation of the apparatus is s stantially as follows:

When treating an emulsion, the valve having been set by its regulating means to hold a given pressure in the receptacl then as the emulson is supplied to the rec tacle by the oil supply pipe 12 the latter be directed down the small compartm through the inlet or feed sleeve 15 by wh the oil is showered onto the adjacent port of the receptacle wall whence it flows do wardly between the wall and the partit 19, any gases being liberated rising to top of this compartment and passing o the upper end of or through the partit 19 and accumulating in the upper end the receptacle 2.

The oil mass, or emulsion in some ca passes beneath the lower end of or thro the partition 19 and thus facilitates the posit or separation of solids such as sand other heavy substances, and the lighter tion of the oil with the gas and the e sion rises into the larger compartment of receptacle and passes upwardly to a l that is determined by the position of pneumatic float 56 The emulsion will compressed to a pressure determined by adjustment of the valve 28 and maintai there, and the emulsion will be broken into its constituent elements thereby. as the oil accumulates it will tend to lift float 56 and through means of the lever 53 rock the rock shaft 51 and through me of its connection to the valve shaft 45 r the latter so as to open or increase the op ing of the oil valve 34 and permit the charge of oil through the outlet member past the valve 34 and to the discharge Thereupon the level of the oil will tend fall until in such action the float 56 in scending will again tend to close or parti close the oil valve 34 and open or increase the opening of the gas valve 26. This alternate action of the automatic valve mechanism will continue so long as oil is supplied by the main oil pipe 13 to the apparatus. It will be understood, however, that the level of the fluid contents of the main chamber of the receptacle 2 will be at all times substantially uniform, and when the apparatus is properly adjusted this level will be maintained at a point well above the mean elevation of the receptacle so as to insure a maximum volume of oil from which the gas may be liberated and yet provide an ample space for the accumulation of the gas.

From the above it will be seen that I have provided a method for separating or facilitating the separation of the gas and oil and separately discharging the same from emulsion; and furthermore have provided a method in which, by maintaining a predetermined pressure in the oil receptacle, the latter is subjected to pressures having the effect of expressing the gaseous content from emulsions, the gaseous constituent in the emulsion being driven from the denser liquids by the increase in the pressure on the oil within the receptacle 2. This, therefore, prevents the loss of the valuable gaseous constituent such as occurs in apparatus in which the oil passes immediately from a well or other source to an apparatus in which it is subject only to atmospheric pressure.

In the event of treating oils that are not emulsified the valve 28 may be omitted and the gas provided with free outlet without pressure or a static head, or pressure may be secured by maintaining a volume of oil in the discharge pipe 35 by providing the latter with a safety extension 35ᵃ. This extension is shown as in the form of a U-shaped pipe stem that is turnable from a horizontal to an upright position about the axis of the pipe 35 and may be temporarily and substantially secured in its angular adjustment by a suitable jack comprising for instance a post 35ᵇ having a sleeve 35ᶜ the upper end of which engages the cross or top section of the siphon tube as at 35ᵈ.

For the purpose of breaking the siphon effect in one branch as 35ᵈ of the tube an inwardly opening check valve 35ᶠ is mounted, preferably, in the bend of the siphonic device so that the oil will not be drawn from the branch to which the pipe 35 is connected but will be maintained therein to prevent drawing out of the gas from the receptacle 2.

Preferably water is maintained in the bottom of the receptacle to a level somewhat above the lower end of the sand sheet or partition or the part therein, so that the incoming supply of oil when passing below the lower end of or through the partition is brought into contact with the water and the attrition between the emulsion ca rapid separation of the gaseous cont the oil and mixture engages the water

It has been found from practical ence in the operation of this apparatu there is an increase in the producti oil from some wells because of the un flow from the separator which preven rapid increment in the quantity of sa dinarily found in wells, and which ment results in the clogging or jammi the well and loss of production unt well is blown.

The sand sheet 19 is, preferably, pro with a window opening that may be co by a cover plate 19ᵃ through which the ber 15 may be applied and removed necessary.

It will be understood that frequentl flow from oil wells is not uniform, th coming in impulses and varying in and pressure, and this occasions the de of sand to a more or less extent, whic quently stops the flow of oil from the

With the use of my apparatus, ho in connection with an oil well, becau the arrangement of parts and the conn between the well and the separating d when a well is flowing a substantially form flow is maintained at all times the delivery of oil from the well is stab and prolonged frequently beyond the of its natural flow, and the impulse prevented also because of the complete lation of the flow from the well. Whe nected with my separator sand is n countered in sufficient quantities to st influence the flow and altogether a uniform and satisfactory delivery fro well is maintained.

It will be further understood that b of the vertical partition within the se ing receptacle and the provision of th arate receiving and separating co ments, the products of the well are del into the smaller receiving chamber a of the agitation of the oil is effected receiving chamber, while the main v of oil is maintained at a stationary lev without agitation in the larger se chamber of the receptacle. Now, this case of oil and gas other than emuls effected without pressure. The tende oil and products is to create pressure the same is agitated in the receptacle a the provision of the partition receptac scribed herein. This agitation and th sequent increase of pressure within t ceptacle is entirely eliminated and th is allowed to flow freely from the gas tion chamber in the minor and uppe tion of the receptacle.

The treatment of emulsions, howev quires a substantial pressure within t ceptacle in order that the emulsion m ken up completely and the constitnent ments thereof liberated and separated.

The pressure is maintained in the recep-le by means of the weight valve 23 which not necessary for the treatment of the ural gas and oil.

at I claim is:

. In an apparatus for separating gas and from oil wells comprising a receptacle, a tical partition arranged in the receptacle terminating short of the top and tom of the receptacle, and means arranged ween the said partition and a contiguous ll of the receptacle for introducing the oil m the well whereby the oil is caused to downwardly and around the lower end the partition and rise on the opposite side l the gas that freely escapes from the oil ng to the top of the receptacle and col-ing above the upper end of the partition, l means including an inlet oil tube ar-ged in the upper part of the receptacle, l tube having a surrounding discharge ve with a deflecting bottom plate for dis-rging the oil toward the receptacle wall.

. In an apparatus for separating gas and from oil wells comprising a receptacle, a a partition arranged in the receptacle terminating short of the top and bottom the receptacle, and means arranged be-en the said partition and a contiguous l of the receptacle for introducing the the well whereby the oil is caused to downwardly and around the lower end he partition and rise on the opposite side the gas that freely escapes from the oil ng to the top of the receptacle and col-ing above the upper end of the partition; l means including an inlet oil tube ar- in the upper part of the receptacle, l tube having a surrounding discharge ve with a deflecting bottom plate for dis-rging the oil toward the receptacle wall; l sleeve having upper and lower outlets gas and oil respectively.

. In an apparatus for separating oil and , a receptacle, having separate discharge ns for the oil and gas, each having an rior valve actuated by a common crank, valves being set for one to open as the er closes and float means within the re-tacle actuated by the height of the liquid, ratively connected to said crank to actu-said valves.

. In an apparatus for separating oil and , the combination of a receptacle having inlet means, separate oil and gas outlet ns comprising pipes extending from the rior of said receptacle, stop cocks ar-ged in said pipes, one to open as the other es and actuated by a common crank, float ns within the receptacle operatively con-ed to said common crank adapted to ate the valves with reference to the ht of the liquid within the receptacle.

5. In an apparatus for separating gas and oil, the combination of a receptacle, a float within the receptacle for actuating gas and oil outlet valves, said float comprising a sealed chamber, two supporting arms of equal lengths pivoted to said float at their proximal ends and pivoted to the wall of the receptacle at their distal ends to hold the chamber in substantially vertical alinement during movement thereof, the pivotal centers of said arms being off-set in vertical alinement at their distal ends, one of said arms having an extension thereon extending through the wall of the receptacle and being operatively connected to an actuating arm of said valves.

6. In an apparatus for separating gas and oil from oil wells, a receptacle having a vertical partition arranged therein and subdividing the interior of said receptacle into two separate chambers arranged for communication near the top and bottom portions thereof, and means in communication with one of said compartments for introducing the oil from a well thereinto, whereby the oil is caused to flow downwardly in said receiving chamber and into the lower portion of said other chamber and to rise on the opposite side of said partition, the gas being liberated from the oil in said receiving chamber and delivered therefrom to and discharged with the gas from the upper portion of said other chamber above the level of oil therein and accumulating therein for delivery to points externally of said receptacle.

7. In an oil and gas separating device, a receptacle for receiving the oil and gas from wells, a vertical partition fixed therein for subdividing the interior of said receptacle into a receiving chamber and a relatively larger settling chamber, said partition being arranged to afford communication near its top and bottom between said chambers, means for delivering a volume of oil and gas from a well into said receiving chamber whereby the gas may be liberated from the oil and discharged into the upper portion of and commonly discharged with the gas in said settling chamber and the oil caused to flow downwardly in said receiving chamber and into and to rise in said settling chamber, and means for regulating the supply and discharge of the oil to and from said receptacle, respectively, whereby the level of the oil in said settling chamber may be maintained at a substantially uniform elevation and substantially above the vertical center of the receptacle, for treating a maximum volume of oil therein.

8. In a gas and oil separating device for oil wells, a receptacle for receiving the oil from a well and having a vertical partition therein subdividing the interior of the receptacle into spearate chambers in communi-

cation at points near the top and bottom, the upper portion of the receptacle serving as a common receiver for the gas from both chambers, and means for supplying a volume of oil to and discharging the same from said receptacle, whereby a maximum volume of oil may be maintained at a substantially uniform level in said receptacle at a plane above the vertical center thereof, the gas liberated therefrom and accumulated above the level of the oil and the heavier constituents caused to settle to the bottom of said receptacle, as described.

9. An oil and gas separator for oil wells including a receptacle having a receiving compartment and a settling compartment arranged for communication at the top whereby the gas may be liberated from the oil introduced into said receiving compartment and discharged into the upper portion of and united with the gas in said other compartment, and also arranged for communication at the bottom whereby the oil may flow downwardly in said receiving compartment and into said other compartment and the other constituents separated therefrom.

10. An oil and gas separator for oil wells including a receptacle having an oil receiving chamber and a settling chamber, a vertical partition therebetween arranged open at the top for providing communication between said chambers, a gas outlet common to both of said chambers and means whereby a substantially uniform level of oil may be maintained in said settling chamber above the vertical center line of said receptacle.

11. An oil and gas separator for oil wells including a receptacle having an oil receiving chamber and a settling chamber arranged for communication near the top and bottom of the receptacle, and provided with a common gas outlet and float controlled means whereby the discharge of the oil from said receptacle is automatically regulated for maintaining a substantially uniform level of oil in the settling chamber, above the vertical center line of the receptacle.

12. An oil and gas separator for oil wells including a receptacle having an oil receiving chamber and a settling chamber, having a common outlet for the gas generated therein, a vertical partition therebetween arranged for affording communication between said chambers near the top and bottom of said receptacle, and float means in the upper portion of said settling chamber for regulating the discharge of the oil, whereby a substantially uniform level of oil may be maintained in said settling chamber above the vertical center line of said receptacle.

13. An oil and gas separator for oil wells including a receptacle having a vertical partition terminating short of the top and bottom thereof, and forming an oil receiving chamber and a settling chamber arranged for communication at the top in botto said receptacle, means for introducin volume of oil from a well into said recei chamber, the oil being delivered from bottom of said receiving chamber into bottom of said settling chamber, the from both of said chambers being un and commonly discharged and means wl by a substantially uniform level of oil be maintained in said settling chamber al the vertical center line of said receptacl

14. An oil and gas separator for oil v including a receptacle having an oil rec ing chamber and a settling chamber ranged for communication between the and bottom of said receptacle, means fo troducing a volume of oil from a well said receiving chamber, the oil being livered from the bottom of said recei chamber into the bottom of said sett chamber, and float controlled means regulating the discharge of oil from the tling chamber, whereby a substantially form level of oil may be maintained in settling chamber above the vertical ce line of said receptacle, the gas being li ated from the oil in said receiving cl ber and being discharged into and from settling chamber.

15. An oil and gas separator for oil v including a receptacle having an oil re ing chamber and a settling chamber ranged for communication near the top bottom of the receptacle, a vertical parti between said chambers, a pivoted float ported in the upper portion of said re tacle for regulating the discharge of th from the settling chamber, whereby a stantially uniform level of oil may be m tained in said settling chamber above vertical center line of said receptacle.

16. An oil and gas separator for oil v including a receptacle having an oil re ing chamber and a settling chamber ranged for communication near the top bottom of the receptacle, a vertical p tion between said chambers, a float ported in the upper portion of said re tacle for regulating the discharge of th from the settling chamber, whereby a stantially uniform level of oil may be m tained in said settling chamber above vertical center line of said receptacle, means for introducing a volume of oil f a well into the top of said receiving cham whereby the gas may be liberated from oil so introduced, and discharged into upper portion of said receptacle, above level of the oil in the settling chamber, a common outlet for the gas from bot said chambers.

17. An oil and gas separator for oil w including a receptacle having a recei chamber therein for the reception of oil its constituents, and a settling chamber

**15,220**

municating with said receiving chamber, a float mounted in the upper portion of said receptacle for regulating the discharge of the oil therefrom, whereby a substantially uniform volume and level of oil may be maintained in said settling chamber at a point above the vertical center of the receptacle.

18. An oil and gas separator for oil wells including a receptacle having a receiving chamber therein for the reception of oil and its constituents, and a settling chamber communicating with said receiving chamber, said receiving chamber and said settling chamber having a common outlet whereby the gas liberated from the oil in both chambers may be commonly discharged, a float mounted in the upper portion of said receptacle for regulating the discharge of the oil therefrom, whereby a substantially uni volume and level of oil may be maintain said settling chamber at a point abov vertical center of said receptacle.

19. An oil and gas separator for oil including a receptacle having a recei chamber and a settling chamber in com cation, a float in the upper portion o receptacle, pivotally supported on the thereof, an oil discharge valve commu ing with said settling chamber and nally mounted on said receptacle, and for operatively connecting said float said valve.

DAVID G. LORRAIN

Witnesses.

J. W. Sheeley,

Luther L. Mack.

| | | | OIL. | | | | GAS. | | | | |
|---|---|---|---|---|---|---|---|---|---|---|---|
| Pressure on trap. | Well Pressure. | Flow plug opening. | Daily production. | Gravity in trap. | Gravity at tank inlet. | Gravity when shipped | Indicated pressure in trap. | Water differential. | Hourly Coefficiant. | Daily gas production (5-lb base, 14.4 lbs. atmos. temp'ture 60/60) | Specific Gravity. |
| Conditions without trap | 360 lbs. | 5/8" | 1161 bl | --- | 29.9 | 29.3 | ---- | ---- | ---- | ------ | 0.83 |
| (Gas escaping in dense, white cloud.) | | | | | | | | | | | |
| 36 lbs. | 380 lb. | 5/8" | 1175 bl | 31.3 | 31.3 | 30.9 | 36 lbs. | 4.5 in. | 1490 | 971,000 cu. ft. | 0.79 |
| (Marked decrease in whitish color of gas.) | | | | | | | | | | | |
| 74 lbs. | 385 lb. | 5/8" | 1205 bl | 31.9 | 31.[illegible] | 31.2 | 74 lbs. | 7.4 in. | 1535 | 947,000 cu. ft. | 0.75 |
| (Nearly all white color in gas eliminated.) | | | | | | | | | | | |
| 177 lbs. | 390 lb. | 5/8" | 1190 bl | 31.7 | 32.2 | 31.1 | 177 lbs. | [illegible].9 in. | 1589 | 1,273,000 cu. ft. | 0.70 |

E 113 - Townsend et al - v - Lorraine

Filed March 23, 1922
Chas. N. Williams, Clerk



Gas discharge lines

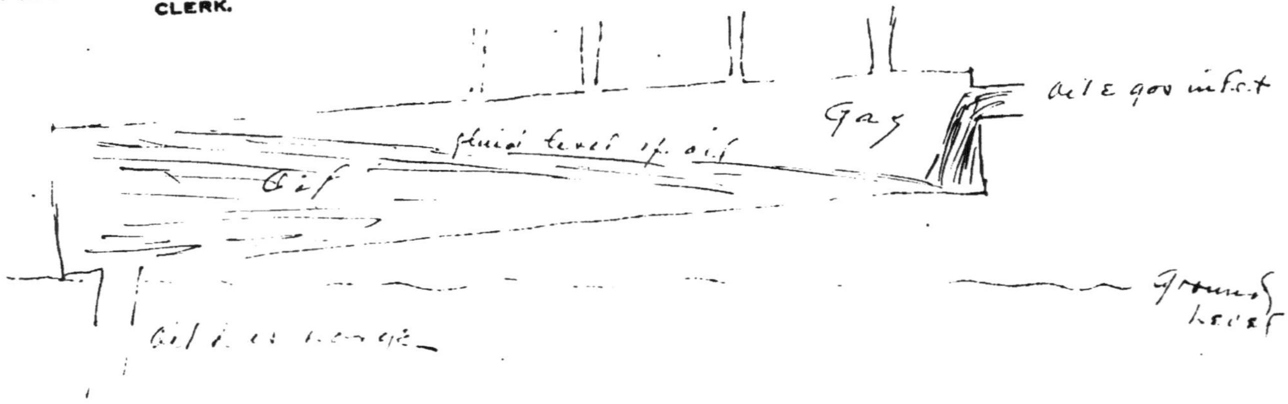

Plff's Exhibit 8-1 & 2

E.113 Eq. Townsend, et al - v - Lorraine

Filed March 22, 1922
Chas N Williams

(1) 8-1

(2)

(3)



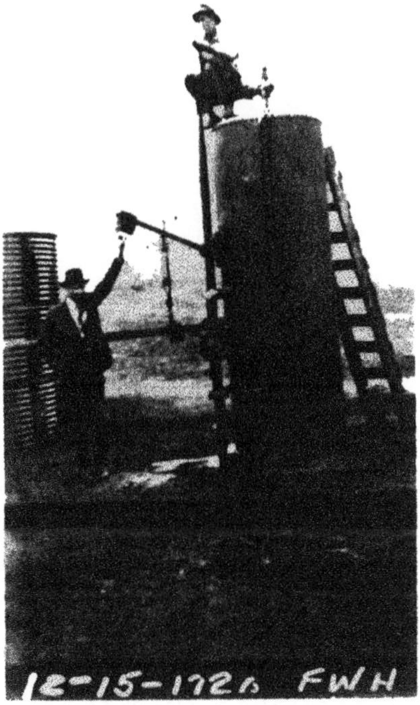


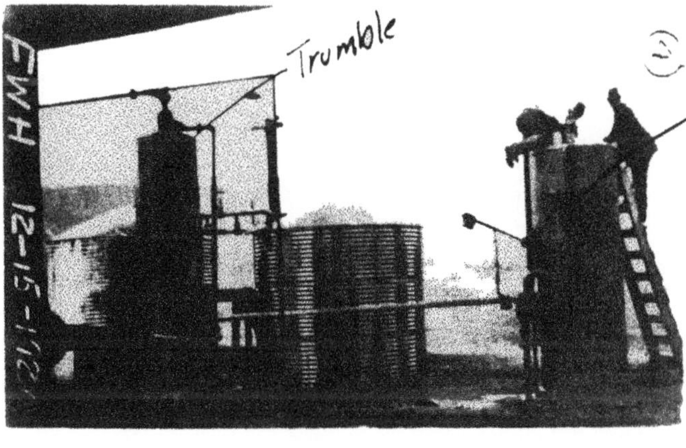

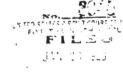

Pltffs Ex 8-1 II

E. 113 Eq.

Townsend, et al

v.

Lorraine

Filed Mch. 22/22

Chas. N. Williams

clerk

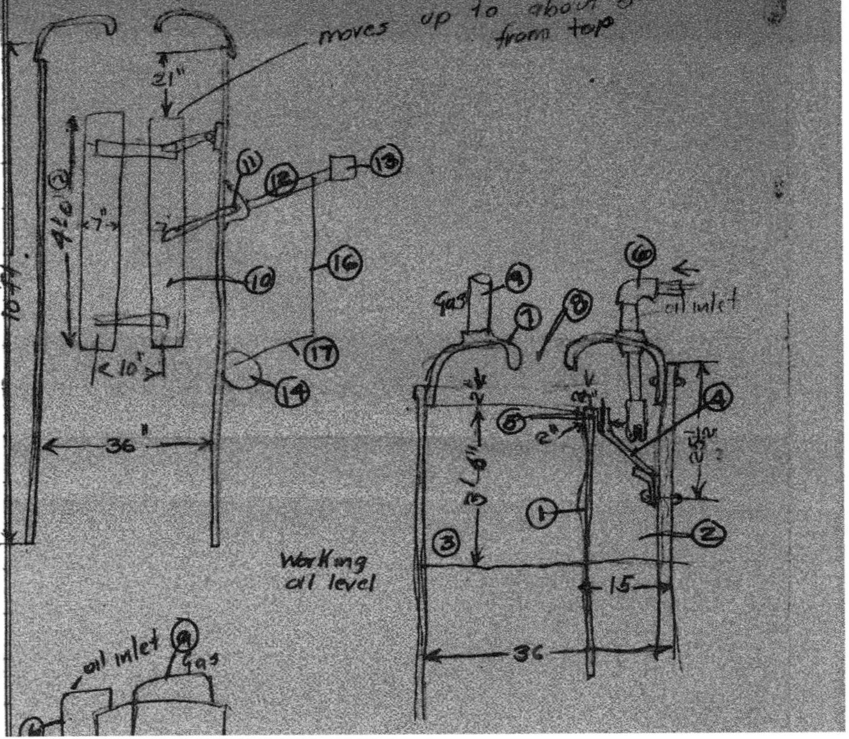

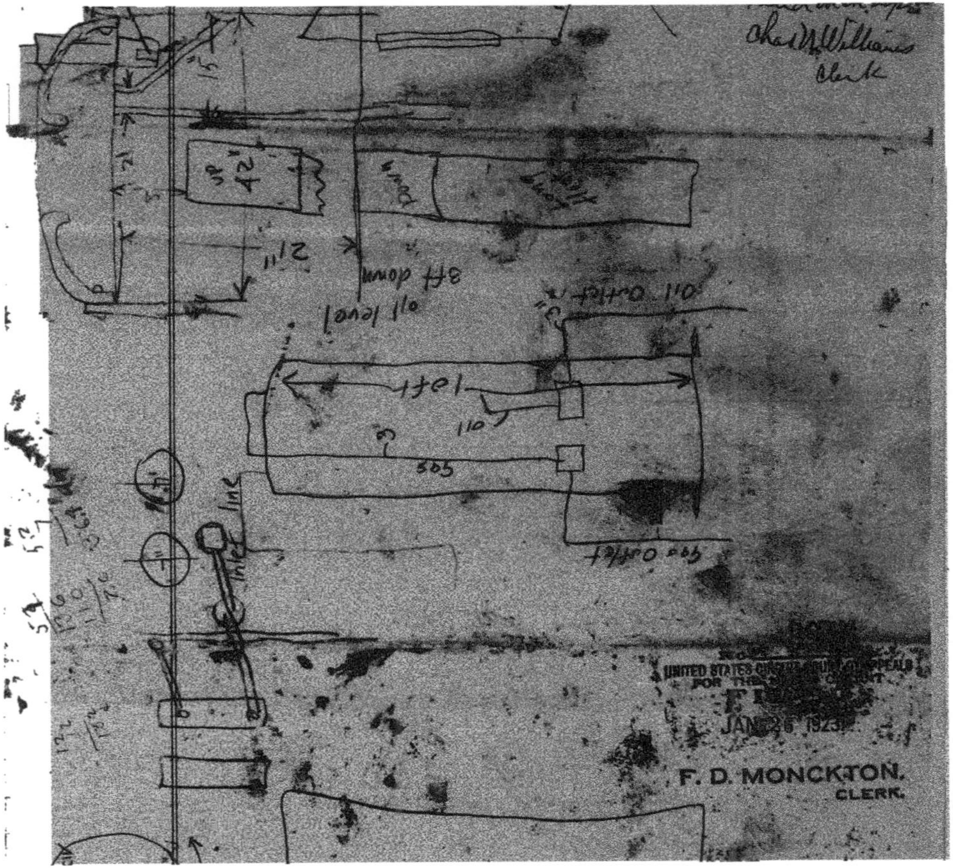


UNITED STATES CIRCUIT COURT OF APPEALS
FOR THE [illegible] CIRCUIT
FILED
JAN 26 1923
F. D. MONCKTON,
CLERK.

Chas W. Williams
Clerk

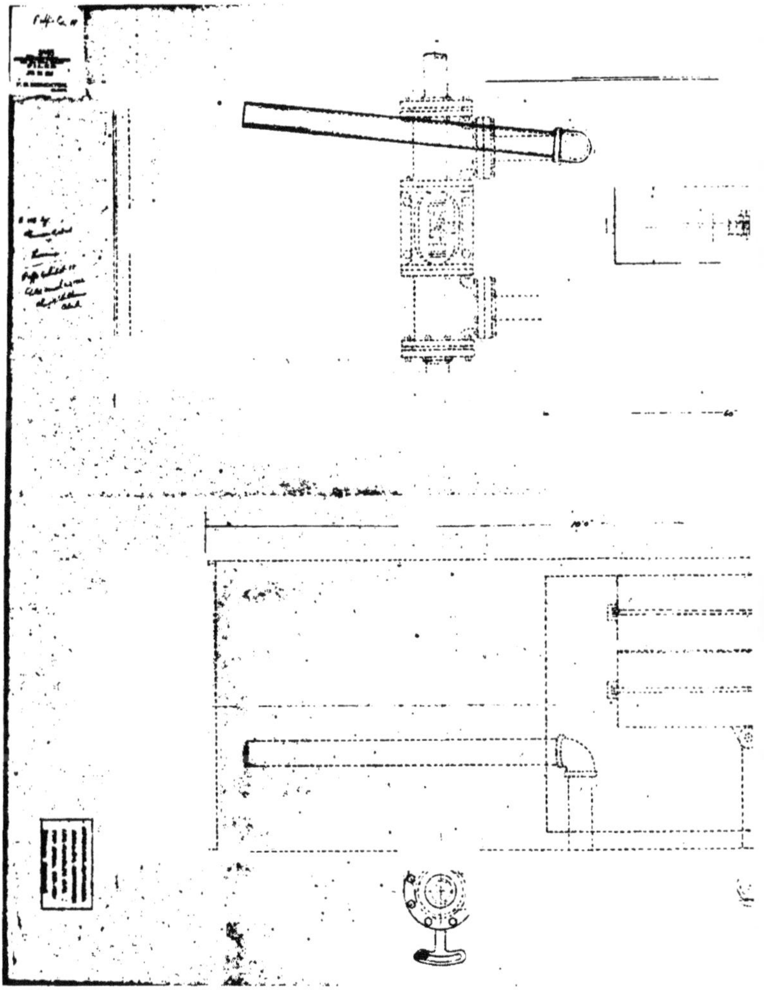

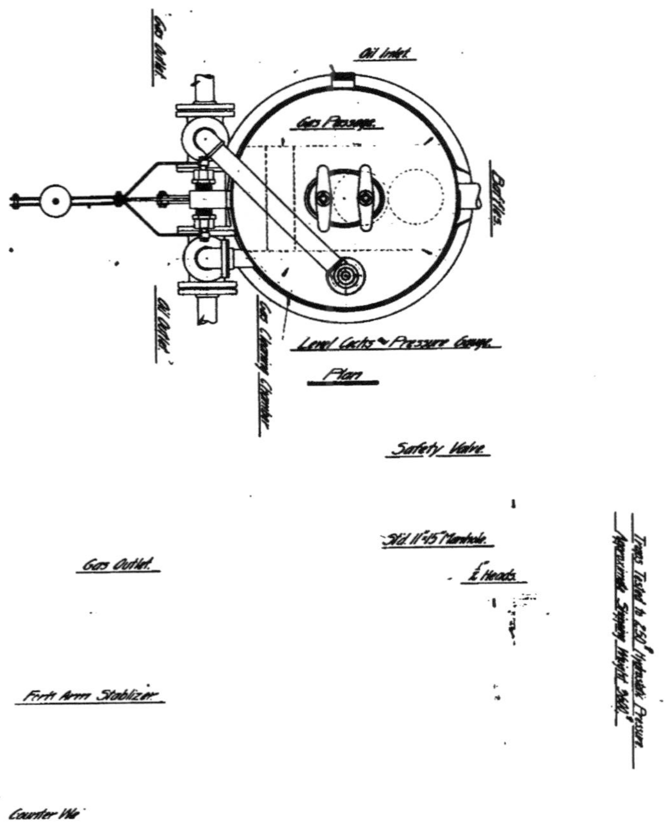
Gas Outlet.
Oil Inlet.
Gas Passage.
Baffles.
Oil Outlet.
Gas Cleaning Chamber
Level Cocks & Pressure Gauge.
Plan
Safety Valve.
Std. 11"x15" Manhole.
Gas Outlet.
Heads.
First Arm Stablizer
Counter Wt.
Traps Tested to 250# Hydrostatic Pressure.
Approximate Shipping Weight 2460#

**Plaintiff's Exhibit No. 13.**

Main 196
132–50

LORRAINE GAS AND OIL SEPARATOR CO.

601 Washington, Building.

Los Angeles.

March 17, 1922

Sold to—General Petroleum Corp.,
Higgins Bldg.,
City.

Your Order Number. 23282.
Our Invoice No. 80.
Terms Net 30 Days

Interest at the Rate of 8 Per Cent Per Annum Will be Charged on All Overdue Accounts.

1. Lorraine Gas and Oil Automatic Separator .................... 1,300.00
Less 7½%.... 97.50

1202.50

Date Mar. 21, 1922—Received. [72]

[Endorsed]: E–133—Eq. Townsend et al. vs. Lorraine. Plffs. Ex. 13. Filed Mch. 23, 1922. Chas. N. Williams, Clerk.

No. 3945. United States Circuit Court of Appeals for the Ninth Circuit. Filed Jan. 26, 1923. F. D. Monckton, Clerk.

**Plaintiff's Exhibit No. 14.**

TRUMBLE GAS TRAP CO.

916 Higgins Building,

Los Angeles, Cal.

December 10, 1920.

PERSONAL.

Mr. Wm. Lacy, President,
Lacy Manufacturing Company,
Suite 601 Washington, Bldg.,
Los Angeles, Calif.

My dear Mr. Lacy:

The Co-Partnership doing business under the name of the Trumble Gas Trap Company is composed of Milon J. Trumble, the inventor of the Trumble Refining Apparatus with which you are undoubtedly fully familiar, A. J. Gutzler and myself, F. M. Townsend. Mr. Gutzler and myself have been associated with Mr. Trumble since 1910 in the development and introduction of the Trumble Refining Patents. An incidental invention of Mr. Trumble's is what is known as CRUDE PETROLEUM AND NATURAL GAS SEPARATORS, for which Letters Patent of the United States was issued on the 11th day of June, 1918, to Milon J. Trumble, Francis M. Townsend and Alfred J. Gutzler, and numbered 1,269,134.

Mr. Gutzler and myself have devoted all of our time to the Trumble matters ever since our first association with Mr. Trumble in 1910, and I believe that you will concede that the general results attained by the Trumble inventions have redounded to the great benefit of Los Angeles manufacturers, as an enormous amount of evaporators, condensers, separators and other appliances employed in the erection of Trumble Refineries have been made both in the shops of the Lacy Manufacturing Company as well as in several of the other shops doing a similiar character of work. I am aware that the amount of business which has been gathered by the Lacy Manufacturing Company from this special branch of industry has amounted to a very large sum. Furthermore, I desire to state that through the influence of the Trumble Refining people, over ninety per cent of the work connected with the manufacturing of Trumble Refining Apparatus has come to the city of Los Anggeles. Much of the materials so manufactured here have been shipped to England, Curacao, Egypt, and Mexico. All of this work is directly traceable to the efforts of Mr. Trumble and his associates.

I desire to call your attention to these facts and to emphasize that if Mr. Trumble has never conceived of his improvements of inventions in oil treating and other devices for handling oils, the manufacturers of Los Angeles would not have had the benefit of the extensive manufacturing with which they have been provided. In other words, I consider, and I believe that you will also agree, that Mr. Trumble has, without doubt, occasioned an enormous benefit to manufacturers in Los Angeles who are equipped for producing sheet metal work and that it is a matter of congratulation that such brains are enabled to place the merits of their inventions before the public so that the public in general can benefit thereby. I believe that an inspection of your books of account will show you that the Lacy Manufacturing Company have been greatly benefited through the manufacturing of Trumble

2.

Mr. Wm. Lacy, President,
Lacy Manufacturing Company,
Suite 601 Washington Bldg.,
Los Angeles, Calif.
December 10, 1920.

Refining devices.

The occasion of my addressing you in the manner above is, that it has lately come to my attention that one, a Mr. Lorraine, has commenced the manufacturing of a so-called Automatic Oil and Gas Separator in your shop in this city and is offering the same in competition with the Trumble Gas Trap which we have introduced to oil producers at great cost and time. Mr. Trumble was the pioneer in conserving the great loss occurred through the waste of natural gas from crude petroleum, and was the first to produce a gas trap which would conserve the light volatile oils and retain them in the crude petroleum while permitting the natural gas to escape comparatively free and dry of liquid hydrocarbons. The patent above mentioned granted to Mr. Trumble and his associates, covers generic features which seem to be absolutely necessary in the construction of a trap that will do work anywhere within the range of practicability.

The trap constructed by Mr. Lorraine in your shops embodies these patented features and is considered by us to be a flagrant and willful infringement upon the rights conferred by the said patent. I am presuming that your attention has never been called to this matter and that you believe you have a full right to manufacture this Lorraine trap, and therefore, I am hereby informing you of the condition of affairs in regard to our rights to the inventions which we claim are infringed by Mr. Lorraine and your company as well as directing your attention to the matter of the threatened interference with the personal business of Mr. Trumble and his associates in the manufacture and sale of a device which fills a want in a field which was unsatisfied prior to the time that Mr. Trumble entered the same and saved to the oil producers a vast sum which otherwise might have been wasted had he not given them the benefit of his inventive skill.

Mr. Lorraine has not originated any new principle in the separation of natural gas from the crude petroleum, but has copied essential features of the Trumble Patent. All of this is covered up under the guise of his having produced a valve for controlling the outlet of oils from the trap, and to which he attributes the great success of his device. This I assert is not true, but on the other hand the degree of success which he does have through his trap is occasioned through the use of the patented features of the Trumble Gas Trap as the particular form of valve outlet for the outflowing oils has no bearing whatsoever upon the means internally arranged for causing the complete separation of gas and oil.

From the many years of acquaintance with your firm, and a somewhat intimate knowledge of your character and business conduct, I believe that you will agree with me that business people engaged in manufacturing enterprises, especially in the city of Los Angeles, will have no disposition to encroach upon one anothers' enterprises even in what might be considered business competition where such encroachment would work to the harm or annoyance of a person like Mr. Trumble who has benefited the manufacturers to the degree that Mr. Trumble has conferred benefits upon manufacturers of

3.

Mr. Wm. Lacy, President,
Lacy Manufacturing Company,
Suite 601 Washington Bldg.,
Los Angeles, Calif.
December 10, 1920.

your class.

I desire to express myself to you personally in this manner so that you will understand my feelings before you have been served with the formal notice of our attorneys that suit would be brought against the Lacy Manufacturing Company in its part in participation of the infringement of the Trumble Patent through its manufacturing of the Lorraine device, as such formal notice will be served upon you in due time within the next few days. We have instructed our attorneys to proceed with the filing of suit for infringement, forthwith, but I wish you to understand that it is without personal feeling by Mr. Trumble or his associates as against the Lacy Manufacturing Company, but is forced upon us through necessity as it has lately come to our hearing that your company is now starting upon the filling of an order for a large number of the Lorraine traps.

I trust that you will receive and consider this matter in the spirit in which it is tendered, and trust that in the future our relations may remain as cordial as in the past.

Yours very truly,
F. M. TOWNSEND.

FMT: ET

[Endorsed]: E.—113—Eq. Townsend et al. vs. Lorraine. Plffs. Exhibit 14. Filed March 23, 1922. Chas. N. Williams, Clerk.

No. 3945. United States Circuit Court of Appeals for the Ninth Circuit. Filed Jan. 26, 1923. F. D. Monckton, Clerk.

**Plaintiff's Exhibit No. 15.**

LACY MANUFACTURING CO.

Plate and Sheet Steel Workers.

Steel Oil Storage Tanks and Riveted Steel Pipe.

Office 601 Washington Building.

Third and Spring Streets.

Phones
Main 196
Home A 3250.

WORKS: Cor. N. Main & Date Streets, Los Angeles, Cal.

All Agreements Are Contingent Upon Strikes, Accidents, Delays of Carriers, and Other Delays Unavoidable or Beyond Our Control. Quotations Subject to Change Without Notice.

Los Angeles, Dec. 13, 1920.

Mr. F. M. Townsend,
Trumble Gas Trap Co.,
916 Higgins Bldg.,
Los Angeles, Cal.

Dear Mr. Townsend:

I am in receipt of yours of the 10th inst., calling my atention to your claim that the oil and gas separator that Mr. Lorraine is selling is an infringement on the Trumble patent, and expressing your opinion that the Lacy Manufacturing Co. does not desire to in any way do an injustice to Mr. Trumble and his associates by being a party to placing on the market any article that would be an infringement upon Mr. Trumble's rights. In reply beg to state that you are right in assuming that we would not care to manufacture any article that we were not legally entitled to, or that the party for whom we were making the article did not have a perfect right to sell. We have made a few of these so-called gas separators for Mr. Lorraine, upon his order and one has been delivered to him, but now since you have called our attention to your claim of infringement on the Trumble patent we would cease making them and call the attention of Mr. Lorraine to your claims and inform him that we shall cease manufacturing any more of these separators until he produces the proper evidence to show that he is not infringing on your patent right.

The writer fully agrees with you in your expressions relative to the advantage to the shops of Los Angeles that the Trumble refining patents have been. We, as well as the other shops in our line, have received many large orders for the construction of Trumble apparatus, which we realize would not have materialized, had not Mr. Trumble devoted his inventive ability to this line.

Yours very truly,

WM. LACY.

WL:HH.

[Endorsed]: E-113-Eq. Townsend et al. vs. Lorraine. Plffs. Ex. 15. Filed March 23, 1922. Chas. N. Williams, Clerk.

No. 3945. United States Circuit Court of Appeals for the Ninth Circuit. Filed Jan. 26, 1923. F. D. Monckton, Clerk.

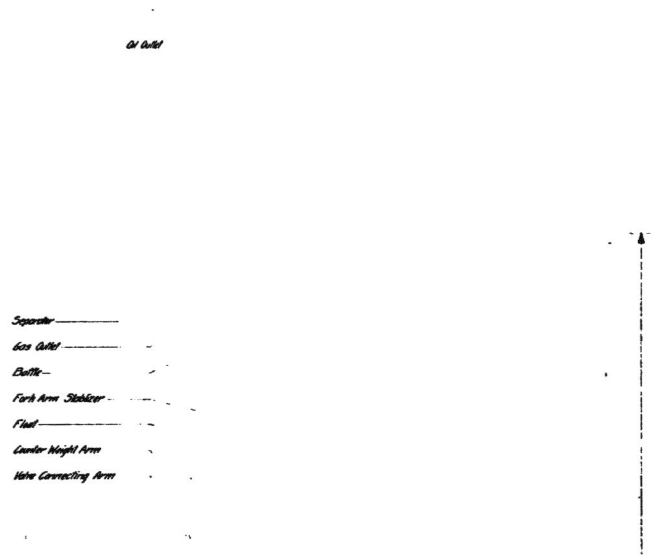
Oil Outlet
Separator
Gas Outlet
Baffle
Fork Arm Stabilizer
Float
Counter Weight Arm
Valve Connecting Arm

E.1138q.
Townsend, et al
v.
Lorraine
Plffs Exhibit 17.
Filed March 3, 1922.
Chas. N. Williams
Clerk

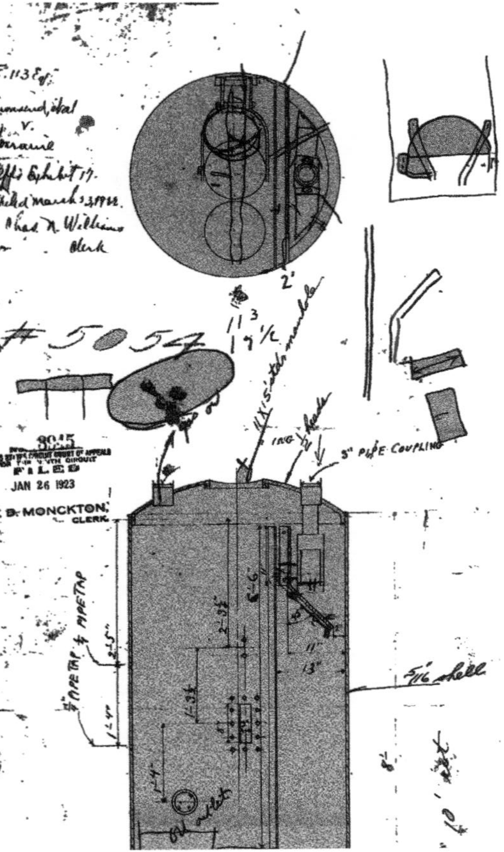

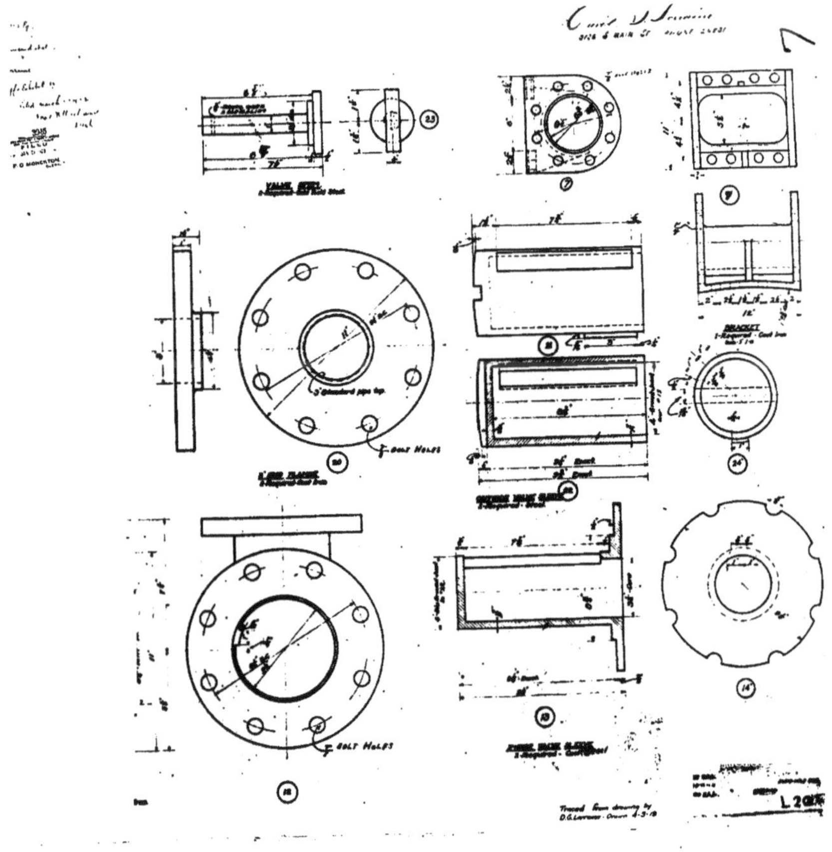
Bolt Holes
Bolt Holes
Bracket
Traced from drawing by

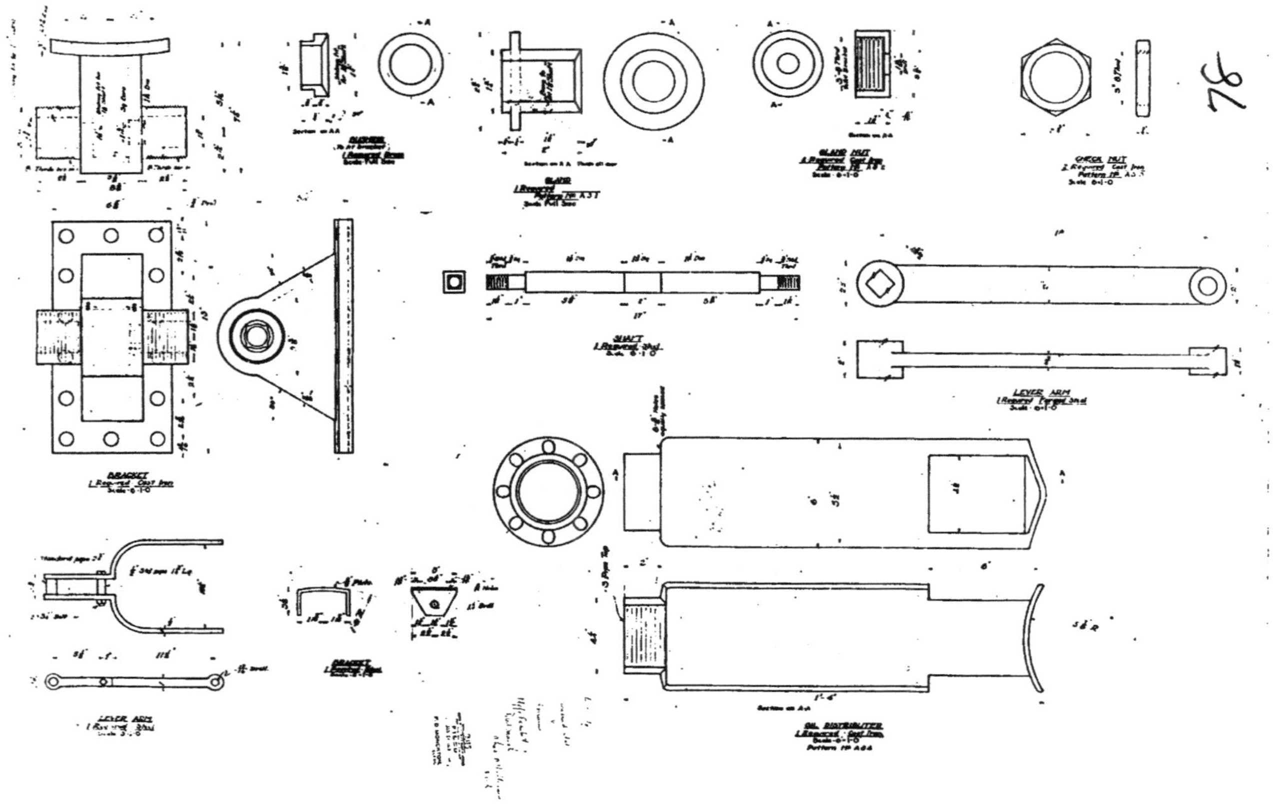

BUSHING
GLAND
GLAND NUT
CHECK NUT
BRACKET
SHAFT
LEVER ARM
BRACKET
LEVER ARM
OIL DISTRIBUTER
Section on A-A
3" Pipe Tap

The 68 year
diplomat, who was officially appointed to the post August 20, has agreed to remain in Low Library "for one year or until a new president is in a position to assume the duties of the office."

Dr. Cordier's appointment came just eight days after Alexander Heard, chancellor of Vanderbilt University, rejected a formal offer by the Trustees for the post.

In another administrative change, Professor of Economics Peter B. Kenen, chairman of the department, has been named University provost. Dr. Kenen will succeed Paul D. Carter, who left Columbia last spring to become vice president and provost of Hamilton College in upstate New York.

It had been expected that the University would be able to find a permanent president before the end of the summer. But, after Chancellor Heard turned down the post, the Trustees asked Dr. Cordier to stay on.

According to one member of the faculty presidential search committee, the panel must now "start from scratch" in drafting a list of potential candidates.

He said that it appears unlikely that a new appointment will be made before June.

Chancellor Heard's decision to reject Columbia's offer took many officials here by surprise. According to reliable sources, the 52-year old political scientist was persuaded to remain at Vanderbilt by Nashville civic and business groups.

At a news conference held on August 12, Dr. Heard stated, "I

(Continued on Page 4).

President A
Library sta
building oc
relative calr

and Surgeons.

The 25-page document, prepared last year, identified some ... at the medical school by race, nationality, political attitudes, and sentiments on union organizing at P&S.

Union officials charged the University administration with preparing the dossiers in an effort to stifle attempts by the union to organize workers at the school.

Employees at P&S will vote in two weeks whether to be represented by 1199, which has successfully organized workers in other divisions of the University, or by the Supporting Staff Organization, a newly-formed union.

Fifteen organizers for 1199 sat in at the office of H. Houston Merritt, dean of the medical school, for almost three hours yesterday. They charged management personnel had intimidated them.

Some Philipino workers contended that hospital officials had threatened them with deportation if they did not stop union organizing.

Douglas S. Damrosch, assistant dean of P&S, yesterday denied knowledge of intimidation on the part of supervisors at the school. He said he plans to look into any charges presented to him.

Sidney von Luther, regional director of 1199, released the compiled dossiers at the sit-in. He said the reports were mailed to him a month ago by an anonymous source.

Mr. von Luther said P&S administrators charged he had stolen the material, but the union leader denied the accusations.

Medical school officials were unavailable for comment on the matter last night.

The typed reports, which are dated June through August, 1968, contain a series of evaluations signed by John J. McNamara, assistant direct of personnel for the University and James J. Dean, an unidentified contributor. Mr. McNamara last night refused to comment on any aspect of the situation.

In the cover letter to the documents, Thomas M. Kerrigan, a attorney in a Park Avenue law firm and a 1928 graduate of the College, notes that it is necessary to "organize the staff at P&S for the pu

(Continued on Page 5)

# Senators Question Subpoena Action

By ROBERT B. STULBERG

The Executive Committee of the University Senate will recommend next week that the Senate curtail the administration's power to divulge privileged information to government sources.

According to members of the executive panel, they will ask that the Senate—not the President—be empowered to set guidelines for release of such data.

It is expected that the 10-man Senate, which was formed last spring, will consider the proposal when it re-convenes next Thursday at its opening session this year.

The committee's proposal is the first substantive measure to be brought before the Senate and marks the first attempt by the body to wrest power from the central administration.

Several Senate representatives initiated the proposed legislation during the summer months, after University officials complied with a government probe into radical activity on campus.

In June, the U.S. Senate Permanent Sub-committee on Investigations, chaired by Sen. John McClellan (D-Ark), subpoenaed Columbia and several other universities for the names and financial records of known campus radicals.

Several weeks later, the central administration, acting on its own, submitted the names and addresses of 29 students and nine non-students who had been accused of participating in two campus protests last spring.

The administration also turned over the students' records of financial aid—records which are normally considered confidential. Most of the persons identified were members of Students for a Democratic Society.

After the administration submitted the requested data, many persons on campus, including several members of the University Senate, scored the University's action.

In a brief, strongly-worded statement, all fifteen senators from the non-tenured faculty criticized President Cordier for complying with the subpoena "without having consulted with the Executive Committee." And in a similar document, sixteen of the 21 student representatives termed the

(Continued on Page 2)

### Publication Notice

*Spectator will commence daily publication with the issue of Thursday, Sept. 25. Ad deadline is 3:00 p.m. Tuesday, Sept. 23.*

### Staff Meeting

*Rumor has it that sophomores and juniors on Spectator will meet Tuesday at 8:00 p.m. in 318 FBH. Those who know the staff well, however, do not put much faith in these reports.*

# Freshmen Endure Orientation Ritual

By JERRY KOPEL

The class of 1973, leaving their homes and their parents behind them, converged on Morningside ... in a ten-day freshman orienta-

F...nding Cu

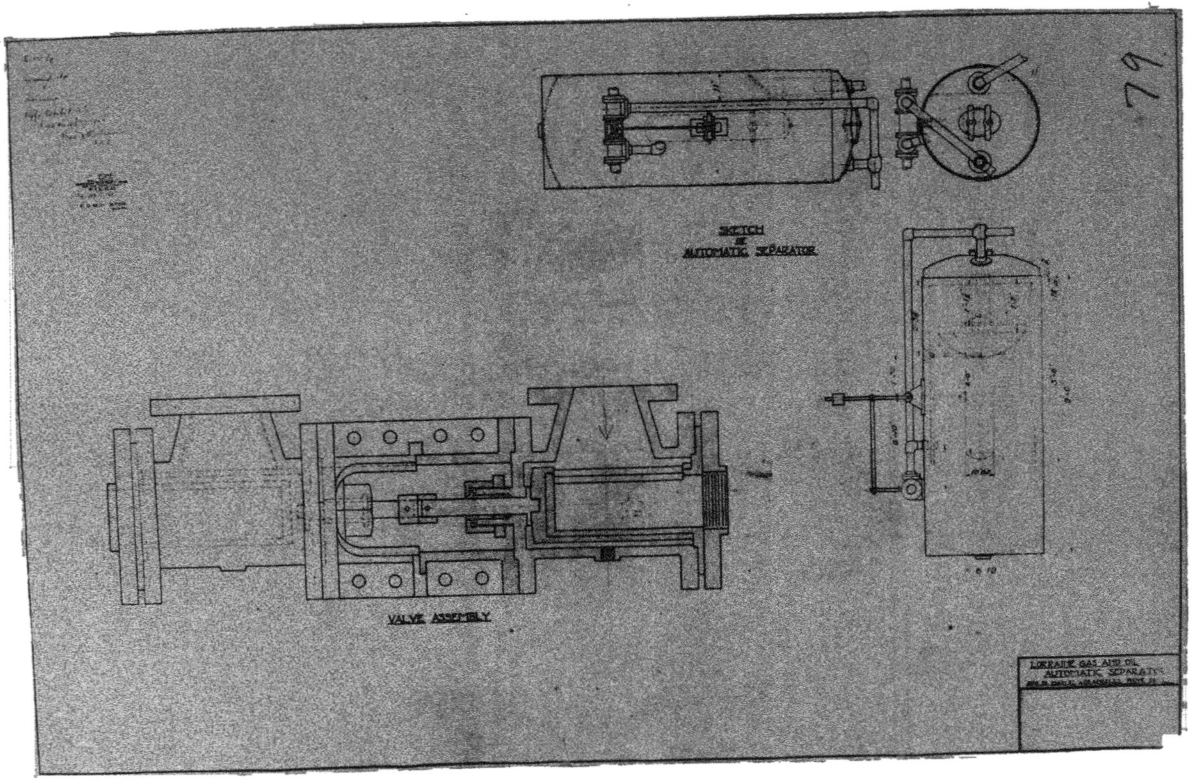
79
SKETCH
OF
AUTOMATIC SEPARATOR
VALVE ASSEMBLY
LORRAINE GAS AND OIL
AUTOMATIC SEPARATOR

**Plaintiff's Exhibit No. 20.**

No. 3945. United States Circuit Court of Appeals for the Ninth Circuit. Filed Jan. 26, 1923. F. D. Monckton, Clerk.

E 113 Ey. - Flammandpdale - V. Lorraine

 Filed March 27, 1922
Chas. N. Williams clerk


# Huntington Beach in California Is Yet a "Blind" Oil Field

LOS ANGELES, CALIF., Feb. 20.—Because of the lack of uniform methods of development, and because the work at the majority of the independent wells is too independent of operations at neighboring wells, few definite facts have been established as to the subsurface conditions in the Huntington Beach field, according to J. B. Case, in the Summary of Operations, California Oil Fields, for November, just published.

Mr. Case devotes considerable space to pointing out the difficutly of assembling accurate information on the subsurface conditions of the field, and his conclusions epitomize his lengthy article as follows:

From the above summary of operations in the Huntington Beach oil field it will be noted that very few definite facts have been established as to subsurface conditions. The lateral extent of the first oil zone, or individual sands within it, is not known. The position of intermediate water has not been located except in two wells. The top of the second oil zone h[as] not been determined.

The investigation discloses that the princi[pal] contributing causes for the lack of informa[tion] may be summed up as follows:

1. The use of the rotary method of drill[ing] with its attendant detrimental factors—us[e of] heavy mud and speed of drilling.

2. The unusual number of wells drilling [si]multaneously over a small acreage.

3. Failure of operating concerns to t[ake] time to make definite tests when the opp[or]tunity presented itself to do so.

4. The large number of poorly financed c[on]cerns, operating on town-lot leases, whose [aim] to produce oil in paying quantities is of s[ec]ondary importance.

5. Cementing of water strings deeper th[an] proposed by operators and approved by [the] department.

6. Cementing through perforations i[n] strings, thereby eliminating certain forma[tions] from production tests. This work was eq[uiv]alent to cementing a water string at a gre[ater] depth than agreed upon.

7. Placing of cement plugs in perfor[ated] casing between depths other than agreed u[pon] and approved.

It would greatly facilitate the obtaining [of] definite data on underground conditions [to] have several test wells drilled for the [sole] purpose of obtaining information, produ[ction] being a secondary consideration. The loc[ation] of these wells should be carefully chose[n in] order to obtain the maximum amount of [in]formation. With data derived from su[ch] wells in which the information is depend[able] a workable plan of development could b[e ar]rived at which would safeguard the inte[rests] and overcome the problems of all, without [in]flicting hardship on anyone.

...for All California Fields for January is as Follows:

| New Rigs | Wells: Drilling | Comp During Month | Abn. During Month | Producing | Prod. Per Day |
|---|---|---|---|---|---|
| .... 4 | 23 | 3 | .. | 2,222 | 18,600 |
| .... 1 | 8 | 1 | .. | 347 | 6,846 |
| ....40 | 84 | 26 | 2 | 2,699 | 118,024 |
| .... 1 | 8 | 3 | .. | 586 | 9,196 |
| .... 1 | 13 | .. | 2 | 1,182 | 34,264 |
| ...... | 11 | .. | .. | 410 | 13,988 |
| .... 1 | 39 | 3 | 2 | 533 | 6,657 |
| ...... | 6 | 1 | .. | 673 | 3,370 |
| ...20 | 129 | 5 | 1 | 1,113 | 88,391 |
| ...17 | 151 | 17 | .. | 66 | 16,206 |
| ...... | .. | .. | .. | 137 | 148 |
| .... 2 | 65 | .. | .. | 9 | 65 |
| ...87 | 537 | 59 | 7 | 9,983 | 315,755 |

...36,184,527 bbls
... 8,625,792 bbls.



## Guiberson Corporation Has New Rotary Tool Joint To[ng]

Los Angeles, Cal., Feb. 20.—A new r[otary] tool joint tong has been designed and pa[tented] by the Guiberson Corporation, and will b[e add]ed to the list of Guiberson specialties fo[r] field use.

The new tong has been in use in Calif[ornia] for sometime, as it is the purpose of the [com]pany to have it thoroughly tested, und[er all] conditions, before offering it in the va[rious] sizes for sale throughout the country.

In spite of the fact that tongs of every [de]scription are patented and marketed to th[e] trade, the new Guiberson tong has 21 cl[aims] made for it, in its patent, claims of new [and] distinct features not heretofore embodied [in any] similar device.

The tong is designed to work automati[cally] and when pushed against the casing it c[loses] automatically and tight, the spring latch [is] so designed that it is impossible for it to [come] loose, until it is released, and at the same [time] it is easy to disengage by hand. The pull [ap]plied to the handle forces a specially desi[gned] "kidney" against the casing wall, and g[ives] a grasp that can't slip.

It is expected by the use of a bushing [the] new tong may be used as any ordinary c[asing] tong, though the tool is specially designed [as] a rotary tool joint tong.

The Guiberson-Mills tong is offered b[y the] same company, and has found a ready de[mand] in all the fields. The new tool is not d[esigned] to take the place of the Guiberson-Mills [tong] but to supplement this tong.

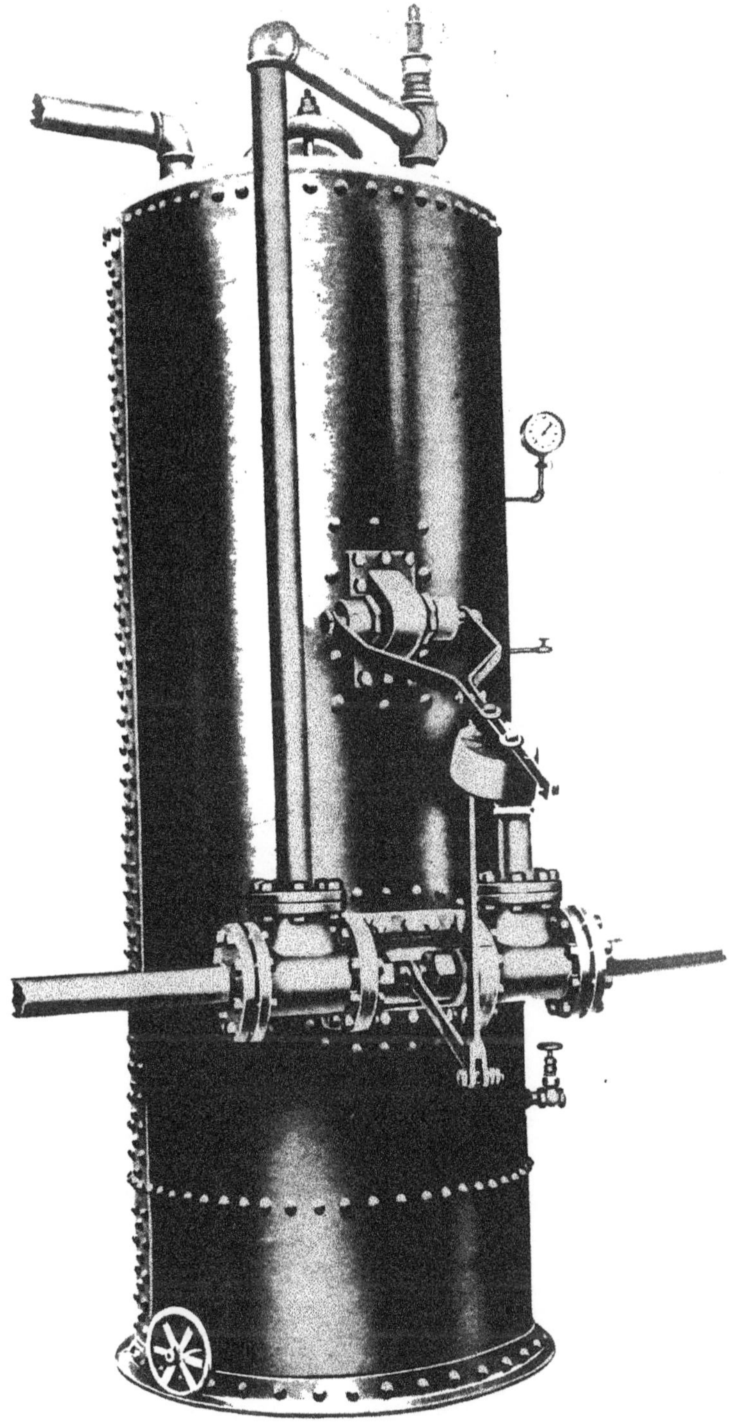

Harris Sketch

Eq.
ind, et al
ie
24
arch 20/22
Williams
lerk

**Plaintiff's Exhibit No. 25.**

[Endorsed]: E–113—Eq. Townsend et al. vs. Lorraine. Plffs. Exhibit 25. Filed March 28, 1922. Chas. N. Williams, Clerk.

No. 3945. United States Circuit Court of Appeals for the Ninth Circuit. Filed Jan. 26, 1923. F. D. Monckton, Clerk.

Fig. 2

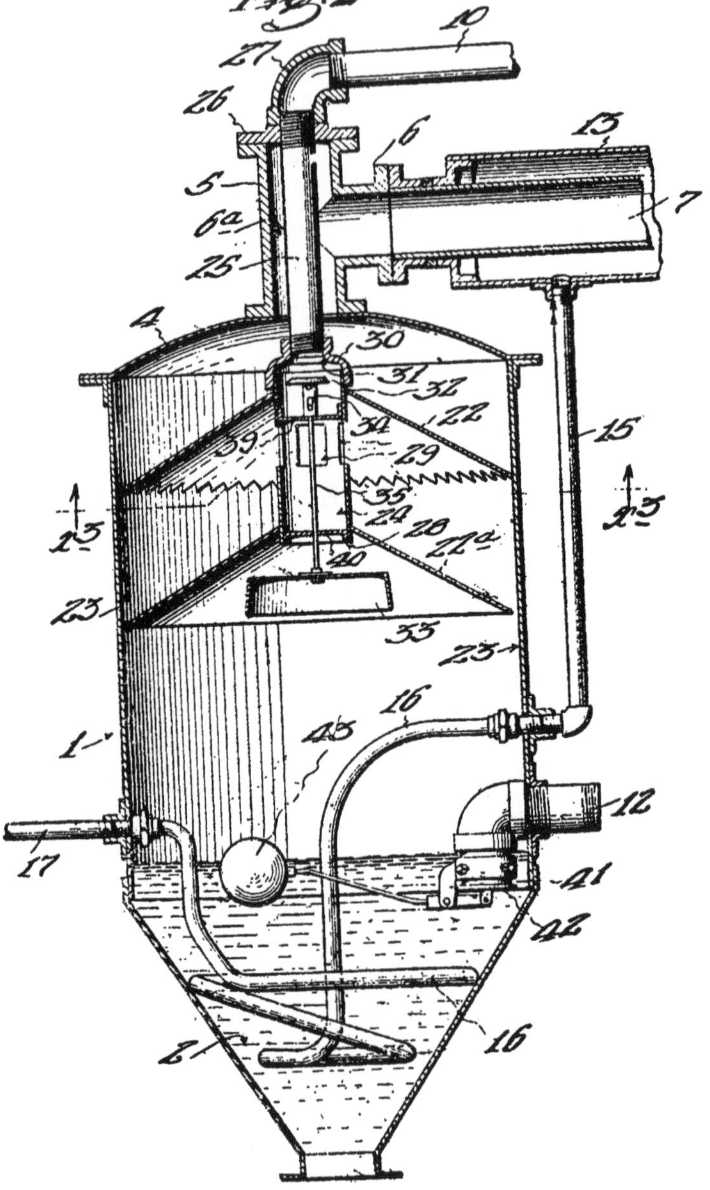
10
27
26
6
13
5
6a
7
25
4
30
31
32
34
22
39
29
15
35
24
28
22a
40
13
23
33
23
16
1
43
12
17
41
42
16
2

**Plaintiff's Exhibit No. 26.**

[Endorsed]: E–113—Eq. Townsend et al. vs. Lorraine. Plffs. Exhibit 26. Filed March 28, 1922. Chas. N. Williams, Clerk.

No. 3945. United States Circuit Court of Appeals for the Ninth Circuit. Filed Jan. 26, 1923. F. D. Monckton, Clerk.

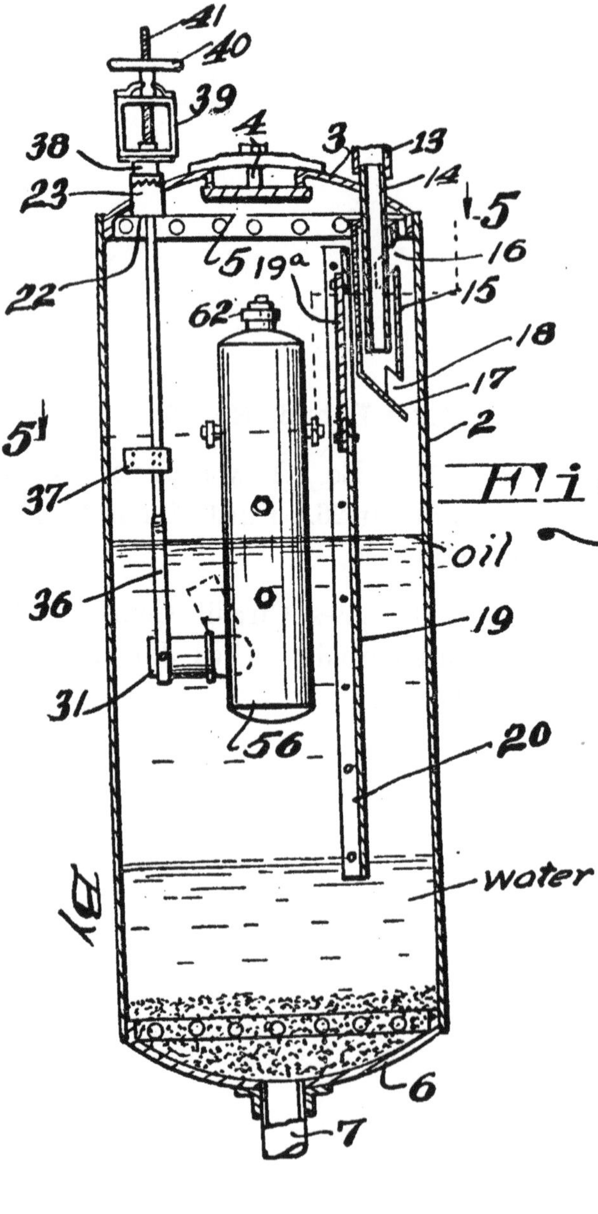
41
40
39
38
23
4
3
13
14
5
16
5
19a
22
15
62
18
17
2
5
37
oil
36
19
31
56
20
water
6
7

Fig. 4

Ex B.

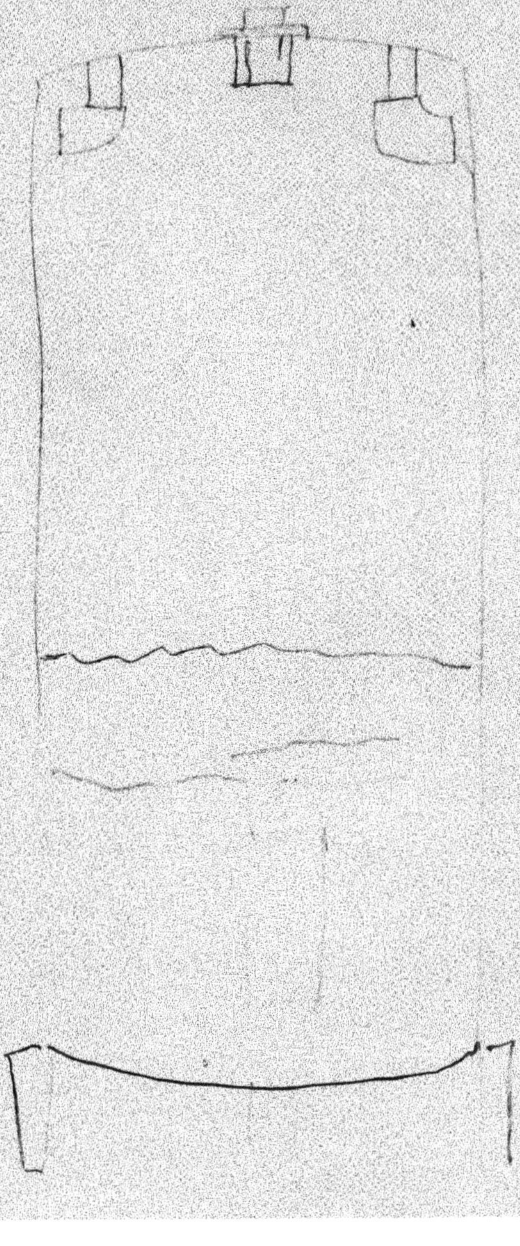

v.
d, etc

d

hibit B.

March 23, 1922
A. N. William
Clerk

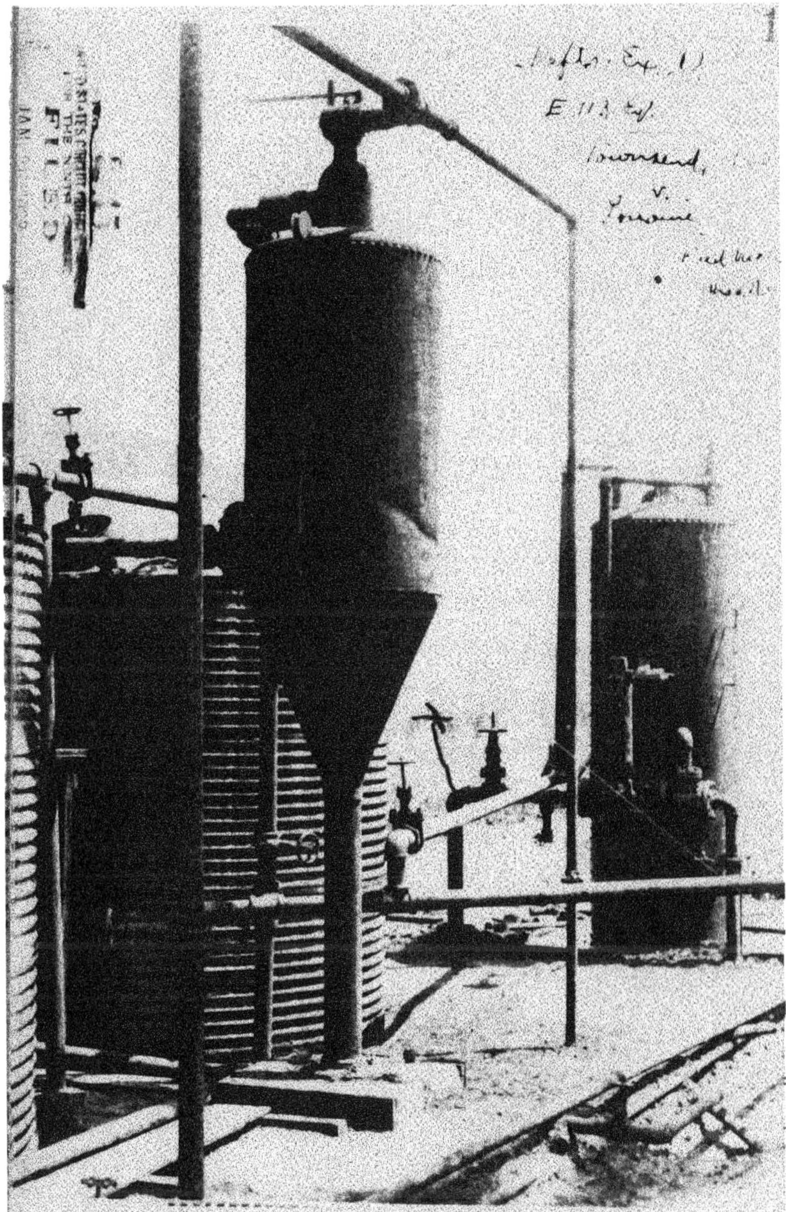

Well #3 Tonner. General Petroluem Oil Co.

**Defendants' Exhibit "E."**

2—390.

UNITED STATES OF AMERICA.

DEPARTMENT OF THE INTERIOR.

UNITED STATES PATENT OFFICE.

To All to Whom These Presents Shall Come, GREETING:

THIS IS TO CERTIFY that the annexed is a true copy from the records of this office of the

Letters Patent of

George L. McIntosh, Assignor of Two-thirds to Kathryn L. Byrne and Roscoe C. Olmsted,

Number 1,055,549, Granted March 11, 1913,

for

Improvement in Mechanism for Separating Gas From Liquids.

IN TESTIMONY WHEREOF I have hereunto set my hand and caused the seal of the Patent Office to be affixed in the District of Columbia this twenty-fourth day of March, in the year of our Lord one thousand nine hundred and twenty-one and of the Independence of the United States of America the one hundred and forty-fifth.

[Seal] M. H. COULSTON,
Commissioner of Patents.

2—362.

No. 1055549.

THE UNITED STATES OF AMERICA,

To All to Whom These presents Shall Come:

WHEREAS GEORGE L. McINTOSH, of Los Angeles, California, has presented to the Commissioner of Patents a petition praying for the grant of letters patent for an alleged new and useful improvement in mechanism for separating gas from liquids,

He having assigned two-thirds of his right, title and interest in said improvement to Kathryn L. Byrne and Roscoe C. Olmsted, both of Pasadena, California, a description of which invention is contained in the specification of which a copy is hereunto annexed and made a part hereof, and he complied with the various requirements of law in such cases made and provided, and

WHEREAS upon due examination made the said Claimant is adjudged to be justly entitled to a patent under the law.

Now therefore these Letters Patent are to grant unto the said George L. McIntosh, Kathryn L. Byrne, and Roscoe C. Olmsted, their heirs or assigns, for the term of Seventeen years from the eleventh day of March, one thousand nine hundred and thirteen, the exclusive right to make, use and vend the said invention throughout the United States and the Territories thereof.

IN TESTIMONY WHEREOF I have hereunto set my hand and caused the seal of the Patent

Office to be affixed at the City of Washington this eleventh day of March, in the year of our Lord one thousand nine hundred and thirteen, and of the Independence of the United States of America the one hundred and thirty-seventh.

[Seal] C. C. BILLINGS,
Acting Commissioner of Patents.

G. L. McINTOSH.

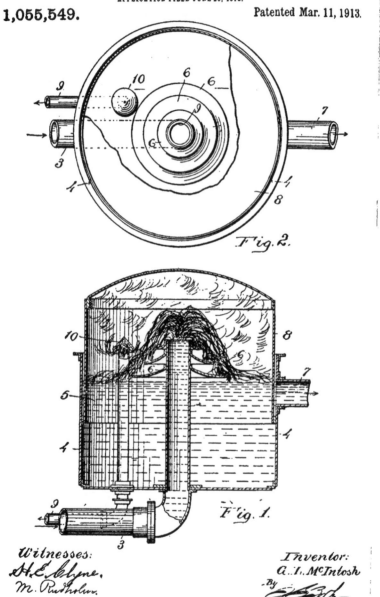

MECHANISM FOR SEPARATING GAS FROM LIQUIDS.

APPLICATION FILED JUNE 20, 1912.

**1,055,549.** Patented Mar. 11, 1913.

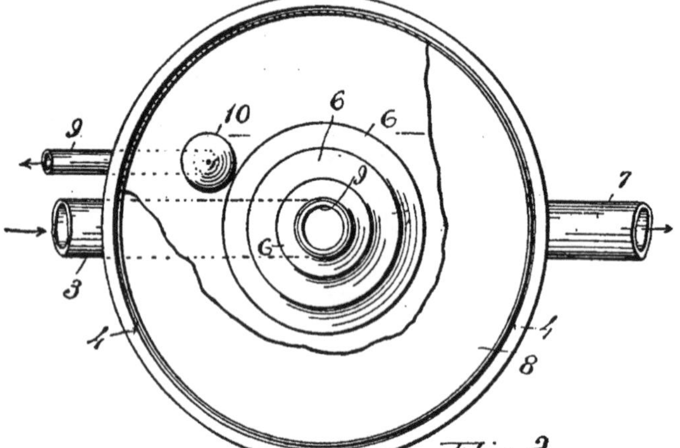


*Fig. 2.*

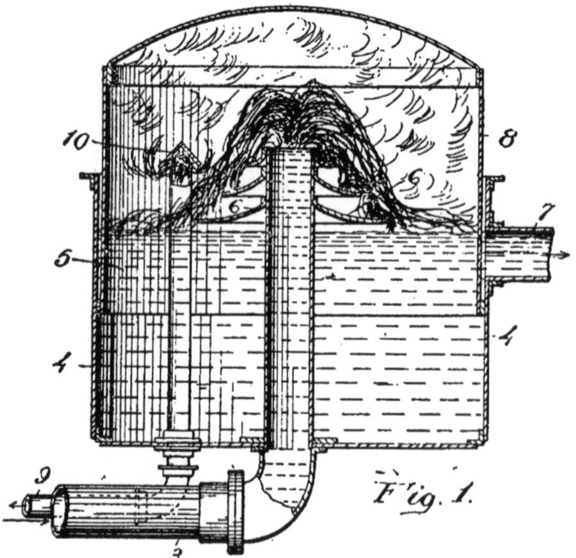


*Fig. 1.*

*Witnesses:*
H. E. Clyne.
M. Rudholm.

*Inventor:*
G. L. McIntosh
By [illegible]
Atty

# UNITED STATES PATENT OFFICE.

ɔORGE L. McINTOSH, OF LOS ANGELES, CALIFORNIA, ASSIGNOR OF TWO-THIRDS TO KATHRYN L. BYRNE AND ROSCOE C. OLMSTED, BOTH OF PASADENA, CALIFORNIA.

## MECHANISM FOR SEPARATING GAS FROM LIQUIDS.

)55,549. Specification of Letters Patent. Patented Mar. 11, 1913.

Application filed June 20, 1912. Serial No. 704,770.

*all whom it may concern:*

3e it known that I, GEORGE L. McINTOSH, itizen of the United States, and resident Los Angeles, in the county of Los Angeles l State of California, have invented cern new and useful Improvements in Mechsm for Separating Gas from Liquids, of ich the following is a specification.

My invention relates to improvements in chanism or apparatus for separating gas m liquids, and it consists of a separating :ice to be used in connection with pump- ; mechanism.

The object of the invention herein is to arate gas from liquids, and to collect the ; into a holder and retain it for use for ting and lighting purposes, as well as to istitute a store of said gas to be utilized · driving gas engines, or for use in other paratuses where gas is employed.

In my present invention I lead the main charge pipe from a pump into a tank of ficient capacity to hold any required quany of water or other liquid, such, for innce, as petroleum, and also to contain a lecting chamber or receiver, as a gas lder. The receiver has a discharge pipe · the liquid, and another pipe extends up o the gas holder, from which the gas colted therein passes through into the heat; or lighting apparatus, or into the cylin· or cylinders of gas engines wherein it is d.

The upper part of the gas discharge pipe provided with two or more flanges of thin tal, or other material, each inclined downrdly, so that the liquid as it issues from discharge pipe passes over the upper end the discharge pipe, and flows down over se inclined flanges of metal, dropping at circumference of each such flange on to next succeeding flange below. In this nner, the liquid flows in thin layers out m the discharge pipe over the succession flanges, so that it is spread over a series thin and extended layers whereby the contained therein in the liquid is freed collected in the gas chamber.

'or the purposes of facilitating the action the pump, as well as for preventing as h as possible of the gas, ascending with liquid into the lift-pipe, from passing the pump, I connect a large discharge · for the gas, coupled at its upper end to acuum pump. The object in this case being to provide as large an area of discharge of the gas as is possible to the vacuum pump, and the discharge from the vacuum pump is connected to the gas pipe, whereby the gas, separated from the liquid passing into the gas holder, is led from the gas holder for lighting and heating purposes, or for driving motive power engines, as will now be set forth in detail.

In the drawings, Figure 1 is a vertical section of the tank and gas holder, for separating gas from liquids. Fig. 2 is a plan, partly in section, of Fig. 1.

In the drawing 3 is the discharge pipe from the pump, which is led through the bottom of a tank 4, and to a sufficient height above the level of the liquid 5, contained therein, to enable the liquid to discharge and trickle down over the inclined flanges 6, which surround the upper end of the pipe. I have shown but three flanges, but a lesser or a greater number of such flanges may be used.

The tank 4 is provided with an overflow pipe 7 for the discharge of the liquid. Within the tank 4 is an inverted gas bell or holder 8, which is air or gas tight, the lower part of which dips to a sufficient depth into the liquid in the tank 4 to prevent leakage or outflow of gas from the bell or gas holder 8, excepting through the gas pipe 9, through which pipe the gas flows to a reservoir or chamber wherein it may be stored for lighting or heating purposes, or to a carbureter for being commingled with air for operating in the cylinders of explosive engines, or for other purposes.

The head of the gas pipe 9, within the gas holder 8, is preferably provided with a hood or shield 10.

It will be understood from the foregoing description that this invention is a separating device, located in the discharge pipe of a pump, which is lifting either water or any other liquid that contains a gas, so that the gas, separated from the liquid, may be used for any purpose for which it is adapted.

I claim:

1. In a mechanism for separating gas from liquids, a tank, a stand pipe in said tank through which liquid is forced into the tank, the stand pipe having below its upper end and supported thereon a plurality of flanges upon or over which the liquid trickles as it is discharged from the stand pipe, an over-

flow pipe, a bell or gas holder located within the tank, a pipe for allowing the escape of gas from the bell or gas holder.

2. In a mechanism for separating a gas from a liquid, a tank, a bell for said tank, a stand pipe centrally located within the bottom of tank and extending upwardly into said tank, a plurality of flanges carried by said stand pipe, said flanges being spaced apart and increasing in size and each lower one being of larger diameter than the one above, an overflow pipe entering the side of said tank, the end of said overflow pipe being flush with the side of said ta draw off pipe entering the bottom tank and extending above the leve liquid.

Signed at the city of Los Angele of Los Angeles State of California, day of April 1912, in the presence nesses.

GEORGE L. McINT

Witnesses:
EDMUND KASOLD,
J. S. ZERBE.



[Endorsed]: E—113—Eq. Townsend et al. v Lorraine. Defendants' Exhibit "E." Filed Marc 27, 1922. Chas. N. Williams, Clerk.

No. 3945. United States Circuit Court of Ap peals for the Ninth Circuit. Filed Jan. 26, 1923 F. D. Monckton, Clerk.

UNITED STATES OF AMERICA,

DEPARTMENT OF THE INTERIOR.

UNITED STATES PATENT OFFICE.

To All to Whom These Presents Shall Come, GREETING:

THIS IS TO CERTIFY that the annexed is a true copy from the Records of this Office of the Letters Patent of

Walter Anderson Taylor,

Number 426,880, Granted April 29, 1890,

for

Improvement in Steam Separators.

IN TESTIMONY WHEREOF I have hereunto set my hand and caused the seal of the Patent Office to be affixed in the District of Columbia this twenty-first day of March, in the year of our Lord one thousand nine hundred and twenty-one and of the Independence of the United States of America the one hundred and forty-fifth.

[Seal] M. H. COULSTON,
Commissioner of Patents.

2—152.

No. 426,880.

THE UNITED STATES OF AMERICA,

To All to Whom These Presents Shall Come:

WHEREAS Walter Anderson Taylor, New Orleans, Louisiana, has presented to the Commissioner of Patents a petition praying for the grant of Letters Patent for an alleged new and useful improvement in Steam Separators, a description of which invention is contained in the specification of which a copy is hereunto annexed and made a part hereof, and has complied with the various requirements of law in such cases made and provided; and

WHEREAS, upon due examination made, the said claimant is adjudged to be justly entitled to a patent under the law;

Now, therefore, these Letters Patent are to grant unto the said Walter Anderson Taylor, his heirs or assigns, for the term of seventeen years from the twenty-ninth day of April, one thousand eight hundred and ninety, the exclusive right to make, use, and vend the said invention throughout the United States and the Territories thereof.

IN TESTIMONY WHEREOF I have hereunto set my hand and caused the Seal of the Patent Office to be affixed, at the City of Washington, this twenty-ninth day of April, in the year of our Lord one thousand eight hundred and ninety, and of the Independence of the United States of America the one hundred and fourteenth.

[Seal] CYRUS BUSSEY,

Assistant Secretary of the Interior.

Countersigned:

ROBERT J. FISHER,

Acting Commissioner of Patents.

(No Model.)

2 Sheets—Sheet 1.

W. A. TAYLOR.

STEAM SEPARATOR.

No. 426,880. Patented Apr. 29, 1890.

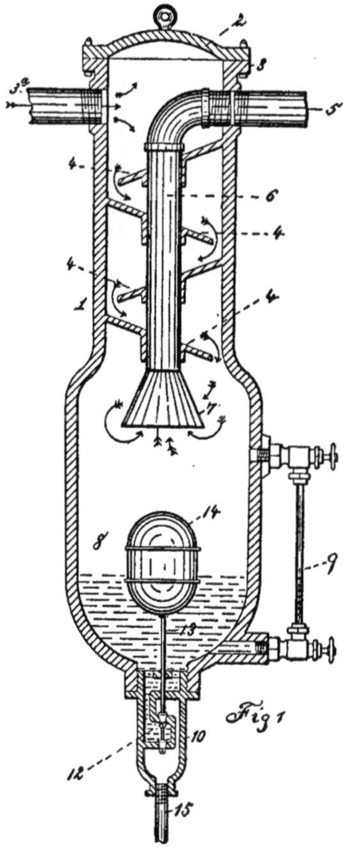


Fig 1

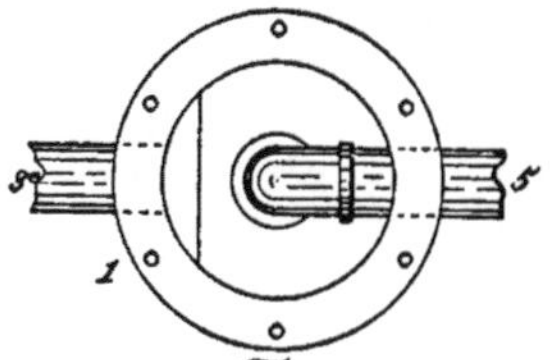


Fig 2

Fig 3

WITNESSES:

Robert Ries.

[illegible signature]

INVENTOR

Walter Anderson Taylor.

By Frederic Cook

Attorney

(No Model.) 2 Sheets—Sheet 2.

W. A. TAYLOR.

STEAM SEPARATOR.

No. 426,880. Patented Apr. 29, 1890.

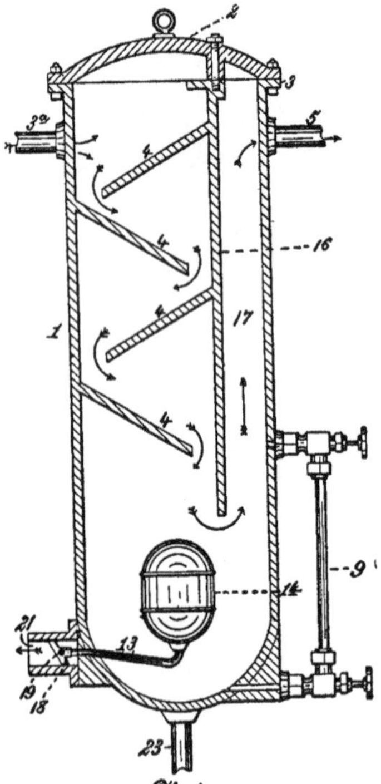

Fig 4

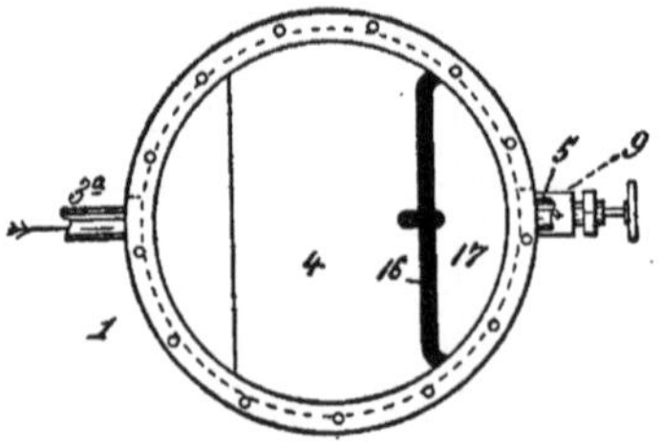

Fig 5.

WITNESSES

Charles De la Croix

Sam B Robeson

INVENTOR

Walter Anderson Taylor

BY Frederic Cook

Attorney.

# UNITED STATES PATENT OFFICE.

WALTER ANDERSON TAYLOR, OF NEW ORLEANS, LOUISIANA.

## STEAM-SEPARATOR.

SPECIFICATION forming part of Letters Patent No. 426,880, dated April 29, 1890.

Application filed November 13, 1889. Serial No. 330,148. (No model.)

 *all whom it may concern:*

Be it known that I, WALTER ANDERSON ·AYLOR, a citizen of the United States, resid·g at New Orleans, in the parish of Orleans ·d State of Louisiana, have invented certain ·w and useful Improvements in Steam-Sepa·tors; and I do hereby declare that the fol·lving is a full, clear, and exact description ·the same, reference being had to the an·nxed drawings, making a part of this specifi·cion, and to the figures of reference marked t·reon.

The object of my invention is to provide a ·vel apparatus for separating the water of c·idensation from live steam and eliminating t·refrom the particles of grease, oil, or other i·purities taken up by the steam in passing f·m the boiler to the steam-chests of the cyl·i·lers. To accomplish this object my inven·tn involves the peculiar features of con·sruction and the combination or arrange·nnt of devices hereinafter described, and s·cifically set forth in the claims, reference b·ng made to the accompanying drawings, i·which—

·igure 1 is a central vertical section of a t·p embodying my invention. Fig. 2 is a p·n view of the same with the cap or cover r·moved. Fig. 3 is a top view of the valve-c·mber. Fig. 4 is a vertical central section s·wing a modified construction. Fig. 5 is a p·n view of Fig. 4, the cover being removed.

·n the said drawings the reference-numeral 1 ·enotes a drum or vessel, preferably of cylin·d·al form, and of small diameter compared w·h its considerable depth. This drum is c·sed at its top by a cover 2, bolted to a f·ge 3, the joint being steam-tight.

·he numeral 3ᵃ denotes the steam-pipe, w·ch enters the drum at at or near the top. F·med or mounted upon the interior face of t· drum are baffle-plates 4, having prefera·b· a downward inclination, and projecting f·n opposite walls of the drum alternately, t·r edges approaching the wall opposite ·t on which they are mounted, but not meet·i·, in order that narrow passages may be ·c for the steam. Any required number of ·se baffle-plates may be used.

·ntering the upper part of the drum above ·baffle plates 4 is a pipe 5, which communicates with a conveyer-tube 6, passing through or by the several plates 4 until its flaring or expanded mouth 7 drops somewhat below the lower plate. The pipe 5 leads to the steam-chest or other point to which the live steam is to be conveyed.

In the lower portion of the drum 1, which may be slightly expanded for the purpose, is the fluid and sediment chamber 8, which is supplied with a water-glass or gage-tube 9. Tapped into or otherwise attached to the bottom of this chamber is a valve-chamber 10, of any suitable construction, in which is arranged a balanced valve 12, carried by a stem 13, which rises into the chamber 8. Upon the valve-stem is mounted a float 14, of suitable size and form, by which the balanced valve may be unseated as the fluid rises in the chamber 8. The lower part of the valve-chamber 10 opens into a pipe 15, which may, if desired, be provided with a suitable cock. As the live steam enters by way of the pipe 3ᵃ, it is compelled to flow downward over the surfaces of the baffle-plates 4, whereby any condensations of vapor or particles of oil or grease carried by it are deposited and caused to adhere to said plates, whence the fluid trickles downward and falls into the chamber 8, while the dry and pure steam enters the mouth of the conveyer 6 and passes to the engine. As the fluid accumulates in the chamber 8, the float is raised and a portion is discharged until the float falls sufficiently to seat the valve. If sediment accumulates in the chamber, the steam-pressure will ordinarily blow it out through the discharge-pipe.

I may substitute the modified construction shown in Fig. 4 in place of that described. In this form of trap the interior of the drum is provided with a partition or plate 16, which is arranged a little to one side of the center, as shown in Fig. 5, where it forms the vertical conveyer 17 for the steam. The baffle-plates 4 extend alternately from the plate or partition 16 and the opposite wall of the drum. It will be seen that this construction is in principle and operation the same as that already described.

In the bottom of the drum is arranged a balanced valve 18, mounted on a pivot 19, and having a stem 13 extending from the outlet 21, in which the valve is pivoted, into the lower part of the drum. On the stem is mounted the float 14. A separate blow-off pipe 23 is arranged in the bottom of the drum and provided with a cock of any suitable form. A water-glass is attached to the drum in the same manner as in the construction described. In this modified form the steam enters, as before, through the inlet-pipe $3^a$ and flows downward over the baffle-plates until it reaches the mouth of the conveyer or passage 17. Passing up through the latter it enters the pipe 5 and is conveyed to the engine.

What I claim is—

1. A steam-purifier consisting of a vertically-arranged closed drum having at its upper end a steam-inlet and a steam-outlet and at its lower end a valved water-discharge, a series of inclined baffle-plates alternately projecting past each other from opposite side portions of the drum and forming a zigzag steam-passage, a vertical steam-conveyer opening at its bottom beneath the lowermost baffle-plate and at its upper end connected with the steam-outlet, and a float sustained by the water in the base of the drum and connected with the valve in the water-discharge, substantially as described.

2. A steam-purifier consisting of a vertically-arranged tight drum having at its upper end a steam-inlet and a steam-outlet, a series of inclined baffle-plates alternately projecting past each other from opposite sides of the drum and each having an orifice, and a steam-conveyer tube extending vertically through the orifices in the baffle-plates, opening at its lower end beneath the lowermost plate and having its upper end connected with the steam-outlet, substantially as described.

3. A steam-purifier consisting of an upright drum having at its top a steam-inlet and a steam-outlet and at its bottom a valved water-discharge and blow-off, a series of baffle-plates alternately projecting past each other from opposite sides of the drum and each having an orifice, a vertical steam-conveyer tube extending through the orifices in the baffle-plates and connected with the steam-outlet, and a float sustained by the water in the base of the drum and connected with the valve, substantially as described.

In testimony whereof I have hereunto subscribed my name in the presence of two witnesses.

WALTER ANDERSON TAYLOR.

Witnesses:
FREDERIC COOK,
JAMES DAVIS.

[Endorsed]: E—113—Eq. Townsend et al. vs. Lorraine. Defendant's Exhibit "F." Filed March 27, 1922. Chas. N. Williams, Clerk.

No. 3945. United States Circuit Court of Appeals for the Ninth Circuit. Filed Jan. 26, 1923. F. D. Monckton, Clerk.

**Defendants' Exhibit "G."**

2—390.

---

UNITED STATES OF AMERICA.
DEPARTMENT OF THE INTERIOR.
UNITED STATES PATENT OFFICE.

To All to Whom These Presents Shall Come,
GREETING:

THIS IS TO CERTIFY that the annexed is a true copy from the Records of this Office of the

Letters Patent of
Arthur W. Barker,

Number 927,476, Granted July 13, 1909,
for
Improvement in Natural-Gas Separators.

IN TESTIMONY WHEREOF I have hereunto set my hand and caused the seal of the Patent Office to be affixed in the District of Columbia this twenty-first day of March, in the year of our Lord one thousand nine hundred and twenty-one and of the Independence of the United States of America the one hundred and forty-fifth.

[Seal] M. H. COULSTON,
Commissioner of Patents.

2—362.

---

No. 927476.

THE UNITED STATES OF AMERICA.

To All to Whom These Presents Shall Come:

WHEREAS Arthur W. Barker, of Fort Pierre,

South Dakota, has presented to the Commissioner of Patents, a petition praying for the grant of Letters Patent for an alleged new and useful improvement in Natural-gas Separators, a description of which invention is contained in the specification of which a copy is hereunto annexed and made a part hereof, and has complied with the various requirements of law in such cases made and provided, and

WHEREAS upon due examination made the said Claimant is adjudged to be justly entitled to a patent under the law.

Now therefore these Letters Patent are to grant unto the said Arthur W. Barker, his heirs or assigns for the term of seventeen years from the thirteenth day of July, one thousand nine hundred and nine, the exclusive right to make, use and vend the said invention throughout the United States and the Territories thereof.

IN TESTIMONY WHEREOF I have hereunto set my hand and caused the seal of the Patent Office to be affixed at the City of Washington this thirteenth day of July, in the year of our Lord one thousand nine hundred and nine, and of the Independence of the United States of America the one hundred and thirty-fourth.

[Seal] C. C. BILLINGS,
Acting Commissioner of Patents.

A. W. BARKER.
NATURAL GAS SEPARATOR.
APPLICATION FILED FEB. 9, 1909.

**927,476.** **Patented July 13, 1909.**

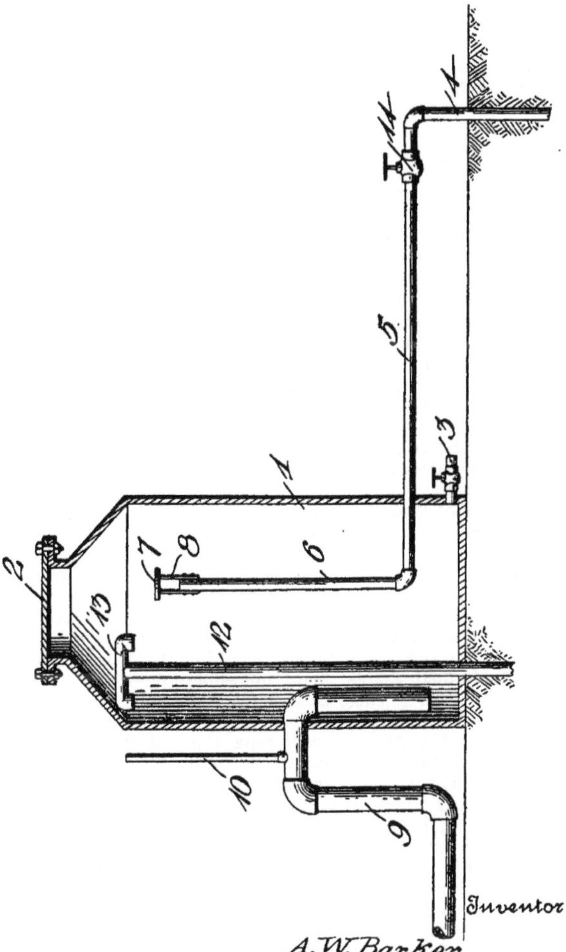

Witnesses
[illegible]
C. H. Griesbauer

Inventor
A. W. Barker
By [illegible]
Attorneys

# UNITED STATES PATENT OFFICE.

ARTHUR W BARKER, OF FORT PIERRE, SOUTH DAKOTA

## NATURAL-GAS SEPARATOR.

No. 927,476. Specification of Letters Patent Patented July 13, 190

Application filed February 9, 1909 Serial No. 477,016.

*To all whom it may concern*

Be it known that I, ARTHUR W. BARKER, a citizen of the United States, residing at Fort Pierre, in the county of Stanley and State of South Dakota, have invented certain new and useful Improvements in Natural-Gas Separators; and I do declare the following to be a full, clear, and exact description of the invention, such as will enable others skilled in the art to which it appertains to make and use the same.

This invention relates to improvements in natural gas separators.

The object of the invention is to provide a device of this character by means of which the gas may be entirely separated from the water flowing from Artesian wells thus saving the gas for use as fuel or for lighting purposes, as well as purifying the water to a sufficient extent to be employed for purposes, other than for human use.

With the foregoing and other objects in view, the invention consists of certain novel features of construction, combination and arrangement of parts as will be more fully described and particularly pointed out in the appended claims.

In the accompanying drawings, is shown a vertical section of my improved gas separator.

Referring more particularly to the drawings, 1 denotes a tank which may be constructed of concrete, sheet metal or any other suitable material and in any suitable size. The upper end of the tank is closed except for a man-hole formed therein, said man-hole being covered by a suitable closing plate 2. In the lower portion of the tank is arranged a drain pipe 3. The Artesian well pipe 4 is connected to the tank by a pipe 5 which enters the tank near its lower end and is provided within the tank with an upwardly projecting discharge tube 6, on the upper end of which is arranged a sprayer 7 which may be of any suitable construction, but which is here shown as preferably consisting of a flat plate which is secured to and spaced above the upper open end of the tube 6 by supporting rods 8.

Connected with the tank 1 is an over-flow pipe 9, the inner end of which extends downwardly in the tank and opens a short distance above the bottom of the tank, as shown. In the highest portion of the pipe outside of the tank is arranged a vent pipe 10 to prevent siphoning. In the tank is also arrange gas discharge pipe 12, said pipe being ranged in the tank in any suitable manne that the open end is at the upper end of tank. The pipe 12 is here shown as exte ing upwardly through the bottom of tank to near the upper end thereof an provided on its upper end with a T head the outer open ends of which are tu downwardly to receive the gas from the per portion of the tank. The outer en the gas conducting pipe 12 may be conne with a pipe line or with a storage tank, shown.

In the operation of the device, the w from the well flows through the pipe 5 tube 6 and is discharged from the tut against the sprayer 7 which breaks the w up into small particles and separates the therefrom. The gas after being thus li ated from the water passes out through T head 13 and gas conducting pipe 12 to pipe line or tank from which it is taken use for fuel or for lighting purposes. water after thus being broken up, passes through the overflow pipe and is conve thereby to a place of use.

The water conducting pipe from the 4 is provided with a regulating valve whereby the supply of water to the tank be controlled. Should the water ente the tank contain more gas than is util or can be carried off by the pipe 12, the p sure of the gas thus accumulating in tank will force the water down below open end of the discharge pipe and will cape through said pipe thus preventing bursting of the tank from an overpress of the gas. As soon as the gas has t liberated itself through the over-flow p the water will again rise therein and remain at a level with the upper portio the over-flow pipe until the gas in the t again reaches a pressure sufficient to f the water down and out of the over-pipe. By this arrangement, the danger of bursting of the tank from an overpres of the gas is entirely eliminated.

By means of a separating device suc herein shown and described, all of the contained in the water may be separ therefrom and the gas utilized for fue lighting purposes, while the water is pur to a sufficient extent to permit the use th of for many purposes.

**927,476**

m the foregoing description, taken in ction with the accompanying draw- the construction and operation of the tion will be readily understood without ing a more extended explanation.

ious changes in the form, proportion he minor details of construction may sorted to without departing from the ple or sacrificing any of the advantages invention as defined in the appended s.

ving thus described my invention, what m as new and desire to secure by Let- 'atent is:

A gas separating device for Artesian comprising a closed tank, a water con- ng pipe to connect said tank with the pipe, a sprayer arranged on the upper f the water conducting pipe to break water discharged therefrom into small les and thereby liberate the gas there- a gas conducting pipe extending into ank, a water over-flow pipe connected e lower portion of the tank, and a vent arranged in said over-flow pipe to prevent the siphoning of the water from the tank.

2. In a gas separating device for Artesian wells, a closed tank, a water conducting pipe to connect said tank to the well pipe, means on the upper end of said water conducting pipe to break up or spray the water discharged therefrom, a regulating valve arranged in said water conducting pipe, a gas conducting pipe extending up through the tank, a T head on the upper end of said pipe, said T head having downwardly projecting open ends to receive the gas from the upper end of the tank, a water discharge pipe arranged in the tank, and extending to near the bottom thereof, and a vent tube arranged in said over-flow or discharge pipe.

In testimony whereof I have hereunto set my hand in presence of two subscribing witnesses.

ARTHUR W. BARKER.

Witnesses:
T. R. STRAIN,
MATHILDE GOLDSMITH.

[Endorsed]: E—113—Eq. Townsend et al. vs. Lorraine. Defendant's Exhibit "G." Filed March 27, 1922. Chas. N. Williams, Clerk.

No. 3945. United States Circuit Court of Appeals for the Ninth Circuit. Filed Jan. 26, 1923. F. D. Monckton, Clerk.

**Defendants' Exhibit "H."**

---

2—390.

UNITED STATES OF AMERICA,

DEPARTMENT OF THE INTERIOR.

UNITED STATES PATENT OFFICE.

To All to Whom These Presents Shall Come,
GREETING:

THIS IS TO CERTIFY that the annexed is a true copy from the Records of this Office of the

Letters Patent of

Eustace Vivian Bray, Assignor of one-half to
Richard S. Haseltine,

Number 1,014,943, Granted January 16, 1912,
for

Improvement in Separators for Removing Gas
from Oil or Other Liquids.

IN TESTIMONY WHEREOF I have hereunto set my hand and caused the seal of the Patent Office to be affixed in the District of Columbia this twenty-fourth day of March, in the year of our Lord one thousand nine hundred and twenty-one and of the Independence of the United States of America the one hundred and forty-fifth.

[Seal] M. H. COULSTON,
Commissioner of Patents.

2—362.

---

No. 1014943.

THE UNITED STATES OF AMERICA,

To All to Whom These Presents Shall Come:

WHEREAS Eustace Vivian Bray, of Coalinga, California, has presented to the Commissioner of Patents a petition praying for the grant of Letters Patent for an alleged new and useful improvement in Separators for Removing Gas from Oil or other Liquids. He having assigned one-half of his right, title, and interest in said improvement to Richard S. Haseltine, of Coalinga, California, a description of which invention is contained in the specification of which a copy is hereunto annexed and made a part hereof, and has complied with the various requirements of law in such cases made and provided, and

WHEREAS upon due examination made the said Claimant is adjudged to be justly entitled to a patent under the law.

Now therefore these Letters Patent are to grant unto the said Eustace Vivian Bray and Richard S. Haseltine, their heirs or assigns for the term of seventeen years from the sixteenth day of January, one thousand nine hundred and twelve, the exclusive right to make, use and vend the said invention throughout the United States and the Territories thereof.

IN TESTIMONY WHEREOF I have hereunto set my hand and caused the seal of the Patent Office to be affixed at the City of Washington this sixteenth dred and thirty-sixth.

[Seal]
C. C. BILLINGS,
Acting Commissioner of Patents.

E. V. BRAY.
SEPARATOR FOR REMOVING GAS FROM OIL OR OTHER LIQUIDS.
APPLICATION FILED MAR. 25, 1911.

**1,014,943.** Patented Jan. 16, 1912.
3 SHEETS—SHEET 1.

Fig 1

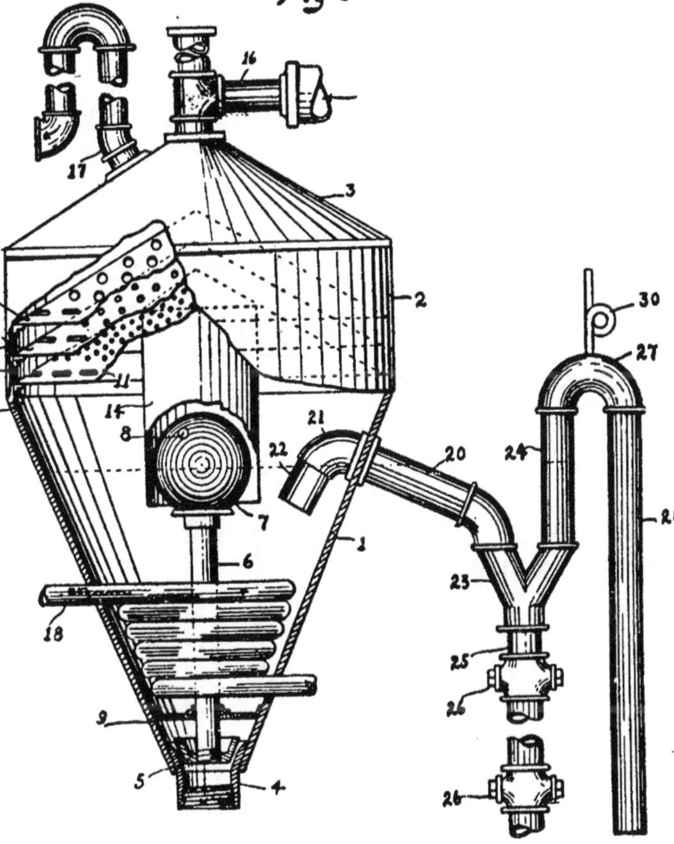


WITNESSES
H.B. Clark,
B.M. Freund

INVENTOR
E.V. Bray
BY
Carlos P. Griffin
ATTORNEY

E. V. BRAY.

SEPARATOR FOR REMOVING GAS FROM OIL OR OTHER LIQUIDS.

APPLICATION FILED MAR. 25, 1911.

**1,014,943.** **Patented Jan. 16, 191**

3 SHEETS—SHEET 1.

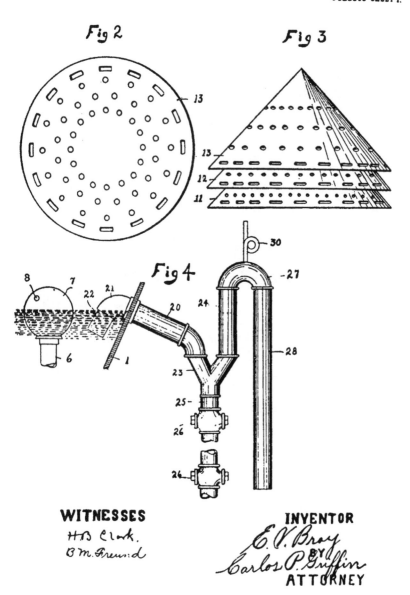

Fig 2

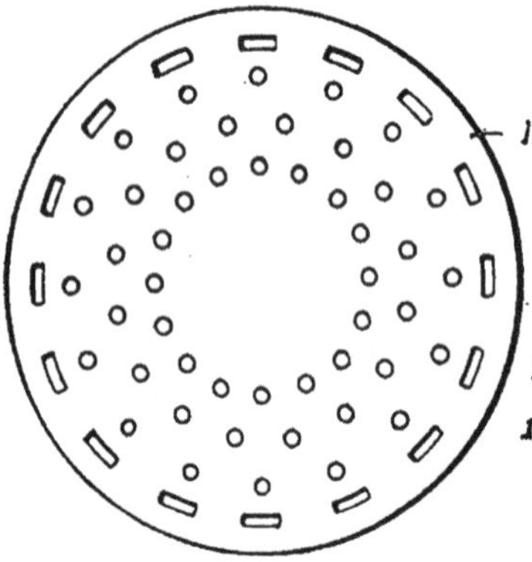


Fig 3

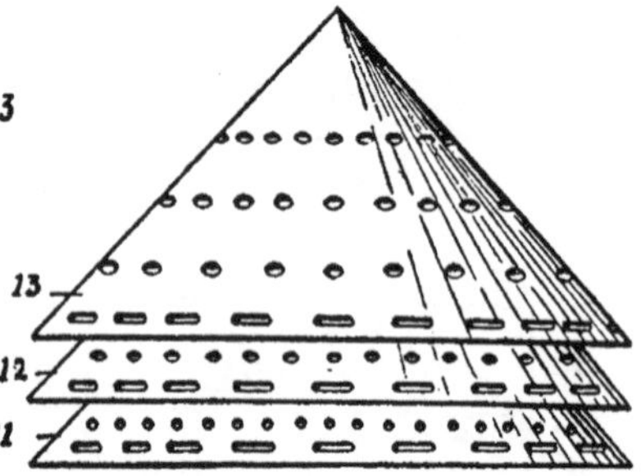


Fig 4

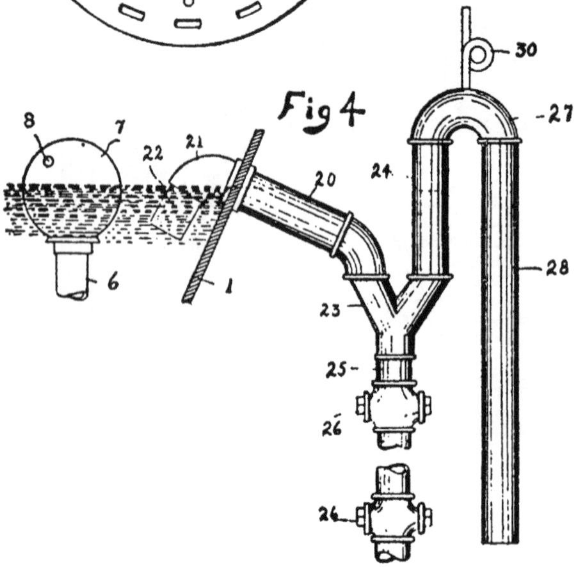


**WITNESSES**

H.B. Clark.

B. M. Freund

**INVENTOR**

E. V. Bray

BY

Carlos P. Griffin

ATTORNEY

ACE VIVIAN BRAY, OF COALINGA, CALIFORNIA, ASSIGNOR OF ONE-
TO RICHARD S. HASELTINE, OF COALINGA, CALIFORNIA.

# SEPARATOR FOR REMOVING GAS FROM OIL OR OTHER LIQUIDS.

,943. Specification of Letters Patent. **Patented Jan. 16, 1912**

Application filed March 25, 1911. Serial No. 616,975.

*ll whom it may concern:*

it known that I, EUSTACE V. BRAY, n of the United States, residing at nga, in the county of Fresno and State alifornia, have invented a new and use- Separator for Removing Gas from Oil her Liquids, of which the following is ecification in such full and clear terms ill enable those skilled in the art to con- t and use the same.

is invention relates to an apparatus for ourpose of the removal and the collec- of gas from crude oil or other liquid.

will be understood by those skilled in rt that crude oil and other liquids as come from the ground and some other ls, contain considerable quantities of ustible gas. This gas is ordinarily per- d to escape and waste. However, the tity of gas in some oils and liquids is onsiderable as to make this a serious e, and to prevent this waste, this appa- has been designed. After the gas is rated from the oil or other liquids and been collected, it may be burned under boiler supplying steam to an oil pump in a gas engine or for heating and ing, or it may be used at any other ble place where power is desired, thus ting a considerable economy in power.

the drawings in which the same nu- l of reference is applied to the same on throughout the several views, Fig- . is a view partly in section of the gas ator, Fig. 2 is a plain view of the top rating screen, Fig. 3 is a side elevation he three separating screens used, and 4 is a view illustrating the overflow

he numeral 1 represents a conical cas- having a cylindrical extension 2, the of which is closed by means of the cy- ical cone 3. At the bottom of the cone ere is a sleeve 4, within which a valve opper 5 seats. The valve 5 is screwed e lower end of a pipe 6, to which pipe cured a float 7, the latter having an ng 8 therciu for a purpose to be ex- ed. Near the bottom of the casing 1, is a guide plate 9 to prevent the valve working out of line with the sleeve 4.

e cylindrical casing 2 is provided with ies of brackets 10, there being a num- of said brackets at different levels to rt the three screens 11, 12 and 13, the latter screen having larger openings therein than the lower screen. The lowest screen has a cylindrical casing 14 secured thereto, which extends downwardly over the float 7, the object being to prevent the incoming oil or liquid from flowing down over the top of the float and accumulating sand and foreign matter thereon, thus preventing its free action.

The oil or liquid is fed into the casing through the pipe 16 at the top of the cone 3, and the gas is removed therefrom through the pipe 17 at the side of the pipe 16. Since heavy fuel oil is more or less viscous, it becomes necessary to provide means to heat the same, a better separation of the gas being effected when the oil is hot. The steam coil also by its heat can cause possibly a greater liberation of gas. This means comprises a coil of pipe 18 at the bottom of the cone 1. In operation the oil or liquid is forced into the apparatus through the pipe 16, one of the openings in said pipe being closed, or both receiving oil or liquid from different sources of supply. The oil or liquid then passes down over the screens and into the cone 1, filling the latter until the float 7 lifts the valve 5 enough to permit such a quantity of oil or liquid to escape as will again seat the valve, or it may be that the valve will be open continuously. However, since the only escape for the gas is through the pipe 17, it may be collected in a suitable receptacle, or may be run to a furnace to be burned, or used in some metallurgical or chemical process, or in the arts.

In order to insure against oil or liquid leaking into the float and interfering with the operation thereof, the pipe 6 is opened to the oil drain pipe, thus permitting the escape of any oil should the float leak, or should the casing 1 be unduly filled, the oil or liquid might run through the opening 8 to the main pipe line.

Since considerable sand or other foreign matter is usually brought along with the oil or liquid, it becomes necessary to provide some means for insuring the continuous operation of the separator should the sand or foreign matter become so great in quantity as to close up the valve 5 and prevent the escape of oil therefrom. This means comprises a pipe 20 connected to the cone 1 and having a downwardly extending elbow 21 and short nipple 22. This pipe

No. 815,407

THE UNITED STATES OF AMERICA,

To All to Whom These Presents Shall Come:

WHEREAS Augustus Steiger Cooper of Berkeley, California, has presented to the Commissioner of Patents a petition praying for the grant of Letters Patent for an alleged new and useful improvement in Separators for Gas, Oil, Water and Sand, a description of which invention is contained in the specification of which a copy is hereto annexed and made a part hereof, and has complied with the various requirements of law in such cases made and provided; and

WHEREAS upon due examination made the said claimant is adjudged to be justly entitled to a patent under the law;

NOW THEREFORE THESE LETTERS PATENT are to grant unto the said Augustus Steiger Cooper, his heirs or assigns for the term of seventeen years from the twentieth day of March, one thousand nine hundred and six, the exclusive right to make, use, and vend the said invention throughout the United States and the Territories thereof.

IN TESTIMONY WHEREOF I have hereunto set my hand and caused the seal of the Patent Office to be affixed, at the City of Washington this twentieth day of March, in the year of our Lord one thousand nine hundred and six and of the Independence of the United States of America the one hundred and thirtieth.

[Seal] F. I. ALLEN,

Commissioner of Patents.

No. 815,407. PATENTED MAR. 20, 1906.

A. S. COOPER.

SEPARATOR FOR GAS, OIL, WATER, AND SAND.

APPLICATION FILED JUNE 14, 1905.

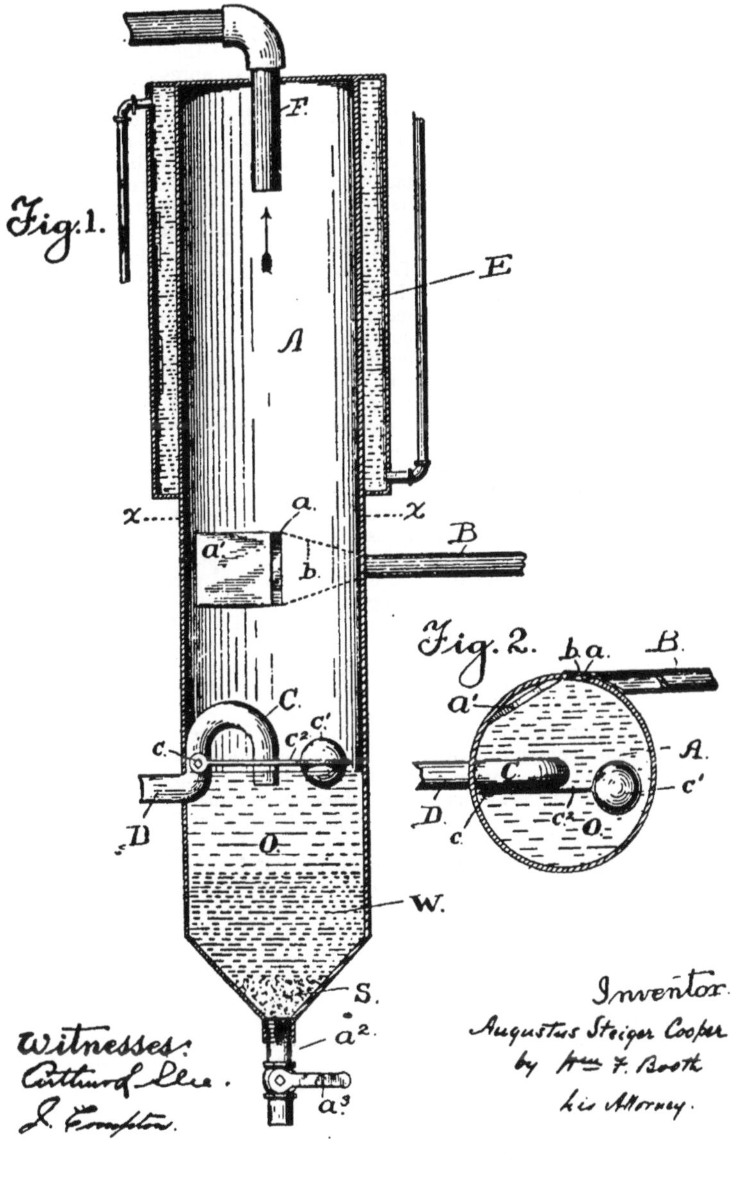

# UNITED STATES PATENT OFFICE.

AUGUSTUS STEIGER COOPER, OF BERKELEY, CALIFORNIA.

## SEPARATOR FOR GAS, OIL, WATER, AND SAND.

815,407. Specification of Letters Patent. Patented March 20, 1906.

Application filed June 14, 1905. Serial No. 265,148.

*l whom it may concern*

it known that I, AUGUSTUS STEIGER R, a citizen of the United States, resid- Berkeley, county of Alameda, State of rnia, have invented certain new and Improvements in Separators for Gas, 'ater, and Sand; and I hereby declare llowing to be a full, clear, and exact de- ion of the same.

invention relates to that class of separa- sed in connection with gas and oil wells ich a separation is effected of the several tuents of the contents issuing from ipe. In that process which I set forth ontemporaneous application for a pat- portion of the gas from the well is used edly as a lift or pulverizer to relieve the f oil and water by raising or assisting to hem. The contents of the discharge- composed of gas, oil, water, sand, and forced into a vessel included in the re system, wherein the constituents parated without affecting the general re in the system and the gas recovered ken to the compressor to be used again t as the lift or pulverizer to relieve the he surplus of the gas being conducted for any useful purpose.

to such a separator that my invention , its object being to efficiently separate liver the gas, oil, water, and sand from a re system used in lifting or pulverizing and water and relieving the well of them. this end my invention consists in the separator which I shall now describe erence to the accompanying drawings, ch—

re 1 is a vertical section of the separa- Fig. 2 is a cross-section on the line *x x* 1.

a cylinder of suitable capacity. In de is made an inlet-opening *a*, leading h the wall at an angle, whereby the rge end *b* of the pipe B from the well is ed to deliver the pipe contents into the er tangentially to its circumference. A earing-plate *a'* is fixed to the inner wall cylinder to receive the material. The end of the cylinder has an outlet $a^2$, with rolling-cock $a^3$.

the cylinder at a predetermined level its bottom issues a pipe C, which con- with the pipe D, which leads to the oil- The pipe C has a cock *c*, which is op- automatically by the float *c'*, connect- h it by an arm $c^2$. Around the upper portion of the cylinder is a water-jacket E, and from the top of the cylinder leads the gas-pipe F to the compressor.

The operation is as follows: Gas, oil, water, silt, and sand are forced from near the bottom of the well by the compressed gas and pass through the pipe B into the cylinder A. Entering the cylinder under pressure tangentially to its circumference, the material is thereby given a rotary motion, which separates the constituents conformably to their specific gravities. The gas rises and passes from the cylinder through pipe F to the compressor. The sand seeks the outer circle and drops down to the bottom, as is indicated by S. The water W lies on the sand, while the oil O lies on the water. When these constituents accumulate to raise the oil to a certain level, the float *c'* rises and opens the cock *c*, thereby permitting the oil to be forced out by the internal pressure in the system through the pipe C to the oil-tank. When enough oil has been forced out to lower the level, the cock *c* is automatically closed by the descent of the float. When the cock $a^3$ is opened, the sand and water may be blown off through the outlet $a^2$. A high pressure of gas may thus be maintained in this cylinder. The water-jacket cools the gas before it is used in the compressor.

Having thus described the invention, what I claim as new, and desire to secure by Letters Patent, is—

1. A separator for gas, oil, water and sand, comprising a cylinder, a means for delivering the material into its side tangentially to its circumference, to impart a rotary motion to said material, whereby its constituents are separated, an exit for the gas at the top of the cylinder, an exit for the sand and water at the bottom of the cylinder, and an intervening exit for the oil automatically controllable by the level of said oil.

2. A separator for gas, oil, water and sand, comprising a vessel, means for imparting to the material fed to it a rotary motion to separate its constituents according to their specific gravities, an exit for the gas at the top of the cylinder, an exit for the sand and water at the bottom of the cylinder, and an intervening exit for the oil, automatically controllable by the level of said oil.

3. A separator for gas, oil, water and sand, comprising a vessel, means for imparting to the material fed to it a rotary motion to separate its constituents according to their specific gravities, an exit at the top for the gas,

2 815,407

an exit at the bottom for the sand and water, an intervening exit for the oil, a cock controlling said last-named exit, and a float connected with the cock to open and close it according to the level of the oil.

4. A separator for gas, oil, water and sand, comprising a cylinder, a means for delivering the material into its side tangentially to its circumference, to impart a rotary motion to said material, whereby its constituents are separated, an exit for the gas at the top of the cylinder, an exit for the sand and water at the bottom of the cylinder, and an intervening exit for the oil, automatically controllable by the level of said oil, by means of a cock operated by a float in said liquid constituent.

In witness whereof I have hereunto set my hand.

AUGUSTUS STEIGER COOPER.

Witnesses:
J. COMPTON,
D. B. RICHARDS.

[Endorsed]: E—113—Eq. Townsend et al. vs. Lorraine. Defts. Exhibit "I." Filed March 27, 1922. Chas. N. Williams, Clerk.

No. 3945. United States Circuit Court of Appeals for the Ninth Circuit. Filed Jan. 16, 1923. F. D. Monckton, Clerk.

**Defendants' Exhibit "J."**

2—390.

---

UNITED STATES OF AMERICA,

DEPARTMENT OF THE INTERIOR,

UNITED STATES PATENT OFFICE.

To All to Whom These Presents Shall Come, GREETING:

THIS IS TO CERTIFY that the annexed is a true copy from the Records of this Office of the

Letters Patent of

Albert T. Newman, Assignor of one-half to Ernest McClure,

Number 856,088, Granted June 4, 1907,

for

Improvement in Water and Gas Separators.

IN TESTIMONY WHEREOF I have hereunto set my hand and caused the seal of the Patent Office to be affixed in the District of Columbia this third day of May, in the year of our Lord one thousand nine hundred and twenty-one and of the Independence of the United States of America the one hundred and forty-fifth.

[Seal] T. E. ROBERTSON,
Commissioner of Patents.

2—362.

---

No. 856,088.

THE UNITED STATES OF AMERICA,

To All to Whom These Presents Shall Come:

WHEREAS Albert T. Newman of Greeley, Kansas, has presented to the Commissioner of Patents a petition praying for the grant of Letters Patent for an alleged new and useful improvement in

Water and Gas Separators,

He having assigned one-half of his right, title and interest, in said improvement, to Ernest McClure, of Greeley, Kansas, a description of which invention is contained in the specification of which a copy is hereunto annexed and made a part hereof, and has complied with the various requirements of law in such cases made and provided; and

WHEREAS upon due examination made the said claimant is adjudged to be justly entitled to a patent under the law;

NOW THEREFORE THESE LETTERS PATENT are to grant unto the said Albert T. Newman and Ernest McClure, their heirs or assigns for the term of seventeen years from the fourth day of June, one thousand nine hundred and seven, the exclusive right to make, use, and vend the said invention throughout the United States and the Territories thereof.

IN TESTIMONY WHEREOF I have hereunto set my hand and caused the seal of the Patent Office

**No. 856,088.** **PATENTED JUNE 4, 1907.**

**A. T. NEWMAN.**

**WATER AND GAS SEPARATOR.**

**APPLICATION FILED AUG. 28, 1906.**

3 SHEETS—SHEET 1.

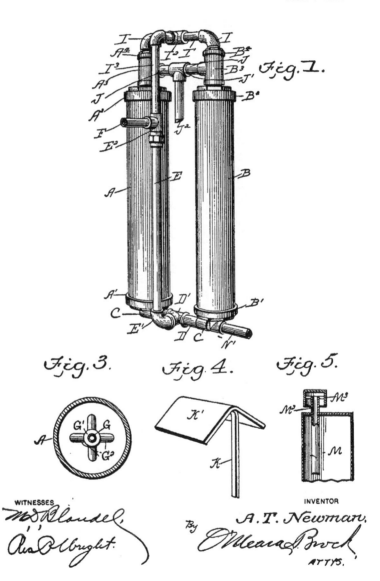

*Fig. 1.*

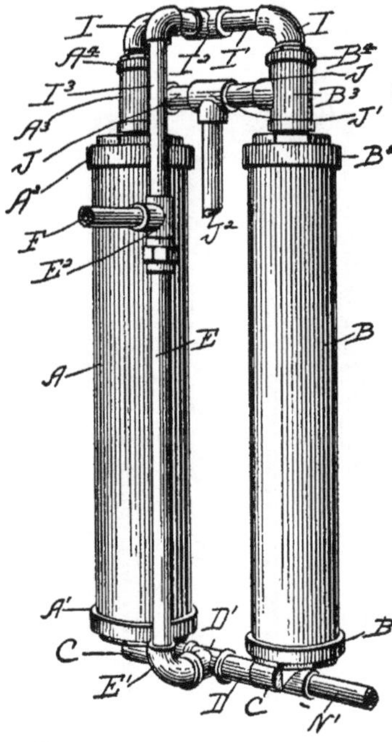


*Fig. 3.* *Fig. 4.* *Fig. 5.*

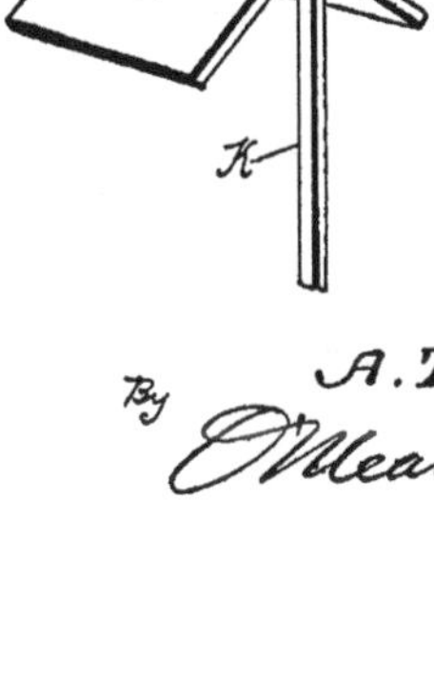

WITNESSES

M. D. Blondel,

Ais P. Albright.

INVENTOR

A. T. Newman,

By O'Meara & Brock

ATTYS.

No. 856,088. PATENTED JUNE 4, 1907.

A. T. NEWMAN.

WATER AND GAS SEPARATOR.

APPLICATION FILED AUG. 28, 1906.

2 SHEETS—SHEET 2.

Fig. 2.

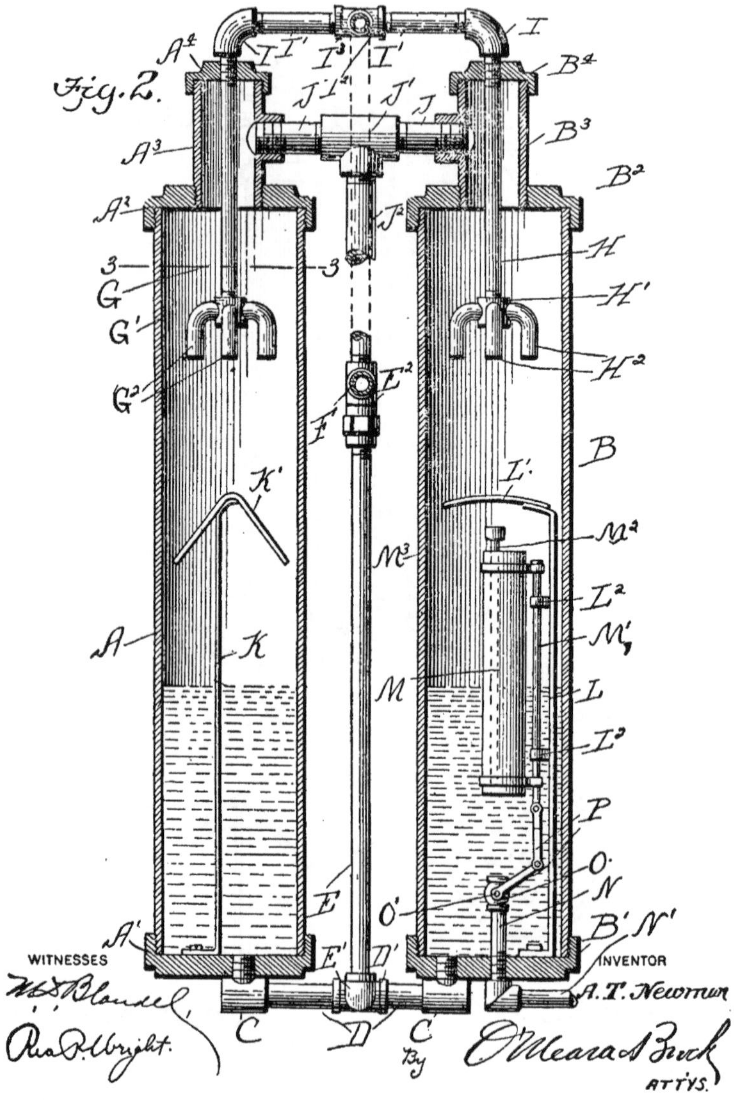

T T. NEWMAN, OF GREELEY, KANSAS, ASSIGNOR OF ONE-HALF TO ERNEST McCLURE, OF GREELEY, KANSAS.

## WATER AND GAS SEPARATOR.

88. **Specification of Letters Patent.** **Patented June 4, 1907.**

Application filed August 28, 1906. Serial No. 332,366.

*hom it may concern:*

nown that I, Albert T. Newman, a the United States, residing at Greele county of Anderson, in the State s, have invented a new and useful ent in Water and Gas Separators, the following is a specification.

vention relates to certain new and provements upon my patent issued 6, 1904, No. 776,753, the object distribute the force of gas, especially ressure so as to eliminate the spray the water chance to collect and the bottom of the tank, and allow ation to be made more perfectly.

this and various other objects in invention consists in the novel feaconstruction, hereinafter fully dend pointed out in the claims.

drawings forming a part of this on:—Figure 1 is a perspective my improved water and gas separa. 2 is an enlarged vertical sectional the separator. Fig. 3 is a transtional view taken on the lines 3—3 . Fig. 4 is a perspective view of e hoods. Fig. 5 is a detail sectional he float.

ng to the drawings A and B indiir of tanks provided with threaded which are secured caps A′, $A^2$, and The caps A′, B′ are provided with openings in which the threaded he **T**-joints C are secured which are n the end of the pipe sections D with a **T**-coupling D′, extending ontally and to which an elbow E′, carrying a pipe E provided with a $E^2$, to which an inlet pipe F is . The caps $A^2$, $B^2$ are provided aded openings in which the lower ends of the pipe sections $A^3$, $B^3$ are hich are provided with caps $A^4$, $B^4$, pper ends provided with threaded through which the threaded ends pes G and H project, which extend the tanks and are provided with H′ carrying downwardly projectd distributing pipes $G^2$, $H^2$. Sethe upper threaded ends of the and H, are elbows I provided with ions I′, carrying a **T**-coupling $I^2$, with a pipe $I^3$, connected at its to the **T**-coupling $E^2$.

The pipe sections $A^3$, $B^3$, are provided with threaded openings in its side in which are secured pipe sections J, carrying a **T**-coupling J′, in which an outlet pipe $J^2$, is secured. Brackets K and L are secured in the tanks A and B, provided with hoods K′, L′ at their upper ends forming reflectors, the hood K being formed inverted **V**-shaped in cross-section. Secured to the bracket L are apertured lugs $L^2$, in which is mounted a rod M′, carrying a hollow float M, which is provided with a tube $M^2$, extending out from the top provided with a cap $M^3$, at its upper end which prevents the same from being filled with water and at the same time to allow the gas to enter the float, thereby equalizing the pressure. Projecting up through a threaded opening in the cap B′, of the tank B is a pipe N provided with a valve O, at its upper end, the stem O′ of which is connected to the bar M′ by links P. The other end of the pipe N is connected to a discharge pipe N′.

The operation of the device is as follows:—The gas is led from the well by the pipe F, into the top and bottom of the tanks A and B through the pipes $I^3$, and E thereby dividing the force of the gas into the tank keeping the agitation in the bottom of the tanks preventing the dirt from settling and blocking up the pipes, and the gas entering the top of the tanks through the pipes G and H, is divided again by the distributers $G^2$, $H^2$, and strikes the hoods K′, L′ preventing the gas from boiling up the water accumulated in the bottom of the tank thus preventing much spray and protects the float M from incoming rush of gas, and as the gas rises, it will pass out the outlet pipe $J^2$. As the water rises in the tank, it will carry the float with it, which will gradually open the valve O allowing a portion of the water to escape out the discharge pipe N and as the float falls, the valve will close gradually.

Having thus fully described my invention, what I claim as new and desire to secure by Letters Patent is:—

1. The combination with a pair of tanks connected together by pipes at their top and bottom, of an inlet pipe connected to said pipes, outlet pipes connected to said tanks and a discharge pipe connected to one of said tanks, for the purpose described.

2. The combination with a pair of tanks, of inlet pipes connected to the top and bot-

tom of said tanks, and an outlet pipe connected to the top of said tanks.

3. The combination with a pair of tanks, of inlet pipes connected to the top and bottom of said tanks, outlet pipes connected to the top of said tanks, a discharge pipe secured in the bottom of one of said tanks provided with a valve and a float connected to said valve, for the purpose described.

4. The combination with a pair of tanks, of pipes connecting the top and bottom of said tank, an inlet pipe connected to said pipes, outlet pipes connected to said tanks, and a discharge pipe provided with a valve arranged in one of said tanks, for the purpose described.

5. The combination with a pair of tanks, connected together at their lower ends, of pipes extending down into said tanks provided with distributers, pipes connected to said pipes and the connecting pipes, an inlet pipe connected to said pipes, and outlet pipes connected to said tanks, for the purpose described.

6. In a device of the kind described, the combination with a pair of tanks, of inlet pipes connected to the top and bottom of said tanks, reflectors arranged in said tanks, outlet pipes connected to said tanks, and a valve controlled discharge pipe, connected to one of said tanks, for the purpose described.

7. In a device of the kind described, the combination with a pair of tanks, of inlet pipes connected to said tanks at their top and bottom, distributers connected to the upper inlet pipes, and reflectors arranged under said distributers, for the purpose described.

8. In a device of the kind described, the combination with a pair of tanks, of inlet pipes connected to the top and bottom of said tanks, distributers connected to the upper inlet pipes in said tanks, reflectors arranged in said tanks under said distributers, outlet pipes connected to said tanks, and a discharge pipe connected to one of said tanks provided with a float operated valve, for the purpose described.

9. In a device of the kind described, the combination with a pair of tanks, provided with reduced upper portions, of inlet pipes connected to the bottom of said tanks, inlet pipes connected to said reduced portion having distributers on their ends, outlet pipes connected to the reduced portion, reflectors arranged in said tanks under said dis ters, a discharge pipe connected to c said tanks, provided with a valve, and ε connected to said valve, for the purpo forth.

10. In a device of the kind describe combination with a pair of tanks pro with reduced upper portions, of pipes necting the lower ends of said tanks, pip tending down into said tank throug tops of the reduced portions, and inlet connected to said pipes and to the pipe necting the lower end of the tanks, pipes connected to the reduced porti said tanks, and a valve controlled discl pipe connected to one of said tanks, f purpose described.

11. In a device of the kind describe combination with a pair of tanks of pipes connected to the top and bott said tanks, outlet pipes connected to t of the said tanks, a discharge pipe exte up into one of said tanks, provided valve, and a float slidably mounted i tank connected to the stem of said val links, for the purpose described.

12. In a device of the kind describe combination with a pair of tanks, of pipes connected to the top and bott said tanks, distributers connected t upper inlet pipes, brackets mounted i tanks provided with deflectors, ape lugs secured to one of said brackets, slidably mounted in said lugs carrying ε a valve controlled discharge pipe con to said tank and links connecting the st said valve to said rod for the purpo scribed.

13. In a device of the kind describe combination with a pair of tanks, of pipes connected to the top and bott said tanks, distributers connected t upper inlet pipes, brackets carrying ( tors arranged under said distributers i tanks, outlet pipes connected to said ta discharge pipe connected to one of said provided with a valve, a float slidably m ed on one of said brackets and links co ing said float to the stem of the valve discharge pipe, for the purpose describ

A. T. NEWM

Witnesses:
CHAS. H. LYON,
W. C. LYON.

[Endorsed]: E—113—Eq. Townsend et al. vs. Lorraine. Defts. Exhibit "J." Filed March 27, 1922. Chas. N. Williams, Clerk.

No. 3945. United States Circuit Court of Appeals for the Ninth Circuit. Filed Jan. 26, 1923. F. D. Monckton, Clerk.

Defts Ex K

E113 Eq.

Townsend, et al
v.
Lorraine

Defts' Ex. K

Filed Mch. 27-1922
Chas. N. Williams
Clerk

5

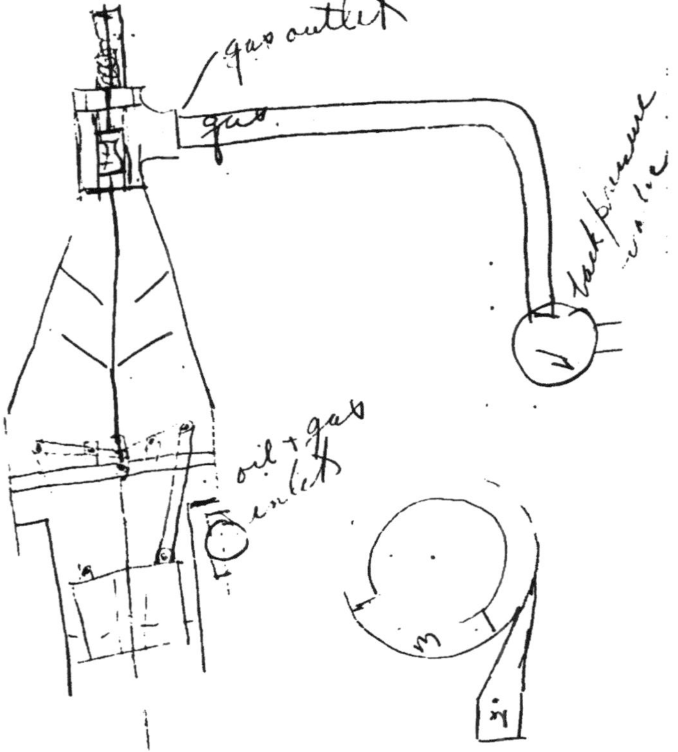

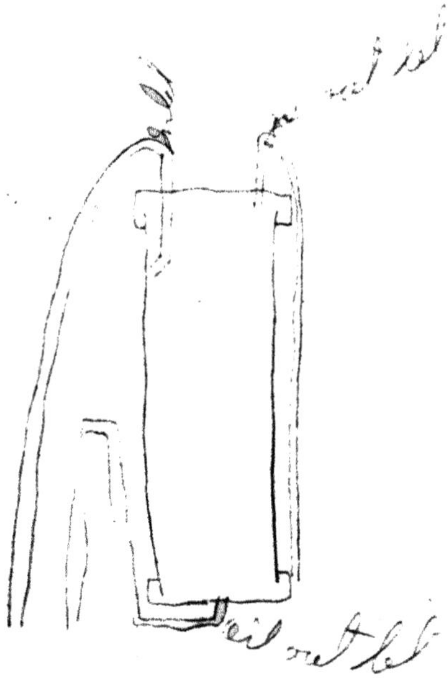

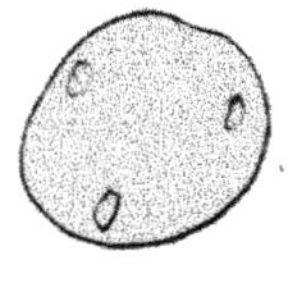

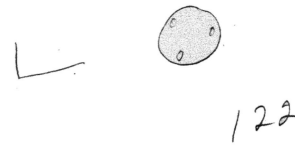

3 Eq.

endy, et al.

V.

aine

Exhibit M

d March 27, 1922
Chas. N. Williams
Clerk



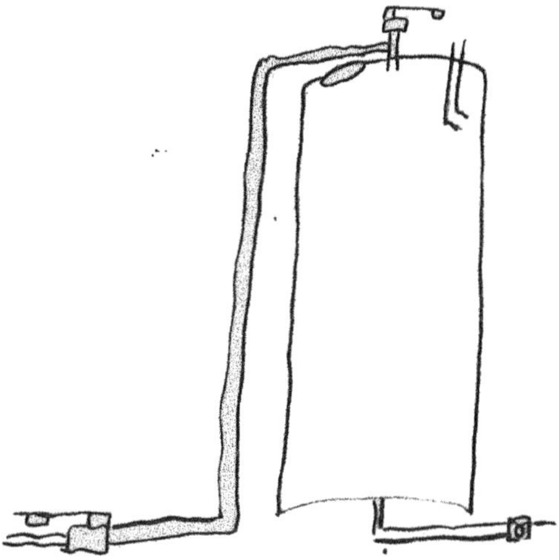

**Defendants' Exhibit "N."**

TRUMBLE GAS TRAP CO.

916 Higgins Building,

Los Angeles, Cal.

June 14, 1921.

Mr. William Lacy,
601 Washington Building,
Los Angeles,
Cal.

Dear Sir:

Referring to the correspondence had with you during the month of December, 1920, relative to the matter of the Lorraine Infringement of the Trumble Gas Trap Patents, I wish to say that I note in yours of December 13th, your statement that you would cease making the Lorraine Gas Trap until he (Mr. Lorraine) produced the proper evidence that he is not infringing upon our patent rights.

We have heard nothing direct from you since that date, but are informed that almost immediately after the receipt of this letter in which you stated that only one Trap had been delivered to Mr. Lorraine, that five additional Traps were sold and placed in the field. We are now informed that you are manufacturing twenty-five of these Traps in your factory here in Los Angeles, and that they will be delivered to Mr. Lorraine for marketing just as soon as it is possible for you to complete them.

It seems that we must presume that you have now elected to become a party to the infringement by Mr. Lorraine, and have deemed it unnecessary to inform us that you were intending to continue such infringement. With this in mind we now find ourselves compelled to look upon you as a most wilful infringer of the Trumble Patent, and as such we think it nothing more than right that The Lacy Manufacturing Company should be made a party to the Infringement Suit now pending against Mr. Lorraine.

2.

Mr. William Lacy,
Los Angeles, Cal. June 14, 1921.

We have noticed that all of the Lorraine Traps so far put into use, are marked with the LACY MANUFACTURING COMPANY plate showing your Company is responsible for their manufacture.

The writer regrets very much the necessity for us to resort to suit for protection against your Company's arbitrary seizing of the rights which we consider are the legal property of Mr. Trumble and his associates.

We shall await a few days in order that you may reply if you so see fit, as we desire to be courteous in the matter.

We are,

Yours very truly,
F. M. TOWNSEND.

FMT. WW.

[Endorsed]: E—113—Eq. Townsend et al. vs. Lorraine. Defts. Exhibit "N." Filed March 27, 1922. Chas. N. Williams, Clerk.

No. 3945. United States Circuit Court of Appeals for the Ninth Circuit. Filed Jan. 26, 1923. F. D. Monckton, Clerk.

---

**Defendants' Exhibit "O."**

LACY MANUFACTURING CO.

Plate and Sheet Steel Workers.

Steel Oil Storage Tanks and Riveted Steel Pipe.

Office 601 Washington Building,

Third and Spring Streets.

Phones Main 196.
Home A 3250.

WORKS: Cor. N. Main & Date Streets.

Los Angeles, Cal.

All Agreements are Contingent Upon Strikes, Accidents, Delays of Carriers, and Other Delays Unavoidable or Beyond Our Control. Quotations Subject to Change Without Notice.

Los Angeles, June 15, 1921.

Mr. F. M. Townsend,
916 Higgins Bldg.,
City.

Dear Sir:

I am in receipt of yours of the 14th inst., in which you refer to my letter to you of December 13, 1920 wherein I made the statement that the Lacy Manufacturing Company would cease making the Lorraine Gas Trap until Mr. Lorraine produced the proper evidence that he is not infringing on your patent rights, and I now beg to inform you that immediately after sending my letter of December 13th, in order to satisfy myself as to your claim that the Lorraine Trap was an infringement of the Trumble Gas Trap patents, I secured the opinion of several able patent attorneys in regard to your claim of infringement and their opinion was absolutely contrary to your contention. After securing these opinions we were satisfied that Lorraine's patent did not in any way conflict with the Trumble patent and we then concluded to fill the orders that Mr. Lorraine had placed with us.

As stated to you before, our policy in the past, as it will be in the future, has been never to infringe upon the legal rights of others, but we are in the business of manufacturing articles of this class and cannot recognize the demands of any one that we cease taking such orders.

The traps that we have made for Mr. Lorraine do not in any way infringe on the Trumble patent and we believe that we shall continue to make them for him as long as he is able to pay for them.

Yours very truly,

LACY MANUFACTURING COMPANY.

WM. LACY.

WL: HH.

[Endorsed]: E—113—Eq. Townsend et al. vs. Lorraine. Defts. Exhibit "O." Filed March 27, 1922. Chas. N. Williams, Clerk.

No. 3945. United States Circuit Court of Appeals for the Ninth Circuit. Filed Jan. 26, 1923. F. D. Monckton, Clerk.

**Defendants' Exhibit "P."**

Patents Trademarks Copyrights Corporations

DR W P KEENE

Attorney and Counsellor in Patent Causes

929 W. P. Story Building

Los Angeles, California

Telephones:
Broadway 1487.
Residence 20730.

December 29, 1920.

Lacy Manufacturing Co.,
601 Washington Building,
Los Angeles, California.

Gentlemen:

Agreeable to your request, I have carefully examined the patent issued to Mr. M. J. Trumble, for Crude Petroleum and Natural Gas Separator, application filed November 14, 1914, numbered 1,269,134, Patented June 11, 1918, and the drawings of the new gas trap invented by Mr. D. G. Lorraine of Los Angeles, California, and I find that there is no infringement of the Trumble patent by the apparatus invented and designed by Mr. Lorraine.

There might exist some doubt as to the third and broadest claim of Trumble, which claims "an expansion chamber having a surface adapted to sustain a flow of oil thereover in a thin body," but this is limited by "pressure maintaining means arranged and adapted to maintain a pressure on one side of the flowing oil."

This pressure maintaining means seems to be only supported in the Trumble patent by the ordinary oil well pump—and it is not understood how the pressure can be maintained on one side of a film of oil,—and as the baffle plate or other spreading means of Lorraine's does not form a thin film and no pressure can normally exist in Lorraine's trap on one side of the flowing oil, this claim is not infringed.

This, as well as the other limited and narrow claims of Trumble, will no doubt have to be interpreted by some court action and judicial authority, before they can be properly construed, and Trumble distinctly shows and describes a velocity reducing means which would reduce pressure.

However, the close proximity of the margin of the first cone to the walls of the chamber might cause a back pressure, but would also interfere with a film on the surface thereof as claimed.

Also, Lorraine merely breaks up the direct flow of his oil stream by striking an inclined baffle plate, the margins being so distant from the sides of the chamber as not to form a film on the walls and a large portion of the baffle being normally immersed in oil and only extending over a small portion of



the total area of the chamber, thus preventing a one-sided pressure as claimed by Trumble.

As to the substitution by Lorraine of other deflecting discharge to break up the flow of oil instead of the baffle plate, all doubt is eliminated as there will be no surface capable of "sustaining a film of oil."

Respectfully,

W. P. KEENE.

[Endorsed]: E—113—Eq. Townsend et al. vs. Lorraine. Defendants' Exhibit "P." Filed March 27, 1922. Chas. N. Williams, Clerk.

No. 3945. United States Circuit Court of Appeals for the Ninth Circuit. Filed Jan. 26, 1923. F. D. Monckton, Clerk.

No. 3945.

IN THE

# United States

# Circuit Court of Appeals,

FOR THE NINTH CIRCUIT.

David G. Lorraine,

*Appellant,*

*vs.*

Francis M. Townsend, Milon J. Trumble and Alfred J. Gutzler, Doing Business Under the Firm Name of Trumble Gas Trap Company,

*Appellees.*

## APPELLANT'S OPENING BRIEF.

WESTALL & WALLACE,
(Ernest L. Wallace and Joseph F. Westall)
By JOSEPH F. WESTALL.
*Solicitors and of Counsel for Defendant-Appellant.*

Parker, Stone & Baird Co., Law Printers, 232 New High St., Los Angeles.

**No. 3945.**

IN THE

# United States Circuit Court of Appeals,

## FOR THE NINTH CIRCUIT.

David G. Lorraine,
*Appellant,*
*vs.*
Francis M. Townsend, Milon J. Trumble and Alfred J. Gutzler, Doing Business Under the Firm Name of Trumble Gas Trap Company,
*Appellees.*

## APPELLANT'S OPENING BRIEF.

## STATEMENT OF THE CASE.

This is an appeal from an interlocutory decree granting an injunction entered by the District Court for the Southern Division of the Southern District of California, which decree and injunction was based upon a finding that Letters Patent No. 1,269,134, granted to Milon J. Trumble, assignor of one-third to Francis M. Townsend, and one-third to Alfred J. Gutzler, (appellees here; plaintiffs in the District Court), on June 11, 1918, for Crude Petroleum and Natural Gas Separators (see Book of Exhibits, page 3, for specifications and drawings of the patent in suit) were valid; and that claims 1 to 4 thereof had been infringed by appellant, David G. Lorraine (defendant in the trial court). Said decree also found that claim 13 of said patent had not been infringed; but this finding being in favor of appellant, and no cross appeal having been taken, no question as to claim 13 is here presented.

## The Patent in Suit.

Inasmuch as the device of the patent in suit (Book of Exhibits, page 3), is quite simple, and its mode of operation easily understandable without special knowledge of the art to which it relates, the following exposition would be superfluous if it did not segregate the relevant matters and throw the spotlight of emphasis upon those things which we contend to be controlling of the issues on this appeal. The Trumble separator consists merely of a closed tank or vessel into which a mixture of oil and gas, under pressure as it flows from an oil well, is conducted by a suitable pipe. Such gas-laden oil is sprayed in at the top of said chamber upon the apex of a cone-shaped plate, and flowing down over the surface of said cone and on to the side walls of the chamber is, during the process of separation, spread out evenly on the plate and the walls of the separator in a thin film. This spreading out permits the gas to become disentangled from the oil. The oil level in the bottom of the chamber is maintained by a float-controlled outlet valve. The separated gas is conducted out of the chamber through a valve controlling the gas outlet. The form of construction of these valves will be found of little materiality on this appeal.

Emphasis is placed by the patentee on the maintenance of pressure in the chamber during the process of separation; thus, at line 120, page 2, of the specification (Book of Exhibits, page 6), Trumble says:

> "In the case of a wet gas, or a gas saturated with an oil of a lighter series, such as gasoline, it is desirable to maintain as high a pressure within the expansion chamber as may be practicable for the reason that a large amount of such light series will thereby be compressed and remain with the oil."

And again at page 3 of the specification, line 30, we find the patentee stating:

> "It will be noted that the action upon the oil while flowing down the wall of the expansion chamber in a thin film under pressure permits the free, dry gas to readily escape therefrom, while the pressure exerted upon the oil surface, backed by the wall of the chamber holds the lighter liquids, such as gasoline, in combination with the oil body, and I desire to be understood as pointing out and claiming this action as being of great benefit to the crude oil derived from the well on account of keeping the gasoline series in combination with the main body of oil."

## Essence of the Trumble Alleged Invention.

When a client places the matter of an application for a patent in the hands of his solicitor, and the art to which it relates happens to be (like in the case at bar as will later appear), an exceedingly well developed one, the patent solicitor, if he is to efficiently serve his client must segregate in his own mind the narrow features of supposed novelty and invention inherent in his client's conception, whether is be a single element, a mode of operation, or a relationship between parts leading to a new result. In other words, he must see and recognize the *spirit* of the supposed invention before he can clothe that spirit with the body of a patent claim. In like manner before a court can intelligently construe a patent claim so as to pass upon issues of validity and infringement, it should disentangle itself from the maze of words and, looking through and around the language of the claims must discover the actual contribution to the art (if it be such), which the claim was designed to protect. Unless it does this, the court cannot intelligently avail itself of such important aids to construction as "the light of the prior art" (Walker on Patents (5th Ed.), page 243,, Sec. 184, and cases); "the position the patentee took before the Patent Office" (decision of this court in Selectasine Patents Co., *et al.*, v. Prestograph Company, 282 Fed. 223); nor is the court in position to construe so as to "cover the real invention" (if there happens to be any invention) (Walker on Patents (5th Ed.), page 239, Sec. 182, and cases);

or "so as to save the invention" (if it should transpire that there has been a valuable contribution to the art); nor can the court interpret the claims so as to give effect to *their spirit and intent* as distinguished from their letter." (Westinghouse v. Boyden, 170 U. S. 537.)

Now what was the essence of alleged invention here in issue? What theory of novelty underlies the grant of the patent in suit? Study and analysis of the specification and drawings of Trumble, and particularly of the proceedings upon which such patent was based (Plaintiff's Exhibit File Wrapper and Contents of Patent Suit; Book of Exhibits, page 9), as well as of the evidence and proceedings upon the trial enables us to answer such questions briefly and positively, and with no fear that the most careful consideration of this court will show our definition to be in any material respect inaccurate. The essence of supposed invention of Trumble consisted of means for spreading the *whole* body of the oil in a *thin film, equally* around and over, not only the surface of an *imperforate* conical spreader-plate, but upon all of the vertical walls of the separator above the oil level and below the periphery of such cone-shaped spreader-plates; and while so spread out on such extended surfaces (necessarily extended in order that *all* the oil might be reduced to a *thin film*) subjecting said oil-film, backed up by such surfaces, to pressure; that is to say, squeezing or pressing the oil-film against the surfaces of the imperforate conical plate and vertical separator walls over which it was designed to flow.

As we shall make clear to the court in our discussion of the prior art and the admissions of the file-wrapper contents of the patent in suit in connection therewith, unless the claims in suit are construed so as to cover only a device in which *all* the oil is spread out in a *thin film of substantially equal thickness* over *all* the separator's vertical walls as well as over an imperforate conical plate or its equivalent, and in which the oil-flow upon such surfaces is sustained, that is to say, in which the oil does not leave such surfaces in any broken-up condition, and does not fall, in whole or in part, in drops or streams—unless the patent is so construed, the patentee will now be granted by judicial construction and decree those things which he expressly disclaimed during the progress of his patent application through the patent office as his contribution to the art, and which are uncontrovertibly shown by the record here to be old; and the public will be deprived of those things which it has purchased from prior inventors in this field at a cost of many seventeen-year grants—in short the public domain will be depleted to award to plaintiffs that to which they are not entitled.

## Errors Assigned and Issues Outlined.

The Assignment of Errors [Transcript of Record, page 549], attacks generally and specifically not only the grant of an injunction, but all findings of the trial court, express or implied, of invention, novelty, utility—in short, validity—of the patent in suit, as well as those findings as to infringement, upon any of which the decree of injunction appealed from might have been predicated.

In the interest, however, of sucinctness and clarity, the following informal explanation of the reasons of this appeal may be helpful: There are in this record fully proven patents admittedly granted long prior to the alleged date of invention of Trumble, responding literally to each of the claims in suit; that is to say, those claims are in all respects correctly descriptive of the disclosures of such prior patents. Now, our last statement might appear at first sight to be equivalent to an assertion that such claims are anticipated. Such, however, is not necessarily the case (although such matter is for the decision of this court); for while the claims do read *literally* on such prior devices, they are susceptible of a very narrow interpretation, and in the light of admissions in the nature of disclaimers in the patent office on the Trumble application—by a close and delicate analysis of the prior art, segregating certain features of novelty of Trumble—(assuming that such exceedingly narrow contribution constitutes patentable invention) and construing the claims so as to cover that only, the court may save such claims from anticipation, but by the same token must necessarily decide in favor of defendant on the issue of infringement; for any interpretation sufficiently narrow to save the Trumble claims in suit from the prior art will exclude defendant's devices as infringements. The dilemma of plaintiff, therefore, is: The claims are either anticipated and void because they literally and correctly describe the disclosures of prior patents of record, or, construing them to read only upon the specific structure of the specification and drawings, there is no infringement.

The court will instantly see that the vital question is: What is the state of the prior art? Most naturally the court will turn to the opinion of the trial court for preliminary light upon that subject. Now, right here we place our finger upon the vital and controlling error complained of on this appeal: The discussion of the prior art is contained in one short paragraph in the trial court's opinion. EVERY STATEMENT IN SAID PARAGRAPH OF THE TRIAL COURT'S OPINION IS RADICALLY MISTAKEN AND ERRONEOUS AS TO THE FACTS. BUT FOR SUCH ERROR THE DECREE APPEALED FROM SURELY WOULD NOT HAVE BEEN ENTERED. The paragraph referred to reads as follows [See Trial Court's Opinion, Transcript of Record, p. 530, near bottom p. 538]:

> "The patents introduced as showing the prior art are readily disposed of. Exhibits 'E,' the McIntosh patent, 'F,' the Taylor patent, 'G,' the Barker patent, and 'J,' the Newman patent, all inject the oil from the well in the form of a spray, having the effect to reduce it to a finely divided condition, and the gas is thus permitted to escape. None of them are provided with baffle-plates except Newman, but the oil does not reach them

except as sprayed upon them, and I think none of these patents contain the element in combination of pressure within the chamber. All these patents are obviated in their evidentiary effect by the restrictions of complainants' specifications and claims as made before the examiner. Exhibit 'H,' the Gray patent, is subject to the same criticism. The oil is there precipitated upon perforated cones, and only slightly if at all, flows down the wall of the chamber or other surface. Exhibit 'I,' the Cooper patent, injects the crude oil tangentially to the wall of the chamber, and causes it to flow down the wall more or less, and the gas escapes upwardly and passes out in a take-off at the top of the chamber. But this patent in combination contains no element of pressure. It must be observed that we are dealing with a combination patenт, and all the elements must be read with reference thereto."

Now, in order that the court may understand the trend of our argument to follow, we point out briefly the errors complained of in the foregoing quoted paragraph from the trial court's opinion:

It is not true that Exhibit "E," McIntosh patent (Book of Exhibits, page 110); "F," Taylor patent (Book of Exhibits, page 115); and "J," the Newman patent (Book of Exhibits, page 136); all inject the oil from the well in the form of a spray, having the effect to reduce it to a finely divided condition. In fact none of them do this. Exhibit "G," the Barker patent (Book of Exhibits, page 121), is the only one that sprays the oil in to the chamber in a finely divided or broken up condition.

with baffle plates except Newman; on the contrary, the fact is that ALL of them except Barker are provided with baffle plates. Further, in *none* of those which are provided with baffle plates is the oil sprayed upon them. Instead of *none* containing the element of pressure within the chamber they ALL, under the admission of defendant's expert as well as the uncontrovertible statements in the specifications themselves and other evidence of record, are shown to contain the element of pressure within the chamber.

It is not true that all or any of these patents are obviated in their evidentiary effect by restrictions of the specifications and claims as made before the Patent Office; for Exhibit "E," McIntosh patent (Book of Exhibits, page 110), for instance, *containing the precise combination of elements of each of the broadest claims of the Trumble patent in issue was entirely overlooked by the Patent Office.* Exhibit "I," the Cooper patent (Book of Exhibits, page 133), *was also overlooked by the Patent Office, and likewise responds to said claims and others.* The same is true of Exhibit "J," Newman patent (Book of Exhibits, page 139). If we neglect to read into the claims in suit the qualifications in the nature of disclaimers of the file-wrapper contents of the Trumble application, Exhibit "H," Bray patent (Book of Exhibits, page 127), is a *complete anticipation of all claims now in controversy.* Such claims read literally upon said Bray disclosure. The same is true of Exhibit "F," the

Taylor patent (Book of Exhibits, page 115), as well as others which will be discussed in our argument, the foregoing being simply intended as a preliminary outline showing the nature of our refutation of the trial court's findings as to the prior art.

The trial court found [Transcript of Record, page 542] that "defendant's patent" (which patent is found in the Book of Exhibits at page 66, being letters patent granted to D. G. Lorraine, reissued Nov. 8, 1921, No. 15,220) infringes claims 3 and 4 of the Trumble patent in suit. Of course, the court did not mean this literally, as, manifestly, the ownership or issuance of a patent does not constitute infringement. Obviously, what the court meant was the making or selling or using of a device constructed in accordance with defendant's patent infringes. Now, as a basis for such finding, the trial court explained in its opinion the construction, as the court understood it, of the device of defendant's said patent [Transcript of Record, page 537]. Here again, a most serious error is to be found. The court's reading of the drawings and description of defendant's patent is vitally mistaken; defendant's patented device does not contain or illustrate or describe any such baffle-plates as the court mentions in connection with its description, nor is oil divided into streams or projected upon any such baffle-plates as set forth in the trial court's opinion. We earnestly believe that if the trial court had fully grasped the meaning of the description and drawings, the decree appealed from would not have been entered.

Another error of sufficient importance to justify brief preliminary notice: In the trial court's opinion [Transcript of Record, middle of page 541], the court said: "Utility has been abundantly proven by the success achieved by plaintiffs' device." If this intended to imply that there is any evidence in the record tending in any degree to show that Trumble contributed anything whatever of value or utility to the art, it is, as we shall later show, clearly erroneous. So far as the evidence discloses, the device illustrated and described in the patent in suit was only useful *insofar as it incorporated means and devices long known and used in the art for identical purposes.* There is no evidence whatever in the record tending to show that any possible difference between the device of the Trumble patent and the prior art, either alone or in combination with other devices as set forth in the claims was in any respect advantageous or had any utility. The only basis of the finding of utility, therefor, was presumption—not evidence.

In the foregoing brief discussion of the trial court's opinion, we have endeavored to point out merely the keystone error of the structure. If in the following argument it shall appear that we have succeeded in doing this, the remainder will fall of its own weight.

## ARGUMENT.

### "Pressure Means," a Vital Element of the Trumble Combination is Disclosed—Not in None (as Found by the Trial Court) But in ALL the Prior Art Patents in Evidence.

In its opinion, after referring repeatedly to the part of the Trumble specification relating to pressure and pressure maintaining means, and quoting particularly [Transcript of Record, page 533], the patentee's explanation of the use, function, and value of maintaining pressure on the thin film of oil on extended surfaces within the separating chamber, the trial court [Transcript of Record, two-thirds down page 535], says:

> "The theory of the patentee is obviously that, pressure being maintained, the dry gas will readily escape from a thin film or body of oil passing down and against a wall or other surface, without at the same time taking off the lighter liquids, such as gasoline, which will yet remain in the crude oil and add to its value."

Having thus emphasized that pressure maintaining means (leaving out of consideration for the moment, the spreading of the oil in a thin film) is, under Trumble's theory of his alleged invention (which theory was sustained by the trial court), a vital and controlling element of his combination, the trial court [Transcript of Record, two-thirds down page 538], dealing with the question of anticipation or limitation by the prior patented art, remarks, "The patents introduced as showing the prior art are readily disposed of," and after discussing several patented structures and their mode of operation, proceeds, "I think none of these patents contain the element in combination of pressure within the chamber," and later in the same paragraph [Transcript of Record, page 539], referring to defendant's Exhibit "I," Cooper patent (Book of Exhibits, page 133), the court concludes:

> "But this patent in combination contains no element of pressure. It must be observed that we are dealing with a combination patent, and all the elements must be read with reference thereto."

Thus, it is obvious that a pivotal point—no doubt the very crux of the trial court's decision as to the pertinence of the prior patented art of record is that, bearing in mind that to establish technical anticipation (not want of invention) the combination of elements of a patent claim in question as an entity, that is to say, the entire combination of each claim, must be disclosed or sufficiently implied in the description or drawings of the prior patent, and that the all-important or vital element of Trumble, namely, pressure maintaining means, is not disclosed in any of the prior art patents, and that, consequently, such art is entitled to little or no weight on questions of validity and interpretation of the claims in suit.

Now, as we have heretofore in our statement of the case set forth, we contend that the trial court greatly erred in "thinking" as it is expressed in the opinion, or assuming or finding that none of these prior art patents disclosed pressure means, for our contention is that ALL *of them, either show or describe specifically such means or clearly imply the same, or under the uncontradicted testimony of plaintiffs' expert during normal operation they all must and do have such pressure means;* and furthermore, it has been recently judicially decided by a court of high authority (whose reasoning we believe will be found unassailable) that pressure in devices of the kind in controversy is a natural incident necessarily always present where gas and oil under normal pressure as it flows from a well is conducted into a closed vessel having as an outlet a pipe line to convey the gas to a place where it is to be used, and cannot of itself or in combination with other elements impart validity to a patent.

The importance and sharpness of the issue thus presented seems to justify its selection for first consideration. If we are correct in our contentions, it will be found that little remains to support the trial court's decision as to the pertinence of the prior art of record —and what little remains can be briefly and conclusively disposed of.

As will no doubt be immediately recognized by the court, unless the mixed oil and gas as it flows from the well is under pressure it will not flow. Indeed the specification of the patent in suit states and assumes this in such expressions, for instance, as at line 20, page 1, of the specifications, where application of the alleged invention is described in connection with "wells where high pressure exists," also at line 21, page 1, of said specification, reading as follows:

> "My invention is also adapted to reduce the velocity and equalize the delivery of oil from wells in which the pressure of gas causes the oil to flow in gushes or by heads."

The source of any pressure in the tank is obviously the well. At Transcript of Record, page 86, this is admitted by plaintiffs' expert, Paul Paine, as follows:

> "Q. Then it finally comes down to a question of what is the pressure of the gas in the well; in other words, you must have pressure of gas in a well before you can get a pressure built up in the trap, must you not?
>
> A. Oh, yes, if there is no gas. If production is coming from the well there must be a pressure to push it in there before one can build up a pressure in the trap."

This pressure from the well normally is considerable. Plaintiffs' expert, Paine, [Transcript of Record, two-thirds down page 74], testified that under certain tests he made, pressure was 380 pounds per square inch at the top of the casing.

The United States Circuit Court of Appeals for the 8th Circuit, Judges Carland and Lewis, circuit judges, and Munger, district judge, in a very recent case decided at the September term, 1922, entitled Standard Oil Company, *et al.,* appellants, v. Oklahoma Natural Gas Company, appellee, reported at 284 Fed. 469, near the top of page 473, found that the usual pressure of natural gas as it flows from the well through pipe lines to the cities for consumption averages 300 pounds per square inch.

The exact amount of pressure, however, is not material for present purposes, for we merely now wish to point out the obvious fact that if the gas under pressure is admitted into a closed vessel, such as plaintiffs' or defendant's devices, and pressure is maintained at the inlet of said vessel, there must be pressure in the vessel. It is also equally clear that if an outlet is provided of less capacity than the inlet there will still be pressure in the vessel. We believe it equally a matter of which the court will take judicial notice that if the diameter of the outlet pipe for gas is equal to the diameter for the inlet pipe (or within bounds even of greater diameter) there would still be pressure in the vessel or separator, because pipe friction would impede the outflow. However, we need not rely upon judicial notice of these matters for the uncontradicted testimony of plaintiffs' expert, Ford Harris, is clear and convincing. Mr. Harris testified [Transcript of Record, near bottom of page 175]:

> "Q. Now, suppose this gas was being carried a long distance, there would be more or less pipe line friction there, would there not?
>
> A. Why, certainly.
>
> Q. And that would have a tendency to cause a back-pressure in the gas trap which would show on the gas gage, would it not, or the pressure gage?
>
> A. Yes.

A. So far as maintaining the pressure on this trap is concerned, it is quite immaterial; that is, a pressure could be maintained either by a valve put directly at the trap or by a valve on the end of the line or perhaps by a contracted opening at the end of the line, which might be an absorption plant, or where it is delivered to the boilers, or anywhere. It is not necessary, to maintain the pressure on the trap, to put the pressure-maintaining valve at the trap; it can be placed at a distance."

It is therefore clear that pressure will always be maintained in a gas and oil separator connected by a suitable pipe to a flowing well, by reason of pipe friction if the outlet is through a pipe line; because of a partial choking of the outlet if there is a contracted orifice or a reduction nozzle or a valve at any place on the line; or if the gas is delivered to an absorption plant or to boilers for use.

Of course, the purpose of separating the gas from the oil is to use it, and in order that it may be used it must be conveyed somewhere, and if it is conveyed there must be a pipe line of some kind and an absorption plant or other apparatus at the end of the line. Thus, the usual and ordinary process of conveying the gas from any kind of gas separator and using it would maintain a pressure in the trap.

The foregoing obvious facts were recognized judicially in the decision of the Circuit Court of Appeals mentioned *supra* (Standard Oil Company, *et al.*, appellants, v. Oklahoma Natural Gas Company, appellee, 284 Fed. 469), when at page 473, the court says:

> "It appears from the evidence in the record that the usual pressure of natural gas as it flows from the wells through the pipe lines to the cities for consumption averages 300 pounds per square inch, and that this pressure is fixed by the requirements of the transportation system. That the pressure is not created or fixed for the purpose of extracting the gasoline content of the natural gas by absorption, but is fixed or created for the purpose of maintaining pressure upon the pipe line system that will transport the gas to the distant markets. It further appears from the evidence that the pressure used on natural gas for the purpose of obtaining gasoline by absorption is whatever pressure that exists on the transportation system where the absorption plant is located.
>
> "The claims of the patent require the natural gas to be under a pressure of not less than about 30 pounds to the square inch above atmospheric pressure, but Saybolt knew from the law of Henry the function that pressure would perform in the absorption of the gasoline vapor from natural gas. The pressure of the natural gas already existed to the knowledge of every one. If the natural pressure of natural gas as it comes from the wells is not sufficient to properly carry the gas to the cities for consumption, then compressors are used to increase the pressure and force the gas to its destination."

At page 475, the court further holds:

> "We have heretofore shown that Saybolt had nothing whatever to do with applying pressure to the natural gas, but simply took the gas at high pressure in the lines as he found it, and that in a gas transportation system a high pressure is used for transportation purposes solely. We do not think that the language above quoted distinguished Saybolt's claims from the prior art, especially when we consider that he had nothing whatever to do with the pressure which existed prior to his entrance into the field."

Pressure thus being necessary, and in fact inherent in the very use which the apparatus is intended to be put, it would seem to be no more necessary to illustrate or describe in the patent drawing or specification of the prior art means for maintaining pressure, such as the long pipe line, an absorption plant, boilers, contracted orifices at any place on the line (Such contracted orifices might be in the form of a valve for shutting down all but a small opening), than it would be for Trumble in his patent to illustrate the bottom of the well and the surrounding source of supply of initial pressure.

It should be borne in mind that Trumble does not claim any particular *amount of pressure* to be maintained within the chamber, nor does he *claim* any specific *means* for maintaining that pressure. Any pressure, variable or constant, above atmospheric pressure, from one ounce to hundreds of pounds would satisfy his claims.

There has not been the faintest suggestion in either evidence or argument on behalf of plaintiff that any novelty or invention resided in the particular means, *per se,* disclosed in the Trumble drawings for maintaining pressure, namely, the weighted safety valve 11; on the contrary, the failure of Trumble to describe or illustrate the construction (except by one small perspective outside view) or mode of operation of this valve (not to speak of his failure to claim it as a separate element) precludes him from so contending; for, as Walker on Patents (5th Ed.), page 220, Sec. 175, quoting from Columbia Motor Car Company v. Duerr & Company, 184 Fed. 893, remarks:

> "A patent is granted for solving a problem, not for stating one. Its description must explain the invention itself, the manner of making it, and the mode of putting it in practice. In the absence of knowledge on these points, the invention is not available to the public without further experiments and further exercise of inventive skill."

Faced by the dilemma of insufficiency of disclosure, Trumble is, therefore, driven to agree with us that the construction, application, use, function, and effect of the ordinary weighted or spring actuated safety or pressure valve, like or similar to this valve 11 (or performing the function stated for that valve—and Trumble only describes the function) was a matter of common knowledge among all those skilled in the art —in fact a matter of judicial notice, being common knowledge—long prior to the alleged date of invention of Trumble (admitted to be not earlier than his application date, Nov. 9, 1914); and to merely show the outside of such a valve and state the purpose for which it was intended to be used was sufficient to enable anyone skilled in the art to select from the prior art a suitable device for such purpose (or, as plaintiffs' expert has suggested, to adopt other pressure maintaining means (the form not being material), such as a pipe line with its friction, a contracted orifice at any place on the line, an absorption plant or other apparatus such as boilers for use of the gas.

Almost any good dictionary published prior to November 9, 1914 (Trumble date of alleged invention), will be found to contain full description and illustration of such forms of pressure valve. Thus for instance, in Funk & Wagnall's New Standard Dictionary (our copy bearing copyright notice of 1913), under "Safety valve," page 2157, is found not only a full description, but also a number of clear illustrations of such form of weighted pressure valve and its mode of construction, and again under "valve," page 2629 of the same authority, is found a reference and description of many other forms of valves which obviously could be used for controlling such pressure, notably a valve described as "back pressure valve."

There can be no doubt of the court's authority to take judicial notice of the meaning of words appearing in the dictionary (see Jones on Evidence, civil cases; thin paper pocket edition, section 130), and indeed the Court of Appeals for the Third Circuit in Werk *et al.* v. Parker, 231 Fed. 121 (Buffington, McPherson and Woolley, circuit judges), in a patent case took judicial notice not only of such matter found in dictionaries, but of encyclopedias, and periodicals, as well as special works relating to the art therein involved. (See also Walker on Patents (5th Ed.), page 570, Sec. 505; Jones on Evidence (thin paper pocket edition) Sec. 128, relating particularly to patent causes.)

That "means to maintain pressure" *per se,* is to be deemed old for the purpose of this proceeding is, moreover, conclusively presumed as a matter of law from the fact that Trumble claimed this element only in combination and not separately. Thus, Hopkins on Patents, page 214, Sec. 137, reads as follows:

> **"Sec. 137, Rule XXXVII.. In a combination claim each element is conclusively presumed as a matter of law to be old, whether old in fact or not.**
>
> The foundation of this rule is, that if, among the elements, there is one which is itself patentably new, it must be separately claimed, or it is dedicated to the public by its inventor's failure to claim it. Thus Judge Baker has said:
>
> 'The failure to claim either one of the elements separately raises the presumption that no one of them is novel.' (Citing Campbell v. Conde Implement Co., 74 Fed. Rep. 745.) And Judge Woods has said: "When a combination is claimed, there arises an implied concession that the elements are old, and not separately patentable." (Citing Hay v. Heath Cycle Co., 71 Fed. Rep. 411-413, 18 C. C. A. 157.)' "

It is, therefore, we submit, to be deemed established, (1) as a presumption of fact arising from Trumble's failure to illustrate or describe the construction of this pressure valve 11; (2) as a presumption of law resulting from his failure to claim it separately as an element (3) as the only horn of the dilemma to sustain the patent on the question of sufficiency of disclosure; (4) from facts of which the court should take judicial notice—that the construction, function, and mode of operation of Trumble's means for maintaining pressure, or any of the equivalents thereof described by plaintiffs' expert, Ford Harris, was old and well known in the art prior to the alleged Trumble invention.

It is elementary in the law of patents that it is not necessary to illustrate or describe those elements or features which in the attempted utilization of the disclosure any mechanic by the exercise of the usual skill of his calling would be able to supply. Thus, Walker on Patents (5th Ed.), page 214, Sec. 174, says:

> "The description of the invention, which forms a part of every specification, is required to set forth that invention, and the manner and process of making and using it, in such full, clear, concise, and exact terms as to enable any person skilled in the art or science to which it appertains, or with which it is most nearly connected, to make and use the same; and in case of a machine, the description is required to explain the principle thereof, and the mode of applying that principle which the inventor believes to be the best." (Citing Revised

Statutes, section 4888; Continental Paper Bag Co. v. Eastern Bag Co., 210 U. S. 405, 1908.)

Quoting Mr. Justice Bradley in Loom Co. v. Higgins, Walker on Patents (5th Ed.), page 215, Sec. 174, says:

> "If a mechanical engineer invents an improvement on any of the appendages of a steam-engine, such as the valve-gear, the condenser, the steam-chest, the walking-beam, the parallel motion, or what not, he is not obliged, in order to make himself understood, to describe the engine, nor the particular appendage to which the improvement refers, nor its mode of connection with the principal machine. These are already familiar to others skilled in that kind of machinery. He may begin at the point where his invention begins, and describe what he has made that is new, and what it replaces of the old."

The contributions of mechanical skill necessary to the utilization of a patented device being thus impliedly contained in every patent description, with much stronger reason it would seem, elements or features not deliberately added by the skilled mechanic to correct or complete the patent description, but resulting as necessary incidents from the placing and connecting of such device in the manner in which it was designed to be applied and operated, should be held to be part of such description.

*Even then if the trial court were correct in its assumption that the element of pressure is not disclosed in any of the prior art patents of record, it would not follow that the value of such evidence on the question of anticipation or as compelling a close and narrow interpretation of the claims would be in any degree lessened. In other words, if pressure means are necessarily implied in a common and ordinary and intended use of the prior art devices, such means need not be expressly described or illustrated.*

The purpose of our thus pressing upon the attention of this court the necessity of taking cognizance of those things which cannot be escaped in the practical utilization of such prior art devices is merely to most glaringly exhibit the fallacy of one of the trial court's most vital conclusions. But we need not rely upon any such necessarily implied pressure means: patents which we shall now proceed to discuss, admittedly granted long prior to any alleged invention of Trumble, *expressly, and unequivocally show or describe such pressure means* in combination with all the other elements of the Trumble claims is controversy.

## COOPER PATENT.

---

**Cooper Patent (Granted Over Eight and One-Half Years Prior to Trumble's Alleged Date of Invention) (Defendant's Exhibit "I", Book of Exhibits Page 133) Which the Trial Court Brushed Aside (Transcript of Record, Page 539) as of Little Pertinence, on the Assumption That it Contained no "Element of Pressure" Distinctly Tells Us That the Device is to be Included in a Pressure System, Mentions Pressure Repeatedly, Finally Stating, "a High Pressure of Gas May Thus be Maintained in This Cylinder." This Patent was Overlooked by the Patent Office. Claims of Trumble Read Literally on This Reference; and the Trial Court was Clearly Mistaken in Implying (if any Such Implication was Intended) That all the Oil was not Spread in a Thin Film on the Walls of the Cooper Expansion Chamber.**

The trial court's entire finding with reference to the pertinence of the Cooper disclosure [Transcript of Record, page 538, and near top of page 539], is as follows:

> "The patents introduced as showing the prior art are readily disposed of. * * * Exhibit 'I', the Cooper patent, injects the crude oil tangentially to the wall of the chamber, and causes it to flow down the wall more or less, and the gas escapes upwardly and passes out in a take-off at the top

of the chamber. But this patent in combination contains no element of pressure. It must be observed that we are dealing with a combination patent, and all the elements must be read with reference thereto."

A brief examination of the very short patent description of Cooper in connection with Fig. 1 of his drawings (Book of Exhibits, page 133), will make clear the construction and mode of operation of this separator. (We also shall exhibit to the court on the argument a model of the device made to the scale of Fig. 1 and have prepared for use an oral argument like models of each of the prior art devices to be referred to.)

Briefly, the mixed oil and gas as it flows from the well enters through the pipe B, the end of this inlet pipe being flattened out to fit a slot-like opening into the Cooper separating or expansion chamber A. The manner in which this inlet pipe opens into such expansion chamber is clearly shown in Fig. 2, which also shows the mode in which the wearing plate a' is disposed with reference to the inlet opening of the chamber. The oil and gas is thus injected, as the trial court correctly found, tangentially on the wall of the chamber; but the trial court is obviously incorrect in the implication (if it is intended as an implication) that possibly less than all of the oil flows down the wall; for clearly all the oil is directed against the wearing plate a' and wall and all *must* be spread upon such wall and if all of the oil is spread on the wall, of course it must flow down the wall. If, however, the mere possibility that some of the oil may leave the wall and fall to the bottom of the chamber (instead of flowing down the sides of the chamber) should exclude or weaken this reference as an anticipation, then, with stronger reason (as the court will see when it considers the question of infringement), *the actual fact* that a very large proportion of the oil is not spread on the wall of defendant's device, but falls to the bottom of the chamber, must conclusively establish that defendant does not infringe.

Does Cooper disclose pressure? At line 21, page 1, of his specification (Book of Exhibits, page 134), Cooper tells us distinctly that his device is "included in a *pressure* system," and at line 22, page 1, of the specification he says that the constituents are separated "without affecting the general *pressure* in the system." At line 30, page 1, of his specification, Cooper says that the object of his invention is "to efficiently separate and deliver the gas, oil, water and sand from a *pressure system.*" At line 59, page 1, of the specification, he says that the gas outlet pipe F leads to a compressor. (Remember that Ford Harris says that a pipeline, boiler, or absorption plant would be equivalent to a valve for the purpose of maintaining pressure.) At line 61, page 1, the Cooper specification discloses clearly that *pressure* must exist by the statement, "gas, oil, water, silt, and sand are *forced* from near the bottom of the well by the compressed gas and passed through the pipe B into the cylinder A." At line 63, Cooper further says: "Entering the cylinder *under pressure* tangentially to its circumference * * *." At line 73, page 1, of the specification, Cooper says:

> "When these constituents (oil, sand and water), accumulate to raise the oil to a certain level the float c rises and opens the cock c, thereby permitting the oil *to be forced out by the internal pressure in the system* * * *."

At line 83, page 1, of his specification, Cooper finally says:

*"A high pressure of gas may thus be maintained in this cylinder."* (Italics ours.)

Certainly, therefore, Cooper contemplated that his device was to be used in a pressure system, and that separation was to take place under pressure. As the court has seen from a consideration of the decision of the Circuit Court of Appeals for the 8th Circuit in the case of Standard Oil Co., *et al.,* v. Oklahoma Natural Gas Company, 284 Fed. 569 (cited and considered *supra*), pressure, being a natural and necessary incident of a gas line, it would be difficult to connect the Cooper separator (or any other separator) to a flowing well and not have pressure in the separating chamber.

Surely, if gas and oil come into this vessel under a normal pressure of 300 to 400 pounds per square inch, it is obvious that the outlet F being of the same diameter as the inlet there would be some pressure in the system (and Trumble, as before pointed out, calls for no particular amount of pressure), even if

Cooper did not repeatedly tell us that it was to be used in a pressure system. If this is not obvious, then the court should be guided by the uncontradicted testimony of plaintiffs' expert, Ford Harris, who states that pipe friction, or an absorption plant, or boilers at the end of the line would maintain pressure. Remember also that back pressure valves, such as safety valves and others illustrated in any old dictionary (of which we have seen the court should take judicial notice without proof, being matters of common knowledge) were old, and their construction, function, and mode of operation well known long before the pretended Trumble invention. Indeed, as we have seen, it is conclusively presumed as a matter of law that such element as pressure maintaining means is old because of Trumble's failure to claim it separately.

Suppose that a skillful mechanic in this art, before November 9, 1914 (Trumble's date of alleged invention) had desired to utilize the Cooper separator, and suppose he found that pipe friction or a compressor, or absorption plant, or contracted orifice, mentioned by plaintiffs' expert Harris did not build up the amount of pressure he desired. Applying the teachings of Cooper with understanding, and referring to those repeated instructions of Cooper to use his device in a *pressure system* and particularly to the statement, "a high pressure of gas may thus be maintained in the cylinder," and being of necessity, as a matter of law, charged with at least the common knowledge found in the dictionary that back-pressure valves of various forms were well known in the art, he would of course, add to the pressure already existing in the device by the employment of an old form of back-pressure valve or by one of the old methods or devices referred to by plaintiffs' expert.

If it was not necessary for the sufficiency of the disclosure of Trumble to more than show the outside of a pressure-valve and indicate its function, surely it was not necessary for the sufficiency of Cooper's disclosure to more than tell us nine or ten times that pressure was to be maintained.

It thus appears that the only reason given by the trial court (namely, because of the assumed absence of pressure in the Cooper separator), for not either declaring void or strictly construing the claims of Trumble in view of the Cooper disclosure, is based upon a mistake of fact.

**Trumble Claims Reading as They do Literally on the Cooper Disclosure, can Only be Saved From a Finding of Invalidity by a Most Strict Construction in the Light of the Specifications and Drawings of Trumble, as Well as of the Position Trumbell Took Before the Patent Office; That is to Say, by an Interpretation Implying or Reading Into the Trumble Claims Limitation not Distinctly Expressed Therein. Under any Interpretation Sufficiently Narrow to Avoid Anticipation by Cooper, Defendant (as Will Later Appear) Does not Infringe.**

Unmistakably, Cooper discloses that oil and gas separators for the identical purposes described by

Trumble, and employing those features decided by the trial court to be of the essence of the alleged Trumble invention, namely, the spreading of the oil in a thin film on the wall of the separator and subjecting such film to pressure as it flowed down such wall, were old seven and one-half years before Trumble even pretended to have invented them.

One of the wisest—and most just because of its wisdom—of rules for the interpretation of patent claims is to determine what if anything a patentee added to the art and then, if possible, to construe his claims so as to cover and protect that contribution. The intelligent application of this rule gives to the patentee the full measure of protection to which he is entitled, and also guards the public against unjust and unauthorized monopoly.

What did Trumble invent, if anything? Certainly, the theory of the trial court as to his contribution cannot be sustained in view of the Cooper disclosure.

Of course, the combination of the claims in suit is the technical definition of Trumble's alleged invention; and the question of anticipation by a prior patent must be determined by comparing the language of those claims with the disclosure of such prior patent. Comparing the Cooper drawings and description with claim 3 of Trumble, for instance, we find that Cooper discloses, "in an oil and gas separator, the combination of an expansion chamber" (Chamber A); "having a surface (wearing plate a' and wall of the chamber A) adapted to sustain a flow of oil thereover in a thin body"; "means for distributing oil on to such surface"; (slotted inlet opening a); "pressure maintaining means arranged and adapted to maintain a pressure ou one side of the flowing oil" (Oil outlet F and pipeline with its friction as described by plaintiffs' expert, Harris, or contracted orifice, valve, or compressor or absorption plant, described as being at the end of the line—in short any admittedly well known means to maintain the high pressure repeatedly spoken of throughout Cooper's specification); "withdrawing means to take gas from the chamber" (the gas outlet) "means for withdrawing old from the chamber" (pipe D and its connections).

The difference between claims 3 and 4 is slight and immaterial to the present comparison. Practically the same reading applies to claim 4.

Claim 1 of Trumble does not call for "means for maintaining pressure" or "pressure means" and a comparison will show that said claim 1, with the exception of the language of the claim requiring the oil and gas to be received in its "upper portion," exactly describes the Cooper separator.

We contend that to raise the oil inlet of Cooper so that instead of being in the middle portion of the expansion chamber, it would deliver the oil a little higher, would not involve invention. The only object of delivering the oil in the upper portion of the chamber is to allow a sufficient distance for it to flow over to effect separation. Patent drawings are not working drawings. (See Walker on Patents, 5th Ed., page

146, last paragraph of section 126 and authorities cited.) Such drawings are only intended to illustrate an invention, and obviously if it were found on actual test of the Cooper device (on account of large volume or pressure of the flowing oil, for instance), that a change of proportions or the relative raising of the oil inlet would improve operation—it would not require inventive genius to do this in view of the full disclosure of the intended action of the oil in the Cooper expansion chamber. The raising of the oil inlet would be a mere matter of degree not affecting the contribution to the art. Furthermore, the idea of injecting the oil in the upper portion of the chamber, in combination with the other elements of the Trumble claims was old, as will be shown by other prior patents, all of which are pertinent in connection with Cooper on this question of interpretation and invention.

The oil inlet of Cooper is very slightly below the middle of the separating chamber. Suppose that after the grant of the Cooper patent someone should have made an application for a patent in which he exactly copied the Cooper drawings and specifications except that he showed and described the Cooper inlet above the middle of the chamber (so that it might be described as being in the upper portion of the chamber as called for by Trumble claim 1), is it conceivable that the Patent Office would have allowed such an application, or if by any chance granted, that any court would sustain the same as involving patentable invention? We think the answer is obvious considering even the Cooper specifications and drawings alone; but when it is further shown that the arrangement of the inlet to inject oil and gas into the upper portion of the chamber was old long before Trumble as shown by numerous other prior patents (for instance, Defendant's Exhibit "E," McIntosh patent (Book of Exhibits, page 110); Defendant's Exhibit "F," Taylor patent (Book of Exhibits, page 115); Defendant's Exhibit "H", Bray patent (Book of Exhibits, page 127); Defendant's Exhibit "G", Barker patent (Book of Exhibits, page 121); Defendant's Exhibit "J", Newman patent (Book of Exhibits, page 139); many of which also contain the combination of the Trumble claims, we submit that any suggestion of possible invention in slightly raising the oil inlet of Cooper is too trifling for even a moment's serious consideration.

Claim 2, not containing the limitation as to the inlet of gas and oil in the upper portion of the chamber, reads literally on the Cooper device, to the same extent as do claims 3 and 4.

## Can Cooper be Avoided as an Antcipation of the Trumble Claims.

At line 33, page 1, of his specification, Trumble tells us (italics ours) that his invention consists *"in the arrangement"* as well as in the combination of parts thereinafter in his specification and claims set forth. In Plaintiffs' Exhibit 2, Trumble File Wrapper and Contents (Book of Exhibits, page 9), in a communication to the Commissioner of Patents, dated March 1, 1915 (Book of Exhibits, middle of page 35), Trumble says:

> "Applicant's specification and claims are now written to define the *peculiar arrangement of applicant's device which is the means for causing the oil to be divided into a thin flowing body* * * *" (Italics ours.)

On the same page plaintiffs also objects to Barker as a reference because he says that in Barker "there *are no means shown for expanding the oil* in such manner that it may flow in a thin film * * *." (Italics ours.) On the same page Trumble likewise objects to Bray as a reference, because the oil flows downwardly in drops or *streams* rather than in a thin film (though Bray (Book of Exhibits, page 127), *does* show spreader cones having very substantial surfaces for the spreading of the oil and directing it upon the walls of the chamber.)

At Book of Exhibits, page 42, Trumble, in order to secure favorable action of the Patent Office, was obliged to define his invention very specifically, which he did as follows (italics ours):

> "Applicant's invention consists of a containing vessel, *an imperforate cone adopted to spread the whole body of the oil to the outer edge of the vessel,* and means for taking off gas from the interior of the cone near the center of the vessel."

With the "arrangement" of his elements thus emphasized— not only in the application proceedings before the Patent Office, but in the Trumble specification as finally allowed—and with his repeated references to the specific means constituting the most vital part of the arrangement for spreading the oil out in a thin film, namely, the spreader cones 22-22a, and bearing in mind that the use of such imperforate cones for spreading the oil in a thin film and spreading it out in such condition equally on all walls of the chamber is the only substantial difference to be found between Trumble and Cooper, we submit that to avoid Cooper as an anticipation (assuming for the moment invention in this infinitesimal contributiôn to the art over Cooper; and also overlooking the fact that in view of the art which we shall later call to the attention of the court, Trumble cannot be given credit for even this minute contribution) the court should construe the element of the Trumble claims, described as "means for spreading the oil over the wall of such chamber." (Substantially, all Trumble claims except claim 4) and the element of claim 4 described as "means within the chamber adapted to cause the oil to flow in a thin body for a distance" as applying strictly to the imperforate spreader cones 22-22a of Trumble or their clear equivalent. While Cooper contains a wearing plate a', this may be held not to be the equivalent of the Trumble cones because it does not spread out and direct equally around and over all the walls of the separator in the manner provided by the cones.

In urging that this is a proper interpretation of the Trumble claims, we remind the court that the law is well settled that the word "means" in a patent claim followed by a statement of function (to use the language of Walker on Patents (5th Ed.), page 138, Sec. 117a:

> "* * * will not include all means, mechanisms, or devices which can perform that function, but only those which are shown in the patent, and their equivalents." (Citing the following cases: Dudley E. Jones Co. v. Munger Mfg. Co., 49 F. R. 64, 1891; Williams v. Steam Gauge & Lantern Co., 47 F. R. 323, 1891; Colts Patent Firearms Mfg. Co. v. Wesson, 127 F. R. 333, 1903; American Can Co. v. Hickmott Asparagus Co., 142 F. R. 141, 1905; Union Match Co. v. Diamond Match Co., 162 F. R. 148, 1908; Continental Paper Bag Co. v. Eastern Paper Bag Co., 210 U. S. 405, 1908.)

In the case of Mossberg *et al.* v. Nutter *et al.,* 135 Fed. 95, the Circuit Court of Appeals for the First Circuit, Judges Colt and Putnam, circuit judges, and Aldrich, district judge, after a careful analysis of the specification and drawings of a patent and of the prior art, on page 99, said:

> "The claims of a patent are to be fairly construed, so as to cover, if possible, the invention, and thus save it, especially if it is be a meritorious one. In approaching a patent, we are to look primarily at the thing which the inventor conceived and described in his patent, and the claims

are to be interpreted with this particular thing ever before our eyes. In confining our attention too exclusively to a critical examination of the claims, we are apt to look at them as separate and independent entities, and to lose sight of the important consideration that the real invention is to be found in the specification and drawings, and that the language of the claims is to be construed in the light of what is there shown and described."

* * * * * * * * *

"The claims, however, must be read in connection with the drawings and specification; and, so construed, we find, in each of these claims, appropriate language descriptive of the vital feature of the Ericson invention. * * * These several phrases, when read with the drawings and specification, were manifestly intended to cover, and do distinctly cover by their language, the primary feature of the Ericson invention. If these terms are susceptible of a broader interpretation, they must by construction be limited to Ericson's actual invention."

In the case of Robins Conveying Belt Company v. American Road Machine Co., 145 Fed. 923, the Court of Appeals for the Third Circuit (Judges Dallas and Gray, circuit judges, and Cross, district judge), the court quoted (page 926) as follows:

> "In a case of doubt, where the claim is fairly susceptible of two constructions, that one will be adopted which will preserve to the patentee his actual invention." (McClain v. Ortmayer, 141 U. S. 419, 425, 12 Sup. Ct. 76, 78, 35 L. Ed. 800.)

> "The court should proceed in a liberal spirit, so as to sustain the patent and the construction claimed by the patentee himself, if this can be done consistently with the language which he has employed." (Klein v. Russell, 19 Wall. 433, 466, 22 L. Ed. 116.) "The claim must be read in the light of the description, and, if it be fairly susceptible of two meanings, that construction must be adopted which will sustain, rather than defeat, the patent." (McEwan Bros. Co. v. McEwan (C. C.) 91 Fed. 787. See, also, Stilwell-Bierce & Smith-Vaile Co. v. Eufaula Cotton Oil Co., 117 Fed. 410, 414, 54 C. C. A. 584; Electric Smelting & Aluminum Co. v. Carborundum Co., 102 Fed. 618, 42 C. C. A. 537; Hogg v. Emerson, 11 How. 587, 606, 13 L. Ed. 824.) "There interpreted, all doubts are dissipated, and we find that the claim requires that the three pulleys be mounted with their axes in the same vertical plane."

In Diamond Patent Company v. S. E. Carr Co., 217 Fed. 401, this court quoted (page 406) approvingly the clear and forceful language of the Court of Appeals for the 3rd circuit in the case of 1900 Washer Company v. Cranmer Co. *et al.*, 169 Fed. 629, at page 632, refuting an argument for the literal reading of a claim, said:

> "This may be true, if we stick in the bark, by looking at the language of the claim, dissociated from the specifications; but no invention can be practically or fairly understood or explained, if such dissociation is absolutely adhered to. As we

> have already shown, the element described in the first claim, as 'means for actuating said lever,' must not be taken to be any means, such as impracticable hand power applied to the lever, but the efficient practical means described in the specifications."

That "the position the patentee took before the patent office" as shown by the file wrapper and contents of his patent application, is to be taken into consideration in construing claims was so very recently decided by this court in Selectasine Patent Co. v. Prest-o-graph Company (Judges Gilbert, Ross and Hunt, Circuit Judges) 282 Fed. 223, that there can surely be no question as to the reasonableness of endeavoring to gather from the arguments and admissions in such file wrapper contents, in connection with the patent specification, the supposed invention, and the real invention (if there be such) to the end that the claims may be construed in the light of such proceedings, for the purpose of protecting the real contribution to the art (if there has been a contribution). So construing the Trumble claims, Cooper does not technically anticipate (although we are still faced by the hair-splitting question of whether or not such claims disclose even a microscopic amount of invention over Cooper) in view of other prior art (particularly in view of patent to Bray, Defendant's Exhibit "H," Book of Exhibits, page 127).

Now, if the foregoing suggestions as to a permissible interpretation of the Trumble claims to avoid

Cooper as an anticipation are adopted, namely, if the court holds with us that there is no anticipation by Cooper because the element described in the Trumble claims, substantially as, "means for spreading the oil over the wall of such chamber" (claims 1 to 3 inclusive), and "means adapted to cause the oil to flow in a thin body" (claim 4), (which a reference to the specification shows to be the spreader cones of Trumble, 22 and 23a) are not disclosed in Cooper; and that the wearing plate a' is not the equivalent of such means because (while, as a matter of fact it does form a surface over which the oil is preliminarily spread in a thin film and directed upon the wall of the separator), it does not spread the oil equally around and over all of the walls of the Cooper separator, then, as the court will see when the question is considered, there is no infringement because, (among other things) defendant's device (like that of Cooper—or even less than Cooper because it contains no wearing plate a') contains no equivalent of such spreader cones.

## McINTOSH PATENT.

**Defendant's Exhibit "E," McIntosh Patent (Granted Nearly Two Years Prior to Trumble's Date of Alleged Invention) (Book of Exhibits, Page 110), Overlooked by the Patent Office on Trumble's Application, Discloses Clearly What the Trial Court Found to Be the Essence of the Alleged Trumble Invention. Moreover, Under the Interpretation Necessarily Placed by the Trial Court Upon the Trumble Claims in Order to Find Infringement by Defendant's Device, McIntosh Discloses a Complete Anticipation. There Is No Question but That McIntosh Spreads the Oil in a Thin Film on Extended Surfaces, i. e., Imperforate Conical Plates, and While so Spread Subjects the Oil Backed by Such Surfaces to Pressure. The Trial Court Greatly Erred in Stating That the Oil in McIntosh Was Injected Into the Separator in the Form of a Spray; on the Contrary, the Oil Is Injected Into the McIntosh Separator in a Less Agitated Condition Than in Trumble's Separator. McIntosh Provides Specific and Very Effective Means to Maintain Pressure. Unless the Trumble Claims Are Construed as Covering Only a Combination in Which the Oil Is Spread Upon Imperforate Cones, and by Them Upon the Walls of the Separator, They Are Anticipated and Void in View of Bray.**

The trial court's findings (transcript of record, near bottom of page 538) with reference to McIntosh patent is as follows:

"Exhibit 'E,' the McIntosh patent [together with several other patents mentioned by the court] all inject the oil from the well in the form of a spray, having the effect to reduce it to a finely divided condition and the gas is thus permitted to escape. None of them are provided with baffle plates except Newman * * * I think none of these patents contain the element in combination of pressure within the chamber. All of these patents are obviated in their evidentiary effect by restrictions of complainant's specification and claims as made before the Examiner."

Every one of the foregoing statements of the trial court is mistaken, as an examination of this reference will conclusively show. (Defendant's Exhibit "E," Book of Exhibits, page 111.)

Referring to the drawings of McIntosh, Fig. 1 shows the separating chamber with its interior construction or arrangement and connections. In this drawing 3 is the inlet pipe through which McIntosh tell us [line 106, page 1 of this specification] the liquid is "forced into the tank" (or separating chamber). Note that the oil rises through this pipe, and therefore always is subject to the restraining influence of gravity to prevent its "spraying" into the chamber. There is no suggestion anywhere in the McIntosh specification or drawings that the oil is "sprayed" into the chamber as stated by the court. Such finding rests entirely upon unwarranted assumption.

The Trumble construction is far more likely to result in spraying the oil into the chamber than that of McIntosh for the reason that the oil under pressure, occupying, as it obviously must, the entire space within the inlet pipe 7 of Trumble is forced to pass through a restricted passage of a T 5, such restriction of the size of the opening into the Trumble separator being caused by the passage of gas outlet pipe 25 through such inlet, gas outlet pipe 25 thus performing a function as to restricting the size of the inlet orifice analogous to that of the needle of a needle valve—lessening the flow in much the same manner as the nozzle of an ordinary garden hose.

There is no such restriction of the volume of the inlet pipe 3 of McIntosh; so that it is at once obvious that any tendency to spray would be greater in the case of the Trumble construction than in that of McIntosh; and that, therefore, the finding of the court that McIntosh sprays the oil into the chamber, while Trumble does not, is mistaken.

McIntosh, at line 34, page 1 (Book of Exhibits, page 111) of his specification says:

> "The upper part of the gas discharge pipe is provided with two or more flanges of thin metal, or other material, each inclined downwardly, so that the liquid as it issues from the discharge pipe passes over the upper end of the discharge pipe, and flows down over these inclined flanges of metal, dropping at the circumference of each such flange on to the next succeeding flange below. In this manner, the liquid *flows in thin layers* out from the discharge pipe over the succession of flanges, so that it is spread a series of *thin and extended layers* whereby the gas contained therein in the liquid is freed and collected in the gas chamber." (Italics ours.)

Thus, it appears that the trial court's statement that this patent does not contain baffle plates (that is, does not contain anything similar to or performing the same function as the Trumble spreader cones 22-22a is erroneous. While McIntosh thus describes the elements provided for spreading the oil in a thin film as "inclined flanges 6," it is plain upon comparing figures 1 and 2 of McIntosh that they are substantially, imperforate spreader cones, the function of which is to spread the oil, as described by McIntosh, in "thin and extended layers" in precisely the manner of Trumble's spreader cones 22-22a.

To maintain pressure, McIntosh provides what he describes (line 81, page 1, of his specification) as "an inverted gas bell or holder 8," which he states is constructed so as to prevent leakage or outflow of the gas except through the gas pipe 9.

Now it requires only an inspection of the drawing of McIntosh to discover that this bell 8 would telescope into the lower portion of the chamber 4, and would drop of its own weight until it rested on the top of the oil inlet pipe 3, and the cover of the gas outlet pipe 10, if it were not for gas pressure. In other words, the weight of the bell 8 is at all times resting upon and exerting a compressive force upon, the gaseous contents of the separator chamber.

It will be noted also that the bell 8 operates as a valve to control the outlet of the oil through the oil outlet 7.

With a full understanding of the nature and mode of operation of the McIntosh separator, it is clear that the following finding of the trial court expressed in its opinion (transcript of record, middle of page 536) is untenable; and that, if such interpretation is placed upon the Trumble patent it must be declared void in view of McIntosh. The trial court says:

> "I am impressed that the patentee is not confined to means of causing the oil to flow down the outer wall of the chamber, but that his patent includes any means that will cause the oil to flow down any surface as well, such as a baffle plate or inner partition or wall, which is reached after the emulsified oil enters the chamber. I think therefore, the patentee is not estopped by the proceedings before the Patent Office to insist upon the broader claims."

The last sentence of the immediately foregoing quotation from the trial court's opinion should be considered in the light of the fact that the Patent Office, as in the case of the Cooper patent and also of the Newman patent (hereinafter to be considered), entirely overlooked this patent to McIntosh and did not cite it at all during the prosecution of the Trumble application.

## Can McIntosh Be Avoided as an Anticipation of the Trumble Claims?

Yes, as in the case of Cooper—by a strict and narrow interpretation of the Trumble claims in the light of the specification, drawings and the proceedings before the Patent Office. In view of the fact, however, that McIntosh, as well as Cooper, discloses what the trial court found (or assumed to be) a contribution to the art by Trumble, namely, the flowing of the oil, as the court says in its opinion, (transcript of record, middle of page 536), "down any surface"—not confining such surface to the outer wall of the chamber—but "any surface * * * such as a baffle-plate or inner partition or wall, which is reached after the emulsified oil enters the chamber,"—in view of such fact why should this court strain to find invention? This method of close and fine interpretation is often necessary where there has been a *valuable contribution to the art*, which it is desired to protect, but where is the contribution of Trumble?

McIntosh, however, may be avoided as an anticipation (if we shut our eyes to the question of invention) by construing the Trumble claims as covering a combination of imperforate spreader cones which spread the oil over the wall of the chamber, so that the oil is first spread upon the cones and by them is later spread upon the walls, and no part of the oil falls to the bottom of the separator without being so spread—but as often before suggested, if this interpretation is adopted, under the same interpretation defendant does not infringe. Manifestly, it would be unjust to this defendant as a representative of the public to adopt one interpretation of the claims for the purpose of avoiding anticipation and another interpretation for the purpose of finding infringement.

## NEWMAN PATENT.

---

**Defendant's Exhibit J, Newman Patent (Book of Exhibits, page 136) (Granted Nearly Seven and a Half Years Prior to the Date of the Alleged Invention of Trumble): As With the Other Patents, the Trial Court Is in Error When It Says (Transcript of Record, Page 538) That Newman Injects the Oil in the Form of a Spray. On the Contrary, the Oil Comes Into the Newman Separator in a Much Quieter Condition Than in Trumble. The Oil Not Coming Into the Chamber in the Form of a Spray Is Not "Sprayed" Upon the Baffle-Plate of Newman as Found by the Court; on the Contrary, It Flows Quietly on the Baffle-Plates $K^1$ and $L^1$, Is Spread by Them (Particularly by Baffle-Plates K ) in a Thin Film, and Is Conducted by Them to the Walls of the Separator Down Which It Flows.**

**Neither Is It True That Newman Does Not Contain the Element of Pressure.**

**This Patent, Like Those to Cooper and McIntosh, Was Overlooked by the Patent Office.**

**All the Trumble Claims in Controversy Read Literally Upon the Newman Disclosure.**

The operation of the Newman separator, is as follows: Fluid and gas approaches the separating chamber through the pipe F, Fig. I of the drawing, conducted by this pipe it passes through a "T" coupling marked E2, and enters the pipe 7. Part of the mixed fluid and gas goes upward and part downward through pipe E. The part which goes upward is again divided at the T coupling $I^2$, and part proceeds by way of each of the pipes I' to the top of the respective chambers A and B, and then descends through the pipes G and H into the separator to the distributing head or pipe H' and G'. From each of these distributing heads are four inlet pipes G2, and four inlet pipes H2.

The portion of the oil which goes downward after reaching pipe E, enters the respective chambers A and B through pipe E and through T-joint C.

Now it will be seen that the diameter of the first mentioned conduit from the well, namely, pipe F, is substantially the same as all the remaining connections from said pipe F to the distributing pipe G2 and H2, and the bottom inlet through the T coupling C. The pipe F will be full of mixed gas and oil and at E2 it will be divided. In the pipe E above E2 there will then be one-half of the amount of oil that was carried by the pipe F. At $I^2$ this one-half will again then be divided. The one-fourth then flowing through I', for instance, will again be divided four times when it reaches the distributing pipe G2. The one-fourth of the contents of pipe F which finally reaches the distributing head H' is divided four times, thus, only one-sixteenth of the original volume flowing through pipe F reaches each of the distributing pipes G2. Manifestly, a volume of oil equal to one-sixteenth of the area of the pipe is only a small trickle. In view, therefore, of the amount of fluid which reaches each of the openings into the chamber, and the comparative size of such openings there is nothing to cause the oil to spray, consequently it must flow quietly upon the conical spreader K1, which reduces it to a thin film and such spreader conducts such film to the wall of the separator.

Newman tells us specifically that one of the objects of his invention, and particularly of those provisions for the repeated dividing of the oil was (line 11, page 1 of the specification) *"to distribute the force of the gas, especially in high pressure, so as to eliminate the spray."* (Italics ours.)

In view of this, how can the finding of the trial court (transcript of record, page 538) that Newman "injects the oil from the well in the form of a spray" be sustained. *That is clearly just what Newman does not do.*

Now, the only other objection of the trial court to Newman as an anticipation, or as requiring a most strict construction of the claims was that he *thought* it did not contain the element of pressure. Upon what this surmize of the court was based does not appear. The Newman specification on this question of pressure is similar to that of Cooper: At line 11, page 1, of his specification, Newman states the object of the invention as being to distribute the *force of the gas especially in high pressure.* At line 70, page 1, Newman specification, he refers to his float and tells us it "thereby equalizes the pressure."

All that we have said regarding pressure under the first head of our argument applies to the Newman device.

It only requires a comparison of the claims of Trumble in controversy (claims 1 to 4, inclusive) with the Newman disclosure to demonstrate that they read *literally* on Newman.

It is clear that if the Trumble claims are to be construed as covering (trial court's opinion, transcript of record, page 536), "any means that will cause the oil to flow down," not only the walls of the chamber, but "any surface as well, such as a baffle-plate" (which interpretation was necessarily adopted by the trial court in order to find infringement), the Trumble patent will be anticipated and void in view of Newman.

### Can Newman Patent Be Escaped as an Anticipation of the Trumble Claims in Suit?

As we have seen, the operation of the Newman separator is in accordance with the principle most emphasized by the trial court as constituting the invention of Trumble: The oil flows quietly into the separator without agitation. It spreads upon the baffle-plate K' in a thin film. In such film-like condition, it is deflected on the walls of the separating chamber; and while flowing over these extended surfaces is subject to pressure. This patent, as before stated was granted over seven and a half years prior to the date of the alleged invention of Trumble. Certainly there was nothing new in either the *principle* of operation or *means* for which Trumble was evidently contending.

*And Trumble's claims in controversy literally describe the Newman device.*

It does not seem possible that the Patent Office would have allowed the Trumble claims if the Newman patent had been discovered in the examiner's search. However, if plaintiffs' counsel is able to point out any patentable difference between Trumble and Newman, we ask the court to consider such difference on the question of infringement.

## NO PRESUMPTION OF VALIDITY.

**Three Most Important Patents Granted Long Prior to the Pretended Invention of Trumble, Namely, the Patents to Cooper, McIntosh, and Newman, Each of Which Contains the Combination of Elements Described in the Trumble Claims; and, What Is of Equal Importance, Each of Which Contains the Alleged Essence of the Trumble Invention, Were Overlooked by the Patent Office Upon the Trumble Application. No Reasonable Theory of Patenable Novelty or Invention Can Be Found Which Would Have Supported the Allowance of the Claims in Controversy Over These References Had They Been Discovered. This Destroys Any Presumption of Validity of the Trumble Claims.**

While we have not strenuously urged invalidity of the Trumble claims in view of the art, either because of anticipation of non-invention (being earnestly of the opinion that our defense of non-infringement was sufficient) we deem emphasis to be justified upon the fact that three references reading literally on the Trumble claims were overlooked by the Patent Office, in view of the finding of the trial court (transcript of record, two-thirds down page 536):

> "I think, therefore, the patentee is not estopped by the proceedings before the Patent Office to insist upon the broader claims,"

—and in view of the further finding (transcript of record, bottom of page 538) reading:

> "All these patents are obviated in their evidentiary effect by the restrictions of complainant's specification and claims as made before the examiner."

Of course, if these patents were not discovered by the examiner, and were not consequently before the Patent Office, it can hardly be said that their effects were obviated by changes made in the specification and claims before the examiner. As we have seen, however, in the discussion of these patents, there were in fact no such restrictions made, otherwise the claims would not read, as they do, literally on the disclosures of these patents. It follows that the quotation by the trial court from the case of National Hollow Brake Beam Co. v. Interchangeable Brake Beam Co. 106 Fed. 693 (transcript of record, page 535) to the effect that there is no estoppel operating against applicant from claiming what is not disclosed by the references cited by the Patent Office, is not in point.

Of course, even if these patents had been discovered and cited by the examiner, the patent claims could not be sustained when they read in letter as well as in spirit upon the disclosure of such prior patents.

As to the effect of failure of the Patent Office to cite references, see Walker on Patents, 5th Ed., Sec. 491, page 557, and cases cited. In American Soda Fountain Co., v. Sample, 130 Fed 145, at the bottom of page 149, the court says:

> "* * * we think the force of that presumption is much diminished, if not destroyed, by the lack of any reference by the examiner to, or consideration of, the "Clark" patents. It does not seem likely that an expert examiner would pass them by, without notice or consideration, if they had been called to his attention. We feel compelled, therefore, to the conclusion, that the first and fifth claims of the patent in suit are invalid for want of patentable novelty."

## BRAY PATENT.

**Defendant's Exhibit "H," Bray Patent (Book of Exhibits, Page 127) Is in All Material Respects the Same Device as That of the Trumble Patent Except That the Conical Spreader Plates of Bray Have Holes in Them. However, an Examination of Figures 2 and 3 of Bray's Drawings Shows That a Large Portion of the Surface of Such Spreader Cones Is Inperforate, and the Oil Flowing Into the Separator First Strikes the Apex of Bray's Cones, Is**

**Spread Out and Flows in a Thin Film for Some Distance, Exactly as Provided for by Trumble, Before It Reaches the First Line of Holes.**

**If the Trial Court Was Correct in Finding That [Transcript of Record, Page 536] Trumble "Is Not Confined to Means of Causing the Oil to Flow Down the Outer Wall of the Chamber," But That His Patent Covers, "Any Means That Will Cause the Oil to Flow Down Any Surface as Well," Clearly the Inventive Idea of Trumble Is Anticipated by Bray.**

**Bray Also Provides for Pressure; and During the Prosecution of the Trumble Application Involving Discussion of the Bray Patent, It Was Never Even Suggested Otherwise.**

**The Trial Court's Understanding of This Reference Is Glaringly Erroneous, for Oil Is Not "Sprayed" Into Bray's Device as Found by the Court; and Bray Does Have Baffle Plates, the Trial Court's Intimation to the Contrary Notwithstanding.**

**The Findings of Infringement Could Only Have Been Arrived at by Overlooking the Extreme Pertinence of Bray as a Reference, and by Losing Sight of the Controlling Admissions Forced From Trumble by Reason of the Citation of Bray in the Proceedings Before the Patent Office.**

We have repeatedly quoted the trial court's findings as to the state of the art (transcript of record, page 538, but for convenience in considering the reference now under discussion, we repeat. The trial court said:

"The patents introduced as showing the prior art are readily disposed of. [Then mentioning several patents, the court finds that they] "all inject the oil from the well in the form of a spray, having the effect to reduce it to finely divided condition, and the gas is thus permitted to escape. None of them are provided with baffle-plates except Newman, * * * and I think none of these patents contain the element of combination of pressure within the chamber. All these patents are obviated in their evidentiary effect by restrictions of complainants' specifications and claims as made before the examiner. Exhibit 'H,' the Bray patent, is subject to the same criticism. The oil is there precipitated upon perforated cones, and only slightly, if at all, flows down the wall of the chamber or other surface."

It only requires a glance at the drawings of Bray and Trumble to recognize the remarkable similarity between these devices. The oil in Trumble enters through pipe 7 at the top, oil enters the Bray device through the pipe 16, also at the top. Upon its entering into the chamber in each case, it falls upon the apex of a spreader cone. Bray's outlet for the gas is at 17 and the outlet for the oil is at the bottom of the chamber controlled by the valve 5, which is actuated by the float 7. There has been no question raised as to the substantial identity of the respective outlets of Trumble and Bray for oil and gas. The only objection to Bray as a complete anticipation of the Trumble claims (as will appear shortly upon our consideration of the file wrapper contents of

Trumble) was that Bray's spreader cones, having holes through them, all of the oil is not conducted by them to the walls of the chamber, but that some of the oil flows through the holes.

In our argument, *supra,* relating specifically to the McIntosh patent, we considered at length the statement of the trial court that the oil was "sprayed" into the separating chamber in a number of prior patents, notably McIntosh and Bray. We called the court's attention to the fact that the reduction of the inlet area of Trumble by the passing through it of the gas outlet pipe 25 would be more likely to "spray" the oil, after the method of operation of a needle valve, than in McIntosh. The same is true of Bray: the area of the inlet "T" attached to pipe 16 of Bray is not diminished, consequently the oil under pressure is much less likely to spray than in the Trumble device. It is, therefore, clear that the trial court was in error in attempting to differentiate between the Trumble and the Bray disclosures on the ground that Bray "sprayed" the oil into the chamber, while Trumble did not.

As to pressure: we have already considered this matter at length under the first heading of our argument. In addition to this we call the court's attention to the fact as shown in the drawings of Bray (Fig. 1) that the diameter of the outlet pipe 17 is less than that of the inlet pipe 16. This, under the uncontradicted testimony of plaintiffs' expert Ford Harris (hereinbefore quoted) would cause pressure in the trap, because, by reason of the restricted outlet orifice the oil coming in under great pressure would not have an opportunity to escape rapidly enough, consequently pressure would be backed up in the trap.

In view of the fact that Your Honors have so recently decided (Selectasine Patent Company, *et al.* v. Prest-o-graph Company, *et al.*, 282 Fed., 223, 224), that this court was,—

> "always in accord * * * that the patentee cannot escape from the position which he took before the Patent Office, and the consequence of not having appealed from the action of the Patent Office,"

—and of the further fact that it is most elementary (Walker on Patents (5th Ed.) Secs. 187 and 187a, page 254 and cases cited) that:

> "Letters patent may be construed in the light of the contemporaneous intention of the inventor and of the Patent Office; and to this end recourse may be had to the files of the application papers to see what changes were made in the description and claims while the application was pending in the Patent Office,"

—it is quite important that we look to the proceedings before the Patent Office as shown by plaintiffs' Exhibit 2; (Book of Exhibits, page 9). Our consideration of the admissions of this exhibit will show that it was only by the closest and narrowest interpretation that the Trumble claims in controversy were allowed over Bray (the Patent Office, as before seen, having overlooked Cooper and McIntosh). Consequently, if consistency be observed in holding to such interpretation, we are confident that the court will agree with us that the finding of infringement cannot be sustained.

Defendant's Exhibit "H," Bray patent (Book of Exhibits, page 127) was first cited by the examiner (Book of Exhibits, page 30) on the rejection of claims 1, 2, 3, 4, 5 and 8 as originally presented.

Only one of these claims (original claim 1) need be referred to. Original claim 1 as rejected on Bray (and afterwards acquiesced in by the applicant) reads as follows:

> "1. The combination in an oil and gas separator, of an expanding chamber having a settling bottom, means for delivering oil and gas to the upper portion of the chamber, means for reducing the oil into a finely divided condition to reduce the tension on the gas contained therein, gas outlet means from the chamber and an oil outlet from the chamber intermediate the settling bottom of the chamber and the gas outlet."

In response to the above action the applicant (Book of Exhibits, page 32), cancelled claims 1, 2 and 3 and inserted in lieu thereof the claims now in controversy which were afterwards allowed, renumbering the remaining claims of the original application and making other changes not material to be here mentioned.

The applicant did not attempt to dispute the accuracy of the examiner's decision that the claim above quoted (namely, claim 1 as originally presented) was found in Bray; on the contrary, in answering the examiner and cancelling the claim, the applicant not only impliedly acquiesced in the action of the examiner, but at Book of Exhibits page 34 of the file wrapper *expressly* acquiesces in the action of the examiner by remarking, "renumbered claim 9 was not been amended for the reason that applicant's attorneys do not believe that it comes within the inventions shown by Bray or Taylor"—thus clearly implying that they *did* believe with the examiner that the cancelled claims did come within the invention of Bray.

At Book of Exhibits page 35, the applicant clearly defines what he considers the essence of his invention as follows:

> "Applicant's specification and claims are now written to define the peculiar arrangement of applicant's device, which is the means for causing the oil to be divided into a thin flowing body while giving off the natural gases carried in combination therewith and applying pressure to the surface of the oil while undergoing this action. This results in maintaining all of the light, volatile liquids in combination with the oil and raising the gravity of the oil when delivered into the receiving tank."

Attempting to forestall any adverse decision of the examiner that the amended claims (being those now charged to be infringed) were not met by the Bray patent under discussion, the applicant says (Book of Exhibits page 35):

> "The patent to Bray shows that the oil after being admitted into the expansion chamber is divided into drops or streams and flows downwardly in that condition."

In Book of Exhibits, page 36, in further effort to distinguish the alleged invention of Trumble from that disclosed in Bray and others, the applicant says:

> "Applicant has discovered that in order to maintain the lighter series of oils in combination with the heavier series when separating the gas, it is necessary to reduce the oil to a thin regularly flowing body, which is not subjected to any breaking up action and to permit the gas to escape therefrom without agitation."

All of the references cited would cause a breaking up of the flowing body of oil, or agitation thereof, and result in the carrying away the light volatile oils with the gas."

In response to the above entitled communication of the applicant the examiner (Book of Exhibits, page 38), says:

> "In this patent to Bray while cones may be shown as perforated, these cones will nevertheless deliver the oil to the wall of the expansion chamber to flow downwardly thereover. This patent is also provided with means for automatically drawing off the heavy residue or deposit as well as means for allowing an overflow of the oil therefrom through the pipe 20. The gas is drawn

In response to this action of the examiner (Book of Exhibits, page 40), the applicant says:

> "We cannot understand the examiner's statement that the cones in the Bray patent will deliver oil to the wall of the expansion chamber. The cones Bray calls screens and they evidently make tight closure with the walls of the chamber. An explanation is requested."

Whereupon, (Book of Exhibits, page 41), the examiner responds:

> "In reconsidering the case in view of applicant's remarks, it is not seen wherein the applicant is justified in saying that the screens make a tight closure with the walls of the chamber. In the patent the casing 2 is provided with a series of brackets 10 to support the screens. This would not look as if the screens make a tight closure with the walls of the chamber. Furthermore, it is said that the oil or liquid passes *down over* the screens and into the cones 1. This would imply that the heavy liquid reaches the casing and passes down its walls. There does not seem to be any authority for assuming that the liquid will do otherwise. While the screens in the patent are perforated it is evident that these perforations are for the passage of the separated gases on their way to the outlet pipe 17, since the heavy liquid would be too thick to pass therethrough."

On the same page of the file wrapper contents, the examiner again rejects the claims here in suit, being claims 1 to 4, inclusive, holding that they were not "patentably distinguishable" from the referennces of record. In response to this last action rejecting the claims, applicant specifically DEFINES HIS INVENTION IN A WAY WHICH UNMISTAKABLY DIFFERENTIATES IT FROM BRAY AND BY THE SAME TOKEN DIFFERENTIATES IT FROM DEFENDANT'S DEVICE NOW CHARGED TO BE AN INFRINGEMENT. At Book of Exhibits, page 42, the applicant responds:

> "Applicant's invention consists of a containing vessel, an *imperforate cone* adapted to spread the *whole* body of the oil *to the outer edge of the vessel,* and means for taking *off gas from the interior of the cone* near the center of the vessel. In the Bray patent there is *no means for spreading the oil,* and even if the examiner holds that screens 11, 12 and 13 spread a portion of the oil, he cannot avoid the conclusion that oil will pass through a screen and that some oil will not be spread." (Most of italics ours.)

Note very particularly that it was not attempted to differentiate the Bray disclosure in any way from that of the claims now in controversy on the ground that Bray did not *disclose means for maintaining pressure.* This is obviously because there are plainly disclosed means in the Bray patent (Fig. 1) for maintaining such pressure, namely, the difference in diameter between the intake pipe 16 and the outlet 17, the inlet being larger than the outlet—which would necessarily build up a pressure. It is clear from this that the patentee was not claiming as the essence of his invention any means for maintaining pressure, because if he had emphasized this particular feature of his device, he would have been met by many references of record which would have shown it to have been old to have applied a valve such as disclosed in Trumble for maintaining pressure to a trap of this character. The argument upon which this patent was allowed over Bray was clearly that the cones of Trumble, performed the *double function* of spreading all the oil in a thin film and of conducting all the oil so spread to the walls of the Trumble separator, while those of Bray had holes in them which impaired the functions of both spreading and conducting.

It follows, therefore, that the elements described substantially in the claims of Trumble as "means within the chamber adapted and arranged to distribute the oil over the walls of the chamber," which (as is made especially clear in the last above quoted applicant's argument before the Patent Office) is an "imperforate cone adapted to spread the whole body of the oil to the outer edge of the vessel,"—is a most important element of the Trumble combination, and the question of its equivalency is to be determined (in accordance with the legal definition of an equivalent) *by all its functions as well as by the manner in which it performs those functions.*

If applicant's argument that Bray's cones are not the equivalent of Trumble's cones because they do not spread *all the oil* out in a thin film and conduct it in such form to the wall of the separator is found valid, it follows necessarily that defendant's devices, *employing no element whatever performing such double function of spreading and conducting are not infringements.*

Look at Fig. 3 of Bray and observe how the center of the upper cone 13 operates to spread the oil in a thin film,—and to spread *all the oil* in such film—in a far more effective manner than is shown in defendant's devices in controversy.

In view of this clear spreading of the oil in a thin film on the imperforate portion of the cone of Bray, how can the trial court's conclusion (transcript of record, middle of page 536) reading:

> "I am impressed that the patentee is not confined to means of causing the oil to flow down the outer wall of the chamber, but that his patent includes any means that will cause the oil to flow down any surface as well, such as a baffle-plate or inner partition or wall, which is reached after the emulsified oil enters the chamber,"

—be sustained? Manifestly, any such breadth of interpretation as here suggested by the trial court would render the claims anticipated in and void in view of Bray; and, furthermore, would totally disregard the close and narrow definition of his alleged invention which Trumble was compelled to adopt before the Patent Office in order to prevail upon the examiner to allow his present claims. Clearly, the examiner allowed the claims because of the interpretation placed upon them by Trumble.

## TAYLOR PATENT.

Defendant's Exhibit "F," Taylor Patent (Book of Exhibits, Page 115): The Trial Court [Transcript of Record, Page 538] Totally Misread and Misunderstood the Taylor Specification and Drawings: First, It Is Not True, as Found by the Trial Court, That in Taylor the Oil From the Well Is Injected in the Form of a Spray; on the Contrary, There Is Nothing Having the Function of a Spraying Device or Which Could by Any Possibility Cause the Oil to Spray, or Which Even Remotely Resembles a Spraying Device, Illustrated or Described by Taylor. Second, It Is Not True as Found by the Trial Court That Taylor Is Not Provided With Baffle-Plates; on the Contrary, the Particular Means for Effecting the Separation in the Taylor Device Consists of Inclined Baffle-Plates Upon Which the Oil Is Spread in a Thin Film; Third, It Is Not True as Suggested by the Trial Court, That Taylor Does Not Contain the Element of Pressure Within the Chamber; on the Contrary It Would Be Impossible to Use This Device as Described by Taylor Without Having the Contents of the Chamber Constantly Under Great Pressure. In Fact, Taylor Distinctly Tells Us (Line 85, Page 1, of His Specification) That Sediment Accumulating in the Bottom of the Chamber Will Ordinarily Be Removed by the Pressure.

**It Is Manifest That the Trial Court Could Not Have Found as It Did [Transcript of Record, Page 536] That Trumble "Is Not Confined to Means Causing the Oil to Flow Down the Outer Wall of the Chamber, but That His Patent Includes Any Means That Will Cause the Oil to Flow Down Any Surface as Well, Such as a Baffle Plate or Inner Partition or Wall, Which Is Reached After the Emulsified Oil Enters the Chamber,"—if It Had Correctly Read and Understood the Taylor Reference; for the Taylor Disclosure Comes Exactly Within This Language of the Court.**

**Taylor Can Only Be Avoided as a Complete Anticipation of the Claims in Suit by the Strictest Possible Interpretation, That Is to Say, by Reading Into the Claims Limitations or Interpretations Suggested in the Proceedings on the Trumble Application, Under Which Interpretation (as Will Later Appear) Defendant Does Not Infringe.**

Taylor (Book of Exhibits, page 115), covers a steam separator, but steam is a gas, and the object of the invention of Taylor as described at line 15, on page 1 of his specification is to separate grease and oil from the steam in passing from the boiler to the steam chests of the cylinders of the engine. It will be noted that the trial court attempted no distinction upon the ground that Taylor did not describe one of the obvious uses to which his device might be put, namely, connection to an oil well to separate gas and oil. Neither did Trumble during the prosecution of his application through the Patent Office attempt any such distinction. The patent to Taylor was cited by the examiner as pertinent and such pertinence was not questioned by Trumble. It has never been even suggested that this Taylor device might not have as ready an application for the separation of oil from hydro-carbon gas as for the separation of oil from water gas.

A mere glance at the drawings is sufficient to understand the operation of the device when it is explained that the oil enters the chamber through inlet pipe 3a. At line 42, page 1, of the Taylor specification, the patentee states that there is mounted upon the interior of the separating chamber "baffle-plates 4, having preferably a downward inclination and projecting from opposite walls of the drum [separating chamber] alternately, their edges approaching the wall opposite that on which they are mounted."

It is obvious that when the oil enters the pipe 3a, it must, as stated by Taylor, line 74, page 1 of his specification, "flow downward over the surface of the baffle plates 4." The gas will pass out through the funnel-shaped opening 7 and through the outlet pipe 6, while the liquid falls to the bottom of the chamber, and its outflow is regulated, as in the case of Trumble by a float-actuated valve. A modified construction of the device is illustrated in Fig. 4 of the Taylor drawings.

The exceedingly narrow interpretation which Trumble was compelled to adopt to secure an allowance over Taylor of the claims now in controversy is shown by the proceedings upon the Trumble application (Plain tiffs' Exhibit, File Wrapper and Contents, Book of Exhibits, page 9). Original claim 8 of the Trumble application (Book of Exhibits, page 25) was rejected upon this patent to Taylor as well as others. Said claim reads as follows:

> "In an oil gas separator, the combination of an expansion chamber, means within the chamber to divide the oil to reduce the surface tension of the oil upon the gas contained therein, an oil outlet from the bottom of the chamber, an oil inlet through the top of such chamber arranged to deliver oil onto the top of the oil and dividing means, and gas outlet means comprising a pipe passing upwardly through the oil inlet supporting the oil dividing means within the expansion chamber."

In response to this action of the Patent Office in disallowing original claim 8, (afterwards renumbered claim 9, the applicant says (Book of Exhibits, page 36):

> "The patent to Taylor is not considered as being adaptable for the use of applicant's invention for the reason that no separation of oil and gas could be effected until both had passed the second baffle from the bottom, as a film of oil started on the upper baffle would flow downward to the lower edge and would fall over in a flowing mass onto the next succeeding plate, and any gas freed in the upper chamber would have to pass through

> such body of oil before it could escape into the chamber formed between the two upper baffles, and this same action would take place at each succeeding fall of oil from the edge of the baffle plates. This would cause an agitation of the oil which would have a tendency to carry off the lighter series of oil with the gas.
>
> Applicant has discovered that in order to maintain the lighter series of oils in combination with the heavier series when separating the gas, it is necessary to reduce the oil to a thin regularly flowing body which is not subjected to any breaking up action, and to permit the gas to escape therefrom without agitation."

There was no contention whatever throughout the file wrapper, that the Taylor patent or any of the others did not show means for maintaining pressure. The argument always was that they did not show means for spreading in a thin film, conveying in such condition all the oil to the wall of the chamber, and allowing it to quietly flow down the wall as contradistinguished from causing it to be violently broken up by contact with various obstructions and falling from a height.

Of course, what the patentee and the Patent Office examiner said during the prosecution of the application is not conclusive as to the facts. A reading and comparison will show that the Trumble claims in controversy (claims 1 to 4 inclusive), correctly describe the disclosure of Taylor. In other words, the differences pointed out by applicant to escape Taylor as an anticipation were not inserted into the claims. In order to save them from anticipation by Taylor, therefore, it manifestly becomes necessary to read the Trumble claims in the light of the interpretation placed upon them in the proceedings on the Trumble application. If this is to be done for the purpose of avoiding Taylor as an anticipation, it should of course be followed when considering the question of infringement.

## BARKER PATENT.

Defendant's Exhibit "G," Barker Patent, (Book of Exhibits, Page 121) was the Only Prior Patent Concerning Which the Trial Court's Finding as to the State of the Art (Transcript of Record, Page 538) was Partially Correct. It is True, as Stated by the Trial Court, That in Barker the Oil Would be Injected in the Form of a Spray; and it is Also True That Barker Does not Provide Baffle Plates; but it is not True as Found by the Court That Barker Does not Contain the Element of Pressure Within the Chamber; on the Contrary, Barker Specifically Provides Means to Prevent the Bursting of the Separating Chamber From Overpressure of Gas.

The Barker Patent is Important as Showing, in Connection With its Discussion on the Trumble Application, the Compulsion Which Trumble

**was Under to Emphasize His "Means for Distributing Oil Over the Wall of the Chamber," (Substantially Claims 1 to 3, Inclusive,) and "Means Within the Chamber Adapted to Cause the Oil to Flow in a Thin Body for a Distance * * * ," (Claim 4) Namely, the Spreader Cones 22 and 22a, as of the Essence of His Supposed Invention Over Barker. (Which Cones are not Employed in Defendant's Device). It is Also of Great Weight as Having Forced an Admission From Trumble on His Application, That if the Oil Falls From the Head in a Solid Body (as it Does in Part, at Least, in Defendant's Device), it is not Within the Spirit of the Supposed Trumble Invention.**

In support of the allowance of the claims now be-re the court (claims 1 to 4) and for the purpose of ticipating any objection of the examiner to their lowance and of differentiating their subject matter om the references that had theretofore been cited, e applicant says (Book of Exhibits, page 35):

> "In the Barker patent there are no means shown for expanding the oil in such a manner that it may flow in a *thin film* but must fall from the head 8 either in a solid body or if accompanied by sufficient pressure in a sprayed or broken up condition." (Italics ours.)

Notice that Trumble does not urge as being of the sence of his invention or supporting patentability of s structure over the prior art the means to maintain pressure within the chamber, nor does he question the fact that such pressure means being old, it would not constitute invention to add the same to the Barker patent if they were not shown therein. The applicant apparently recognized that the Barker patent showed that the tank operated under a pressure, and had pressure maintaining means, as the patentee, beginning line 86, page 1, of the Barker specification (Book of Exhibits, page 122), says:

> "Should the water entering the tank contain more gas than is utilized or can be carried off by the pipe 12, the pressure of the gas thus accumulating in the tank will force the water down below the open end of the discharge pipe and will escape through said pipe, thus preventing the bursting of the tank from an over-pressure of the gas. As soon as the gas has thus liberated itself through the overflow pipe, the water will again rise therein and will remain at a level with the upper portion of the overflow pipe until the gas in the tank again reaches a pressure sufficient to force the water down and out of the overflow pipe. By this arrangement, the danger of the bursting of the tank from an overpressure of the gas is entirely eliminated."

## Estoppel by Proceedings in the Patent Office.

The Trumble alleged invention was obliged to squeeze itself into exceedingly diminutive proportions in order to squirm through the meshes of the Patent Office machinery; it tried, in combination with the other elements, to have the oil in a "finely divided condition" (original claim 1, Book of Exhibits, page 22), but this was found too broad, whereupon it contracted to "peculiar arrangement which is the means for causing the oil to be divided into a thin flowing body." (Book of Exhibits, page 35.) This "thin flowing body" was, in view of Cullinan (Book of Exhibits, page 35), disclaimed as a "solid stream" and in view of Bray (same page reference) as a series of "drops or streams"; nor could it, in view of Barker (same page reference), be construed "as a solid body" or to cover a "sprayed" or "broken up condition." In fact, it was necessary to narrow "thin flowing body" to "thin film," (Book of Exhibits, pages 31, 35 and 36) which "thin film" is thus shown to have no breadth of interpretation whatever. If this "thin film" was broken or interrupted by falling from one baffle-plate to another (Taylor, Book of Exhibits, page 36), it did not come within the spirit of the alleged invention, nor, if the gas on one side of the film broke through to the other side (same reference); for such falling from baffle to baffle or passing through of bubbles of gas would "agitate" the oil, and if the oil were agitated it was not the Trumble alleged invention. Furthermore, this thin, unbroken, unagitated film must regularly flow (Book of Exhibits, page 36) and should not be subject to "breaking up" or agitation.

Let it be noticed particularly that not even a suggestion is contained throughout the file wrapper contents that pressure might have been lacking in any one of the prior patents. The reason for this was that, obviously, all showed pressure.

But even with several of the partitions between meshes of the Patent Office screen broken by the failure of the Examiner to discover and cite the patents to Cooper, McIntosh or Newman as a bar to the grant, this timid and diminutive "alleged invention" was still too large to wiggle through, so Trumble was obliged to squeeze it still more by defining it as a device in which *all the oil* must be so spread (Book of Exhibits, page 42 to the *"outer edge of the vessel"*; and furthermore, all such oil must be so spread in such thin, unagitated, unbroken film to such outer edge by means of a *"imperforate cone."* (Book of Exhibits, page 42.)

This was the insignificant aspect of the Trumble alleged invention, when it sought for a hole to get through the Patent Office; but now, like the genie in the vase of the Arabian Nights entertainment, it has grown and expanded until it has become the flow under pressure of even a small part of the oil, whether in a film or not, on any surface, or to any extent on such surface, without regard to whether it is agitated, broken up, sprayed, or regularly flowing, or whether it is conducted to the outer edge of the vessel by anything resembling an imperforate cone or not. Its present great size is simply—surface and pressure. As the genie threatened his liberator, the fisherman, with the loss of his life, so this pigmy grown to giant stature by the decree of the District Court, dominating the entire art, threatens its creator with the loss of those things for which the public has paid the price of many seventeen-year grants.

Accordingly, the finding of the trial court that the patentee is not estopped by the proceedings in the Patent Office (said court not having found anything about the proceedings which *should* have taken place in the Patent Office on the Cooper, McIntosh and Newman patents, but which through inadvertence did not take place) does not seem to us to be consistent with reason.

The limitations, explanations, interpretations, and attempted differentiations contained in the Trumble file wrapper contents were put there by the applicant to avoid prior references—to define his supposed invention so as to avoid them—to make clear to the public what Trumble did not deem within the spirit of his proposed claims. Surely, they must have *some* weight. *Certainly, if they had not been made, the Examiner would not have granted the patent.*

Our prior argument has shown that these limitations were reasonable and necessary. If the Patent Office be deemed to have, notwithstanding these references, granted claims which, literally construed, seem to read on such references, then it is clear that such literal interpretation was not intended by the parties to the patent contract. This court, we urge, should not permit an interpretation which would have the effect of nullifying them. Surely, this court had these considerations in mind and was not mistaken as to their application in the case of Selectasine Patent Co., *et al.*, v. Prest-o-graph Company, *et al.*, 282 Fed. 223, where it held that a "patentee cannot escape from the position which he took before the Patent Office."

## Defendant's Patented Device Misunderstood by the Trial Court.

In its opinion, transcript of record, page 542, the trial court found that defendant's patent infringes claims 3 and 4, in suit. This finding is based upon a description by the trial court [Transcript of Record, page 537] of what the court evidently understood to be the construction of defendant's said patented device. As we have intimated in our statement of the case, this exposition of the trial court reflects a serious failure to grasp the true significance of the description and drawings of the device in question. As a foundation for a correct decision we must, of course, thoroughly understand the construction and mode of operation of the mechanisms under consideration; we shall therefore, by pointing out the error of the trial court in this regard, endeavor to explain the construction of the Lorraine patented separator.

In attempting to describe the device of said Lorraine patent, the trial court [Transcript of Record, page 537] says:

> "Defendant's patent, referred to in counsel's brief as Model 1, has an inner partition set away from the wall on one side more than one-third the distance of the diameter of the chamber, and extending below the oil level. To this partition, at some distance from the top of the chamber, is attached a baffle-plate extending downward on an incline of perhaps 45 degrees, and to within an inch and a half or two inches from the wall for the entire segment cut off by the partition. The

oil inlet, consisting of a pipe, extends downward to within a short distance of a baffle-plate. The pipe has two openings, so that the stream of oil is divided and projected on the baffle-plate in two directions laterally. The device is provided with a gas take-off above the partition and one from underneath the baffle-plate, to permit gas to pass off eventually from the upper portion of the major chamber."

It only requires a brief but careful examination of defendant's patent (Book of Exhibits, page 66) to make clear the error of the foregoing description by the trial court.

The court was obviously in error in stating that the inner partition was set away from the wall on one side "more than one-third of the distance of the diameter of the chamber." The partition mentioned by the court is numbered 19, shown in Figs. 4 and 5 of said Lorraine patent (Book of Exhibits, page 67), Place a pair of dividers on these figures of drawings and it will immediately be seen that the partition is exactly one-fourth of the diameter from the wall.

The court also says:

> "To this partition, at some distance from the top of the chamber, is a baffle-plate extending downward on an incline of perhaps 45°, and to within an inch and a half or two inches from the wall for the entire segment cut off by the partition."

This is a mistake. There is no baffle-plate whatever attached to the partition; and it will be noted that the drawings of the patent are not to scale and so it is impossible to given dimensions in inches.

Instead of being as described by the court, the construction is as follows: Referring to Fig. 4 of the Lorraine patent (Book of Exhibits, page 67), described line 71, page 2, Lorraine specification (Book of Exhibits, page 69), oil enters through the connection 13 from the supply pipe 12 (supply pipe shown in Figs. 1, 2, and 3), and goes downward through the pipe 14. This pipe 14 has around its lower end a sleeve or hood 15, provided at its upper portion with a gas outlet 16. The lower end of this sleeve 15 has an inclined bottom 17, below the open mouth 18 formed in the side of the sleeve. Now, it will be noted that such inclined bottom 17 is of very slight dimensions outside of the inlet passage. (Of course, the oil does not discharge into the separating chamber until it reaches the mouth.) A very gross error of the trial court is that of assuming that such inclined bottom 17 is "attached to the partition," for plainly it is not, being part of the sleeve 15. Furthermore, and very important, it does not extend from the wall "for the entire segment." On the contrary, it is only very slightly (and at its lower side only) larger than the sleeve 15 of which it forms a part. In Fig. 5 of the Lorraine drawings (Book of Exhibits, page 67) is shown the outline in plan of this inclined bottom of the sleeve. In this figure, the sleeve 15 is clearly indicated, and the little triangular corners extending out from each side of the sleeve 15 indicate the relative area of such inclined bottom.

It is thus seen that by far the greatest area of surface of this inclined bottom is covered by the end of the inlet pipe and sleeve and that the extent of its surface within the chamber, upon which the oil might spread, is negligible.

This area is obviously very small as compared with the entire area of the segment of the separator formed by the partition 19. The court was clearly mistaken in assuming that this small inclined impingement plate occupied area nearly equal to this segment.

The trial court was also greatly mistaken in stating (latter part of the immediately preceding quotation) that the "pipe has two openings, so that the stream of oil is divided and projected on the baffle-plate in two directions laterally." There is nothing in the drawings or description of the patent in question to support such statement of the trial court.

## Defendant's Patented Device Does not Infringe.

The oil, as we have seen, comes in under great pressure from the well. The supply pipe 12 will be presumably filled with oil under this great pressure. The oil is thus injected violently and partly falling and partly being forced, strikes inclined bottom 17. Now obviously, this inclined bottom is too small to spread all the oil out in a thin film. Why, patents to McIntosh, Cooper, Newman, and Bray all spread the oil out on surfaces far more extended than this diminutive inclined bottom 17. Cooper's wearing plate a, for instance, is of larger comparative dimensions and was so designed and placed that *all* the oil must be spread on the walls of the chamber, but inclined bottom 17 of defendant's patented device being of approximately triangular shape, only a part of the oil is deflected to the side wall, while there is obviously a very large part which shoots off the sides of the plate and falls to the bottom of the chamber without striking any of the walls of the separator.

Remember, in view of the Bray patent, discussed *supra,* the patentee distinctly in his file wrapper disclaimed a construction in which part of the oil was not spread on the walls of the chamber. Trumble had to do this in view of the disclosure of Bray. Remember, Trumble said his invention consisted in causing the oil to flow in a "thin film," as distinguished from a "broken up" or "agitated" condition. Barker had a small plate similar to the inclined bottom of the Lorraine patent, but Trumble took the position before the Patent Office that such plate would break up or atomize the oil. This is also obviously true of the Lorraine patent. There can be no doubt but that a much larger proportion of the oil, after striking the inclined bottom of sleeve 15, will fall off such plate laterally and will not strike the wall of the separator at all, but will be in a broken up, agitated condition, which Trumble again and again emphasizes throughout his application papers as not his invention. Trumble stated to the Patent Office that his invention consisted of "the peculiar arrangement," which was the "means for causing the oil to be divided into a thin flowing body," namely, not "sprayed," "broken up," not in "drops or streams," not "in a solid stream," not "irregularly flowing body," and not a device which does not spread the whole body of the oil to the outer edge of the vessel in such thin film, and by use of an imperforate baffle-plate.

We submit that, clearly, defendant's patented device does not infringe.

## Construction of Model 2 Described.

The only error we see in the trial court's description of this Model 2 is contained in the implication that, like the court's erroneous description of Model 1 (defendant's patented device), the partition cutting off a segment of the chamber is set away from the wall to which it is nearest about one-third or more the diameter of the chamber. This is not correct. The line of the segment formed by the partition was not more than *one-fourth* of the diameter of the chamber from the nearest wall.

In describing this Model 2, the trial court [Transcript of Record, bottom of page 537] said:

> "Model 2 contains a like partition to that described in Model 1. The oil inlet consists of a pipe extending into the side of the minor chamber, supplied with what is called a nipple, bell-shaped, to allow the oil to spread when discharged into the chamber. The nipple is set at an angle with and extended within proximity of the inner wall, the effect of which is, when the oil is discharged into the chamber, to carry part of it down the inner partition wall, part down the outer wall, at and near the intersection of the inner with the

outer wall, and part of it down by gravity *without reaching either wall.* The device is provided with a gas take-off above the nipple. This sufficiently describes the models to make the application later. I may add, further, that the nipple in the model in evidence is machined off on one side to sit closely against the partition wall. Defendant says this was done through mistake in setting the nipple, the machine having allowed it to extend too far inwardly. If this is true, it only shows how easy it is to set the nipple in without discovery. But we are dealing with the model in evidence, which complainants say infringes their patent."

In order that this description of the trial court may be clear, on the two immediately following pages of this brief we set forth as Figs. 1 and 2 the constructions in question.

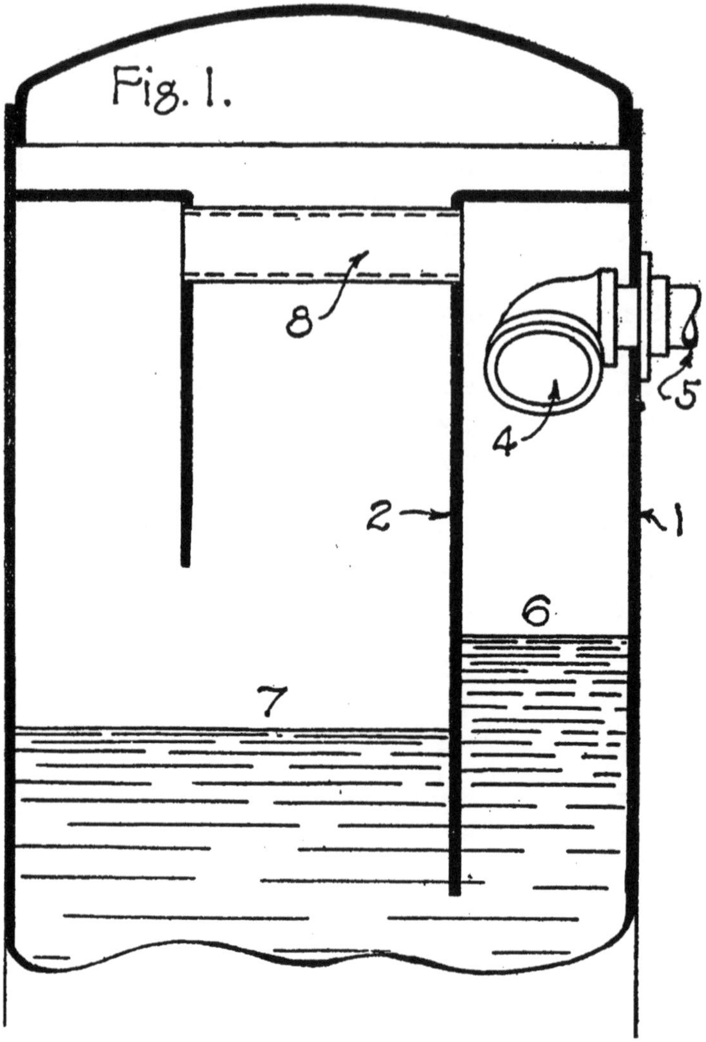
Fig. 1.
8
4
5
2
1
6
7

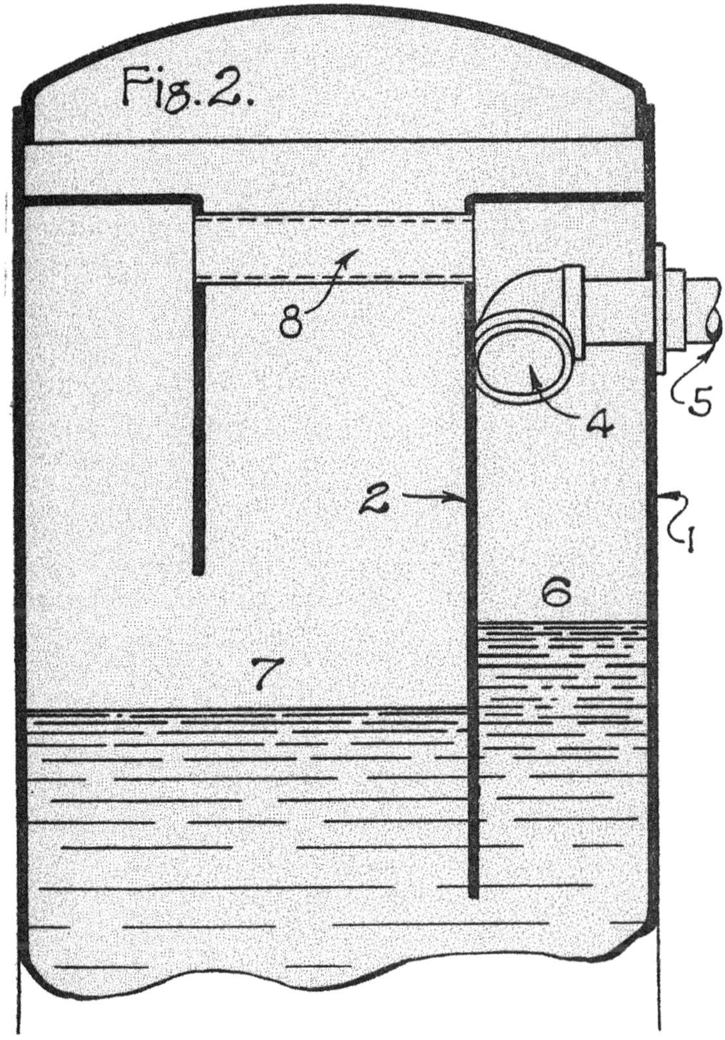


Fig. 1 represents the device of Model 1 as it was usually put out, while Fig. 2 represents the same construction with the so-called nipple machined off, as described by the court, so as to "sit closely against the partition wall." These drawings have been carefully *made to scale*, and correctly represent the constructions referred to. In Figs. 1 and 2 the wall of the separator is indicated by 1, the partition by 2. and the discharge opening into the chamber by 4. The oil supply pipe leading from the well and connected to the oil inlet 4 is indicated by 5. The oil level in the portion of the separator called the minor chamber by the court, is indicated by 6, and the level of the oil in the portion of the separator called by the court the major chamber, is indicated by 7. Opposite to the partition 2 is a short partition and extending between the partitions is a pipe 8. Gas which arises from the oil in the minor chamber passes through the pipe 8 across the separator behind the short partition, then downwardly below the short partition and upwardly through a gas outlet in the upper portion of the chamber. The gas outlet or take-off means is not of any special pertinence.

### Defendant's Model 2 Does not Infringe.

It would seem that a clear understanding of the construction and purpose of the device of this Model No. 2, in the light of the prior descriptions of many other similar devices in the record, especially including defendant's patented device, would render supererogatory any further description of the operation of the device illustrated in the foregoing Figs. 1 and 2. It would seem obvious that the oil coming through the inlet opening 4 *must* in large part fall to the bottom of the separator without striking the walls at all. Indeed, the trial court distinctly so found, stating (near the top of page 538 of the transcript of record) that part of the oil descends "by gravity *without reaching either wall.*" And again, at transcript of record page 541, the trial court says that the oil is injected into this Model 2 only "in part at least against the partitions." The court further states [same page reference of the Transcript of Record]:

> "While part of the oil is reduced to a spray and falls by gravity to the settled fluid below, its action does not obviate the objectionable feature of a part flowing down the partition and a part down the wall."

Now, it would seem as clear as the noonday sun that unless the trial court had entertained the erroneous impression that the prior art contains nothing pertinent to the claims, how could such finding of infringement be arrived at? There is a much larger surface in the center of Bray's cone over which the oil is spread in a thin film and down which it flows, and the only difference Trumble was able to point out to differentiate his supposed invention from Bray was that in Bray's device some of the oil would fall to the bottom of the separator without being spread in a thin film over the walls of the separator, by the action of imperforate cones. To find Model 2, either with the so-called nipple machined off or otherwise, an infringement, surely disregards entirely the limitations of the file wrapper contents of the Trumble application. If the patent in suit is construed so that any device in which the oil even *partly* strikes the wall of the chamber is an infringement, then every patent which we have mentioned or discussed in this brief is a complete anticipation. How, for instance, under such interpretation, can we escape the Cooper patent in which all the oil is directed in a ribbon-like stream upon the walls? Certainly, McIntosh spreads all the oil out in a thin film on extended surfaces of baffle-plates and under pressure.

Again, overlooking for the moment the broad general rule by which the *spirit* of the supposed contribution to the art controls the letter of the claim, as the rule is expressed in, for instance, Westinghouse v. Boyden, 170 U. S. 537, where the Supreme Court said:

> "We have repeatedly held that a charge of infringement is sometimes made out though the letter of the claims be avoided * * *. The converse is equally true"—

ignoring for the moment this broad principle of interpretation, let us look to the letter of the claims.

Claims 1 to 3, inclusive, of Trumble found to be infringed contain the elements described, substantially (Claim 2, for instance) as "means within the chamber adapted and arranged to distribute the oil over the wall of the chamber in a downwardly flowing film." This same element is referred to in Claim 4 as "means adapted to cause the oil to flow in a thin body for a distance."

It is Hornbook law that the omission of a single element of those mentioned in the combination of a claim from a defendant's device defeats a charge of infringement. Thus, in Pittsburgh Meter Co. v. Pittsburgh Supply Co., 109 Fed. 644, 651, the court said:

> "A claim for a combination is not infringed if any one of the described and specified elements is omitted, without the substitution of anything equivalent thereto."

Baker, Judge, in Adams v. Folger, 120 Fed. 260, states the same law in slightly different language as follows:

> "If a patentee claims eight elements to produce a certain result, when seven will do it, anybody may use the seven without infringing the claim; and the patentee has practically lost his invention by declaring the materiality of an element that was in fact immaterial."

Of course, it follows from this law that we must find some equivalent of each element of the claims of Trumble alleged to be infringed.

Now, it is equally a matter of settled law that the term "means" in a claim, followed by a statement of function, does not extend the scope of the claim in the least—the patentee is, nevertheless, limited to the precise element to which this broad language relates, and its obvious mechanical equivalents, and the term "mechanical equivalent" has a definite technical meaning.

Referring to the word "means" followed by a statement of functions, Walker on Patents (5th Ed.), page 138, Sec. 117a, says:

> "But such general language will not include all means, mechanisms, or devices which can perform that function, but only those which are shown

> in the patent and their equivalent,—citing many cases to which we refer the court."

The same author (Walker on Patents (5th Ed.), page 541, Sec. 358) says:

> "It is * * * safe to define an equivalent as a thing which performs the same function, and performs that function in substantially the same manner, as the thing of which it is alleged to be an equivalent."

We, therefore, turn to the specification and drawings of Trumble, as well as to his file wrapper contents, to discover what the parties to this patent contract meant when they used the language "means to distribute the oil over the wall of the chamber," etc. We have seen that Trumble defines this "means" very specifically, in connection with the statement of what he supposed he actually added to the art, as "an imperforate baffle-plate adapted to spread the whole body of oil to the outer edge of the vessel," *i. e., distribute the oil equally around and over all the walls of the chamber.* Manifestly, there is no such element in defendant's Model No. 2. This element is described as being *within* the chamber. The oil does not reach the chamber until it is discharged from the opening in the so-called bell-shaped nipple, and upon entering into the chamber falls in large part to the bottom of the chamber, only incidentally striking or splashing on the walls. We, therefore, submit that defendant's Model No. 2, either with the so-called nipple set against the partition or away from the partition, does not infringe.

## One Other Construction Made by Defendant Urged in the Trial Court to be an Infringement, but Upon Which the Trial Court did not Pass.

There was evidence concerning a separator of defendant, known as Tonner No. 3 trap. The trial court does not mention this separator, and the only findings of the court from which appeal is taken was, first, that defendant's patent infringes Claims 3 and 4; and second, that his Model No. 2 infringes Claims 1 to 4, inclusive, both of which findings we have, under immediately preceding headings of this brief, fully considered.

There is no cross-appeal in this case, and the *failure* of the trial court to find the trap known as Tonner No. 3 as an infringement is not before the court. Accordingly, the only reason that we mention this Tonner No. 3 separator is because the court evidently confused it with the device illustrated and described in defendant's patent. The court's description of what it supposed was defendant's patented device as shown and described in the Lorraine specification and drawings, comes much nearer to describing this Tonner No. 3 separator than it does defendant's patent. On the following page, in Fig. 3, we illustrate this separator.

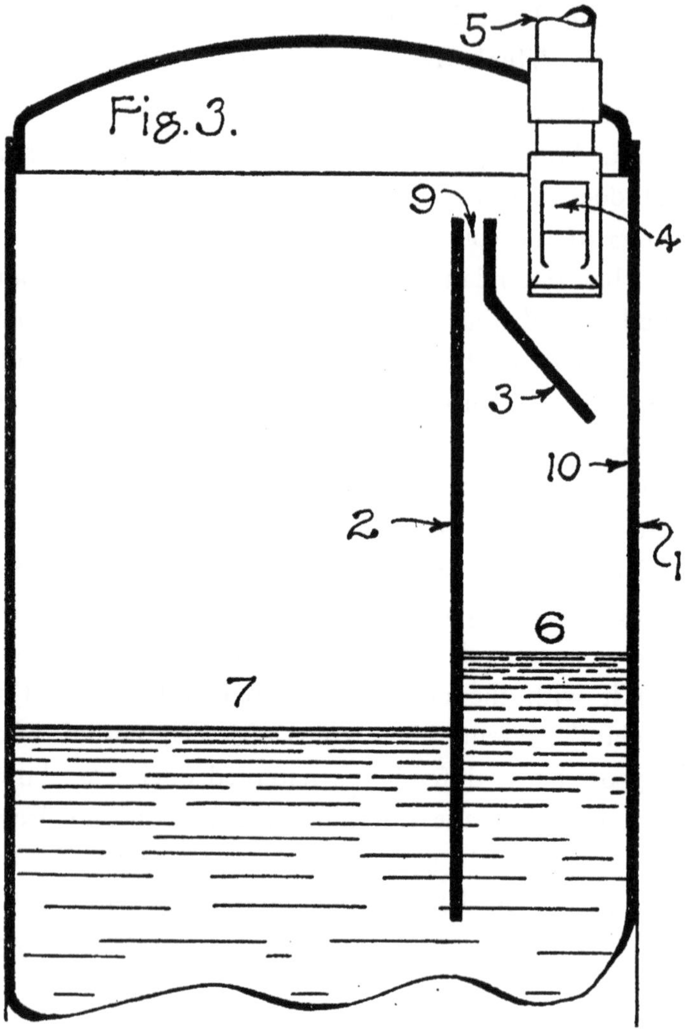


As with preceding Figs. 1 and 2, this Fig. 3 is drawn to scale and correctly represents the distance of the partition 2 from the wall of the separator 1, and also of the downwardly extending plate 3 from the wall of the separator 1. It also correctly illustrates the disposition of the inlet 4 and its opening into the chamber of the separator. It will be unnecessary to describe this construction completely by reference to the figures of the drawings, because such description is obvious from a consideration and understanding of Figs. 1 and 2, *supra.* The same description applies with exceptions which we now note: The part 3 extends entirely across the segment cut-off by the partition 2 having the same relative distance from the wall 10 at all points. The pipe has two openings, so that the stream of oil is divided and projected against this inclined plate 3 in two directions laterally. Of course, it strikes against and splashes off the plate and impinges upon the side of the separator; the distance from the opening to the plate 3 being relatively considerable. There is not a quiet, placid flow of the oil into this device. No infringement should be found for the reasons heretofore stated in discussing Figs. 1 and 2, and the further argument is to be noted that this device contains the objectionable feature upon which Trumble relied in differentiation of his invention from Bray, that the oil falls in drops or streams from the inlet. The angle of this plate 3, which is correctly shown in Fig. 3, is too great to permit a spreading out in a thin film and its distance from the wall shows that its design is not to permit any oil film to spread quietly over the wall, but to permit it to fall, in large part at least, violently to the bottom of the chamber. The oil is thus sprayed and broken up and agitated and is not spread in a thin film to the outer edge of the vessel. Manifestly, much of the oil will not touch the walls, even by splashing upon them. We submit this construction does not infringe, although we believe its consideration irrelevant to this appeal.

Respectfully submitted,

WESTALL & WALLACE,
(Ernest L. Wallace and Joseph F. Westall)
By JOSEPH F. WESTALL.
*Solicitors and of Counsel for Defendant-Appellant.*

No. 3945.

IN THE

# United States

# Circuit Court of Appeals,

FOR THE NINTH CIRCUIT.

| | |
|---|---|
| David G. Lorraine, *Appellant,* *vs.* Francis M. Townsend, Milon J. Trumble and Alfred J. Gutzler, Doing Business Under the Firm Name of Trumble Gas Trap Company, *Appellees.* | Suit on Trumble Patent No. 1,269,134 |

## BRIEF FOR PLAINTIFFS-APPELLEES.

Frederick S. Lyon,
Leonard S. Lyon,
Frank L. A. Graham,
*Solicitors for Plaintiffs-Appellees.*

Parker, Stone & Baird Co., Law Printers, 232 New High St., Los Angeles.

**No. 3945.**

IN THE

# United States
# Circuit Court of Appeals,

FOR THE NINTH CIRCUIT.

---

| | |
|---|---|
| David G. Lorraine, *Appellant,* vs. Francis M. Townsend, Milon J. Trumble and Alfred J. Gutzler, Doing Business Under the Firm Name of Trumble Gas Trap Company, *Appellees.* | SUIT ON TRUMBLE PATENT No. 1,269,134 |

## BRIEF FOR PLAINTIFFS-APPELLEES.

---

## STATEMENT OF THE CASE.

This cause comes before this court on appeal from the District Court for the Southern District of California, the interlocutory decree therein finding that United States Letters Patent No. 1,269,134 granted to Milon J. Trumble, Francis M. Townsend and Alfred J. Gutzler, plaintiffs-appellees, on June 11th, 1918, for

---

Italics appearing herein may be regarded as ours.

crude petroleum and natural gas separators, the invention of Milon J. Trumble, one of the plaintiffs-appellees, are valid, claims 1, 2, 3 and 4 infringed, and granting an injunction.

The patent is for a "Crude Petroleum and Natural Gas Separator."

Petroleum oil as it comes from a well generally contains more or less gas which must be eliminated before the oil is suitable for commercial purposes or storage. The product of a well consists of a mixture comprising a liquid which under ordinary conditions is always a liquid; a fixed gas which under ordinary conditions is always a gas; and an intermediate product, principally gasoline. Under ordinary well pressures, this intermediate product of gasoline remains in a liquid state, but, if the well pressure be released, will tend to become vaporized.

It has been and is the practice to pass the oil as it comes from the well through a suitable feed-line and into storage tanks, which tanks are under atmospheric pressure. During the passage of the oil from the well to the storage tanks, it has been and is common to separate the gas from the oil in order that the gas may be saved and utilized, and in order that the oil may be passed to the tanks free from gas. The device or apparatus for accomplishing this purpose is called in the art a "gas trap." The device of the patent in suit is a gas trap, but it was designed with a new object in view. Prior to the Trumble invention, the designers and users of gas traps had in mind only separating or extracting the fixed gas from the oil, the fate of the gasoline being ignored. As a necessary incident to the operation of the prior art gas traps, the intermediate mixture of gasoline was to a large extent separated from the oil and carried away with the gas. The prior gas traps made no provision to prevent this loss of the most valuable part of the oil. Until taught differently by the Trumble invention, the art proceeded on the theory that the loss of a considerable portion of the gasoline was a necessary incident to the separation of the fixed gas from the oil.

Trumble recognized the enormous loss to the oil industry growing out of this uncontrolled dissipation of gasoline incidental to the separation of the fixed gas from the oil. He set about producing a gas trap which would effectively separate the fixed gas from the oil, but which would prevent the gasoline from passing out of the oil with the fixed gas, and which would maintain the gasoline in the oil. The invention of Trumble, in its broadest aspect, was the striking discovery made by him that if a mixture of oil, gas and gasoline be maintained under pressure in the gas trap until after the separation of oil and gas had occurred, that the pressure will have a diametrically opposite effect on the gasoline and on the fixed gas. That is to say, the pressure will, on the one hand, maintain the gasoline in the oil in its liquid form, and, on the other hand, will squeeze out the fixed gas from the oil and effect a clean-cut separation. This was the introduction of an entirely new and unexpected principle in gas traps. Prior to the invention of Trumble, such teaching as there was relating to the effect of pressure on a gas trap, led to the natural belief that pressure on a gas trap would seriously impede the separation of the fixed gas from the oil. It was an elementary principle of physics, as recognized by the so-called Law of Henry, that the solubility of a gas in a liquid increases in accordance with any increase of pressure. Naturally, it would be supposed, from the Law of Henry, that the more pressure put on a gas trap, the less efficient would be the separation of the gas from the oil, which, of course, is a primary prerequisite of a gas trap. Trumble was the first to recognize that pressure could be applied to a gas trap for the purpose of holding gasoline in the oil without interfering with the separation of the fixed gas from the oil.

The introduction to the art by Trumble of this new principle, based upon his discovery of the opposite effects of pressure on gasoline and fixed gas, was recognized by the industry to be of great commercial value. At the time of the Trumble invention, as today, gasoline is the most valuable portion of the oil. The Trumble invention disclosed for the first time how to prevent the loss and dissipation of gasoline theretofore considered necessarily incidental to the separation of the fixed gas from the oil. The Trumble invention saves for the user of the trap, the value of the gasoline that would otherwise pass away from the oil with the fixed gas. Oil is ordinarily bought and sold on a sliding scale, the price increasing as the gasoline content of the oil increases. The operator's saving from the use of the Trumble invention appears in the increased value of his oil.

Upon the introduction of the Trumble trap, the industry recognized and appreciated the enormous saving that could be agorded by the use of the Trumble invention. Trumble gas traps went into widespread and general use, displacing other gas traps, and became, and are today, the standard gas trap in use in the oil fields of the United States and foreign countries. The evidence shows that at the time of the trial of this case, 583 Trumble traps had been sold for use in California, Texas, Louisiana, Arkansas, Oklahoma, Wyoming, and foreign countries. The importance of these traps to the oil industry may be appreciated from the testimony of the witness Paine, to the effect that one of these traps alone saved $125 a day (over $45,000 a year) for the Honolulu Oil Company, at Taft, California. [R. 117.]

Mr. Paine, a director of the Union Oil Company of California, and a man highly skilled in this art, testified as to the advantage of a Trumble trap employed by him, as follows:

"The ultimate effect therefore of the trap was first to conserve and utilize about one million feet per day of gas worth $50 per day, an increase in the value of the oil of about $60 per day, and an increase in actual oil saved from the well of about 40 barrels per day, having the value of, at that time, of about $15 per day." [R. 79.]

The scientific theory or basis of the operation of the Trumble trap, and the reason why pressure, as discovered by Trumble, has a diametrically opposite effect on fixed gas and gasoline in a gas trap, is explained by Mr. Paine as follows:

"That is due to a rather deep question of physics that has come into importance in connection with the manufacturing or increasing of gasoline from natural gas, the principle of 'partial pressures' as it is called, which is this: If there are some vapors of gasoline in a gaseous state in gas and these vapors are in a comparatively small proportion, such vapors if alone not mixed with other gases may condense at comparatively low pressure. One can have a gas for instance which is a gas vapor which is condensable at 10 pounds pressure if applied to it. Now, if that gas is mixed with other gases which are practically non-condensable in proportion of 5 per cent, say a 10-pound or 20-pound pressure then applied to the gas will not condense those vapors, but the pressure must be according to increase in the ratio of this dilution. If it were present there at 5 per cent then the pressure would have to be increased twenty times that which was required to condense it when it is alone and not intermingled with these non-condensable gases." [R. p. 80.]

That is to say, as long as the gasoline is kept as a liquid in the oil and not permitted to vaporize, a comparatively small pressure, say a pressure of 20 pounds per square inch, will hold this gasoline in the oil without any danger of its becoming a vapor. If, on the other hand, the gasoline is allowed to vaporize and to mix with the fixed gases, pressures as high as 200 pounds per square inch must be resorted to to condense them back into a liquid. It is thus possible, by maintaining a comparatively small pressure on the interior of the Trumble trap, to simultaneously prevent the vaporization of the gasoline without materially affecting the fixed gas in the trap, since the fixed gas requires a very large pressure to effect any appreciable condensation thereof. While it is now possible for Mr. Paine to point out the reasons for the success of the Trumble invention, there is not evidence in the prior art to indicate that this was understood or appreciated prior to Trumble's practical application of this then unknown result and his disclosure of the same to the art.

A further advantage of the Trumble trap is emphasized by Mr. Paine in the clean-cut separation of fixed gas from the oil that is effected by the Trumble trap. This is important, as pointed out by Mr. Paine, because, if this separation is not clean-cut, and a portion of the fixed gas remains in the oil that is passed to the storage tank, the fixed gas (being always a gas) will thereafter percolate out of the oil and rise from the oil in the storage tank and will carry with it gasoline vapor. This is explained by Mr. Paine in his testimony herein as follows:

"The reason I will give for that will be this: That at the higher pressure, still higher pressure maintained upon that well certain gases were retained in the oil which went into the solution in the oil. Now, these gases were always gases. They were not gasoline vapors that had been retained in the oil in virtue of maintaining this higher pressure on the trap, and those gases which are fixed gases and are incondensable under the ordinary agreement of pressure or temperature, escaped from the oil in the shipping tank, and it is altogether possible—I observed it on occasion—that the escaping of those gases carried along with them small quantities of gasoline vapor, so that I have had instances, similar instances where, through the maintaining of an unduly high pressure we arrived at quality of oil which was lower than that which would be obtained in a pressure range of from 75 to 100 pounds held on the trap." [R. 76, 77.]

Some years after the invention of Trumble, and after the Trumble gas trap had become standard in the oil fields of this country, the defendant, David G. Lorraine, invented a valve adapted for installation on a trap of the Trumble type. The defendant was familiar with the Trumble trap, and went to the owners of the Trumble trap to interest them in installing his valve on Trumble traps. The owners of the Trumble trap were not interested in the Lorraine valve, because the valve used on the Trumble trap was entirely satisfactory and they saw no reason for changing the same. Finding that he could not interest the owners of the Trumble trap in his valve, the defendant, Lorraine, proceeded to appropriate the Trumble invention, making only immaterial mechanical variations in an endeavor to escape the Trumble patent, but employing in the trap made by the defendant, the fundamental principles discovered by Trumble and all of the advantages of the Trumble invention. Thereupon, the owners of the Trumble patent brought this suit against defendant for infringement of the Trumble patent.

The bill of complaint is in ordinary form for infringement of a patent, and prays for an injunction and an accounting. The answer of the defendant does not deny or place in issue the validity of the Trumble patent. At the trial, defendant conceded the validity of the Trumble patent. [R. 61.] The only defense was one of non-infringement; but it conclusively appeared at the trial of the case that the devices made and sold by defendant differed only in immaterial mechanical variations from the device illustrated in the drawings of the Trumble patent, and employed and appropriated to the fullest extent the invention made by Trumble.

The case was tried before the Honorable Charles E. Wolverton, sitting in the Southern District of California by special designation, and the trial consumed five days in open court. During the trial, Judge Wolverton heard the testimony of the various witnesses who testified concerning the construction, the mode of operation and the principle of the devices involved, and the advantages and results attained by the Trumble invention. During the trial, Judge Wolverton made a trip to West Alhambra, California, and attended a demonstration showing the operation of one of the infringing devices that had been sold by the defendant. In the early stages of the trial, the defendant strenuously denied that he employed any means for placing pressure in his traps and contended, therefore, that he did not infringe upon the Trumble patent. As the trial went on, and after Judge Wolverton had attended the demonstration at West Alhambra and had heard the testimony of an official of the General Petroleum Corporation who had purchased and used a number of defendant's traps, it became obvious that this position of defendant was untenable; and finally counsel for defendant stipulated that pressure was maintained in the defendant's traps.

The case was submitted on briefs to Judge Wolverton, and the latter took the case and briefs to Portland for consideration. In the brief for defendant submitted to Judge Wolverton, the defendant stated:

"At the outset of this argument we wish to state that, for the purposes of this case, the defendant is willing to admit there is more or less pressure in its separators or traps during actual operation; so there is no need to discuss this question further.

"The only defense relied upon by this defendant in this case is that of non-infringement."

In the answer to the original bill, defendant alleged, in paragraph 12, that he had then pending an application for letters patent describing and covering the gas trap complained of by the original bill of complaint herein, which paragraph 12 further states:

"And this defendant states that the device for separating natural gas from crude petroleum which he has made is made according to the specifications and claims made and so allowed in his said application for letters patent, and for which letters patent will be issued in due time, and which said device the defendant states does not, in any manner whatsoever, infringe the alleged patent of the plaintiffs."

Subsequent to the filing of this answer, and prior to the trial of the case, Letters Patent No. 1,373,664 (Book of Exhibits, pp. 60-65) were granted to the defendant upon the application referred to in defendant's answer. At the trial, counsel for defendant admitted that patent No. 1,373,664 was granted on the application referred to in defendant's answer [R. 52], and that it showed the device defendant was making and selling. It should, however, be noted that defendant's patent No. 1,343,664 has been reissued as Reissue No. 15,220, shown in Book of Exhibits, pages 66-74. Subsequent to the filing of the bill of complaint herein, plaintiffs discovered that the defendant had altered the construction of his device, and on the Saturday preceding the opening of the trial before Judge Wolverton, plaintiffs caused to be purchased from defendant one of defendant's latest devices (Book of Exhibits, p. 83). The construction of this trap is shown on page 81, Book of Exhibits. At the hearing, upon the suggestion of Judge Wolverton [R. 146], plaintiffs obtained leave to and filed a supplemental bill bringing before the court defendant's latest construction or type of device so purchased and charging infringement thereby. [R. 28.] The defendant filed an answer to this supplemental bill, denying that defendant's modified device infringed upon the Trumble patent, but again failing to contest the validity of the Trumble patent. [R. 42, 43.]

The case was submitted to Judge Wolverton on March 18, 1922, and on September 11, 1922, Judge Wolverton filed a written opinion [R. 530] in detail considering and passing upon all of the contentions made before him by the defendant. This opinion is reported in 283 Fed. Rep. 806. At the outset of his opinion, Judge Wolverton states:

"The defendant does not question the validity of complainant's patent, but claims that he does not infringe, for two reasons: * * *." [R. 530.]

In his opinion, Judge Wolverton refers to the original form of defendant's device, which form was alleged in the answer of the defendant to conform to defendant's patent No. 1,373,664, as "Model No. 1." Defendant's modified device, presented by the supplemental bill herein, and shown on page 81, Book of Exhibits, is referred to by Judge Wolverton as "Model No. 2." The opinion of Judge Wolverton concludes as follows:

> "I find the defendant's patent infringes claims 3 and 4 of complainants' patent, that his model No. 2 infringes claims 1 to 4 inclusive, and that claim 13 is not infringed."

Upon the filing of the opinion, counsel presented to Judge Wolverton a draft of a decree, which decree was duly entered by Judge Wolverton on September 26, 1922. [R. 543-547.] The decree herein follows the opinion of Judge Wolverton, and adjudges:

"That the validity of said patent was not denied or put in issue by defendant in the above case; that said letters patent are good and valid in law, particularly as to claims 1, 2, 3 and 4 thereof;" [Par. 1, R. 543.]

"that the apparatus made and sold by defendant referred to in this case as Defendant's Model No. 1, and described in reissued letters patent of the United States, No. 15,220 granted November 8, 1921, to defendant, infringes upon said claims 3 and 4 of plaintiff's said patent, and that the apparatus made and sold by defendant referred to in this case as Defendant's Model No. 2, infringes upon said claims 1, 2, 3 and 4 of plaintiff's said patent both when made with the nipple machined off on one side to sit closely against the partition wall as illustrated in said Model No. 2, and when made without such machining or setting, as defendant claims such devise was intended to be constructed;" [Par. 2, R. 545.] and directs an injunction and an accounting accordingly.

The assignments of error filed herein by defendant [R. 549] are prolix and totally disregard that the validity of the Trumble patent was conceded by the defendant at the trial of this case. Error is assigned, that the District Court erred "in failing and refusing to find and decree that said Letters Patent No. 1,-269,134, granted on June 11, 1918, to plaintiffs, was null and void in law for want of patentable invention over the prior art."

It is a fundamental principle of equity practice that points not raised before the trial court cannot be considered by the Appellate Court:

Bell v. Bruen, 1 How. 169;
Alviso v. U. S., 8 Wall. 337;
National Bank v. Commissioners, 9 Wall. 353;
Rogers v. Ritter, 12 Wall, 317;
Klein v. Russell, 19 Wall. 433;
Supervisors v. Lackawanna Co., 93 U. S. 619;
Wilson v. McNamee, 102 U. S. 572;
Wood v. Weimer, 104 U. S. 786;
U. S. v. Amer. Bell. Telephone Co., 106-7 U. S. 224.

Appellant's opening brief herein proceeds in total disregard of the issues before the District Court, and in total disregard of the admissions and testimony upon which Judge Wolverton decided this case and made his decree herein. The principal contentions made in appellant's opening brief are not based upon the record in this case, but are apparently based upon the theories of the present counsel for appellant as to what he thinks the facts ought to be, totally disregarding the evidence before Judge Wolverton. This case is before this court for determination upon the record made and presented, and we regret that most of the consideration that this court will have to give to this case will necessarily be taken up by a discussion of contentions made by appellant without regard to the record herein. For example, Judge Wolverton considered all of the evidence and explicitly found as a fact that the maintenance of pressure on a trap for the purpose above stated, was novel and original with Trumble. There was no testimony given, or claim to the contrary made, before Judge Wolverton. In appellant's opening brief, present counsel for appellant now apparently urge that every prior patent for, and every prior gas trap, must have employed this pressure principle. Appellant's opening brief in no manner suggests where such contention is supported by any testimony in this case. The unwarranted character of such a contention is established by the testimony given by defendant himself, that gas traps may be operated with a pressure above atmospheric pressure, at atmospheric pressure, or even under a vacuum. [R. 379.] Indeed, defendant, at the early stages of the trial of this case, testified that he did not intend that pressure should be used in his trap. [R. 298.] What, then, becomes of the contention in appellant's opening brief that all prior art gas traps necessarily operated under pressure? The testimony in the record is directly to the contrary. For example, the witness Paine, a director of the Union Oil Company, testifies that when he was operating the Honolulu Oil Company at Taft, California, and the proposition of installing a Trumble trap was first presented to him, he hesitated, because he did not know what effect the use of pressure would have, and he first operated the Trumble trap without any pressure and gradually increased the pressure from day to day and felt his way along to see what the effect would be. [R. 72.] Mr. Paine in his testimony gave figures showing the saving resulting from the use of the pressure principle in the Trumble trap, as distinguished from running the same trap without the pressure principle, thereby concretely establishing the enormous benefit of such pressure principle. [R. 76.]

Before taking up the more detailed portions of the case, the court should be fully informed as to the nature of the invention and the part which it has played in the commercial art. Judge Wolverton held that "utility has been abundantly proven by the success achieved by complainant's device." This court has repeatedly held that such commercial success is evidence of invention and entitles the patent to favorable consideration and interpretation. It is our position that the Trumble patent is for a broad and generic invention and introduces to the art a new and controlling principle, providing a device operating by a novel mode of operation and productive of results and answering a purpose prior to Trumble unthought of and not understood. Trumble was the first to conceive and provide means for separating fixed gas from oil without dissipating a substantial portion of the gasoline content of the oil. The Trumble invention provides a device which is not only a gas and oil separator, but is at the same time *a gasoline saver.* In this, Trumble was a pioneer, for he gave to the art a totally new thought.

## The Trumble Invention.

We have stated that, in its broadest aspect, the Trumble invention comprises the discovery or conception that a gas trap, if maintained under pressure, will prevent the loss of gasoline without interfering with the separation of the gas from the oil. The advantages of maintaining pressure on a gas trap having been conceived, it was necessary for Trumble to go further and devise means whereby the advantage of the pressure principle could be availed of in a gas trap. Obviously, if the mixture from the well was merely shot rapidly through a tank, the pressure feature would be of little benefit, because it would have small opportunity to effect its action on the gasoline and gas. Trumble perceived that the passage of the mixure through the gas trap must be at a reduced speed, and under such conditions, as would give the pressure element the necessary time to effect its action; in other words, the mixture as it comes from the well must be slowed down in the trap to give the pressure plenty of time to act. Trumble further saw that, to obtain the maximum benefit of the pressure principle, it was necessary to spread out or distribute the oil in order that the pressure element would have the maximum opportunity to act. He appreciated that if the mixture as it passed through the trap was kept in one bulk or body, the gas entrained in the oil must necessarily travel for a relatively long distance through the oil to escape, and the separation would be incomplete and hampered. It is the same proposition as attempting to wring water out of a bathing-suit,—for everyone knows that the most efficient way to separate or wring water out of a bathing-suit is to divide the suit up into small portions and squeeze each portion individually.

A narrower aspect of the Trumble invention is the provision by Trumble of means to slow down and spread out the oil in its passage through the trap, thereby permitting a maximum pressure effect and the readiest separation of the gas from the oil.

The main advantage of the Trumble invention is set forth in the specification of the Trumble patent as follows:

"My invention is also effective in maintaining the lighter gravity series of the crude oil in combination with the heavier series of the oil, thereby producing from oil wells a product of lighter gravity than where the oil and gas is permitted to separate on exposure to ordinary atmospheric conditions."

At the close of the specification of his patent, Trumble takes particular care to state that he desires to be understood as pointing out and claiming the advantage and action of the pressure principle in gas traps, as follows:

"It will be noted that the action upon the oil while flowing down the wall of the expansion chamber in a thin film under pressure permits the free, dry, gas to readily escape therefrom, while the pressure exerted upon the oil surface backed by the wall of the chamber holds the lighter liquids, such as gasoline, in combination with the oil body, and I desire to be understood as pointing out and claiming this action as being of great benefit to the crude oil derived from the well on account of keeping the gasoline series in combination with the main body of the oil."

There can be no doubt that the Trumble patent clearly expresses the intent to cover the broad aspects of the Trumble invention. Having specifically set forth his invention in its broad aspect, and having stated that he wanted to be understood to claim the full benefit thereof, Trumble proceeded in his patent to describe the one form of device in which his invention could be embodied. The Trumble invention resides in the striking discovery made by Trumble that pressure, if maintained in a gas trap, would maintain the gasoline in the body of the oil without interfering with the separation of the gas. The device illustrated in the Trumble patent is not the Trumble invention, but is merely one embodiment thereof. As said by the Supreme Court in Smith v. Nichols, 21 Wall. 112:

> "A patentable invention is a mental result. * * * The machine, process or product is but its material reflex and embodiment."

Before proceeding in his patent to describe a material embodiment of his invention, Trumble was careful to point out that:

"The accompanying drawings illustrate my invention and *one method* of its application:"

The illustrative embodiment of the Trumble invention described in the Trumble patent is best represented in figure 2 of the patent drawings. (Book of Exhibits, p. 51.) The device consists principally of a tank of expansion chamber designated by numeral 1. Oil is admitted to the expansion chamber 1 through a pipe 7 leading from the casing head 10 of a well (illustrated in figure 1 of the patent drawings, Book of Exhibits, p. 50.) As the oil enters at the top of the expansion chamber 1, it comes in contact with a spreading or distributing surface 22 which is of cone or umbrella shape. The passage of the incoming mixture is retarded by the cone and the oil, distributed into a relatively thin body, is directed to the inner surface of the shell of the expansion chamber 1. The mixture passes downwardly through the expansion chamber in a thin body, and during such passage it is acted upon by pressure, the gas readily escaping from the oil, and the gasoline being maintained in the oil by the pressure. The oil flows into the bottom of the expansion chamber 1 and passes therefrom through the outlet 12, the level of the oil in the chamber being maintained by a valve 41 controlled by a float 43. The gas passes upwardly from the expansion chamber to the under side of the cones and out the pipe 10. The gas outlet pipe 10 is provided with a pressure regulating valve 11 (illustrated in Figure 1 of the patent drawings; Book of Exhibits, p. 50) to maintain on the trap the desired pressure of the well. To regulate such pressure to any desired degree, the pressure regulating valve 11 is provided with an adjustable weight.

Comparing the particular device illustrated in the Trumble patent drawings with the broad aspects of the invention of Trumble, and remembering that Trumble explicitly pointed out that his patent drawings illustrate only "one method" of embodying his invention; we find that the device of the patent drawings provides means for maintaining pressure within the chamber (the pressure regulating valve 11 controlling the natural gas pressure of the well), and provides means for slowing down and spreading out the oil to afford an opportunity for the pressure to act thereon (the cones 22 and 22a). Trumble is careful in his patent specification to state that the cones 22 and 22a are merely examples of means for dividing or spreading the oil, stating:

"Oil dividing means are provided interior of the expansion chamber, *such as* cones 22 and 22a."

This statement is further evidence that Trumble appreciated and intended to claim the full benefit of the broadest aspect of his invention, and that the arrangement illustrated in his patent drawings was by way of example only.

The intent of Trumble to cover the broad aspect of his invention is further evidenced by the claims of his patent. Claim 4 is one of the claims held by Judge Wolverton in this case to have been infringed, both by the original and modified types of defendant's infringing trap referred to by Judge Wolverton as Model 1 and Model 2. Claim 4 reads as follows:

"In an oil and gas separator, the combination of an expansion chamber, means for delivering oil and gas into the chamber, means for maintaining pressure within the chamber, means for drawing oil from the chamber, and means within the chamber adapted to cause the oil to flow in a thin body for a distance to enable the gas contained and carried thereby to be given off while the oil is subjected to pressure."

It is apparent from a reading of this claim, that Trumble intended to avail himself of a the full benefit of his invention, and that the Patent Office, in granting the claim, recognized and approved of such intent. Analyzing the claim into its elements, it will be found that none of these elements are restricted specifically, and that each of the elements expressly avails itself of various equivalents. For example, the first element is "an expansion chamber," without limitation as to any particular type of expansion chamber. The next element is "means for delivering oil and gas into the chamber," likewise unlimited as to the form or details thereof. The next element consists of "means for maintaining pressure within the chamber," and such element is obviously not limited to any particular form or detail of valve or other mechanism. The next element of the claim is "means for drawing oil from the chamber," and by its terms contemplates and includes various specific arrangements. The last element of the claim is "means within the chamber adapted to cause the oil to flow in a thin body for a distance to enable the gas contained and carried thereby to be given off while the oil is subjected to pressure." This last element could not more explicitly or accurately convey to the reader the crux of the Trumble invention. It embodies the slowing down and spreading of the oil to permit the operation of the pressure maintaining principle. Coupled with the statement above quoted from the Trumble specification, wherein Trumble states that the cones 22 and 22a are "oil dividing means," the intent of Trumble to avail himself of the full benefit of the various and sundry specific arrangements that could be provided for this purpose, is made manifest. It is absurd to contend that Trumble's retarding and spreading invention is limited to any particular form of surface. Any number of surfaces are adaptable.

Claims 1, 2 and 3 of the Trumble patent, also held by Judge Wolverton to have been infringed, are directed in varying language to the broad aspect of the Trumble invention. The differences in phraseology between these various claims will appear best from a reading thereof, and, for our purpose here, need not be repeated as the differences between these claims are unimportant in this case.

The Trumble invention is of a fundamental and pioneer character, for it introduces to the art a new principle productive of a result never before attained. It is not limited to specific forms or details of construction. The Trumble patent claims are clearly and definitely drawn, with a breadth of scope adequate to protect Trumble in his invention. The Trumble invention, and the scope to be accorded to the Trumble patent, may be paraphrased by the words of the Supreme Court in *Winans v. Denmead,* 15 How. 340, as follows:

> "Its substance is a new mode of operation, by means of which a new result is obtained. It is this new mode of operation which gives it the character of an invention, and entitled the inventor to a patent; and *this new mode of operation* is, in view of the patent law, *the thing entitled to protection.*"
>
> (14 Law. Ed. 721.)

In *Winans v. Denmead, supra,* the Supreme Court further said:

> "There being evidence in the case tending to show that other forms do in fact embody the plaintiff's mode of operation, and, by means of it, produce the same new and useful result, the question is, whether the patentee has limited his claim to one out of the several forms which thus embody his invention.
>
> "Now, while it is undoubtedly true that the patentee may so restrict his claim as to cover less than what he invented, or may limit it to one particular form of machine, excluding all other forms, though they also embody his invention, yet such an interpretation should not be put upon his claim if it can fairly be construed otherwise, and this for two reasons:
>
> "1. Because the reasonable presumption is, that, having a just right to cover and protect his

whole invention, he intended to do so. (Haworth v. Hardcastle, Web. P. C. 484.)

"2. Because specifications are to be construed liberally, in accordance with the design of the Constitution and the patent laws of the United States, to promote the progress of the useful arts, and allow inventors to retain to their own use, not anything which is matter of common right, but what they themselves have created. (Grant v. Raymond, 6 Pet. 218; Ames v. Howard, 1 Sumn. 482, 485; Blanchard v. Sprague, 3 Id. 535, 539; Davoll v. Brown, 1 Wood & M., 53, 57; Parker v. Haworth, 4 McLean, 372; Le Roy v. Tatham, 14 How. 181, and opinion of Parke, Baron, there quoted; Neilson v. Harford, Web. P. C. 341; Russell v. Cowley, Id. 470; Burden v. Winslow, decided at the present term, 15 How.)"

(14 Law Ed. 721.)

Judge Wolverton, after a full consideration of all the evidence in the case, concluded that the Trumble invention introduced a new principle to gas traps and obtained a new result:

"It is argued that the principle of subjecting oil to pressure, for the purpose of keeping lighter hydrocarbons in solution in the oil while the dry gas constituent separates from the body of the oil, is old, but this overlooks the theory of complainants that they have discovered a more efficient way of separating the gas from the oil, whereby a greater proportion of oil value is secured than had theretofore been derived by the use of any trap in existence or previously operated." [R. 541.]

Judge Wolverton also clearly recognized the character of the Trumble patent as a combination patent:

"It must be observed that we are dealing with a combination patent, and all the elements must be read with reference thereto." [R. 539.]

Under these circumstances, the Trumble patent is entitled to a generous interpretation and protection against the use of mechanical equivalents thereby to cover and protect the patentee in the new mode of operation and principle discovered and introduced by him. As said by this court in *Smith Cannery Machines Co. v. Seattle-Astoria I. W.,* 261 Fed. 85, at page 88:

> "Where a combination patent marks a distinct advance in the art to which it relates, as does the appellant's invention here, the term 'mechanical equivalent' should have a reasonably broad and generous interpretation, and protection against the use of mechanical equivalents in a combination patent is governed by the same rules as patents for other inventions. Imhaeuser v. Buerk, 101 U. S. 647, 25 L. Ed. 945."

## Infringement.

The truth as to the mode of operation of defendant's infringing trap is best obtained from the description thereof contained in the patent applied for by defendant before this litigation arose. In his answer filed in this case, defendant has stated that his device "is made according to the specifications and claims" now embodied in his patent. [R. 14.] The defendant's patent No. 1,373,664 contains a full disclosure of the construtcion of defendant's infringing trap "Model 1'" and of the mode of operation thereof, and is illustrated in figure 4 of the drawings of defendant's patent reproduced on page 61, Book of Exhibits. This trap comprises a tank or expansion chamber 2 to which oil from a well is introduced at the top of the expansion chamber through a pipe 14. Upon entering the expansion chamber 2, the progress of the oil is retarded by an inclined baffle-plate 17, which slows down the oil and distributes the same upon the inner surface of the shell of the expansion chamber 2. The oil passess to the bottom of the expansion chamber and leaves the tank by means of a pipe 33 controlled by a valve 34 operated by a float 56 as shown in figure 1, Book of Exhibits, page 60. The gas passes up to the top of the expansion chamber 2 and out through a gas pipe variously numbered 23, 25 and 27 in figure 1. Pressure is *maintained* in the trap by means of a gas pressure regulating valve 28, shown in figure 3, Book of Exhibits, page 60, which is exactly the same in construction and operation as the valve 11 of Trumble, even having an adjustable weight thereon.

The identity of mode of operation between the defendant's infringing trap and the Trumble trap, and the full extent to which the infringing device appropriates the invention of Trumble in all its aspects, is immediately apparent from a reading of the specification of the Lorraine patent. Indeed, the Lorraine patent extolls the advantages of maintaining pressure in the trap as follows:

"As above mentioned there is maintained in the receptacle a gas pressure as determined by the adjustment of the pressure regulating valve 28 * * *" (Book of Exhibits, p. 63, lines 111 to 114.)

"A further object of the invention is to provide means for maintaining a pressure continuously on the oil or emulsion as it is continuously admitted to the separating chamber so that the emulsion may be subjected to sufficient pressure to express the gaseous constituent of the emulsion from the mixture." (Book of Exhibits, p. 62, lines 30 to 37.)

"From the above it will be seen that I have provided a method for separating or facilitating the separation of the gas and oil and separately discharging the same from emulsions; and furthermore have provided a method in which, by maintaining a predetermined pressure in the oil receptacle, the latter is subjected to pressures having the effect of expressing the gaseous content from emulsions, the gaseous constituent in the emulsion being driven from the denser liquids by the increase in the pressure on the oil within the receptacle 2. This, therefore, prevents the loss of the valuable gaseous constituent *such as occurs in apparatus in which the oil passes immediately from a well or other source to an apparatus in which it is subject only to atmospheric pressure.*"* (Book of Exhibits, p. 64, lines 41 to 63.)

———*N. B. *How* can appellant reconcile its brief with this recognition that prior traps operated without pressure?

We cannot conceive of a fuller or more explicit demonstration that the Lorraine trap, in theory and in mode of operation, incorporates the very crux and broadest aspect of the Trumble invention. The defendant in his patent recognizes the value of the pressure principle first conceived by Trumble, and expressly recognizes the diametrically opposite effects in a gas trap of pressure upon gasoline and gas. It should be borne in mind that the defendant's patent was applied for many years after the Trumble trap had become standard in the oil fields of this country, and *pro tanto* vouches for the utility and advantage of the novel principle first conceived of and incorporated in a gas trap by Trumble and fully described in his patent.

Passing to the narrower aspect of the Trumble invention, it is found from the Lorraine patent specification that defendant's infringing device coincides even with the narrower aspect and appropriates the narrower as well as the broader advantages of the Trumble invention. It will be remembered that the narrower aspect of the Trumble invention was the conception of slowing down and spreading out the mixture in order that the pressure might effectively operate thereon. In the Lorraine patent specification, it is stated that:

"the oil is showered onto the adjacent portion of the receptacle wall whence it flows downwardly between the wall and the partition 19, any gases being liberated rising to the top of this compartment and passing over."

Here is identity with the narrower aspect of the Trumble invention, to-wit, the slowing down and spreading out of the oil mixture in order that the pressure may act thereon to separate the gas from the oil and maintain the gasoline in the oil.

In appellant's opening brief, pages 95 to 97, inclusive, counsel shows and describes a form of trap called by him the Tonner pressure trap. A comparison of the trap on page 96 of appellant's opening brief with figure 4 of the Lorraine patent, found on page 61 of the Book of Exhibits, shows that the so-called Tonner trap is, in all substantial particulars, the trap shown in the Lorraine patent. Moreover, as counsel ingeniously states on page 95, the court's description of defendant's patented device as shown and described in the Lorraine specification and drawing, comes much nearer to describing the Tonner No. 3 separator than does the exact drawings of defendant's patent. This Tonner trap was treated by all parties and the court as being the device described in defendant's patent. The Tonner trap, for all purposes identical with the construction illustrated in the defendant's patent, was referred to in the lower court as "Model 1." In the brief filed before Judge Wolverton, counsel for defendant stated:

"Plaintiffs have brought in two different models of defendant's device, each of which differs materially from the other in construction and action. We will, therefore, discuss each of these separately, and will designate the one described as being located on well No. 3 of the Tonner lease as "Model No. 1," and the one which was used in the demonstration, as "Model No. 2."

The device referred to as "Model No. 2" is the modified form of defendant's device presented by the supplemental bill herein, and illustrated in Plaintiffs' Exhibit No. 11. (Book of Exhibits, p. 81.)

## Model No. 1.

In his opinion herein, Judge Wolverton describes the infringing device "Model No. 1" as follows:

"Defendant's patent, referred to in counsel's brief as Model 1, has an inner partition set away from the wall on one side more than one-third the distance of the diameter of the chamber, and extending below the oil level. To this partition, at some distance from the top of the chamber, is attached a baffle-plate extending downward on an incline of perhaps 45 degrees, and to within an inch and a half or two inches from the wall for the entire segment cut off by the partition. The oil inlet, consisting of a pipe, extends downward to within a short distance of a baffle-plate. The pipe has two openings, so that the stream of oil is divided and projected on the baffle-plate in two directions laterally. The device is provided with a gas take-off above the partition and one from underneath the baffle-plate; all to pass off eventually from the upper portion of the major chamber." [R. 537.]

The inclined baffle-plate employed in this device to slow down and spread out the incoming liquid mixture, is illustrated at 17 in defendant's patent. (Book of Exhibits, p. 61.) The evidence in the case is clear,

Judge Wolverton found, that means were employed in this model 1 type of trap to maintain a pressure on the trap. The defendant's patent illustrates the gas pressure regulating valve 28 for the gas outlet pipe. (Book of Exhibits, p. 60.) It was also demonstrated to Judge Wolverton that a gas outlet valve 26 shown in pipe 25 in figure 1, Book of Exhibits, page 60, constituted a further means for maintaining a pressure on this trap. [R. 371.] Mr. McLaine, the director of production for the General Petroleum Company, having charge of the Tonner lease, testified as follows:

"Q. (By Mr. L. S. Lyon): After the adjustment what pressure, if you remember, was maintained in that trap?

A. I believe it was in the neighborhood of 28 pounds, as nearly as I can remember." [R. 512.]

Judge Wolverton's conclusion as to the infringing trap "Model No. 1" is as follows:

"The defendant's trap, Model No. 1, infringes, in that the baffle-plate furnishes a surface down which the oil flows, with pressure against the oil, by which the gas escapes from the oil and passes out of the chamber by the take-off." [R. 541.]

The description in appellant's opening brief of the splashing of the incoming oil in Model No. 1, is not supported by any evidence in the case whatsoever, and is the volunteered imagination of appellant's counsel. Judge Wolverton heard the witnesses, observed the devices, and his finding of fact is controlling.

"In such a case, the trial court having the advantage of seeing and especially examining the material which it is claimed infringes, an Appellate Court, without such advantage, will not disturb the conclusion reached, unless it appears clearly that the finding is against the obvious weight of the testimony."

*Diamond Patent Co. v. Webster Bros.,* 249 Fed. 155, at 158.

'The court heard the evidence of the witnesses, and had before it alleged infringing devices, as well as the ring-shaped blanks out of which appellant makes them. The evidence of the witnesses was contradictory, and, in so far as the decree is predicated thereon, we cannot disturb it."

*Blettner v. Gill,* 251 Fed. 81 at 82.

**Model No. 2.**

The modified form of defendant's infringing device is described by Judge Wolverton in his opinion herein as follows:

"Model 2 contains a like partition to that described in Model 1. The oil inlet consists of a pipe extending into the side of the minor chamber, supplied with what is called a nipple, bell-shaped, to allow the oil to spread when discharged into the chamber. The nipple is set at an angle with and extended within proximity of the inner wall, the effect of which is, when the oil is discharged into the chamber, to carry part of it down the inner partition wall, part down the outer wall, at and near the intersection of the inner with the outer wall, and part of it down by gravity without reaching either wall. The device is provided with a gas take-off above the nipple." [R. pp. 537, 538.]

Model No. 2 is illustrated in Plaintiffs' Exhibit 11 (Book of Exhibits, p. 81.) The defendant was asked how many traps like "Model No. 1" he had made, and answered, "Just one." [R. 253.] The defendant testified that the modification of his trap from Model No. 1 to Model No. 2, comprises taking out the baffle 17 [R. 253], and in substituting for the form of inlet and baffle shown in defendant's patent, a reducing elbow, and that otherwise "the valves and all the mechanism hasn't been changed." [R. 257.] To settle a conflict which had arisen in the testimony as to the purpose and result attained by the reducing elbow in the Model No. 2 type of infringing device, on March 24, 1922, Judge Wolverton attended a demonstration exhibiting the operation of a Model No. 2 trap. This trap had been sold a few days before by the defendant, and was cut open with an acetylene torch in order that Judge Wolverton might observe the operation within the trap. At the conclusion of the demonstration, Judge Wolverton questioned the defendant as to what had been observed at the demonstration, and compelled the defendant to admit that the effect of the reducing elbow was to widen or spread out the incoming stream, causing part of the oil to flow down the inner partition and part to flow down the wall of the trap.

"The Court: There was oil thrown over here though?

Mr. Lorraine: On the wall, yes, Your Honor.

The Court: On the outer wall?

"Mr. Lorraine: Yes, sir; I admitted that on the vertical partition it was thrown on the wall.

* * * * * * * *

The Court: What proportion of it ran down the partition and what proportion down the wall, as you observed it?

Mr. Lorraine: I should say it was nearly equally divided." [R. pp. 358, 359.]

The finding of fact made by Judge Wolverton in view of the opportunity afforded him to personally examine the internal operation of one of the Model No. 2 infringing devices, cannot be assailed by defendant on this appeal.

*Diamond Patent Co. v. Webster Bros., supra.*

Judge Wolverton's opinion concludes as follows as to Model No. 2:

"So of defendant's device, Model 2, the oil is injected in part at least, against the partition, as well as against the chamber wall, so that it flows down thereon with pressure on the moving oil, from which the gas escapes. While part of the oil is reduced to a spray which falls by gravity to the settled fluid below, its action does not obviate the objectionable feature of a part flowing down the partition and a part down the wall. I am of the opinion also that defendant's trap will likewise infringe with the nipple constructed, as he claims it should be, according to drawings and specifications." [R. pp. 541, 542.]

The latter sentence of Judge Wolverton refers to the fact that in the Model 2 device observed by him at West Alhambra, the defendant had machined off the edge of the reducing elbow in order that the reducing elbow should fit closely against the inner partition. The defendant asserted that he did not intend to so position the reducing elbows in his Model No. 2 traps. This would, of course, be immaterial if true, and Judge Wolverton so held. Judge Wolverton followed an elementary principle of patent law in ruling that:

"While part of the oil is reduced to a spray which falls by gravity to the settled fluid below, its action does not obviate the objectionable featue of a part flowing down the partition and a part down the wall."

It is well settled that infringement resides in the appropriation of a part of a patented invention, although the whole is not taken.

30 Cyc. 977, note 15.

> "Clearly, the mere interposition of squeezing surfaces in the defendant's machine is not a substantial difference, but one purely formal; and, if it is, it is none the less an infringement. The law is well settled that infringement is not avoided by dividing an integral element of a patent machine into distinct parts so long as the function and operation remain substantially the same; and the same rule applies as to the joinder of two elements into one integral part accomplishing the purpose of both, and no more, so long as the same results are accomplished. *The impairment of the function of a part of a patented structure*

*by omitting a portion will not avoid infringement, nor will a mere change in form where the principle of operation is preserved and appropriated.* Winans v. Denmead, 15 How. 330, 342, 14 L. Ed. 717; Nathan v. Howard, 75 C. C. A. 97, 143 Fed. 889, and numerous cases there cited."

*Manton-Gaulin Mfg. Co. v. Dairy Machinery & C. Co.,* 238 Fed. 210, at 215.

Infringement of claim 4 of the Trumble patent by both Models No. 1 and No. 2, is apparent from a comparison of the claim with the structures of the infringing devices. Claim 4 of the Trumble patent has the following elements:

(1) An expansion chamber;

(2) Means for delivering oil and gas into the chamber;

(3) Means for maintaining pressure within the chamber;

(4) Means for drawing oil from the chamber;

(5) Means within the chamber adapted to cause the oil to flow in a thin body for a distance to enable the gas contained and carried thereby to be given off while the oil is subjected to pressure.

The wording of this claim is clearly broad enough and comprehensive enough to include both Models Nos. 1 and 2. It cannot be denied that both models of the infringing device embody each of the elements of this claim. There is no difference between Model No. 1 and Model No. 2 in regard to the embodiment of the elements of the Trumble claim, except that the two models differ in the particular form of the "means within the chamber adapted to cause the oil to flow in a thin body for a distance to enable the gas contained and carried thereby to be given off while the oil is subjected to pressure." In Model No. 1, this element comprises the inclined baffle-plate, while in Model No. 2, this element comprises the reducing elbow. The functions of the inclined baffle-plate and the reducing elbow are the same and coincide exactly with the function of the cone 22 of the Trumble patent. In this manner, both models of the infringing device appropriate the narrower aspect of the Trumble invention which resides in Trumble's conception of slowing down and spreading out the oil to give an effective oportunity for the action of the pressure maintained in the trap.

Notwithstanding the fact that the validity of the Trumble claims has been conceded in this case and is not in issue, and notwithstanding the clear wording of the claims, appellant's opening brief contains a discussion under the title "Estoppel by Proceedings in the Patent Office." A reading of the discussion in appellant's opening brief under that title, demonstrates that appellant cannot find anything in these claims of the Trumble patent that does not apply fully and directly to both forms of the infringing device. Appellant relies solely upon statements made by way of argument to the Patent Office during the prosecution of the application for the Trumble patent, *which argu-*

*ment is not reflected in the claims as granted by the Patent Office.* Since the claims held to be infringed in this case were actually granted in their present form by the Patent Office without amendment, and since infringement of these claims follows from their precise language, any proceedings leading up to these claims are immaterial in this case. This is not a case where it was necessary to place a limitation in a claim in order to obtain its allowance, and, thereafter, the patentee attempts to escape such limitation. Here we accept the claims as they are written, without any attempt to enlarge their scope. Mere argument to the Patent Office, if not reflected in the claims, is immaterial. A leading case by this court on this point is:

> *Fullerton Walnut Growers' Ass'n v. Anderson-Barngrover Mfg. Co.*, 166 Fed. 443,

in which the patent covered a certain process of bleaching nuts, involving the addition of a weak acid to a certain solution and plunging the nuts into the solution. In the Patent Office, the attorney for the applicant in his arguments "insistently pointed to the fact that the applicant's invention differed from all prior bleaching processes in that the nuts are plunged into the bleaching solution *at the very instant of adding a weak acid thereto.* (P. 451.) The patent as granted contained a claim which was not limited to plunging the nuts into the bleaching solution at the instant of adding the acid. As the scope of the invention was not so limited, and as the claim was not so limited, the court held that the argument made to the Patent Office was immaterial, stating:

> "The question arises whether those claims are to be limited in their scope by the arguments or admissions made by the patentee's attorney in the proceedings in the Patent Office. It is true that where an applicant presents to the Patent Office a claim, which is rejected as being anticipated by prior patents or publications, and in consequence thereof *he amends his claim* so as to avoid such anticipation, he may not thereafter contend that the claims are to receive the construction to which they would have been entitled if such limitation or restriction had not been inserted. *But in the present case there was no amendment or restriction of the claims.* They were allowed in the terms in which they were originally formulated.
>
> "In the light of these authorities, it is clear that the claims of the patent, unambiguous as they are, are to be interpreted according to the meaning of their own terms, *and are not to be controlled or limited by any argument or representation made in the Patent Office by the applicant's attorney as to the scope of the invention or the features wherein it differs from the prior art.*"

To the same effect is the decision of the Circuit Court of Appeals for the 2nd Circuit in

> *Auto Pneumatic Action Co. v. Kindler & Collins*, 247 Fed. 323, at 328,

in which the court said:

> "so far as we know there is no decision which goes further than to hold that, where the appli-

cant has assented to *changes in a claim* upon a reference in the Patent Office, he may not, by subsequent construction, resort to the elements which he has thus abandoned. *We are far from being willing to establish a rule that arguments made in the Patent Office by the applicant to the examiner are to be taken as a measure of his patent.* We read the claims as they are written, like the language of any other formal statement drawn up as the final memorial of the parties' intentions, and we decline to consider what was said arguendo during the passage of the case through the Patent Office, or any other of the preliminary negotiations which the patent itself was intended to subsume."

American Caramel Co. v. Thomas Mills & Bro., 149 Fed. 743, 747;

Boyer v. Keller Tool Co., 127 Fed. 130-134;

A. G. Spalding & Bro. v. Wanamaker, 256 Fed. 530-3,

and many other cases to the same effect.

Claims 1, 2, 3 and 4 of the Trumble patent, the claims held by Judge Wolverton to be infringed in this case, were allowed as filed. (Book of Exhibits, p. 32). No amendments were ever made to claims 1, 2, 3 or 4, and no claims involving the Trumble principle of maintaining pressure were rejected by the Patent Office and cancelled. Judge Wolverton carefully considered the contention made by appellant as to the effect of the arguments made in the application for the Trumble patent. In his opinion, Judge Wolverton rules as follows on this matter. R. 532:

"Turning to the file-wrapper showing the proceedings before the examiner, claim 1 as made in the application contains the element 'means for reducing the oil into a finally divided condition to reduce the tension on the gas contained therein.' Claim 2, 'Oil dividing means arranged in the expanding chamber to reduce the oil to a thin film-like condition.' Claim 3, 'gas freeing means consisting of means to reduce the oil to a thin film arranged within the expanding chamber;' and claim 4, 'A cone arranged near the top of such chamber to receive the incoming oil and spread it over the wall of the chamber in a thin film-like form.'

"When the application came to the examiner, claims 1, 2 and 3 were each rejected on the application on the patents of Barker and Bray, and 4 on patent of Bray. The action of the examiner induced the petitioner to add the following to his specifications:

" 'It will be noted that the action upon the oil while flowing down the wall of the expansion chamber in a thin film under pressure permits the free, dry gas to readily escape therefrom, while the pressure exerted upon the oil surface backed by the wall of the chamber holds the lighter liquids, such as gasoline, in combination with the oil body, and I desire to be understood as pointing out and claiming this action as being of great benefit to the crude oil derived from the well on account of keeping the gasoline series in combination with the main body of oil.'

"Also to cancel claims 1, 2, and 3, and to insert claims 1, 2, 3 and 4 as now contained in the issued patent.

"The examiner again rejected claims 1 to 4 inclusive, on patent of Bray, and 5 to 13, inclusive, were held not to patentably distinguish from Bray, and accordingly were rejected. In response to these objections, the applicant added claims 13 and 14 as now contained in the patent. As presented, the examiner again rejected claims 14 and 15, being claims 13 and 14 in patent, also claims 1 to 13, inclusive, as not to patentably distinguish from references of record. The applicant replied to the action of the examiner, stating among other things, that the 'applicant's invention consists of a containing vessel, an imperforate cone adapted to spread the whole body of the oil to the outer edge of the vessel, and means for taking off gas from the interior of the cone near the center of the vessel'; this to distinguish from the Bray patent. He says, further: 'Moreover Bray does not take off his gas below his screens, and the claims of Trumble are quite specific in stating that the gas is taken off inside the cone.'

"The matter coming again before the examiner, on reconsideration, all the claims were allowed as contained in the patent. Claim 9 (original claim 8) was rejected as met by patent of Bray, and has been eliminated from the patent.

"A patentee, where he is required by the rulings of the Patent Office to modify and restrict his claims, to obviate anticipation by previous patents, is by the limitations he thus imposes upon such claims, and where the patent is for a combination of parts, his claims must be limited to a combination of all the elements which he has included in his claims as necessarily constituting that combination. Phoenix Caster Co. v. Spiegel, 133 U.

S. 360, 368; New York Asbestos Mfg. Co. v. Ambler Asbestos A. C. C. Co., 103 Fed. 316. And it was said in Roemer v. Peddie, 132 U. S. 313, 317:

" 'When a patentee, on the rejection of his application, inserts in his specification, in consequence, limitations and restrictions for the purpose of obtaining his patent, he cannot, after he has obtained it, claim that it shall be construed as it would have been construed if such limitations and restrictions were not contained in it.'

"See also, National Hollow Brake-Beam Co. v. Interchangeable Brake-Beam Co., 106 Fed. 693, 714, where the court adds:

" 'But this is the limit of the estoppel. One who acquiesces in the rejection of his claim because it is said to be anticipated by other patents or references is not thereby estopped from claiming and securing by an amended claim every known and useful improvement which he has invented that is not disclosed by those references.'

"Two thoughts were uppermost with the patentee in making the changes indicated: First, to avoid the objection with reference to Barker and Bray with means for reducing the oil into a finely divided condition; and, second, to confine the oil in its flow down a wall or surface with *maintained pressure* meanwhile. The theory of the patentee is obviously that, pressure *being maintained,* the dry gas will readily escape from a thin film or body of oil passing down and against a wall or other surface, without at the same time taking off the lighter liquids, such as gasoline, which will yet remain in the crude oil and add to its value.

"The limitation and restriction which the patentee has imposed upon his patent must be gathered from his addition to his specifications and the claims which were finally approved by the examiner. He says in the added specifications that the free, dry gas readily escapes, while the pressure exerted upon the oil surface, backed by the wall of the chamber, holds the lighter liquids in the oil body. In his claims, however, he asserts a broader scope for his invention, as in claim 3, which comprises 'the combination of an expansion chamber having a surface adapted to sustain a flow of oil thereover in a thin body, means for distributing oil on to such surface, pressure-maintaining means arranged and adapted to maintain pressure on one side of the flowing oil.' All this was approved and allowed by the examiner.

"Construing the whole together, the added specifications and the claims, I am impressed that the patentee is not confined to means of causing the oil to flow down the outer wall of the chamber, but that his patent includes any means that will cause the oil to flow down any surface as well, such as a baffle-plate or inner partition or wall, which is reached after the emulsified oil enters the chamber. I think therefore, the patentee is not estopped by the proceedings before the Patent Office to insist upon the broader claims."

## The Prior Art.

In view of the fact that defendant has conceded and has failed to deny the validity of the claims of the Trumble patent in issue, and in view of the fact that these claims are certainly broad enough to include both Model 1 and Model 2 of the infringing device, we fail to perceive any force or materiality in considering the prior art. However, almost all of appellant's opening brief is devoted to construing the prior art patents in an endeavor to show by some process of implication they contain and disclose something, which Judge Wolverton has expressly found is not contained therein, and something which no witness in the case has in any manner attempted to testify is contained therein. The prior patents referred to in appellant's opening brief are mere paper patents, and there is no evidence showing that any successful device has ever been made in accordance with any one of them.

> "The novelty of an invention is not negatived by a prior useless process or thing, nor is anticipation made out by a device which might, with slight modification, be made to perform the same function. The invention must have been complete, and capable of producing the result. One should not be deprived of the results of a successful effort merely because some one else has come near it."

*Diamond Patent Co. v. S. E. Carr. Co.,* 217 Fed. 400, at 405. (9th C. C. A.)

If there was any incidental use of pressure in any of these prior patents, it was accidental, not recognized by the patentee, and no disclosure thereof made to the public. The record in this case shows that there was no trap in public use in the art at the time Trumble made his invention, that operated on the principle of maintaining pressure, and that Trumble's ntroduction to the art of that principle was immediately recognized as of controlling importance and benefit. The prior patents, to which a great portion of the appellant's opening brief is devoted, were thoroughly considered by Judge Wolverton. As has already been pointed out, it is not seen wherein they have any bearing on the claims in the Trumble patent, inasmuch as the validity of the patent is admitted, and inasmuch as the language of the claims is plain and requires no interpretation to include both forms of the infringing device. Trumble was the first to suggest that pressure *be maintained* on a trap to hold the gasoline in the oil and express the fixed gas from the oil. We challenge appellant to point out this suggestion in any of the prior patents referred to in appellant's brief. The ultimate effect of the argument made in appellant's opening brief is that necessarily, in the use of the prior devices, pressure was to some extent maintained. This is not sufficient, under the law, to constitute an anticipation of the Trumble invention.

A controlling decision precisely in point was rendered by the Supreme Court on February 19, 1923, in the case of

> *Eibel Process Co. v. Minnesota & Ontario Paper Co.,* (not yet published).

The patent in that case before the Supreme Court was on a paper-making machine, and the invention consisted in lifting the feed end of the machine so that gravity assisted the machine in transporting the paper pulp during the operation of the machine. It was argued to the Supreme Court that machines had been so positioned, in the prior art, as to necessarily give the same effect. The Supreme Court sustained the patent, reversed the Circuit Court of Appeals, and said:

> "It is contended on behalf of the defendant that whether Barrett and Horne perceived the advantage of speeding up the stock to an equality with the wire, yet the necessary effect of their devices was to achieve that result and therefore their machine anticipated Eibel. In the first place we find no evidence that any pitch of the wire, used before Eibel, had brought about such a result as that sought by him, and in the second place if it had done so under unusual conditions, accidental results, not intended and not appreciated, do not constitute anticipation. Tilghman v. Proctor, 102 U. S. 707, 711; Pittsburgh Reduction Company v. Cowles Electric Co., 55 Fed. Rep. 301, 307; Andrews v. Carman, 13 Blatchford 307, 323."

The controlling facts referred to by the Supreme Court in the above case are exactly in accord with the facts in this case. In the first place, there is no evidence in the record before this court that the pressure appellant seeks to imply in the prior art patents, "brought about such a result as that sought by him" (Trumble). In the second place, if any such result did accrue, it was "not intended and not appreciated," and "does not constitute anticipation."

In spite of the obvious fallacy in fact and law, of the contentions urged by appellant as to the prior patents, we will briefly discuss the same in order that the court may not be misled by the wealth of misinterpretation placed thereon by counsel. We regard the following points as elementary patent law.

*Point One.* Novelty is not negatived by any prior patent or printed publication unless the information contained therein is full enough and precise enough to enable any person skilled in the art to which it relates, to perform the operation or make the thing covered by the patent sought to be anticipated. Walker on Patents (5th Ed.) Sec. 57 citing:

> Seymour v. Osborne, 11 Wall. 516;
> Cawood Patent, 94 U. S. 704;
> Downton v. Milling Co., 108 U. S. 466;
> Eames v. Andrews, 122 U. S. 66.

Each of the prior patents referred to in appellant's opening brief can be scanned with the most careful scrutiny without finding therein any appreciation or disclosure of Trumble's discovery, that by maintaining pressure on a gas trap, the gasoline can be maintained in the oil and the fixed gas expressed from the oil.

*Point Two.* Novelty is not negatived by anything beneficially incapable of the function of the subject of the patent, even though apparently similar thereto. Walker on Patents (5th Ed.) Sec. 65, citing:

> Crown Cork & Seal Co. v. Ideal Stopper Co., 123 Fed. 666;
> Kirchberger v. Am. Acetylene Co., 124 Fed. 764;
> Dececo Co. v. Gilchrist Co., 125 Fed. 293;
> Farmers' Mfg. Co. v. Spruks Mfg. Co., 127 Fed. 691.

There is no evidence before this court indicating that any of the prior patents discussed in appellant's opening brief could be used to perform the function of the Trumble patent, even though some of these devices might be similar in appearance to the Trumble device.

*Point Three.* Novelty is not negatived by any prior accidental occurrence or production, the character and function of which was not recognized until later than the date of the patented invention sought to be anticipated thereby. Walker on Patents (5th Ed.), Sec. 67, citing:

> Tilghman v. Proctor, 102 U. S. 711;
> Ranson v. New York, 1 Fisher 256;
> Pelton v. Waters, 1 Bann. & Ard. 399;
> Andrews v. Carman, 2 Bann. & Ard., 277.

Appellant's opening brief attempts to interpolate into the prior patents discussed by him by a process of implication, *the maintenance of pressure.* There is not an iota of evidence in this case supporting this implication. The implication is as unsupported as is the representation made in appellant's opening brief that all prior gas traps necessarily operated by the maintenance of pressure. This is absolutely contrary to the record. Indeed, the defendant's patent expressly recognizes the existence of gas traps "in which the oil passes immediately from a well or other source to an apparatus in which it is subject only to atmospheric pressure." (Book of Exhibits, p. 64, lines 55-63.) It will be found, from a careful scrutiny of each of the prior patents discussed by defendant's counsel, that none of them in any way appreciate or rely upon the maintenance of pressure as a means of maintaining the gasoline in the oil and as a means of expressing the fixed gas from the oil, and each of these patents will be found to have a different mode of operation. It has been settled in many cases by this and other courts that "a device which does not operate on the same principle cannot be an anticipation."

> Los Alamitos Sugar Co. v. Carroll, 173 Fed. 280, at 284 (9th C. C. A.), citing:
> Western Electric Co. v. Home Telephone Co., 85 Fed. 649;
> Dederick v. Cassell, 9 Fed. 506;
> Pattee v. Moline Plow Co., 9 Fed. 821;
> Fuller v. Yentzet, 94 U. S. 288, 24 L. Ed. 103;
> Topliff v. Topliff, 145 U. S. 156, 36 L. Ed. 658;
> Robinson on Patents, Vol. 1, Sec. 282;
> Walker on Patents (4th Ed.) Sec. 62.

*Point Four.* Novelty is not negatived by anything which was neither designed, nor apparently adapted, nor actually used, to perform the function of the thing covered by the patent, though it might have been made to perform that function by means not substantially different from that of the patented invention. Walker on Patents (5th Ed.), Sec. 68, citing:

> Topliff v. Topliff, 145 U. S. 161;
> Carnegie Steel Co. v. Cambria Iron Co., 185 U. S. 422;
> Knickerbocker Co. v. Rogers, 61 Fed. 297;

Kinnear & Sager Co. v. Capital Sheet-Metal Co., 81 Fed. 492;

Bowers v. San Francisco Bridge Co., 91 Fed. 410.

The applicability of the last cited rule to this case is demonstrated by the absence, in each of the prior patents, of any appreciation or suggestion of maintaining the gasoline in the oil by pressure, and none of these prior patents in any manner recognize or instruct in the saving of gasoline.

We will now discuss the prior patents in order.

### Cooper Patent.

In this patent, it is stated:

> "In that process *which I set forth in a contemporaneous application for a patent,* a portion of the gas from the well is used repeatedly as a lift or pulverizer to relieve the well of oil and water by raising or assisting to raise them. The contents of the discharge pipe, composed of gas, oil, water, sand, and silt, is forced into a vessel included in the pressure system, wherein the constituents are separated without affecting the general pressure in the system and the gas recovered and taken to the compressor to be used again in part as the lift or pulverizer to relieve the well, the surplus of the gas being conducted away for any useful purpose."

(Book of Exhibits, p. 134, lines 14 to 28.)

Attention is called to the fact that this statement refers, not to any disclosure in the Cooper patent itself, but to *a contemporaneous application which is not before the court,* and of whose nature we are entirely ignorant. In addition, this patent states, at lines 83 and 84:

> "A high pressure of gas may thus be maintained in this cylinder."

We call attention to the fact that no method is shown for maintaining this high pressure of gas, and that this patent falls within all of the rules previously cited; that is to say, it falls within the rule that the information contained in the patent is not sufficient to perform the process referred to in the patent. It falls within the rule that anything beneficially incapable of the function of a patent is not an anticipation. It falls within the rule that any advantages flowing from the use of pressure were unrecognized and accidental and therefore not an anticipation. It falls within the rule that anything neither designed nor apparently adapted nor actually used to perform the function of the patent is not an anticipation. There is no suggestion in the patent that maintenance of pressure would hold the gasoline in the oil and express the fixed gas from the oil. Certainly the Cooper patent falls within the rule even if it used pressure, of being a prior accidental occurrence or production, the character and function of which was not recognized until the Cooper patent was resurrected by counsel in this case. *It is clear that the device of the Cooper patent does not operate by the mode of operation first disclosed in the Trumble invention and there is no suggestion of or reliance upon this mode of operation.* On the contrary, the Cooper patent expressly states that the mode of operation is by means of *centrifugal separation.* That the mode of operation of the Cooper device is based on centrifugal operation, and not on pressure, is stated in the Cooper patent as follows:

> "Entering the cylinder under pressure tangentially to its circumference, the material is thereby given a rotary motion, which separates the constituents conformably to their specific gravities."

It is obvious, from a study of the Cooper patent, that he was particularly seeking to separate water and sand from oil, and that he used this rotary or centrifugal machine for this purpose. Instead of suggesting the pressure mode of operation conceived by Trumble, the centrifugal action of the Cooper device would dispel and lead in the opposite direction from Trumble's invention.

In conclusion, the Cooper patent stands in this case as follows:

(a) as a paper patent which never contributed anything practical to the art;

(b) as a patent for a device which operated upon the principle of centrifugal separation;

(c) as a patent which neither discloses nor recognizes the maintenance of pressure in a gas trap to maintain the gasoline in the oil and express the fixed gas from the oil;

(d) as a patent which merely mentions pressure in conjunction with and as an incident to centrifugal separation;

(e) as a patent which shows no means for maintaining pressure; and,

(f) as a patent in which the mixture of oil and gas is rapidly whirled and moved and is not slowed down and spread out to enable the same to be effectively acted upon by pressure.

## McIntosh Patent.

(Book of Exhibits, p. 110).

This patent shows a gas trap, but it is entirely free from any reference whatsoever to the maintenance of pressure. In appellant's opening brief, counsel states that it requires only an inspection of the drawing to discover certain things, namely, that the bell 88 is movable up and down and that it acts as a valve for the pipe 7. There is nothing whatsoever in the patent to indicate that any such action takes place. There is certainly nothing to indicate that McIntosh ever thought of or appreciated that his device could be used to retain gasoline in the oil and to express gas therefrom, and the device of McIntosh has its own mode of operation, distinct and different from the Trumble mode of operation. The McIntosh patent falls within all the rules of law stated above.

## Newman Patent.

(Book of Exhibits, p. 136).

There is nothing in this patent to indicate that any pressure whatsoever is used therein. It is apparently not a gas trap at all; indeed, it is stated to be a water and gas separator. It was evidently not used in connection with oil; and the teachings of the Newman patent would be of no benefit to anyone seeking to solve the problem solved by Trumble.

The Newman patent clearly falls within the rules of law previously stated.

## Taylor Patent.

(Book of Exhibits, p. 115).

This patent does not show a gas trap. It shows a steam separator, that is to say, a device for taking grease or oil *out of steam.* It shows no means for *maintaining pressure*, and it was certainly not designed, nor is it adapted, nor has it ever been used, for the purpose of the Trumble invention. It is true that the patentee states that "if sediment accumulates in the chamber, the steam-pressure will ordinarily blow it out through the discharge-pipe." This does not indicate that there was any pressure *maintained* in the chamber, but merely that the steam was allowed to blow through and that it carried any sediment with it. There was certainly no *means* for maintaining pressure, even if pressure were momentarily built up due to the presence of sediment. Certainly, the Taylor patent contains nothing which would in any way assist one confronted by the problem which confronted Trumble at the time he made his invention.

**Barker Patent.**

(Book of Exhibits, p. 121).

This patent does not disclose a gas trap. The invention relates to improvements in natural gas separators.

> "The object of the invention is to provide a device of this character by means of which the gas may be entirely separated *from the water* flowing from Artesian wells, thus saving the gas for use as fuel or for lighting purposes, as well as purifying the water to a sufficient extent to be employed for purposes, other than for human use."
>
> (Book of Exhibits, p. 122, lines 14-21.)

Obviously, the Barker device was not designed or adapted or ever used for the purposes of the Trumble invention. There was no means shown in the Barker device for spreading incoming water so as to allow the escape of gas therefrom, and any pressure means which may be involved in the patent is merely a pressure relieving means, and not a means for *maintaining pressure.* It is very evident that the Barker patent is not only incapable of the function of the Trumble device, but it is also evident that there is nothing in the Barker patent which would in any way lead to the solution of the problem so ably solved by Trumble.

**Bray Patent.**

(Book of Exhibits, p. 128).

The misinterpretation of the patents in the appellant's opening brief reaches its climax in the interpretation of the Bray patent. After quoting from the testimony of an expert, who stated that if the gas was carried a long distance, there would be some pipe friction, and who said that this might have a tendency to cause a back-pressure, and who further stated that if the gas was being delivered to an absorbing plant, *which had a* reduction nozzle or some other means of reducing the flow, it would cause a back-pressure,—counsel for appellant proceeds to interpret this testimony as meaning that a short open pipe will maintain pressure in a gas trap. Referring to page 127 of the Book of Exhibits, the court will note that the Bray patent is provided with a pipe 17 which is open at the outer end, and which is connected into the top of the trap. This, appellant maintains, is a pressure *maintaining* means. This is the ultimate in the art of interpretation. To support this most absurd interpretation of a mechanical principle, the appellant's counsel is forced to a plain misstatement of facts. He states that the "diameter of the outlet pipe 17 is less than that of the inlet pipe 16. Appellant's counsel fails to mention that both oil and gas enter through the pipe 16, and that the oil is taken out through the pipe 20, which is much larger in area than the pipe 16, the gas being allowed to escape through the pipe 17. Counsel, in addition, is absolutely wrong in his statement that the pipe 17 is smaller in area than the pipe 16. A careful calipering of the drawing indicates that this pipe is of the same size as the pipe 15 within any limits of mechanical reproduction or measurement. Of course the relative size of an inlet and an outlet would be immaterial in any event, unless the inlet be filled to capacity.

We regret that it is necessary for us to so contradict the assertions made by counsel for appellant, but this arises through present counsel for appellant attempting to make contentions before this court in total disregard of the evidence considered by Judge Wolverton. None of these contentions are supported by the record and testimony before Judge Wolverton. They are certainly not supported by the facts relied upon by counsel. The novel contentions injected into this case by counsel for appellant, and unsupported in any manner by any testimony in this case, should have little weight on this appeal. Defendant had full opportunity to make any such contention before the court below, and, although testimony was given as to all of these patents, no support was developed for the contentions here made. Furthermore, all of these patents were before the Patent Office, and the ruling of the Patent Office in granting the Trumble patent is contrary to all of the contentions made by counsel for appellant. The Trumble patent is *prima facie* valid, and the burden is upon defendant to show proof in the record invalidating the same. These proofs are not in the record, and their place cannot be taken by mere argument of new counsel.

Counsel for appellant has explained at some length the case of *Standard Oil Co. v. Oklahoma Natural Gas Co.,* reported in 284 Fed. 469. This suit was upon a patent which is not of record in this case. The patent therein relates to a process of *extracting* gasoline from natural gas, and has nothing whatever to do with a gas trap. Extracting gasoline from natural gas is an entirely different matter, and, in many respects, quite the opposite from preventing *the escape* of gasoline during a separation of natural gas. The patentee in that case had no problem of *maintaining gasoline in oil* and of so doing without hindering the separation of gas from the oil. The patentee in that case did not have to meet the problem of an effective separation of gas from oil in quantity. He was merely attempting to extract gasoline from *gas,* whereas Trumble was confronted with the problem of preventing gasoline from ever leaving the oil and of so doing without hindering the separation of the gas from the oil. The evidence in the Standard Oil case showed that a similar process to that disclosed in the patent there had been operated to recover hydro-carbons from artificial gas, and the only apparent novelty resided in performing the same process on a gas of "natural origin on the way from its underground source to its places of consumption." The court held:

> "We are satisfied that Saybolt discovered nothing new; that he simply applied an old and well-known process to a new use, which produced no new result or an old result in a better or easier way." (284 Fed. 478.)

The facts are entirely different in the Trumble case. Trumble did produce a new result, and also produced an old result in a better way. The new result produced by Trumble was saving gasoline and *maintaining* it in the oil during the separation of the fixed gas. The old result that Trumble produced in a better way was the effective and clean separation of the fixed gas from the oil. Trumble discovered something new—in fact, he had a totally new object in view. Trumble not only introduced a new mode of operation, but he did so for a new object, and thereby produced a new result by the application of a new principle. There is no identity between the Saybolt patent (which is not in the record here, and which was considered in the Standard Oil case, *supra*), and the Trumble patent. There is no testimony in this case in any way asserting or establishing anticipation by the Saybolt patent, and no such issue is raised by the pleadings herein. In fact, the distinction between the Saybolt process and the Trumble gas trap may be briefly described by stating that Saybolt disclosed that by the use of pressure he could *take gasoline out* of gas as it flowed from a gas well; whereas, Trumble found that by the use of pressure he could *hold gasoline* in a mixture of oil and gas without hindering the separation of the gas from the oil. Instead of following the teachings of the prior art, Trumble branched out on new territory and proceeded contrary to the information then contained in the art. If Trumble had blindly followed the prior art, he would have relieved the pressure on his gas trap in order that the separation of the gas from the oil would not be hindered. There is no evidence in this case concerning the Saybolt patent; but we might state, for the benefit of the court, that the Saybolt process could not be applied to an oil well, but is applicable only to a gas well; and the opposite state of facts is true as to the Trumble trap.

## Conclusion.

In conclusion, it is respectfully submitted that defendant has infringed each of claims 1 to 4 of the Trumble patent in suit, as found by Judge Wolverton in this case; and particularly claims 3 and 4 by model 1 and claims 1 to 4 by model 2 of the infringing devices. The record clearly shows that Trumble made an important and fundamental addition to the gas trap art, of the character and kind that the patent laws of this country were designed to reward and protect. As said in *Topliff v. Topliff*, 145 U. S. 156:

> "The object of the patent law is to secure to inventors a monopoly of what they have actually invented or discovered, and it ought not to be defeated by a too strict and technical adherence to the letter of the statute, or by the application of artificial rules of interpretation."

The decree entered by Judge Wolverton was according to law, according to equity and good conscience, and in accordance with the facts, and should be affirmed.

FREDERICK S. LYON,
LEONARD S. LYON,
FRANK L. A. GRAHAM,
*Solicitors for Plaintiffs-Appellees.*

No. 3945.

IN THE

# United States
# Circuit Court of Appeals,

## FOR THE NINTH CIRCUIT.

David G. Lorraine,
*Appellant,*

*vs.*

Francis M. Townsend, Milon J. Trumble and Alfred J. Gutzler, Doing Business Under the Firm Name of Trumble Gas Trap Company,
*Appellees.*

## APPELLANT'S REPLY BRIEF.

Westall and Wallace,
Ernest L. Wallace,
Joseph F. Westall,
By Joseph F. Westall,
*Solicitors and of Counsel for Defendant-Appellant.*

Parker, Stone & Baird Co., Law Printers, 232 New High St., Los Angeles.

**No. 3945.**

IN THE

# United States
# Circuit Court of Appeals,

FOR THE NINTH CIRCUIT.

---

| | |
|---|---|
| David G. Lorraine, *Appellant,* | |
| *vs.* | |
| Francis M. Townsend, Milon J. Trumble and Alfred J. Gutzler, Doing Business Under the Firm Name of Trumble Gas Trap Company, *Appellees.* | |

## APPELLANT'S REPLY BRIEF.

---

### Argument of Our Opening Brief Ignored.

While the opening page of brief for plaintiff-appellees carries a main heading "Statement of the Case," there is in appellees' brief no title, "Argument." In fact all other headings throughout the brief appear as subdivisions of plaintiffs-appellees' "Statement of the Case." This error of classification, however, need not confuse as an examination of the context shows that the argument proper begins with the first sentence beginning at the 5th line from the bottom, page 4, of plaintiffs-appellees' brief—at any rate that is where we must begin it, for we find ourselves in almost total disagreement with the statements of counsel thereafter following as to the nature, scope and alleged importance of Trumble's asserted contribution to the art.

Preliminarily, as a general indicator of direction, it should be noted, that nearly four-fifths (66 of the 83 pages) of the argument of appellant's opening brief is devoted to a consideration of the art prior to Trumble in order that the court might have fully before it *evidence and reasoning* in support of our contention that the most that Trumble can claim to have added to the art (whether the addition was of even the smallest practical value by no means granted) was the idea of *spreading the whole body of the oil upon its entry into the chamber in a thin film, unbroken and unagitated, over not only an imperforate conical spreader plate, but by such spreader plate equally around and upon the walls of the expansion chamber, so that no part of the oil would fall to the bottom of the chamber without being so spread.*

When an argument cannot be squarely met, a common method is to ignore it: barely one-ninth of plaintiffs-appellees' brief (a scant seven pages out of sixty-four) are devoted to any attempted refutation of the argument which we so elaborately developed in our opening brief. We are confident that further analysis by this court, in the light of plaintiffs-appellees' brief will show that the reason for this was that our argument on the prior art was found to be unanswerable.

## The Principal Fallacy of Plaintiffs-Appellees' Brief.

It is the fallacy most to blame for the fatal errors of individuals and nations since the world began—*unsupported assertion.* Disregarding the clear showing of the evidence to the contrary (which we pointed out and dwelled upon at length in our opening brief), beginning near the bottom of page 4 of plaintiffs-appellees' brief, we find an opening of two and one-half pages relating to the nature, scope, and importance of Trumble's asserted discovery. Not a single reference to the record in support of counsel's assertions is there found, but important and even vital questions, are *brazenly begged,* namely, that Trumble had "a new object in view," "what the prior art designers of traps had in mind"; that in prior art traps "gasoline was carried away with the gas;" that with the prior art devices "loss of a considerable portion of the gasoline was a necessary incident to the separation;" that Trumble made a "striking discovery" which was "a new and unexpected principle in gas traps;" etc.

Eagerly turning these pages from page 4 of plaintiffs-appellees' brief, we search *for some reference to the record* cited as support for these important and far-reaching assertions. We find the first of such references about the middle of page 7, in support of counsel's statement that the "industry recognized and appreciated the enormous saving" that could be afforded by the use of the Trumble trap; that such traps went into "widespread and general use, displacing other traps, and became the standard gas trap in use in the oil fields in the United States and foreign countries," that is to say, *throughout the world.* Now what is the support in the record cited for these last mentioned assertions of counsel? Why, counsel states that 583 gas traps have been sold from the time of the alleged invention of the Trumble trap (November 9, 1914), to the beginning of this suit, namely, March 22, 1922—nearly seven and one-half years—an average of *only less than seventy-eight traps a year* THROUGHOUT THE WORLD. Now, this, clearly, is *no* support for such assertion, for the reason, in the first place, we do not know how many traps are required for use throughout the world. However, beginning with the second paragraph of page 4 of plaintiffs-appellees' brief, counsel state "petroleum oil as it comes from the well generally contains more or less gas which must be eliminated before the oil is suitable for commercial purposes." Thus, under counsel's statement, practically all wells, in order to produce oil suitable for commercial purposes must have some device for separating the gas, that is to say, must have a gas trap. There appears to be no evidence in the record of how many wells are in operation throughout the world, but anyone who has even glanced over just one of the large oil fields, or has even seen a picture of an oil field will readily have an inkling of the preposterously insignificant proportion than an average of *only seventy-eight gas traps per year* bears to the total number of wells in operation.

Thus, the statement, that the use was "widespread," "general" or "standard" or that such traps displaced other traps, as asserted by counsel, carries on its face its own refutation.

Of course, even if a large proportion of the gas traps used throughout the world were traps made in accordance with the Trumble specifications and drawings, this would by no means be conclusive as to the value of any infinitesimal contribution of Trumble to the art; for it might well be that the sales were due to persistent advertising, to the fact that the Trumble trap was of recognized value *only insofar as it incorporated devices long and well known in the prior art.* Dealing with a necessary and staple article, if Trumble was a leading manufacturer, aggressively advertising his product, assiduously copying those things demonstrated to be of value in the art, he would, of course, sell many traps. *Someone* would have to supply the large demand, and it might as well be Trumble. However, the admittedly pitifully small actual output of Trumble traps testifies louder than mere words to the fact that Trumble after seven and one-half years' effort has failed miserably as a competitor with manufacturers of other gas traps throughout the world.

We submit that the immediately foregoing discussion furnishes a most impressive warning against unsupported statements of counsel. If the first citation to the record so grossly fails as support, how can we place any confidence whatever in the preceding two pages in which possible evidential support is not even suggested?

**Counsel's Assertion (Middle of Page 7, Plaintiffs-Appellees' Brief) That One of the Trumble Traps Saved Over Forty-Five Thousand Dollars a Year for the Honolulu Oil Company Is Misleading, in That It Implies That Such Saving Was the Result of the Employment Only of the Trumble Alleged Invention. The Witness Paine in so Testifying Was Not Comparing the Operation of the Trumble Trap With Any Other Trap, but Merely Testifying That the Amount Saved Was the Result of the Employment of a Gas Trap as Compared With the Employment of No Gas Trap Whatever.**

An even more striking example of the zeal of counsel which overreaches is found beginning at the middle of page 7 of appellees' brief, where counsel asserts:

"The importance of these traps to the oil industry may be appreciated from the testimony of the witness Paine, to the effect that one of these traps alone saved $125.00 per day (over $45,000 a year) for the Honolulu Oil Company at Taft, California."

Counsel's reference [R. 117] immediately following this statement is erroneous: the correct reference seems to be the last part of answer at R. 79.

Now, the immediately foregoing statement of counsel and the quotation and discussion of the witness Paine's testimony which follows (beginning page 7 of plaintiffs-appellees' brief) is obviously set forth by counsel for plaintiffs-appellees as proof that Trumble's contribution to the art resulted in this great saving. The argument is implied that if Trumble had produced a device which leads to such wonderfully advantageous results it must have great merits, and that this court should go to great lengths to reward such merit by such an interpretation of the Trumble claims as will avoid anticipation and find infringement.

The impression that one gains from a reading from counsel's said discussion is *that the Trumble trap did something that prior traps could not and did not accomplish, namely, the saving of $125.00 per day for the Honolulu Oil Company. As we shall immediately show, any such impression is entirely mistaken and should be guarded against. Paul Paine was not comparing the Trumble trap with any other trap.*

In effect, Paine's testimony is simply that the use of a gas trap saved $125.00 per day over the use of *no* gas trap. In other words, the Paine testimony merely supports an argument that a gas trap was needed and its employment by the Honolulu Oil Company profitable. Thus, at R. 70, Paine testifies that he installed the first Trumble gas trap in 1915; that previous to this he had noted the operation of such a trap on the adjoining property.

At R. 71, he says, that prior to the installation of said Trumble Gas Trap *the gas was allowed to escape into the atmosphere and was entirely lost.* The witness then proceeds to testify [R. 71] as to general conditions at the well, and as to the effect of pressure, the substance of which is merely that the use of pressure increased the value of the oil by keeping in the oil the lighter series. If Paine had been conversant with the art prior to the time he made his first observation, he would have known what the Circuit Court of Appeals for the 8th Circuit in Standard Oil Company v. Oklahoma Natural Gas Company, 284 Fed. 469—cited in our opening brief and quoted from at page 21—when in speaking of the Saybolt patent granted in 1911, many years before Trumble found:

> "Saybolt knew from the law of Henry the function that pressure would perform in the absorption of gasoline vapor from natural gas. The pressure of natural gas already existed to the knowledge of every one."

It is clear that the evidence, authority, and argument of our opening brief are not met by counsel—they are ignored.

Again we find at the middle of page 9, plaintiffs-appellees' brief, the following statement:

"While it is now possible for Mr. Paine to point out the reasons for the success of the Trumble invention, there is no evidence in the prior art to indicate that this was understood or appreciated prior to Trumble's practical application of this then unknown result and his disclosure of the same to the art."

As we have seen, Mr. Paine did not point out any reason for the alleged success of the Trumble pretended invention which *was not equally a reason for the success of any of the prior art traps in evidence.* It is not true that there is no evidence in the prior art to indicate that this was not understood or appreciated as asserted by counsel. As we have shown the court in discussing the prior art patents in our opening brief—take the McIntosh patent only as an instance,—McIntosh distinctly provided for reducing the oil to a thin film and subjecting it while so spread out to pressure, stating in his specification that the gas is thereby "freed and collected in the gas chamber." (See quotation from the McIntosh specification, bottom of page 48, of appellant's opening brief, and particularly the last line of such quotation.)

We deny the assertions of counsel at page 10 of plaintiffs-appellees' brief, as to the invention by Lorraine of a "valve adapted for installation on a trap of the Trumble type." We notice that counsel has cited no reference to the record where such assertions are supported. What Lorraine invented is fully shown and described in his letters patent beginning at page 66 of the Book of Exhibits. The Lorraine patent covers expressly—not a valve or valves as stated by counsel—but "an Oil, Gas and Sand Separator."

At plaintiffs-appellees' brief near the top of page 12 counsel have fallen into error concerning the proper interpretation of the testimony relating to pressure in defendants' device. The fact is that the evidence first related to a trap put out by defendants which had not been equipped with a pressure valve; later other testimony was introduced relating to another trap which did contain the pressure valve, whereupon counsel for defendant stipulated to the fact.

At plaintiffs-appellees' brief, page 17, we find counsel insisting that pressure was novel and original with Trumble, stating (plaintiffs-appellees' brief, bottom page 16), "Judge Wolverton considered all the evidence and explicitly found as a fact that the maintenance of pressure on a trap for the purpose above stated was novel and original with Trumble." Now, in the first place, the word "explicitly" overstates the finding of the trial judge. The fact is that Judge Wolverton was *very uncertain* in expressing himself in this regard. He did *not find*, he only states that he *thought* [Tr. of Rec., bottom p. 538, Court's Opinion] that the prior art patents contained no element of pressure. In this surmise we have seen in our opening brief the trial judge was in error.

Again, counsel's statement near the top of page 17, plaintiffs-appellees' brief, is reckless and mistaken where they assert that there was no testimony given or claim made before the trial court to the effect that the use of pressure in the trap was not novel or original with Trumble; for at Record, page 426, for instance, the witness Trout on behalf of defendant-appellant testifying concerning a prior art trap expressly describes and marks on a drawing produced by him the back pressure valve, which Mr. Lyon suggested should be marked "4." Again at Record, page 448, the witness Swoap, testifies to the use of a trap in 1897, in which pressure was employed together with the other elements in controversy.

However, the question of whether or not there was any *testimony* in the case to the effect that pressure was old before Trumble is a false issue, because our main contention was that *prior patent documents, not necessarily testimony* established the antiquity of such pressure in gas traps. So conclusive was the showing of these prior patents, that we have not squandered the court's time by pointing out the testimony of Trout and Swoap as to the state of the art. Counsel was careful to use the word "testimony" and not "evidence," but even then were mistaken in view of the testimony of Trout and Swoap above mentioned.

At the bottom of page 17, plaintiffs-appellees' brief, the fact adverted to that the witness Paine "felt his way along to see what the effect [of pressure] would be," merely emphasizes the fact admitted by Mr. Paine, that he had never had any experience with gas traps before he installed the first Trumble trap in 1915. About all he knew was that such things existed, but he had never experimented with them.

## Counsel's Attempt to Define the Alleged Trumble Invention.

At plaintiffs-appellees' brief, page 19, counsel state: "The Trumble invention comprises discovery or conception that a gas trap, if maintained under pressure will prevent the loss of gasoline without interfering with the separation of the gas from the oil."

If counsel are correct then the alleged invention is *not patentable subject matter.* Only arts, *i. e.,* processes, machines, manufactures, or compositions of matter, and their improvements are patentable. (Sec. 4886, R. S. U. S.) The discovery that if a gas trap were maintained under pressure gasoline would be saved does not belong to the classes of things for which patents may be legally granted.

Of course, this is merely one of those subtle attempts to get away from the subject-matter of the claims. It is elementary that a patentee is bound by his claims. (Walker on Patents, 5th Ed., Sec. 176.)

After the smoke of patent litigation has somewhat cleared away, and it is seen just what the state of an art is, it is not permissible to gather from the debris a few hazy generalities and reconstruct them into a patent claim.

Counsel's argument under the heading, "The Trumble Invention" (Plaintiffs-Appellees' Br., p. 19) most clearly shows the fact to be as we have stated in our opening brief (p. 79) namely, that after its escape through the Patent Office it is now attempted to construe the Trumble patent as covering broadly only surface and pressure. In other words, the claims if so rewritten would be: "a separator in which the oil is spread to any degree whatever upon any surface and subjected to pressure."

Assume, for the sake of argument, that none of the prior art patents *did* mention or describe pressure. This would not be determinative of the question of infringement, for Trumble in the proceedings before the Patent Office distinctly limited himself to imperforate spreader plates for spreading all the oil to the outer edge of the vessel without breaking up or agitation, and causing it to flow wholly down the walls of the separator. Irrespective, therefore, of the question of pressure, defendant-appellant does not infringe; for he does not use this spirit of invention as there defined.

On page 20 plaintiffs-appellees' brief, a so-called "narrower aspect" of the Trumble alleged invention is mentioned, namely, to "slow down and spread out the oil in its passage through the trap." Note how counsel have here overlooked the question of infringement. In defendant-appellant's device there is no slowing down of the oil. A large part must drop to the bottom of the separator, without being retarded or spread.

In order to show the strength of our position, let counsel for appellees define the pretended invention of Trumble as they please—let them rewrite the claims if they choose—then compare with the prior art, and it will always be found that if such claims are broad enough to include defendant's devices they will be anticipated. *In other words, an invention is not mere words and cannot be constructed of thin air.*

The so-called "broadest aspect" of the alleged Trumble invention is distinctly found in Cooper, McIntosh, Newman, and other patents of record; the alleged "narrower aspect" is found in all prior art patents discussed in our opening brief.

### The Court's Duty to the Public.

There seems to be a theory pervading plaintiffs-appellees' brief, that because defendant-appellant in the trial court did not rely upon defenses attacking validity, that the way is thereby open to totally disregard the prior art, and to stretch the claims to almost any length. Now, of course, the public interest as well as private interests are represented by defendant-appellant. The right of the public to buy its gas and oil separators which employ those things which are old without the imposition of a royalty, is fully as much entitled to the protection of this court as the right of a patentee. If a patent is valid a patentee stands in the same light as the grantee of a tract of land from the government. Merely because there has been a grant, is no reason why we should totally disregard its metes and bounds. The grantee should have exactly what the spirit and intent of his instrument of title covers and no more. If it is clear that the public owns adjoining property care should be taken not to place such an interpretation upon the grant as will appropriate public domain.

Even if counsel were correct in their statement (Plaintiffs-Appellees' Br. middle of p. 21), that Trumble made a "striking discovery * * * that pressure if maintained in a gas trap, would maintain the gasoline in the body of the oil without interfering with the separation of the gas" (which is certainly not true as shown by the evidence) this would no more entitle Trumble to a disregard by this court of the metes and bounds of the grant of his patent as construed in the light of the prior art than it would entitle any public benefactor whose claim for public land was before the court for adjudication to a disregard of the law.

The assertions of counsel as to the "great value," "striking discovery," etc., of Trumble, should not, however, confuse the court. The fact is that Trumble's alleged contribution to the art is so small that one has difficulty in finding it. Trumble did not invent pressure. Trumble did not first apply pressure in a gas trap. He did not first devise means for slowing the flow of the oil. He did none of those things asserted by counsel.

## Defendant-Appellant's Devices Do Not Infringe.

We are pleased to note that the accuracy of our description (Defendant-Appellant's Br., p. 81 *et seq.*) of defendant's devices is not questioned in plaintiffs-appellees' brief. As we stated in appellant's opening brief, the drawings were made to scale, and we strenuously endeavored to set forth a clear description which could be accepted by counsel for plaintiffs-appellees. In this it appears we have succeeded. We regret, however, to note that there is a slight question as to our description of the *operation* of the so-called Model No. 1, found in plaintiffs-appellees' brief at the bottom of page 34 where counsel state: "The splashing of the incoming oil in Model No. 1, is not supported by any evidence in the case whatsoever and is the voluntary imagination of appellant's counsel." In this criticism counsel for plaintiffs-appellees have fallen into error, *as it was distinctly found by the trial court* that the part of the oil in model No. 2 [Tr., bottom of p. 537; quoted Defendant-Appellant's Op. Br., top of p. 87]—which is the same as model No. 1, except that the nipple is closer to the wall—descended by gravity *without reaching either wall.* Furthermore, as we point out near the top of defendant-appellant's opening brief, page 91, the court at transcript of record, page 541, upon this question of splashing said: "while part of the oil is reduced to a spray and falls by gravity to the fluid below, its action does not obviate the objectionable feature to a part flowing down the partition and a part down the wall." It is therefore clear that counsels' only criticism of our description of both the construction and mode of operation is based upon error.

In the discussion under the head of "Infringement" (Plaintiffs-Appellees' Br. p. 28 *et seq.*), counsel, ignoring the limitations by the prior art, as well as the admissions in the Trumble proceedings before the Patent Office; overlooking or disregarding completely the fact that under the interpretation suggested the patent in suit would be anticipated and void, urges not that the specific claims with the equivalents recognized by the Patent Office had been appropriated by the defendant, but that what in effect are *two new claims,* namely, reconstructions, labeled, respectively, "broadest aspect" and "narrowest aspect" had been adopted by defendant. Under such interpretation, if defendant uses any means for retarding the flow of the oil or any degree of pressure either separately or in combination it infringes. Of course, as we have repeatedly pointed out any such interpretation disregards the showing of the prior art which we so elaborately considered in our opening brief as well as the admission of the Trumble file wrapper contents as to the spirit and scope of the alleged invention.

Let it be noted that at the bottom of page 33 plaintiffs-appellees' brief, while counsel do not question the accuracy of our description of the diminutive atomizing device, *i. e.,* the inclined bottom of sleeve 18, which is numbered 17 in defendant's patent (Book of Exhibits, p. 67; described in our Opening Brief, p. 83), they do by the use of the term "baffle-plate" suggest that it might perform the function of Trumble's spreader cone and might spread out and slow down the oil, so as to bring it within the so-called "narrowest aspect" of the Trumble patent. If this small atomizing bottom 17 is a baffle-plate, then the small plate of Barker, for instance, is also a baffle-plate and anticipates. Both of these plates perform the same function.

In considering the small portion of the testimony quoted by counsel at the top of page 37, plaintiffs-appellees' brief, it must be borne in mind that the court, after considering all the evidence, found with regard to this model No. 2 [Court's Opinion, Tr. of Rec., bottom of p. 537; quoted in our Opening Brief, top of p. 87] that the part of the oil fell to the bottom "without reaching either wall"; and again at transcript of record, page 541; quoted in our opening brief, near the top of page 91, that part of the oil is reduced to a spray and falls by gravity to the settled fluid below. There is no question, therefore, that *all of the oil was not spread upon any surface within the chamber* in defendant's devices, and it is also clear that the oil was agitated, sprayed, or broken up.

### Alleged Validity of the Trumble Patent.

At page 40, plaintiffs-appellees' brief, counsel state:

"Notwithstanding the fact that the validity of the Trumble claims has been conceded in this case and is not in issue, and notwithstanding the clear wording of the claims, appellant's opening brief contains the discussion under the title of "Estoppel by Proceedings in the Patent Office."

In view of the fact that the trial court, in its opinion [Tr. of Rec, p. 536, ⅔ down the page] said:

"I think, therefore, the patentee is not estopped by the proceedings before the Patent Office to insist upon the broader claims"—and in view of the fact that, notwithstanding the statement that validity was not in issue, the court *did* distinctly decree that the patent was valid—we fail to understand counsel's criticism. Certainly, the proceedings before the Patent Office which we emphasized, especially in view of the Bray patent discussed in our opening brief should estop Trumble from asserting *any possible breadth of construction;* and, more than that, should preclude *Trumble from placing an interpretation upon his claims which is different from the interpretation he had to place upon them before the Patent Office to secure the allowance of his patent.*

## The Prior Art.

The fallacy of counsel's argument as to the lack of pertinence of the prior art is apparent from a consideration of the very first sentence of their discussion under this head (Plaintiffs-Appellees' Br. p. 47) where it is said:

"In view of the fact that defendant has conceded and has failed to deny the validity of the claims of the Trumble patent in issue, and in view of the fact that these claims are certainly broad enough to include both model 1 and 2 of the infringing device, we fail to perceive any force or materiality in considering the prior art."

Why, nothing is more elementary in the law of patents than that the prior art, even though not pleaded is admissible. Thus, Dunbar v. Myers, 94 U. S. 187, holds:

> "Proof of the state of the art is admissible in equity cases, without any averment in the answer touching the subject, and in actions at law, without giving the notice required when evidence is offered to invalidate the patent. It consists of proof of what was old and in general use at the time of the alleged invention; and may be admitted to show what was then old, or to distinguish what is new, or to aid the court in the construction of the patent."

Why should the court desire to know what was old or to distinguish what was new, unless it was to construe the claims in the light of the prior art, *so as to avoid an interpretation which would trench upon the public domain?* Again, why did the trial court consider the prior art and why was evidence of the prior art admitted without objection, if it was not pertinent? We discover the *spirit* as distinguished from the *letter* of the claim by a study of a file wrapper and contents as well as by a consideration of the prior art. Having determined what was actually new (if anything) we construe the claims if possible to cover that novelty, applying it narrowly, if necessary, to the particular form of means and devices shown in the patent drawings. This gives to the patentee all he is entitled to and it protects the public in the use of those things which are old.

In the light of counsels' opening statement under this title, it is easy to understand why they have accorded so little consideration to the main subject of appellant's opening brief.

In the middle of page 49, plaintiffs-appellees' brief, counsel suggest a false issue, as follows:

"We challenge appellant to point out this suggestion [the suggestion that pressure will hold gasoline in the oil] in any of the prior patents referred to in appellant's brief."

It is quite elementary that it is not necessary for a patentee to mention all the uses, benefits, and advantages resulting from his device or process in order to cover and secure those advantages. Thus, Walker on Patents (5th Ed.), p. 217, Sec. 175, says:

> "An inventor need not explain in his description, or know in point of fact, what laws of nature those are which cause his invention to work (citing St. Louis Stamping Co. v. Quinby, 4 Bann. & Ard. 195, 1879; Haffcke v. Clark, 46 F. R. 770, 1891; Dixon-Woods Co. v. Pfeifer, 55 F. R. 395, 1893; Jemolin Co. v. Harway, etc., Mfg. Co., 138 F. R. 54, 1905; and many other cases) nor is a patent void on the ground that the principle of the invention is not fully understood; or if understood by anyone, not understood alike by all (citing many cases) and was not understood correctly by the inventor himself (citing many cases) or that all its advantages and possibilities were not understood by or known to the inventor. (Citing many cases.) Neither is any description insufficient in the eye of the law on account of any mere errors it may be

> found to contain, where those errors would at once be detected and their remedies be known, by any person skilled in the art, when making specimens of the invention set forth, or when practicing that invention, if that invention is a process (citing many cases) nor where such errors consist in mistaken statements of immaterial facts (citing many cases); nor where such errors relate to the degree of efficiency of the invention (citing cases) nor where they consist in deficient description of the functions of some parts of a machine. (Citing cases.) Nor need a description state every use to which the described and claimed invention is applicable, in order to cover every such use. (Citing many cases.)"

In other words, patents are granted for *mechanical devices, means, or things* and not for the setting forth of a catalogue of supposed advantages.

## Cooper Patent.

(Discussed Plaintiffs-Appellees' Br. p. 54 *et seq.*) Counsel have failed to meet our argument on this patent. Not only does Cooper repeately mention pressure, not alone in connection with the prior patents to which he refers, but in the very patent in evidence. Cooper says that his device is to be used in a pressure system. Pressure systems, as we have seen, were old. As we pointed out in our opening brief, it was no more necessary for Cooper to illustrate the means for maintaining pressure (which Trumble himself merely mentions and illustrates prospectively on the outside, but does not describe) than it was for either Cooper or Trumble to illustrate the source of initial pressure at the bottom of the well. Remember, also, that plaintiffs-appellees' expert, Ford Harris, has testified that an absorption plant at the end of the line, or boilers, or other place of use (quoted page 20, our opening brief) would cause pressure in the trap. Cooper distinctly tells us (line 68, page 1, of his specification—Book of Exhibits, p. 134): "the gas rises and passes through the pipe F to the compressor." The compressor, under Ford Harris' testimony would back up pressure. Furthermore, Harris says (quotation, bottom page 19, appellant's opening brief) that pipe friction would back up pressure. Cooper's statement, therefore, that the gas passes *through a pipe* to the compressor, under the testimony of Harris indicates a pressure means in addition to the compressor.

Counsel's criticism (Plaintiffs-Appellees' Br., top of p. 56) that Cooper operates on a different principle, namely, centrifugal separation, is unmerited. Furthermore, it does not affect the pertinence of Cooper for the reason, that mere *addition* to a patented thing does not change its patentable character nor avoid anticipation or infringement. This rule, we believe, is obvious, but see Walker on Patents (5th Ed.), Sec. 338, 347, 367. Cooper spreads the oil in a thin film on the wall, at the same time subjecting it to pressure. Now, if he gives to the oil a swirling motion upon its entry into the chamber, thereby utilizing an additional principle as an aid to separation, he has notwithstanding *disclosed the subject matter claimed by Trumble.* Trumble surely cannot, *by foregoing one of the advantages of Cooper* and utilizing less than the complete Cooper disclosure, claim novelty and invention in the part appropriated.

There is no evidence as stated by counsel that Cooper was "a mere paper patent," and the argument of the witness Paine (plaintiffs' expert) in support of the value of a gas trap, namely, that it saved $125.00 per day, proves that Cooper did contribute something of value to the art. The Cooper device does not operate on the principle of centrifugal action. Such action is merely incidental and additional to the spreading of the oil on the wall of the separator in a thin film and subjecting it to pressure. Cooper does distinctly and repeatedly recognize the use of pressure with the device. Means for maintaining pressure in Cooper, counsel's assertion to the contrary notwithstanding, are shown, mentioned, and described in Cooper.

## McIntosh Patent.

Counsels' half page (Plaintiffs-Appellees' Brief, p. 57) of reply to our argument on this patent speaks eloquently of an inability to meet the clear showing of this evidence. The argument that McIntosh does not mention or describe the effect of pressure, *i. e.*, the advantages resulting therefrom, is met clearly by the quotation from Walker on Patents (5th Ed.), Sec. 175, which we have set forth *supra,* to the effect that *uses* or *advantages* need not be mentioned to be covered by a patent.

The suggestion that McIntosh does not enlarge upon the operation of his inverted bell as a means for maintaining pressure, is met by the fact that no one could examine the drawing and description of this patent without instantly understanding the function performed by the bell. Remember, too, that a patent description is addressed to those skilled in the art. We urge that even one unskilled in this art could understand the clear showing of this McIntosh description and drawing. Moreover, Trumble is in no position to question the sufficiency of the disclosure of McIntosh on the subject, for he does not describe the construction of his pressure valve, he only mentions its function and has illustrated merely the outside view of the device. Certainly McIntosh is as clear as Trumble as to the feature of maintaining pressure.

### Newman Patent.

Counsels' suggestion (Plaintiffs-Appellees' Brief, p. 58) that there is nothing in the Newman patent to indicate that pressure is used therein, ignores the fact that Newman at line 11, page 1 of his specification, states that the object of his invention was to distribute *the force of the gas especially in high pressure,* and the fact that at line 70, page 1, Newman's specification, referring to the float, he tells us that it "thereby equalizes the pressure."

Again we point out there is a pipe line from the Newman separator, which under plaintiffs' expert Harris' testimony, will back up pressure in the trap.

## Taylor Patent.

It is just as reasonable for counsel to assert that Taylor shows no means for maintaining pressure as it would be for us to point out that Trumble illustrates no means for producing the initial pressure from the well—namely, the bottom of the well.

Steam, of course, *is a gas;* and thus, the Taylor device is for the purpose of separating gas and oil. On the Trumble application it was adjudicated by the Patent Office (by its citation by the examiner) to be pertinent art, and Trumble did not object to its consideration, and raised no point that it could not be used for the separation of the products of an oil well. Of course, such a device could not be placed in position between the boiler and steam chest of an engine without having constant boiler pressure within the chamber.

The true difference between Taylor and Trumble was properly pointed out in the discussion before the Patent Office. That difference only related to the breaking up or agitation of the oil. We submit that conclusively this patent requires the adoption of the interpretation placed by Trumble upon his claims before the Patent Office. Under such interpretation it is clear that defendant-appellant does not infringe.

## Barker Patent.

(Discussed Plaintiffs-Appellees' Br., p. 59.) Counsel opens discussion by the statement, "This patent does not disclose a gas trap." This is clearly a mistake. The patent is entitled "NATURAL GAS SEPARATOR." True, the separation of gas from water is more emphasized by Barker than that of other products, but counsel evidently overlooks the fact that the Trumble devise also mentions *as a first object* (Trumble's specification, p. 1, line 13) the separation of gas from not only oil, *but water*.

During the prosecution of the Trumble application through the Patent Office, when this Barker patent was cited, it was not objected that it was not pertinent. On the contrary, Trumble at that time recognized it as a proper citation and did not attempt to distinguish it except on the ground of structural differences. Thus, both the Patent Office and Trumble agreed that Barker was a pertinent citation, and the present suggestion of counsel is totally inconsistent with such accord before the Patent Office.

Again counsel is mistaken in asserting "there was no means shown in the Barker device for spreading the incoming water so as to allow escape of gas therefrom." There is indicated by Barker a very similar atomizing plate (namely, plate 7 of the Barker patent) to the bottom 17 of the sleeve 18 of the Lorraine patent. Both of these devices have the same effect. If the Barker device is not capable of the function of the Trumble separator, as stated by counsel, then defendant's devices, which in large part atomizes and sprays the oil without spreading it upon any wall or surface within the chamber, are not infringements, as they, too, must be incapable of performing the function of the Trumble separator.

**Bray Patent.**

Counsel places much emphasis upon the fact that Bray provides in addition to the pipe 17 (which notwithstanding counsel's assertion to the contrary is of smaller diameter than the inlet pipe 16) an auxiliary oil outlet pipe 20. However, counsel fail to take into consideration the fact that *the oil level shown by the dotted line of Fig. 1 of Bray completely covers the outlet through pipe 20. There is, therefore, a locked space in the chamber above the oil level which normally closes this auxiliary outlet 20.* True, some pressure, when the same is excessive, may be relieved through the pipe 20, but nevertheless, obviously, there will be considerable pressure above the oil level before the same becomes so excessive as to force the oil upward and out through the pipes 21-22. It surely takes *some* pressure to thus force the oil out through the auxiliary outlet. Remember again, that Trumble provides *for no specific amount of pressure,* and even the slightest pressure within the chamber over atmospheric pressure will meet the Trumble claims.

Counsel's discussion of the case of Standard Oil Company v. Oklahoma Natural Gas Company (Plaintiffs-Appellees' Br., p. 62) is based upon error.

Is not the Trumble device for forcing the lighter series, namely, among other things, gasoline, from the natural gas into the oil? Trumble like Saybolt sought to prevent the escape of the gasoline with the gas. We submit the principle is precisely the same in both the Trumble and Saybolt patents. Trumble *takes the gasoline out of the gas* and puts it in the oil, or he prevents it from escaping from the oil. It is totally immaterial which. Obviously, no one can tell just what proportion of gasoline is squeezed from the gas into the oil, or is prevented from leaving the oil in any of these pressure separators, like Cooper, McIntosh, Newman, Trumble, and others.

## Conclusion.

We agree with counsel in the applicability of the quotation from Topliff v. Topliff, 145 U. S. 156:

> "The object of the patent law is to secure to inventors a monopoly of what they have actually invented or discovered * * *."

Trumble actually invented or discovered nothing but the specific form of his imperforate conical spreader plate and means for spreading the oil in a thin film upon the plate and equally around and upon the walls of the separator in an unagitated and unbroken up and unsprayed condition. He did not invent simply pressure and the flowing of the oil, either separately or in combination. To construe his patent so broadly would confer upon him by judicial construction and decree that which he distinctly disclaimed as his invention before the Patent Office, and would grant to him that which clearly belongs to this defendant as a member of the public.

We submit, that the decree appealed from should be reversed and the cause remanded with directions that the bill be dismissed for want of equity at appellees' costs.

Respectfully submitted,

WESTALL AND WALLACE,
ERNEST L. WALLACE,
JOSEPH F. WESTALL,
By JOSEPH F. WESTALL,

*Solicitors and of Counsel for Defendant-Appellant.*

No. 3945.

IN THE

# United States
# Circuit Court of Appeals,

FOR THE NINTH CIRCUIT.

David G. Lorraine,
*Appellant,*
*vs.*
Francis M. Townsend, Milon J. Trumble and Alfred J. Gutzler Doing Business Under the Firm Name of Trumble Gas Trap Company,
*Appellees.*

## PETITION FOR REHEARING.

Westall and Wallace,
Ernest L. Wallace,
Joseph F. Westall.
California Bank Bldg., 629 So. Spring St.,
Los Angeles, Cal.
*Solicitors and of Counsel for Appellant.*

Parker, Stone & Baird Co., Law Printers, 232 New High St., Los Angeles.

# Circ

---

David

Franc
ble a
ness
ble

---

*To t*

n
nan
a r
of

**No. 3945.**

IN THE

# United States
# Circuit Court of Appeals,

FOR THE NINTH CIRCUIT.

---

No. 3945.

David G. Lorraine,
*Appellant,*
*vs.*
Francis M. Townsend, Milon J. Trumble and Alfred J. Gutzler Doing Business Under the Firm Name of Trumble Gas Trap Company,
*Appellees.*

## PETITION FOR REHEARING.

*To the Honorable Judges of the United States Circuit Court of Appeals for the Ninth Circuit:*

Your petitioner, David G. Lorraine, appellant above named, respectfully petitions this Honorable Court for a rehearing of the appeal herein, and that the opinion of this court filed herein June 4, 1923, in so far only as it directs the District Court to enter a decree enjoining defendant-appellant from manufacturing and selling or using any device infringing the Trumble patent in suit and particularly from making or selling the construction of Tonner No. 3 trap—be reconsidered and reversed and that the decree of the District Court appealed from be wholly reversed at appellee's costs and for grounds therefor allege—

## Brief Statement of Grounds for Rehearing.

That this court erred in matter of law apparent upon the record and arising upon questions which were not argued at the original hearing in that—

A. The record clearly shows that there was only a single infringing device of the only form of device (namely, Tonner No. 3 Trap) found to infringe the Trumble patent in suit, made by appellant.

B. The record further establishes that such single infringing device was made experimentally only and no prior actual knowledge on the part of defendant appellant of appellee's rights in the premises was attempted to be proven.

C. The record further shows that such single infringing device made by appellant was unsuccessful or unsatisfactory, and appellant discontinued its further manufacture with no intention whatever to continue.

D. There being only such single infringement, there is no jurisdiction in equity to support the grant of an injunction or further proceedings on accounting or otherwise in the trial court, in that—

(1) The infringement was trifling and therefore beneath the dignity of a court of equity to consider.

(2) There was no reason or occasion on the part of appellees to fear any further trespass, in that—

(a) The single infringement was unsatisfactory and experimental only.

(3) An action at law to recover damages would have furnished adequate relief without recourse to equity;

(4) There was and is no ground to support any form of equitable relief—either of injunction or accounting.

E. The Tonner No. 3 trap was in fact not an infringement, in that—

(1) The device of the Trumble patent in suit has been found or assumed by the court to be useful and successful;

(2) The Tonner No. 3 trap, not being satisfactory could not constitute an infringement of a successful device.

## ARGUMENT.

As to the general ground for rehearing above stated, Foster Federal Practice (6th Ed.) Vol. II, Sec. 445, bottom of page 2172, says:

> "Unless the judge acts on his own motion, a rehearing will be granted only for errors of law apparent upon the record and arising upon questions which were not argued at the original hearing, or upon newly discovered evidence of such a character that it would have authorized a new

trial in an action at law." (Citing Daniel v. Mitchell, 1 Story, 198; Jenkins v. Eldredge, 3 Story, 299; Emerson v. Davies, 1 W. & M. 21; Tufts v. Tufts, 3 W. & M. 426; Giant P. Co. v. California V. P. Co., 5 Fed. 197; Swann v. Austell, 257 Fed. 870.)

## There Was Only One Infringing Device Ever Made by Appellant.

At transcript 253 (middle of page) appears the following testimony:

"Mr. Baggs: Now, Mr. Lorraine, I will ask you to state how many traps like the one you have just described as being located on the Tonner lease No. 3 you put out or built? A. Just one."

On cross-examination, the witness Lorraine also testified [Tr. of Record, p. 314]:

"Q. Now, didn't you testify yesterday that the only trap you had ever made that had a baffle in like the Tonner trap was the Tonner trap, and you put it up and found it wasn't any good and took it out.

A. Yes, sir; that is the only one I have made like that Tonner baffle."

## The Single Infringing Device Was Not Satisfactory in Operation.

At transcript, page 253, *et seq.*, (beginning at the middle of the page) appellant Lorraine testifies:

"Q. After you had put that one out what did you do with reference to other models?

A. Why, we took this baffle out entirely, and I think Mr. Lacy there has the record. He can tell just how many we built without that baffle in there at all, but we still had this incoming baffle like that, but we didn't have this here deflector on the bottom (indicating).

Q. That is, you mean to say you didn't have this baffle at all or do you mean to say that it was not—

A. We used this vertical partition the same, just the same, but we abandoned the use of this baffle here and just [210] used this divider in the top.

Q. And there was no baffle there, the oil struck nothing, then, after passing out here (indicating)?

A. No, sir.

Q. After it passed out here it struck nothing?

A. Well, it may be some of it shot over this way with a big gas force (indicating).

Q. But it just dropped down?

A. Yes.

Q. Now, why did you change from that form with the baffle-plate to the form without the baffle-plate?

A. Well, we found there were several reasons. We already decided not to use this baffle as we found that it was no good in there in that position as it held the oil up too high here (indicating). As soon as I installed the trap and put the trap into operation why I told Mr. Burrows and Mr. Swope that I wished that that baffle was out of there as it held the oil up too high on that side of the vertical partition.

Q. And you took it out?

A. Not that one. We left that one in.

Q. The Court: That was held up because the space between the baffle-plate and the oil was not large enough to let the oil pass through, was that the reason for it?

A. Well, that might have been possible, but then it shot the froth and the foam, some of it over the top, and we had to carry our oil level too low on this side to make a complete separation of the oil when the oil reached the storage tank.

Q. Well the baffle-plate in there was the cause of the oil shooting over the top?

A. That was it.

Q. Caused it to pile up there and it didn't have room to pass down by the wall, is that it?

A. Well, yes, it held up the froth.

Q. I see.

A. I would say that is the reason; yes, sir."

**The Law Clearly Is That Equity Has no Jurisdiction in the Case of a Single Infringement Where no Threat to Continue to Infringe Is Proven. The Remedy at Law Is Adequate.**

In the complaint, paragraph VIII, transcript of record, page 6, and in the supplemental complaint [Tr. p. 28] defendant is charged. not only with infringement, but also *with the intent and threat to continue to infringe.* This charge implies an acquiesence on the part of plaintiffs in the soundness of the general rule of law which requires something more than proof of a single sale or a single manufacture as a basis for equitable relief. The answer specifically denies, not only the infringement and continuance thereof, but also the allegation of threat and intent to continue infringement. The burden is thus placed upon plaintiffs to prove circumstances justifying apprehension of future trespasses which could be the only logical basis for the grant of injunctive relief.

Hopkins on Patent page 787. says:

"A single sale, as Judge Lacombe has said, 'may, in connection with other proof, be persuasive evidence of other sales, and convincing proof of an intention to sell whenever the opportunity of doing so without detection is presented.' (Citing Lever Bros. Ltd., v. Pasfield, 88 Fed. Rep. 484.)"

Hopkins also quotes Judge Coxe in Hutter v. De Q. Bottle Stopper Co., 128 Fed. Rep. 283, as follows:

"A single sale made in circumstances which indicate a readiness to make other similar sales upon application is sufficient to make out a *prima facie* case.' It must be borne in mind that unless *prima facie* case indicates that a substantial recovery of profits may be had, the court may refuse to order an account; (citing Ludington v. Leonard, 127 Fed. Rep. 155, 62 C. C. A. 269) notwithstanding the general rule that the right to an account of profits is incidental to the right to an injunction. (Citing Stevens v. Gladding, 58 U. S. 447, 15 L. Ed. 155.)

"What has been said is entirely consistent with Judge Lurton's doctrine. 'that a single infringement by making and selling a single infringing machine would not justify the interposition of a court of equity for the purpose of restraining further infringement by the making and sale of other infringing machines, if it appeared clearly that there was no reason to apprehend any further infringement.' (Citing Johnson v. Foos Mfg. Co., 141 Fed. Rep. 73, 72 C. C. A. 123.)

The case of Globe-Wernicke Co. v. Brown & Besly (C. C. A. 7th Cir.) 121 Fed. 91, was an action to enjoin defendant from using the plaintiffs' patent imprint on letter files made by defendant. It appeared that defendant, which was a manufacturer of such files, was given an order by a customer for a lot of files to be made according to a sample submitted, which sampel had been made by plaintiff and had been marked patented because plaintiff in making it had used a patented pin. Defendant made up the order without this pin, but without the knowledge of defendant's officers, its employees put the patent imprint on the files so made. It was never used except the once. The court said:

> "Treating what was done as a trespass upon appellant's rights, there was a plain and adequate remedy at law, for the evidence fails to sustain the allegation of threatened continuation and irreparable injury. There was no error in dismissing the bill for want of equity."

In the case of Woodmanse etc. Co. v. Williams (C. C. A.) (6th Cir.) 68 Fed. 489-492, the court said:

> "The ground upon which a court of equity will take cognizance of a suit for an infringement of a patent is the relief through an injunction. There is nothing so peculiar to a suit for damages and profits for infringement of a patent as will, independently of some recognized ground of equitable jurisdiction, justify a court of chancery in assuming jurisdiction. It must appear that the legal remedy at law is inadequate, and if the case is one in which equitable relief by injunction is inappropriate, as where the patent has expired, or where the circumstances are such as to justify a court in

> refusing equitable relief, the suit will not be entertained for the mere purpose of an account of past damages and profits."

In Edison Phonograph Co. v. Hawthorne Etc. Co., 108 Fed. 630, the court held that where all acts of infringement charged are in the past, and there are no allegations that their continuance is threatened or intended, no injunction will issue, saying:

> "The jurisdiction of equity in the case of infringement of letters patent exists only when the bill states facts upon which the right to some form of equitable relief may properly rest. For infringement merely, the remedy at law is complete and adequate."

In Plotts v. Central Oil Co. (C. C. A. 9th Cir.) 143 Fed. 901, it was held that where defendant had made but one infringing machine and had not used it, but had offered to pay royalty on it, and no intention of defendant to continue its use in violation of plaintiffs' rights appeared, the bill was properly dismissed.

In the case of General Electric Co. v. Pittsburg, Etc., Co., 144 Fed. 439, it was held that where defendant admits it was infringing without knowledge of plaintiffs' rights, but had ceased doing so, and had no intention of resuming a preliminary injunction would not issue.

We submit that the foregoing authorities strictly follow the basic principles of equity jurisprudence and should be followed.

All but a trifling amount of the record in this case is devoted to controversy relating to forms of devices found not to infringe. The record shows that the whole business of defendant-appellant was based upon the manufacture and sale of such non-infringing devices. Certainly, if the proper interpretation of the Trumble patent in suit as indicated by this court in its opinion had been understood and adopted by the trial court, jurisdiction in equity could not have been sustained. The infringement proven is trifling. The remedy at law is adequate. Even in an action at law, surely the unsuccessful nature of the experiment found by this court to constitute infringement (the making and selling of the single Tonner No. 3 Trap) could not have justified more than an award of nominal damages.

We submit that no injunction should be awarded against the appellant under the circumstances, and that the mandate of this court should direct the trial court to enter a decree dismissing the bill for want of equity at the costs of appellees.

David G. Lorraine,
*Appellant-Petitioner,*
By Westall and Wallace,
Ernest L. Wallace,
Joseph F. Westall,
*Solicitors and of Counsel for Appellant.*

We hereby certify that the foregoing petition for rehearing is in our judgment well founded and that it is not interposed for delay.

Westall and Wallace,
Ernest L. Wallace,
Joseph F. Westall.

1331 **No. 3945.**

IN THE

# United States
# Circuit Court of Appeals,
FOR THE NINTH CIRCUIT.

David G. Lorraine,
*Appellant,*
*vs.*
Francis M. Townsend, Milon J. Trumble and Alfred J. Gutzler Doing Business Under the Firm Name of Trumble Gas Trap Company,
*Appellees.*

## PETITION FOR REHEARING.


Frederick S. Lyon,
Leonard S. Lyon,
Frank L. A. Graham,
*Solicitors for Plaintiffs-Appellees.*


Parker, Stone & Baird Co., Law Printers, 232 New High St., Los Angeles.

**No. 3945.**

IN THE

# United States Circuit Court of Appeals,

FOR THE NINTH CIRCUIT.

David G. Lorraine,
*Appellant,*

*vs.*

Francis M. Townsend, Milon J. Trumble and Alfred J. Gutzler Doing Business Under the Firm Name of Trumble Gas Trap Company,
*Appellees.*

## PETITION FOR REHEARING.

*To the Honorable Judges of the United States Circuit Court of Appeals for the Ninth Circuit:*

The plaintiff-appellees, FRANCIS M. TOWNSEND, MILON J. TRUMBLE and ALFRED J. GUTZLER, doing business as the TRUMBLE GAS TRAP COMPANY, believing themselves aggrieved by this court's decision filed June 4, 1923, come now and respectfully petition this court for a rehearing upon the following grounds:

Italics appearing hereinafter may be deemed ours.

## Both Parties Concur in Asking Rehearing.

We have been served with copy of petition of defendant-appellant for rehearing of this case. It is evident that the decision of this court is not considered an equitable one by either party, that a rehearing should be granted, and that the decision should be restated.

## The Trumble Patent Is Admittedly Valid.

In considering this case, since the court did not declare any claim to be invalid, the court evidently did not lose sight of the fact that claims 1, 2, 3 and 4 of the Trumble patent sued on in the court below are shown by the record to have been repeatedly stipulated and admitted by the defendant to be valid, in the court below, and that the only question before the court is the interpretation or scope to be given the claims.

We think the court has been led to take an erroneous position as to certain points of fact and law as follows:

## First: The Court Was Misled as to the Proceedings in the Patent Office, in Regard to the Trumble Patent, and as to the Law Relating Thereto Reversed Its Own Settled and Established Practice.

This is a most important point and one upon which the whole decision rests.

It cannot be said that any of the claims of the Trumble patent are ambiguous or uncertain but on the contrary, although fairly broad, they are most definite.

This point can be settled by a mere reading of the claims and needs no argument from us. Claim 4, which we will hereinafter use as an example to save extended discussion, reads as follows:

"4. In an oil and gas separator, the combination of an expansion chamber, means for delivering oil and gas into the chamber, **means for maintaining pressure** within the chamber, means for drawing oil from the chamber, and means within the chamber adapted to cause the oil to flow in a thin body for a distance to enable the gas contained and carried thereby to be given off while the oil is subjected to pressure.

Can it be said that this claim is *ambiguous* or *uncertain?* It seems to us most definite and certain.

It is well settled that a patentee who acquiesces in the rejection of claims by the Patent Office is thereby estopped from later asserting that the allowed claims should be construed as equivalent to the rejected ones. See Cole v. Ed. G. Hookstratten Cigar Co., 250 Fed. Rep. 629; W. F. Schultheiss Co. v. Phillips, 264 Fed. Rep. 971; Selectasine Patents Co. v. Prest-O-Graph Co., 282 Fed. Rep. 223. In all these cases *this* court definitely relied on and restated this doctrine.

*But this doctrine has no application to the file wrapper of the patent in suit.* Claim 4, for example, was inserted by amendment of March 15, 1915, p. 31, Exhibits. It did not supersede broader claims which were cancelled but was in itself broader than original claim 7, p. 24, Exhibits, the only original claim which contained pressure regulating means in the gas outlet and which was later allowed and appears in the patent as claim 8.

At the same time that claim 4 was inserted the specification was amended, p. 31, Exhibits, to further bring out the novel feature of the Trumble invention already well stated in the specification, i. e., the pressure regulating means or **means for maintaining pressure** which was used to force the dry gas out of the oil and hold the gasoline in. This shows that Trumble appreciated and at that time claimed broadly the substance of claim 4. The Patent Office *never* rejected any claim on the grounds that **pressure maintaining means** was old in the art.

This amendment was *admitted* by the Patent Office *and the claims were allowed.* Claim 4, so far as *structure* is concerned was much broader than any claims previously in the case. There was no acquiescence in the rejection of any claim containing a **pressure regulating means** and no estoppel.

The solicitor for Trumble was, however, met by a rejection as to structure claimed in *claim 1* on the patent to Bray and he met this rejection with an argument as to the structural differences, and not as to the fundamental feature of pressure since the examiner had not contended and could not contend that **means for maintaining pressure** were old in the art. In other words the solicitor met the examiner's rejection with an argument on that rejection and naturally did not go outside that rejection to argue points that were not in issue. With all the evidence in this case before us we can *now* see that the argument of the solicitor did not state *all* the grounds upon which Trumble might have relied. It met the rejection, however, and *it was successful.* The court, in considering the Trumble file wrapper, evidently thought that since Trumble's solicitor did not state *all* the available arguments to overcome the Bray patent before the Patent Office, we are thereby precluded from presenting to this court other arguments which are amply supported by the evidence.

In the absence of any estoppel of the patentee by acquiescence it would seem to us that this court should proceed to do justice on the *facts,* unhampered by any limitations due to the solicitor's mere argument or to the failure of the solicitor to fully present to the Patent Office *all* the arguments in the applicant's favor. If Trumble's solicitor had presented to the Patent Office the facts urged here and had *lost,* before the Patent Office, we might have been estopped. What he did, however, was to present a *much weaker* argument before the Patent Office upon which he *won.* Certainly this does not preclude our using the stronger argument here, since it is clearly supported by the evidence.

The Supreme Court has well enunciated the true doctrine as follows:

> "While not allowed to revive a rejected claim, by a broad construction of the claim allowed, yet the patentee is entitled to *a fair construction* of the terms of his claim as actually granted."

Huppel v. United States, 179 U. S. 77, 45 L. Ed. 95.

So also other Circuit Courts of Appeal have spoken definitely on this point as follows:

> "It is of the essence of the rule * * * that * * * the estoppel does not extend to a matter not stated in the objection or disclosed by the reference."

Vrooman v. Penhallow, 179 Fed. Rep. 297.

> "The rule on this topic * * * is to the effect that, in order that the proceedings in the Patent Office should operate as a waiver or estoppel, they must relate to the pith and marrow of the alleged improvement, and be understandingly and deliberately assented to. This rule has been many times approved by the federal courts."

United States Peg Wood, S. & L. B. Co. v. B. F. Sturtevant Co., 125 Fed. Rep. 384.

See also:

National Hollow Brake Beam Co. v. Interchangeable Hollow Brake Beam Co., 106 Fed. Rep. 714;

Stead Lens Co. v. Kryptok Co., 214 Fed. Rep. 375;

New York Scaffolding Co. v. Whitney, 224 Fed. Rep. 462;

J. L. Owens Co. v. Twin City Separator Co., 168 Fed. Rep. 259.

It is not necessary, however, to go to either the Supreme Court or to other Circuit Courts of Appeal for a complete exposition of the law on this point. This court itself has handed down decisions in which the rule is clearly and definitely enunciated. In one case decided in 1908, this court in a very thorough and sweeping analysis of the law, after carefully reviewing various authorities, said:

> "In the light of these authorities, it is clear that the claims of the patent, unambiguous as they are, are to be interpreted according to the meaning of their own terms, and are not to be controlled or limited by any argument or representation made in the Patent Office by the applicant's attorney as to the scope of the invention or the features wherein it differs from the prior art."

Fullerton Walnut Growers Association v. Anderson-Barngrover Mfg. Co., 166 Fed. Rep. 452.

So, also *this* court in another decision said:

> "But it has been held, in effect, by this court that, where the claims allowed are not uncertain or ambiguous, the courts should be slow to permit their construction of the patent actually granted and delivered to be affected or controlled by alleged interlocutions between the Patent Office and the claimant."

Selectasine Patents Co. v. Prest-O-Graph Co., 267 Fed. Rep. 845.

Citing:

Fullerton W. G. Assn. v. Anderson Barngrover Mfg. Co., 166 Fed. Rep. 443;

Westinghouse v. Boyden Power Brake Co., 170 U. S. 582.

We respectfully submit that the court's decision in this case is almost wholly based on the assumption that Trumble limited his claims by the proceedings in the Patent Office. This is an erroneous assumption as to the facts as is evident from a short study of the Trumble file wrapper. The paragraph beginning at the top of page 9 of the decision and upon which the whole decision rests is not only based upon an erroneous assumption of facts but also is an absolute reversal of the law as established by the Supreme Court case, the three cases in other circuits and the two cases of *this* court above cited.

It cannot be said that claim 4 is ambiguous or uncertain or that it is susceptible of various interpretations. It is definite and certain, admitted by the defendant-appellant to be valid, and most certainly infringed by defendant-appellant's devices.

**Second: The Court Was Misled Into Holding That There Was No Claim Made, That the Physical Law Involved in the Patent in Suit Was New but That This Law Was Well Known, the Court Citing Standard Oil Co. v. Oklahoma, 284 Fed. 469-472.**

This is a serious error as to the facts.

The Saybolt patent referred to in this cited decision did not relate to a gas trap or a device for *taking gas out of oil*, but to a device for an exactly opposite purpose, that is, a device for causing a portion of the *gas to be absorbed by oil.* Saybolt used pressure, not to facilitate the *escape* of gas from oil, but for the purpose of forcing a portion of the natural gas to be *absorbed* by an oil. Saybolt was seeking *absorption;* Trumble was seeking *release.* The teaching of the Saybolt patent would in fact lead one to believe that the Trumble gas trap would be inoperative, as Saybolt used pressure for an entirely different purpose and obtained an exactly opposite result to Trumble. Moreover, since counsel for defendant cited the Standard Oil v. Oklahoma decision he also knew of the Saybolt patent, but he wisely did not put it in evidence. It is, therefore, to be assumed, as is a fact, that the Saybolt patent is not pertinent to the issues herein. The court was undoubtedly misled by counsel's argument on this most essential point, and this error has led to serious inconsistencies and further error, as will be later pointed out. Trumble *did* discover and apply a new principle in the art, namely, that pressure **maintained** on a gas trap would facilitate the escape of the dry gas therefrom. This was a novel and unexpected result not foreshadowed in any way in the prior art. It was not based on the principle of absorption used in the Saybolt patent, but on the principle of *partial pressures* which was unknown at the time Trumble made his invention, as clearly shown by the evidence. [See Record, p. 80.]

Trumble's object was to save gasoline, the saving of gas being of minor importance, and no prior patent was directed to this object, all of the prior patents being directed to saving the gas and incidentally wasting the gasoline.

Third: **The Court Was Also Misled as to the Facts and Assumed That the Combinations Claimed by Trumble Were Old in the Art.**

For convenience in discussion, we can consider claim 4 which contains (in addition to certain old elements) two significant elements *in combination,* that is

(a) **Means for maintaining pressure** within the chamber;

(b) Means within the chamber adapted to cause the oil to flow in a thin body for a distance to enable the gas contained and carried thereby to be given off while the oil is subjected to pressure.

Note that means (a) is a **means for maintaining pressure** and is not met by showing that due to exterior causes pressure *might* at times be partially or temporarily caused in a trap.

Of the prior gas traps found in the art, *not a single trap shows the combination of elements (a) and (b)*, and further in no case is any trap in the prior art provided with any structure suited to *maintain* pressure.

The Trumble patent is not a process patent but an apparatus patent claiming a *combination* of specific mechanical elements. One of these elements, namely, the **pressure maintaining means** *(b) is not found in the art.* The court has *assumed* that it is present in certain prior patents but the patents themselves are silent on this point.

By dividing his oil into a thin body and *maintaining* a pressure on it, Trumble accomplished certain very wonderful results, as stated in his patent specification and as clearly shown by the evidence.

The Cooper patent No. 815,407, p. 133, Exhibits, does not show or describe any "**means for maintaining pressure.**" It is true Cooper says that *"in a contemporaneous application"* he shows means, namely, a compressor, for forcing gas from a gas trap *into a* *well,* but this is the application of pressure to gas going down into a well and not to the gas trap. Whatever Cooper may have shown in a *contemporaneous application* is not evidence here and since the defendant did not put this contemporaneous application in evidence, we are justified in assuming it is not of any importance in this case.

It is also true that Cooper says that high pressure *may* be maintained in the cylinder, but he does not say that it *is* maintained and neither shows nor described any **means for maintaining this pressure.** His gas outlet line F is free from even a valve. If it had an ordinary manually operated valve it would still not have **means for maintaining pressure,** this means in the Trumble patent being the automatic "gas pressure regulating valve" 11. The valve 11 of Trumble *automatically maintains* pressure under all conditions of gas and oil flow of the well. If the flow increases, it opens. If the flow diminishes, the valve 11 closes to the necessary degree to maintain a predetermined pressure. It is a **means for maintaining pressure.** The defendant has a valve exactly like it in the valve 28 of his patent, p. 66, Exhibits. No such valve or anything approaching it is either shown or described in the prior art.

The court apparently entirely lost sight of the fact that the Trumble patent is *a combination* apparatus patent and that elements entirely absent from the prior art cannot be imported or read into combinations, from which they are absent, for the purpose of anticipation.

McIntosh also has no means for maintaining pressure. His gas pipe 9 is entirely free from valves. There is not one word in his patent which indicates that he intended **to maintain pressure** therein. Even accepting the court's assumption that pressure *might* be present, we are still far short of finding any **means for maintaining pressure.** In other words, we do not find Trumble's claimed combination of elements.

Bray is far from showing **pressure maintaining means.** In fact he shows an overflow pipe 20 equipped with a *vacuum breaker* 30. His gas outlet pipe 17 is free from valves and the trap is vented through the pipe 28 to the open air.

None of the prior patents except Cooper *speak* of pressure and he is *vague* and *indefinite.* Certainly none of them show or describe actual *means* for accomplishing Trumble's new and useful result, i. e., the combination of the elements (a) and (b).

Fourth: **The Court Having Been Led by Counsel to Assume Certain Elements to Be Found in the Prior Art in the Combination Claimed by Trumble, Resolved All Doubts in Relation Thereto Against the Plain Words of the Patent and the Evidence. This Is an Error of Law and Is Contrary to the Well Established Principles of This Court.**

This court has, in the past, always construed patents fairly and in accordance with their plain words. For example, see:

Letson *et al.,* v. Alaska Packers' Ass'n, 130 Fed. Rep. 129;

Beryle v. San Francisco Cornice Co. v. Hickmott Asparagus Canning Co., 137 Fed. Rep. 86, and

Los Angeles Art Organ Co. v. Aeolian Co. *et al.,* 143 Fed. Rep. 880.

These are all Ninth Circuit cases in which this court has refused to construe patents otherwise than in accordance with their *obvious and plain language.*

In this case the court has made certain assumptions. First, it has assumed that Trumble limited his patent by the proceedings in the Patent Office. Second, it has assumed that the principle of the Trumble gas trap was old. Third, it has assumed that **means for maintaining pressure** on such a trap was old in the art. None of these assumptions are supported by the evidence, but are based wholly upon inference. The court has then, basing its findings on these assumptions, placed the narrowest structural construction on claims which specify in the broadest terms *"means"* for accomplishing a result.

In so doing the court has, in effect, reversed its previous fair, liberal and constructive policy.

The Fifth Circuit in a case published June 21, 1923, says:

> "Where anticipation is relied on as a defense it should be clearly proved, and, in cases of reasonable doubt, the doubt should be resolved in favor of the patent attacked."

Atlantic, Gulf, & Pacific Co. v. Wood, 288 Fed. Rep. 154.

Citing:

Coffin v. Ogden. 18 Wall 120 21 L. Ed. 821;

Victor Talking Machine Co. v. Duplex Phonograph Co., 177 Fed. 248;

Simonds Rolling Machine Co. v. Hathorn Mfg. Co., 93 Fed. 958.

Fifth: **The Court Was Misled Into Ignoring the Great Weight of the Evidence as to Utility in the Trumble Invention.**

This was a wide departure from the settled practice of this court, and one which sets a dangerous and insidious precedent.

The evidence shows that the Trumble gas trap made enormous savings in the oil industry due to the combination of the dividing means (a) and the pressure maintaining means (b). For example, one trap alone saved $125.00 per day and Trumble sold 583 traps. There was no claim made or evidence introduced to show that any trap in the prior art could have made this saving, and in fact no prior trap could have made it. Trumble has conserved enormous quantities of gasoline and oil. His patent is not a mere paper patent. It is an invention that made enormous savings and is now making enormous savings for the oil industry.

The court has evidently not considered this but has given the patent an interpretation which limits it to definite structure and which absolutely ignores the true nature of the Trumble invention and the great public benefit that has accrued therefrom.

In so doing, the court has again widely departed from its previous well settled policy.

The court is referred to the following cases:

- H. J. Heize Co. v. Cohn, 207 Fed. Rep. 547;
- Morton v. Llewellyn *et al.*, 164 Fed. Rep. 693;
- Stebler v. Riverside Heights Orange Growers Ass'n *et al.*, 205 Fed. Rep. 735;
- Hyde v. Minerals Separation, Limited *et al.*, 214 Fed. Rep. 100;
- Sherman Clay Co. v. Searchlight Horn Co., 214 Fed. Rep. 86;
- Majestic Development Co. v. Westinghouse Elec. & Mfg. Co., 276 Fed. Rep. 676.

These are Ninth Circuit cases in which this court has given weight to the proven utility and important commercial results flowing from the invention before it.

The court should also not lose sight of the fact that long before the defendant-appellant appropriated the Trumble combination, that Trumble had made and sold hundreds of his gas traps and therefore occupied a dominating position in the commercial field.

**Sixth: The Court Has Been Misled to Such a Degree That It Has Issued Directions to the Lower Court Which Cannot Be Consistently Followed or Interpreted by the Lower Court Without Clarification by This Court.**

The directions must be explained and amplified by this court to allow the lower court to take intelligent action thereon.

The court has said, as to the Trumble patent, that

it wishes "to adopt a construction of which it is fairly susceptible."

Element (b) of claim 4 reads as follows:

"means within the chamber adapted to cause the oil to flow in a thin body for a distance to enable the gas contained and carried thereby to be given off while the oil is subjected to pressure."

This claim was and is conceded to be valid by the defendant-appellant, and, by implication, by this court.

The court says, page 3 of decision, that this claim (with the others) must be interpreted to mean:

"a structure where the whole *body* of the crude oil is spread *equally* in a *thin film* upon the conical spreader plates and upon the entire chamber wall intermediate between them and the pool level." (The italics are the court's.)

Note that the claim is entirely silent on the following points:

(a) It does not say that the *whole* body of oil is effected.

(b) It does not say that it is *spread equally*.

(c) It does not say anything about a *thin film*.

(d) It is not limited in language to *conical spreader plates*.

(e) It says nothing about spreading the oil on the *chamber wall.*

The court directs the District Court "to enter a decree interpreting the Trumble patent in harmony with the views herein expressed."

How can the District Court logically enter a decree holding the patent *valid* and also holding that the plain words of element (b), claim 4, mean all that this court says it does?

How can the District Court ignore the plain wording of this claim and read into it *five* limitations not found therein?

It seems to us that the court must reconsider and rewrite its decision in such form that it can be acted upon. Certainly no District Court can enter a decree "in harmony" with the present decision.

It should be noted that our position on this point is concurred in by defendant-appellant who shows in his petition for rehearing that the decision must be amended to permit a consistent decree by the District Court.

Respectfully submitted,

FRANCIS M. TOWNSEND,
MILON J. TRUMBLE,
ALFRED J. GUTZLER,
*Appellees-Petitioners.*

By FREDERICK S. LYON,
LEONARD S. LYON,
FRANK L. A. GRAHAM,
*Attorneys.*

I hereby certify that I have examined the foregoing petition, and in my opinion it is well founded; that the case is one in which the prayer of the petitioner should be granted by this court; and that the petition is filed in good faith and not for the purpose of delay.

FRANK L. A. GRAHAM,
*Of Counsel for Plaintiff-Appellee.*

Lightning Source UK Ltd.
Milton Keynes UK
UKHW010810110119
335238UK00010B/1082/P

9 780266 824589